Excel

SUCCESS ONE® HSC*

MATHEMATICS

Past HSC Papers & Worked Answers 1996–2017

PLUS

Topic Index of Past HSC Questions

PASCAL PRESS

ISBN 978 1 74125 669 7

Pascal Press
PO Box 250
Glebe NSW 2037
(02) 8585 4044
www.pascalpress.com.au

Publisher: Vivienne Joannou
Commissioning and series editor: Mark Dixon
Project editor: Rosemary Peers
Edited by Aaron Butler and Rosemary Peers
Answers checked by Peter Little
Cover design by Michael Sherman
Page layout and typesetting by lj Design (Julianne Billington), DiZign Pty Ltd and Monoset Typesetters
Cover photo by Janie Barrett, Fairfax Syndication
Printed by Vivar Printing/Green Giant Press

Contents

1996 HSC 2 Unit Mathematics
Examination Paper **1**
Worked Answers 12

1997 HSC 2 Unit Mathematics
Examination Paper **19**
Worked Answers 30

1998 HSC 2 Unit Mathematics
Examination Paper **37**
Worked Answers 48

1999 HSC 2 Unit Mathematics
Examination Paper **57**
Worked Answers 68

2000 HSC 2 Unit Mathematics
Examination Paper **75**
Worked Answers 87

2001 HSC Mathematics
Examination Paper **95**
Worked Answers 107

2002 HSC Mathematics
Examination Paper **116**
Worked Answers 129

2003 HSC Mathematics
Examination Paper **138**
Worked Answers 151

2004 HSC Mathematics
Examination Paper **163**
Worked Answers 174

2005 HSC Mathematics
Examination Paper **184**
Worked Answers 197

2006 HSC Mathematics
Examination Paper **207**
Worked Answers 221

2007 HSC Mathematics
Examination Paper **233**
Worked Answers 246

2008 HSC Mathematics
Examination Paper **257**
Worked Answers 270

2009 HSC Mathematics
Examination Paper **280**
Worked Answers 295

2010 HSC Mathematics
Examination Paper **308**
Worked Answers 325

2011 HSC Mathematics
Examination Paper **336**
Worked Answers 351

2012 HSC Mathematics
Examination Paper **360**
Worked Answers 378

2013 HSC Mathematics
Examination Paper **386**
Worked Answers 401

2014 HSC Mathematics
Examination Paper **411**
Worked Answers 426

2015 HSC Mathematics
Examination Paper **434**
Worked Answers 451

2016 HSC Mathematics
Examination Paper **461**
Worked Answers 475

2017 HSC Mathematics
Examination Paper **485**
Worked Answers 501

Reference Sheet **511**

PAST HSC EXAMINATION PAPERS

The past HSC Examination papers contained in this publication have been reproduced under licence from the NSW Education Standards Authority (NESA) in whom copyright is vested.

NESA takes no responsibility for errors in the reproduction of past HSC Examination papers contained in this publication.

NESA was the first publisher of each examination paper in the year indicated on the first page of each examination paper.

WORKED ANSWERS

The worked answers contained in this publication are examples of answers which the authors believe would score full marks.

They are not necessarily ideal or model answers, nor are they the only answers which would score full marks. They are not endorsed by NESA.

Worked answers for the 1996–2003 HSC Examination papers were written by Keith Saines.

Worked answers for the 2004–2005 HSC Examination papers were written by Lynne Knapman and Berra Mossemenear.

Worked answers for the 2006–2009 HSC Examination papers were written by Berra Mossemenear, Graeme Downward, Bronwyn Opferkuch and Robert Russell.

Worked answers for the 2010–2017 HSC Examination papers were written by Allyn Jones.

THE HSC MATHEMATICS EXAMINATION

From 2001, HSC Mathematics was the new name for the former HSC 2 unit Mathematics course.

Since 2012 the HSC Mathematics Examination paper has consisted of two sections. Section I has 10 objective-response questions worth 10 marks in total. Section II has six questions worth 15 marks each. Each of these six questions consists of a number of short-answer parts.

There is reading time of 5 minutes and working time of 3 hours.

Students should note that marks will be shown for each part of each question and should allocate their time accordingly.

HIGHER SCHOOL CERTIFICATE EXAMINATION

1996
MATHEMATICS
2/3 UNIT (COMMON)

Time allowed—Three hours
(Plus 5 minutes' reading time)

DIRECTIONS TO CANDIDATES

- Attempt ALL questions.
- ALL questions are of equal value.
- All necessary working should be shown in every question. Marks may be deducted for careless or badly arranged work.
- Standard integrals are printed on page 12.
- Board-approved calculators may be used.
- Answer each question in a *separate* Writing Booklet.

QUESTION 1. Use a *separate* Writing Booklet.

Marks

(a) Find the value of $13^{-1 \cdot 2}$ correct to two significant figures. **2**

(b) Factorise $2x^2 + 3x - 2$. **2**

(c) Rationalise the denominator of $\dfrac{4}{\sqrt{5}+2}$. **2**

(d) Simplify $\dfrac{2}{3} - \dfrac{x-1}{4}$. **2**

(e) Find a primitive function of $6 - x^{-3}$. **2**

(f) Graph the solution of $|x+2| \leq 3$ on a number line. **2**

QUESTION 2. Use a *separate* Writing Booklet.

Marks

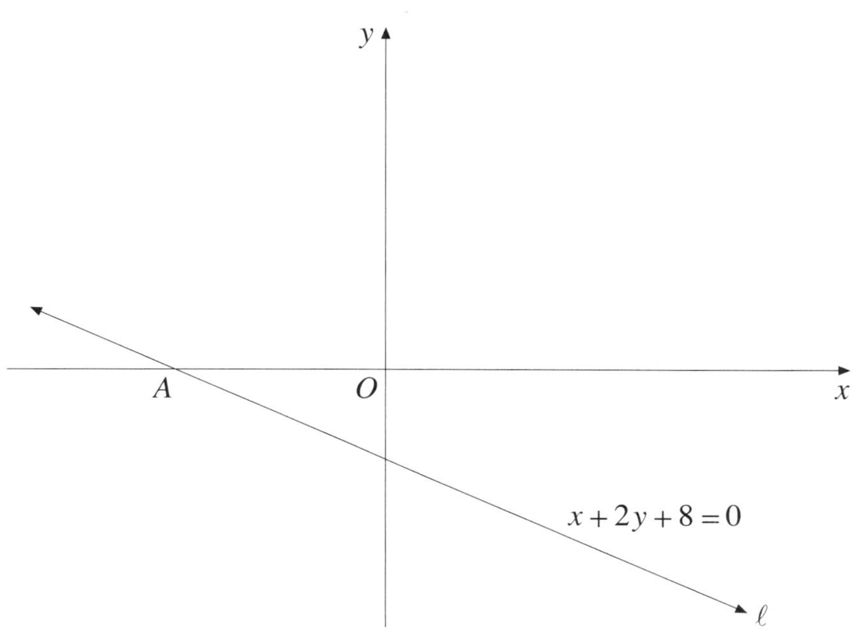

The line ℓ is shown in the diagram. It has equation $x + 2y + 8 = 0$ and cuts the x axis at A.

The line k has equation $y = -\frac{1}{2}x + 6$, and is not shown on the diagram.

Copy or trace the diagram into your Writing Booklet.

(a) Find the coordinates of A. **1**

(b) Explain why k is parallel to ℓ. **1**

(c) Draw the graph of k on your diagram, indicating where it cuts the axes. **2**

(d) Shade the region $x + 2y + 8 \leq 0$ on your diagram. **1**

(e) Write down a pair of inequalities which define the region between k and ℓ. **2**

(f) Show that $P(8, 2)$ lies on k. **1**

(g) Find the perpendicular distance from P to ℓ. **2**

(h) $Q(4, -6)$ lies on ℓ. Show that Q is the point on ℓ which is closest to P. **2**

QUESTION 3. Use a *separate* Writing Booklet.

Marks

(a) Differentiate: **5**

(i) $\sqrt{x}$

(ii) xe^{2x}

(iii) $\cos^2 x$.

(b) A layer of plastic cuts out 15% of the light and lets through the remaining 85%. **3**

(i) Show that two layers of the plastic let through 72·25% of the light.

(ii) How many layers of the plastic are required to cut out at least 90% of the light?

(c) Find $\int \sec^2 6x \, dx$. **1**

(d) Evaluate $\int_1^{e^3} \frac{5}{x} \, dx$. **3**

QUESTION 4. Use a *separate* Writing Booklet.

Marks

(a) **3**

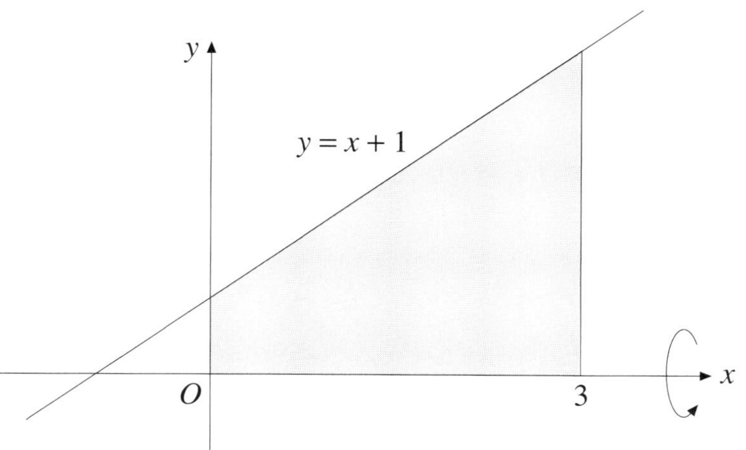

The region which lies between the x axis and the line $y = x + 1$ from $x = 0$ to $x = 3$ is rotated about the x axis to form a solid. Find the volume of the solid.

(b) Let $f(x) = \sqrt{25 - x^2}$. **9**

(i) Copy the following table of values into your Writing Booklet and supply the missing values.

x	0	1	2	3	4	5
$f(x)$	5·000		4·583			0·000

(ii) Use these six values of the function and the trapezoidal rule to find the approximate value of

$$\int_0^5 \sqrt{25 - x^2}\,dx.$$

(iii) Draw the graph of $x^2 + y^2 = 25$ and shade the region whose area is represented by the integral

$$\int_0^5 \sqrt{25 - x^2}\,dx.$$

(iv) Use your answer to part (iii) to explain why the exact value of the integral is $\dfrac{25\pi}{4}$.

(v) Use your answers to part (ii) and part (iv) to find an approximate value for π.

QUESTION 5. Use a *separate* Writing Booklet.

Marks

(a) The graph of $y = f(x)$ passes through the point (1, 3), and $f'(x) = 3x^2 - 2$. Find $f(x)$. **2**

(b) Solve the equation $u^2 - u - 1 = 0$, correct to three decimal places. **2**

(c) **6**

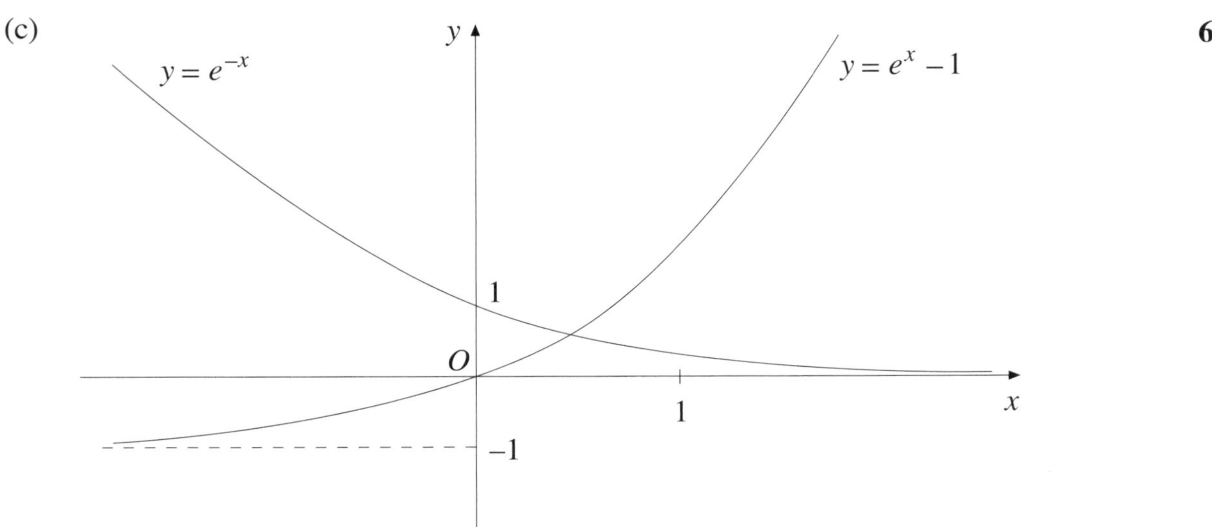

The diagram shows the graphs of $y = e^x - 1$ and $y = e^{-x}$.

(i) Find the area between the curves from $x = 1$ to $x = 2$. Leave your answer in terms of e.

(ii) Show that the curves intersect when

$$e^{2x} - e^x - 1 = 0.$$

(iii) Use the results of part (b) with $u = e^x$ to show that the x coordinate of the point of intersection of the curves is approximately 0·481.

(d) For all x in the domain $0 \le x \le 4$, a function $g(x)$ satisfies $g'(x) < 0$ and $g''(x) < 0$. **2**

Sketch a possible graph of $y = g(x)$ in this domain.

QUESTION 6. Use a *separate* Writing Booklet.

Marks

(a) Lee takes some medicine. The amount, M, of medicine present in Lee's bloodstream t hours later is given by **7**

$$M = 4t^2 - t^3, \text{ for } 0 \le t \le 3.$$

(i) Sketch the curve $M = 4t^2 - t^3$ for $0 \le t \le 3$ showing any stationary points.

(ii) At what time is the amount of medicine in Lee's bloodstream a maximum?

(iii) When is the amount present increasing most rapidly?

(b) **5**

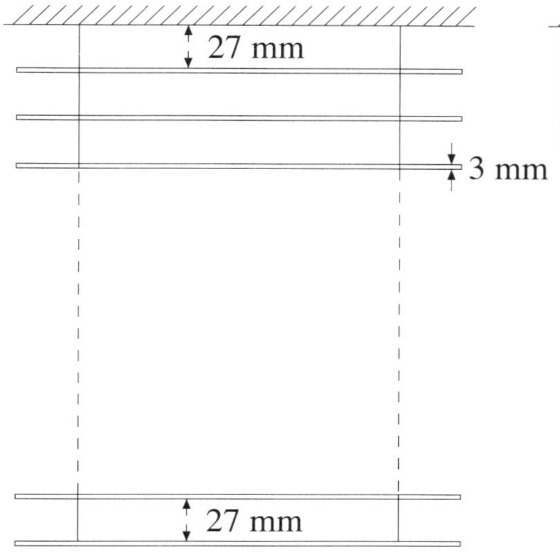

27 mm

3 mm

27 mm

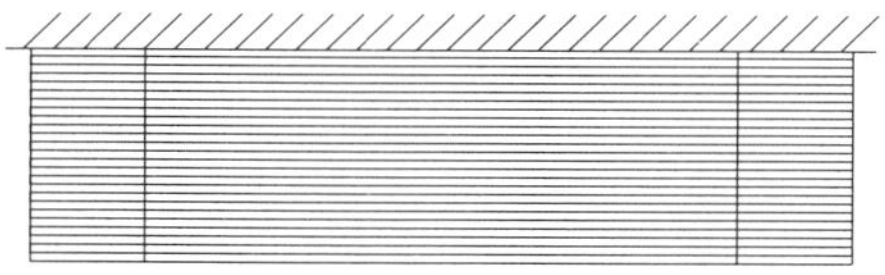

A venetian blind consists of twenty-five slats, each 3 mm thick. When the blind is down, the gap between the top slat and the top of the blind is 27 mm and the gap between adjacent slats is also 27 mm, as shown in the first diagram.

When the blind is up, all the slats are stacked at the top with no gaps, as shown in the second diagram.

(i) Show that when the blind is raised, the bottom slat rises 675 mm.

(ii) How far does the next slat rise?

(iii) Explain briefly why the distances the slats rise form an arithmetic sequence.

(iv) Find the sum of all the distances that the slats rise when the blind is raised.

QUESTION 7. Use a *separate* Writing Booklet. **Marks**

(a) (i) Sketch the graph of $y = 2\cos x$ for $0 \le x \le 2\pi$. **5**

(ii) On the same set of axes, sketch the graph of

$$y = 2\cos x - 1 \text{ for } 0 \le x \le 2\pi.$$

(iii) Find the exact values of the x coordinates of the points where the graph of $y = 2\cos x - 1$ crosses the x axis in the domain $0 \le x \le 2\pi$.

(b) **7**

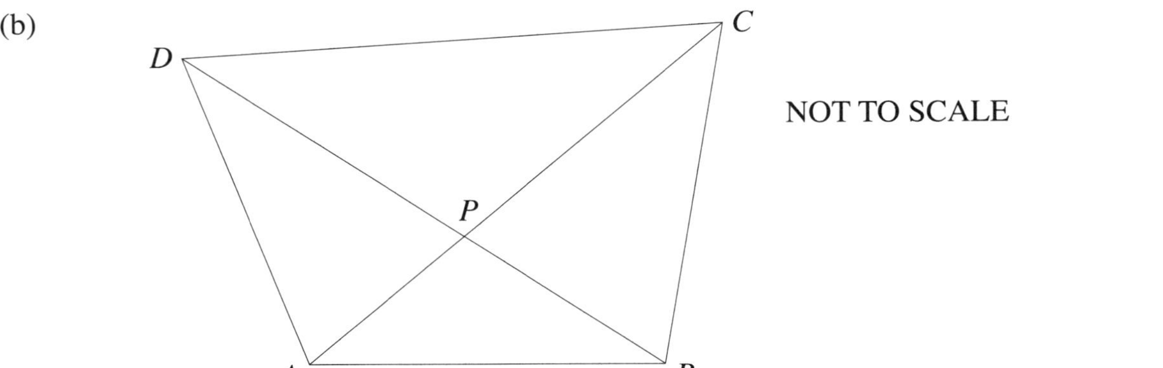

ABCD is a quadrilateral. The diagonals *AC* and *BD* intersect at *P*. $AD = BC$ and $AC = BD$.

Copy the diagram into your Writing Booklet.

(i) Show that triangles *ABC* and *ABD* are congruent.

(ii) Show that triangle *ABP* is isosceles.

(iii) Hence show that triangle *CDP* is isosceles.

(iv) Show that *AB* is parallel to *CD*.

QUESTION 8. Use a *separate* Writing Booklet.

Marks

(a) Students studying at least one of the languages, French and Japanese, attend a meeting. Of the 28 students present, 18 study French and 22 study Japanese. **4**

(i) What is the probability that a randomly chosen student studies French?

(ii) What is the probability that two randomly chosen students both study French?

(iii) What is the probability that a randomly chosen student studies both languages?

(b) **8**

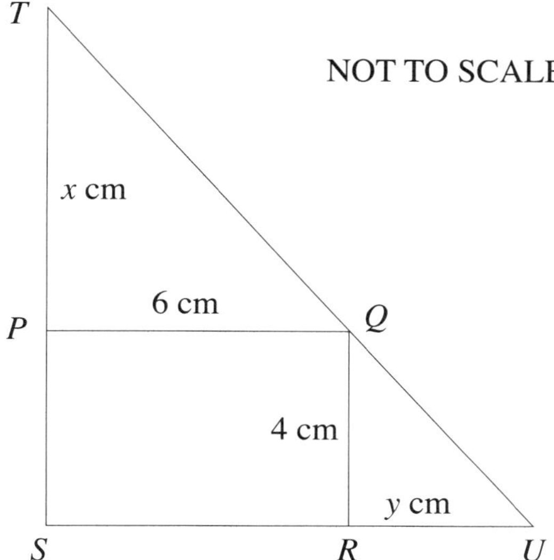

$PQRS$ is a rectangle with $PQ = 6$ cm and $QR = 4$ cm. T and U lie on the lines SP and SR respectively, so that T, Q, and U are collinear, as shown in the diagram. Let $PT = x$ cm and $RU = y$ cm.

(i) Show that triangles TPQ and QRU are similar.

(ii) Show that $xy = 24$.

(iii) Show that the area, A, of triangle TSU is given by

$$A = 24 + 3x + \frac{48}{x}.$$

(iv) Find the height and base of the triangle TSU with minimum area. Justify your answer.

QUESTION 9. Use a *separate* Writing Booklet. **Marks**

Two particles P and Q start moving along the x axis at time $t = 0$ and never meet. Particle P is initially at $x = 4$ and its velocity v at time t is given by $v = 2t + 4$.

The position of particle Q is given by $x = 1 + 3\log_e(t+1)$. The diagram shows the graph of $x = 1 + 3\log_e(t+1)$.

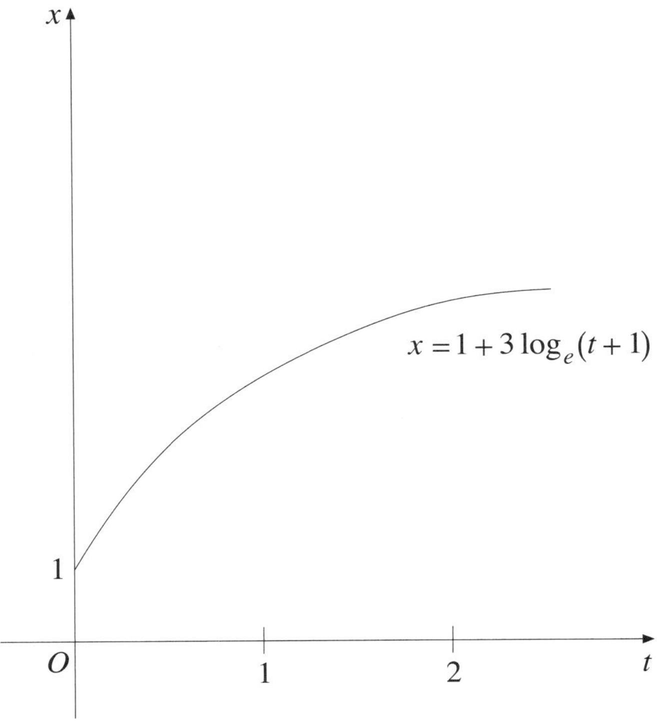

(a) Find an expression for the position of P at time t. **2**

(b) Copy the diagram into your Writing Booklet and, on the same set of axes, draw the graph of the function found in part (a). **2**

(c) P and Q are joined by an elastic string and M is the midpoint of the string. Show that the position of M at time t is given by **1**

$$x = \tfrac{1}{2}\left[t^2 + 4t + 3\log_e(t+1) + 5\right].$$

(d) Find the time at which the acceleration of M is zero. **3**

(e) Find the minimum distance between P and Q. **4**

QUESTION 10. Use a *separate* Writing Booklet.

Marks

(a) (i) On the same set of axes, accurately draw the graphs of **5**

$$y = \sin x \text{ and } y = \tfrac{2}{3}x \text{ for } 0 \le x \le \pi .$$

(ii) Find the gradient of the tangent to $y = \sin x$ at the origin.

(iii) For what values of m does the equation

$$\sin x = mx$$

have a solution in the domain $0 < x < \pi$?

(b) **5**

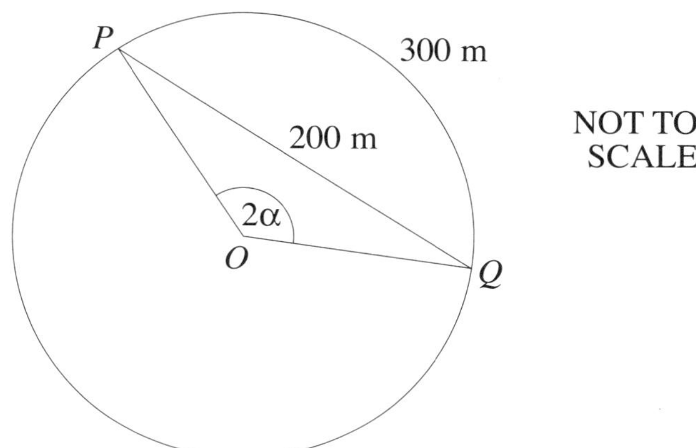

The diagram shows a circle centre O with points P and Q on the circle. The angle $POQ = 2\alpha$, the length of the chord PQ is 200 metres, and the length of the arc PQ is 300 metres as shown.

(i) Show that $\sin \alpha = \dfrac{2\alpha}{3}$.

(ii) Use your graphs from part (a) (i) to find an approximate value for α.

(iii) Hence find the size of $\angle POQ$, and also find the radius of the circle.

(c) A and B are two points 200 metres apart. For what values of ℓ is it possible to find a circular arc AB of length ℓ metres? Justify your answer. You may use the results from parts (a) and (b). **2**

1996 Higher School Certificate Worked Answers

QUESTION 1

(a) $13^{-1.2} = 0.04605\ldots$ (by calc.)
$= 0.046$ to two significant figures.

(b) $2x^2 + 3x - 2 = (2x - 1)(x + 2)$.

(c) $$\frac{4}{\sqrt{5}+2} = \frac{4}{\sqrt{5}+2} \times \frac{\sqrt{5}-2}{\sqrt{5}-2} = \frac{4\sqrt{5}-8}{5-4} = 4\sqrt{5} - 8.$$

(d) $$\frac{2}{3} - \frac{x-1}{4} = \frac{4(2) - 3(x-1)}{4 \times 3} = \frac{8 - 3x + 3}{12} = \frac{11 - 3x}{12}.$$

(e) $$\int (6 - x^{-3})dx = 6x - \frac{x^{-2}}{-2} + C = 6x + \frac{x^{-2}}{2} + C = 6x + \frac{1}{2x^2} + C.$$

(f) $|x + 2| \leqslant 3$
$\therefore -3 \leqslant x + 2 \leqslant 3$
$\therefore -5 \leqslant x \leqslant 1$

-5 1 x

QUESTION 2

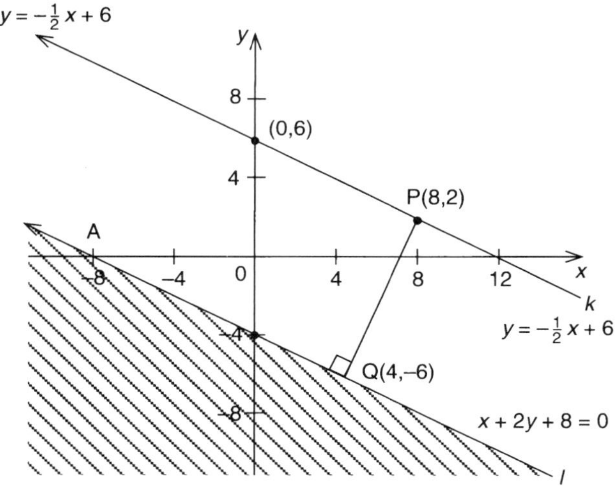

(a) At A, $y = 0$
$\therefore x + 0 + 8 = 0$
$\therefore x = -8$

$\therefore$ Coordinates of A are $(-8, 0)$.

(b) Line k: $y = -\frac{1}{2}x + 6$

Gradient $= m_1 = -\frac{1}{2}$.

Line l: $x + 2y + 8 = 0$
i.e. $2y = -x - 8$
$y = -\frac{1}{2}x - 4$

Gradient $= m_2 = -\frac{1}{2}$

Since $m_1 = m_2$, the lines are parallel.

(c) $y = -\frac{1}{2}x + 6$

$\therefore m_1 = -\frac{1}{2}$, y-intercept is 6.

(d) See diagram.

(e) $x + 2y + 8 > 0$

$y + \frac{1}{2}x - 6 < 0$

(f) Substituting (8, 2) into equation:

LHS $= 2$
RHS $= -\frac{1}{2}(8) + 6$
$= -4 + 6$
$= 2$
$=$ LHS

$\therefore$ (8, 2) lies on line k.

(g) $P \equiv (8, 2)$ line l: $x + 2y + 8 = 0$

$$d = \left|\frac{Ax_1 + By_1 + C}{\sqrt{A^2 + B^2}}\right| = \left|\frac{1(8) + 2(2) + 8}{\sqrt{1^2 + (2)^2}}\right| = \left|\frac{8 + 4 + 8}{\sqrt{1 + 4}}\right| = \left|\frac{20}{\sqrt{5}}\right|$$

$= \dfrac{20}{\sqrt{5}} \times \dfrac{\sqrt{5}}{\sqrt{5}}$

$= \dfrac{20\sqrt{5}}{5}$

$= 4\sqrt{5}$ units.

(h) $PQ = \sqrt{(8-4)^2 + (2-(-6))^2}$

$= \sqrt{(4)^2 + (8)^2}$

$= \sqrt{16 + 64}$

$= \sqrt{80}$

$= 4\sqrt{5}$ units

This is the perpendicular distance of P from l.
$\therefore$ Q is the point on l closest to P.

QUESTION 3

(a) (i) $\dfrac{d}{dx}(\sqrt{x}) = \dfrac{d}{dx}(x^{\frac{1}{2}})$

$= \dfrac{1}{2}x^{-\frac{1}{2}}$

$= \dfrac{1}{2x^{\frac{1}{2}}}$

$= \dfrac{1}{2\sqrt{x}}.$

(ii) $\dfrac{d}{dx}(xe^{2x}) = \text{x}.\dfrac{d}{dx}(e^{2x}) + e^{2x}.\dfrac{d}{dx}(x)$

$= x.2e^{2x} + e^{2x}.1$

$= 2xe^{2x} + e^{2x}.$

(iii) $\dfrac{d}{dx}(\cos^2 x) = \dfrac{d}{dx}[(\cos x)^2]$

$= 2(\cos x)^1 . \dfrac{d}{dx}(\cos x)$

$= 2\cos x(-\sin x)$

$= -2\cos x \sin x.$

(b) (i) Two layers let through
$85\% \times 85\%$

$= \dfrac{85}{100} \times \dfrac{85}{100}$

$= 0.7225$ (by calc.)

$= 72.25\%.$

(ii) Cut out at least 90% means let through 10% or less

$\therefore \left(\dfrac{85}{100}\right)^n \leqslant 10\%$

$\left(\dfrac{85}{100}\right)^n \leqslant 0.1$

$n \log(0.85) \leqslant \log(0.1)$

$n \geqslant \dfrac{\log(0.1)}{\log(0.85)}$

$n \geqslant 14.1681\ldots$ (by calc.)

$\therefore n = 15$

$\therefore$ 15 layers are required.

(c) $\displaystyle\int \sec^2 6x\,dx = \frac{1}{6}\tan 6x + C$

(d) $\displaystyle\int_1^{e^3} \frac{5}{x}\,dx = [5\ln \text{x}]_1^{e^3}$

$= 5(\ln e^3 - \ln 1)$

$= 5(3\ln e - 0)$

$= 5 \times 3$

$= 15.$

QUESTION 4

(a) $V = \pi \displaystyle\int_0^3 y^2\,dx$

$= \pi \displaystyle\int_0^3 (x+1)^2\,dx$

$= \pi \left[\dfrac{(x+1)^3}{3 \times 1}\right]_0^3$

$= \pi \left[\dfrac{(x+1)^3}{3}\right]_0^3$

$= \dfrac{\pi}{3}[(x+1)^3]_0^3$

$= \dfrac{\pi}{3}(4^3 - 1^3)$

$= \dfrac{\pi}{3}(64 - 1)$

$= \dfrac{63\pi}{3}$

$= 21\pi$ units^3

(b) (i)

x	0	1	2	3	4	5
$f(x)$	5.000	4.899	4.583	4.000	3.000	0.000
	y_0	y_1	y_2	y_3	y_4	y_5

(ii) $\displaystyle\int_0^5 \sqrt{25 - x^2}\,dx$

$\doteqdot \dfrac{b-a}{2n}[y_0 + 2(y_1 + y_2 + y_3 + y_4) + y_5]$

$= \dfrac{5-0}{2 \times 5}[5.000 + 2(4.899 + 4.583 + 4.000 + 3.000) + 0.000]$

$= \dfrac{5}{10}[37.964]$

$= 18.982$ (by calc.)

(iii)

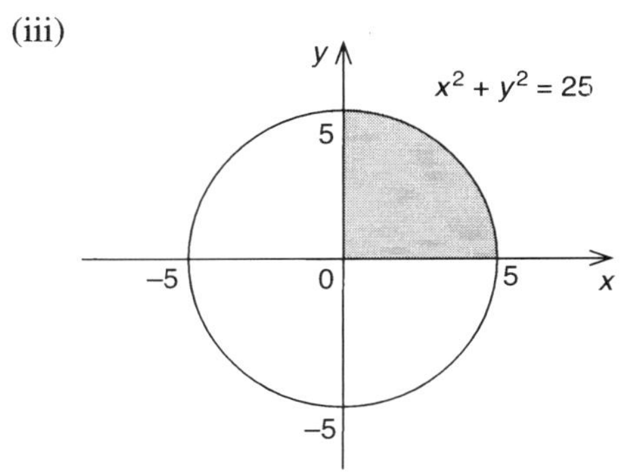

(iv) Area of circle $= \pi r^2$

$$\therefore \int_0^5 \sqrt{25 - x^2}\, dx = \frac{\pi r^2}{4} \quad \text{(quarter of circle)}$$
$$= \frac{\pi(5)^2}{4}$$
$$= \frac{25\pi}{4}.$$

(v) From parts (ii) and (iv):

$$\frac{25\pi}{4} \doteqdot 18.982$$
$$\pi \doteqdot 18.982 \times \frac{4}{25}$$
$$\pi \doteqdot 3.037\ldots \quad \text{(by calc.)}$$
$$\pi \doteqdot 3.04.$$

QUESTION 5

(a) $f'(x) = 3x^2 - 2$

$$f(x) = \int (3x^2 - 2)\,dx$$
$$= \frac{3x^3}{3} - 2x + C$$
$$\therefore f(x) = x^3 - 2x + C$$

Now $y = f(x)$ passes through $(1, 3)$

$$\therefore 3 = 1^3 - 2(1) + C$$
$$3 = 1 - 2 + C$$
$$\therefore C = 4$$

$$\therefore f(x) = x^3 - 2x + 4$$

(b) $u = \dfrac{-b \pm \sqrt{b^2 - 4ac}}{2a}$

$$= \frac{-(-1) \pm \sqrt{(-1)^2 - 4(1)(-1)}}{2(1)}$$
$$= \frac{1 \pm \sqrt{1 + 4}}{2}$$
$$= \frac{1 \pm \sqrt{5}}{2}$$
$$= \frac{1 + \sqrt{5}}{2} \text{ or } \frac{1 - \sqrt{5}}{2}$$
$$= 1.6180\ldots \text{ or } -0.6180\ldots \quad \text{(by calc.)}$$
$$= 1.618 \text{ or } -0.618 \text{ to 3 decimal places.}$$

(c) (i) $A = \displaystyle\int_1^2 (e^x - 1)\,dx - \int_1^2 e^{-x}\,dx$

$$= \int_1^2 (e^x - 1 - e^{-x})\,dx$$
$$= \left[e^x - x + e^{-x}\right]_1^2$$
$$= (e^2 - 2 + e^{-2}) - (e^1 - 1 + e^{-1})$$
$$= e^2 - 2 + \frac{1}{e^2} - e + 1 - \frac{1}{e}$$
$$= \left(e^2 + \frac{1}{e^2} - e - \frac{1}{e} - 1\right) \text{ units}^2$$

(ii) The curves intersect when

$$e^x - 1 = e^{-x}$$
$$\therefore e^x - 1 = \frac{1}{e^x}$$
$$\therefore e^{2x} - e^x = 1$$
$$\therefore e^{2x} - e^x - 1 = 0.$$

(iii) From part (b), curves intersect at 1.618, but $u = e^x$

$$\therefore e^x = 1.618$$
$$x = \log_e 1.618$$
$$\therefore x \doteqdot 0.481 \quad \text{(by calc.)}$$

(d)

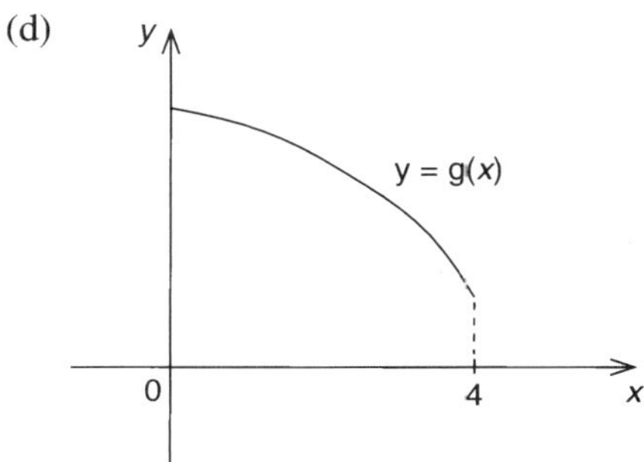

QUESTION 6

(a) (i) $M = 4t^2 - t^3$

$$\frac{dM}{dt} = 8t - 3t^2$$
$$= t(8 - 3t)$$
$$\frac{d^2M}{dt^2} = 8 - 6t$$

Stationary points occur when

$$\frac{dM}{dt} = 0$$
$$t(8 - 3t) = 0$$
$$t = 0 \text{ or } \tfrac{8}{3}$$

When $t = 0$, $M = 0$

and $\dfrac{d^2M}{dt^2} = 8 > 0$

$\therefore$ A minimum turning point at $(0, 0)$.

When $t = \frac{8}{3}$, $M = 4\left(\frac{8}{3}\right)^2 - \left(\frac{8}{3}\right)^3$

$$= 4\left(\tfrac{64}{9}\right) - \tfrac{512}{27}$$
$$= \frac{4 \times 64 \times 3}{27} - \frac{512}{27}$$
$$= \frac{256}{27}$$
$$= 9\tfrac{13}{27}$$

and $\dfrac{d^2M}{dt^2} = 8 - 6\left(\frac{8}{3}\right)$

$$= 8 - 16$$
$$= -8$$
$$< 0$$

$\therefore$ A maximum turning point at $\left(\frac{8}{3}, 9\frac{13}{27}\right)$

End point of domain:
When $t = 3$, $M = 4(3)^2 - (3)^3$
$= 36 - 27$
$= 9$

$\therefore$ End point is (3, 9).

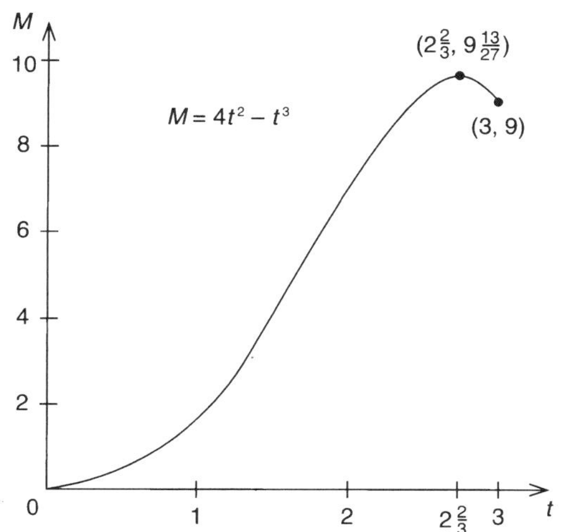

(ii) Maximum occurs when $t = \frac{8}{3}$
$= 2\frac{2}{3}$ hours.

(iii) The amount is increasing most rapidly when the slope of the curve is the greatest,

i.e. when $\frac{d^2M}{dt^2} = 0$

$$8 - 6t = 0$$
$$6t = 8$$
$$t = \frac{8}{6}$$
$$= \frac{4}{3}$$
$$= 1\tfrac{1}{3} \text{ hours.}$$

(b) (i) Since there are 25 gaps between the top and the last slat,

Rise $= 25 \times 27$ mm
$= 675$ mm

(ii) Rise $= 24 \times 27$ mm
$= 648$ mm.

(iii) First slat moves 27 mm.
Second slat moves 54 mm.
Third slat moves 81 mm.

This is an arithmetic sequence with $a = 27$ and $d = 27$

(iv) $S_n = \frac{n}{2}(a + l)$

$n = 25$, $a = 27$, $l = 675$

$$S_{25} = \frac{25}{2}(27 + 675)$$
$$= \frac{25}{2}(702)$$
$$= 8775 \text{ mm.}$$

QUESTION 7

(a) (i) (ii)

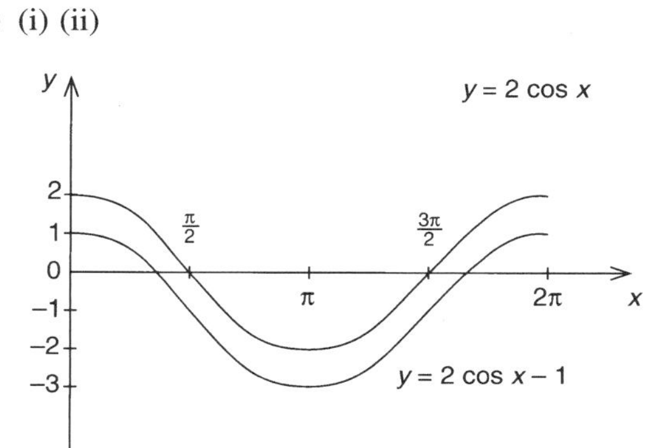

(iii) $y = 2\cos x - 1$ crosses the x-axis when $y = 0$

$\therefore 2\cos x - 1 = 0 \quad 0 \leq x \leq 2\pi.$

$$2\cos x = 1$$
$$\cos x = \frac{1}{2}$$
$$\therefore x = \frac{\pi}{3} \text{ or } \left(2\pi - \frac{\pi}{3}\right)$$
$$= \frac{\pi}{3} \text{ or } \frac{5\pi}{3}.$$

(b)

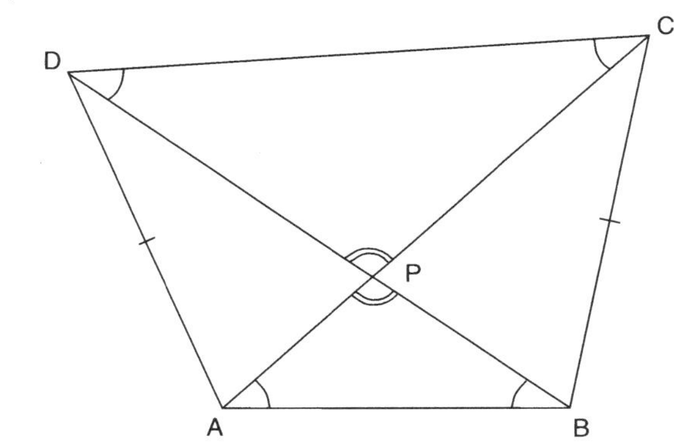

(i) In Δ's ABC and ABD,

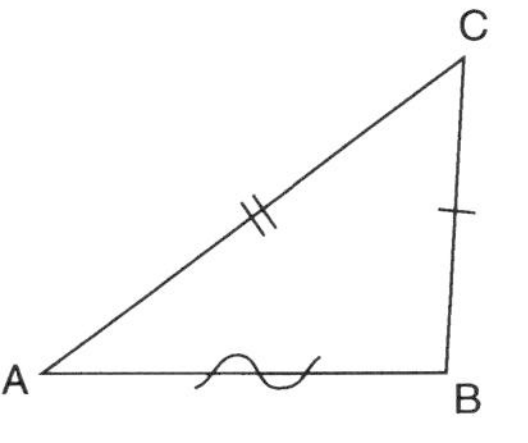

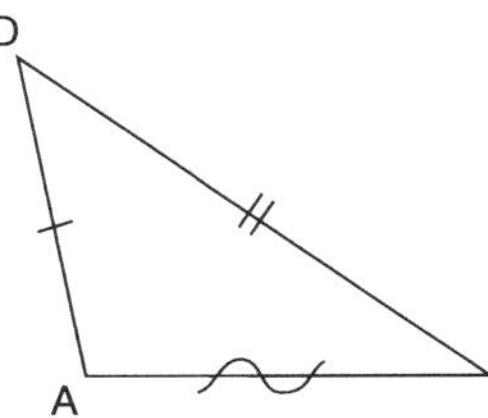

AB is common
$AC = DB$ (given)
$BC = DA$ (given)

$\therefore \Delta ABC \equiv \Delta ABD$ (SSS)

(ii) $\therefore\ \angle CAB = \angle DBA$
(corresponding angles of congruent Δs)
$\therefore\ \angle ABP$ is isosceles (base angles equal)

(iii) Since ΔABP is isosceles,
$\therefore\ AP = BP$
Now $AC = BD$ (given)
$\therefore\ PC = PD$
$\therefore\ \Delta CDP$ is isosceles (two equal sides)

(iv) Now $\angle APB = \angle DPC$
(vertically opposite)

and ΔAPB and ΔCDP are both isosceles triangles with equal base angles.

$\therefore\ \angle PDC = \angle DCP = \angle PBA = \angle BAP$
$\therefore\ \angle PDC = \angle PBA$
$\therefore\ DC \parallel AB$ (alternate angles equal)

QUESTION 8

(a) Let x be the number of students studying both French and Japanese.

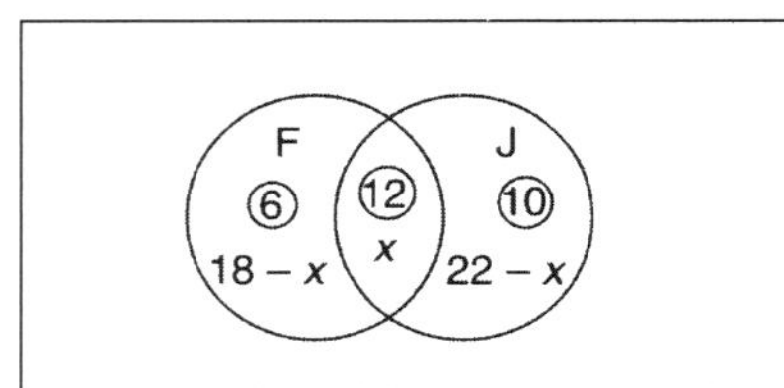

Now $18 - x + x + 22 - x = 28$
$40 - x = 28$
$x = 12$

(i) $P(F) = \frac{18}{28}$
$= \frac{9}{14}$.

(ii) $P(FF) = \frac{18}{28} \times \frac{17}{27}$
$= \frac{17}{42}$.

(iii) $P(\text{Both}) = \frac{12}{28}$
$= \frac{3}{7}$.

(b) (i) In Δ's TPQ and QRU:

$\angle TPQ = \angle QRU = 90°$
($PQRS$ a rectangle)
$\angle TQP = \angle QUR$
(corresponding, $PQ \parallel SU$)
$\therefore\ \Delta TPQ \parallel\!| \Delta QRU$ (equal angles)

(ii) Since $\Delta TPQ \parallel\!| \Delta QRU$,

$$\frac{TP}{QR} = \frac{PQ}{RU}$$
$$\frac{x}{4} = \frac{6}{y}$$
$$\therefore\ xy = 24 \ (\therefore\ y = \frac{24}{x})$$

(iii) $\therefore\ A = \frac{1}{2} \times SU \times TS$
$= \frac{1}{2}(6 + y)(4 + x)$

but from (ii) $y = \frac{24}{x}$

$$\therefore\ A = \frac{1}{2}\left(6 + \frac{24}{x}\right)(4 + x)$$
$$= \frac{1}{2}\left(24 + 6x + \frac{96}{x} + 24\right)$$
$$= \frac{1}{2}\left(48 + 6x + \frac{96}{x}\right)$$
$$= 24 + 3x + \frac{48}{x}$$

(iv) $A = 24 + 3x + 48x^{-1}$

$$\frac{dA}{dx} = 3 - 48x^{-2}$$
$$= 3 - \frac{48}{x^2}$$
$$\frac{d^2A}{dx^2} = 96x^{-3}$$
$$= \frac{96}{x^3}$$

Stationary points occur when

$$\frac{dA}{dx} = 0$$
$$3 - \frac{48}{x^2} = 0$$
$$3 = \frac{48}{x^2}$$
$$x^2 = 16$$

$\therefore\ x = 4$ ($x = -4$ is ignored as x is a positive length)

When $x = 4$

$$\frac{d^2A}{dx^2} = \frac{96}{64} > 0$$

$\therefore$ A minimum value occurs when $x = 4$

$\therefore$ Height $= x + 4$
$= 4 + 4$
$= 8$ cm

Base $= 6 + \frac{24}{x}$
$= 6 + \frac{24}{4}$
$= 12$ cm.

QUESTION 9

(a) $v = 2t + 4$

i.e. $\frac{dx}{dt} = 2t + 4$

$\therefore x = t^2 + 4t + C$

When $t = 0, x = 4 \therefore C = 4$

$\therefore x = t^2 + 4t + 4$

(b)

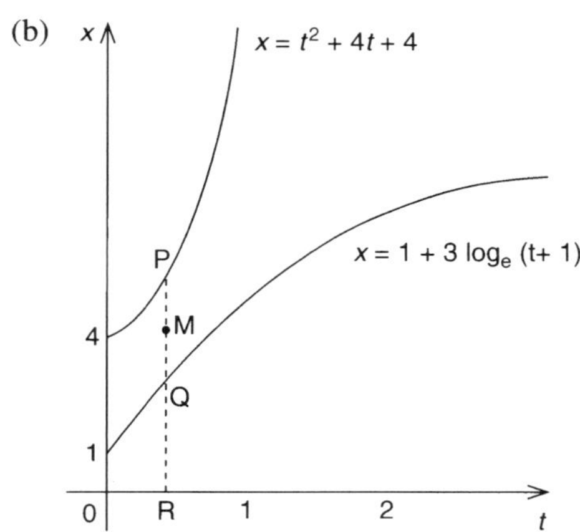

(c) The position of M is given by

$\frac{PR + QR}{2}$

i.e. $x = \frac{1}{2}[t^2 + 4t + 4 + 1 + 3\log_e(t + 1)]$

$= \frac{1}{2}[t^2 + 4t + 3\log_e(t + 1) + 5]$

(d) $v = \frac{dx}{dt} = \frac{1}{2}\left[2t + 4 + \frac{3}{t + 1}\right]$

$= \frac{1}{2}[2t + 4 + 3(t + 1)^{-1}]$

$a = \frac{d^2x}{dt^2} = \frac{1}{2}[2 - 3(t + 1)^{-2}]$

$= \frac{1}{2}\left[2 - \frac{3}{(t + 1)^2}\right]$

When $a = 0$:

$\therefore \frac{1}{2}\left[2 - \frac{3}{(t + 1)^2}\right] = 0$

$2 - \frac{3}{(t + 1)^2} = 0$

$2(t + 1)^2 - 3 = 0$

$2(t + 1)^2 = 3$

$(t + 1)^2 = \frac{3}{2}$

$t + 1 = \pm\sqrt{\frac{3}{2}}$

$\therefore t = \sqrt{\frac{3}{2}} - 1$

($-\sqrt{\frac{3}{2}} - 1$ is ignored as it is negative)

(e) $D = (t^2 + 4t + 4) - (1 + 3\log_e(t + 1))$

$= t^2 + 4t - 3\log_e(t + 1) + 3$

$\frac{dD}{dt} = 2t + 4 - \frac{3}{t + 1}$

$= 2t + 4 - 3(t + 1)^{-1}$

$\frac{d^2D}{dt^2} = 2 + 3(t + 1)^{-2}$

$= 2 + \frac{3}{(t + 1)^2}$

Stationary points occur when

$\frac{dD}{dt} = 0$

$2t + 4 - \frac{3}{t + 1} = 0$

$(2t + 4)(t + 1) - 3 = 0$

$2t^2 + 6t + 4 - 3 = 0$

$2t^2 + 6t + 1 = 0$

$t = \frac{-6 \pm \sqrt{(6)^2 - 4(2)(1)}}{2(2)}$

$= \frac{-6 \pm \sqrt{28}}{4}$

These are both negative times which do not exist.

$\therefore$ The minimum distance occurs when $t = 0$ (as the curves are diverging)

i.e. $D = 4 - 1$

$= 3$

QUESTION 10

(a) (i)

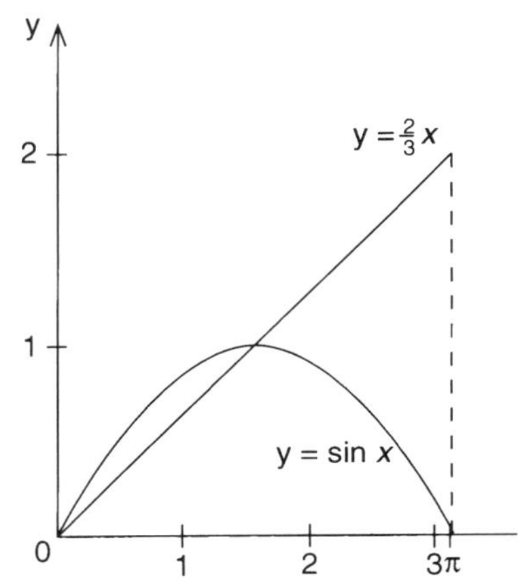

(ii) $y = \sin x$

$\frac{dy}{dx} = \cos x$

When $x = 0$, $\frac{dy}{dx} = \cos 0$

$= 1$

$\therefore$ Gradient of tangent at the origin is 1.

(iii) $0 < m < 1$

(b) (i)

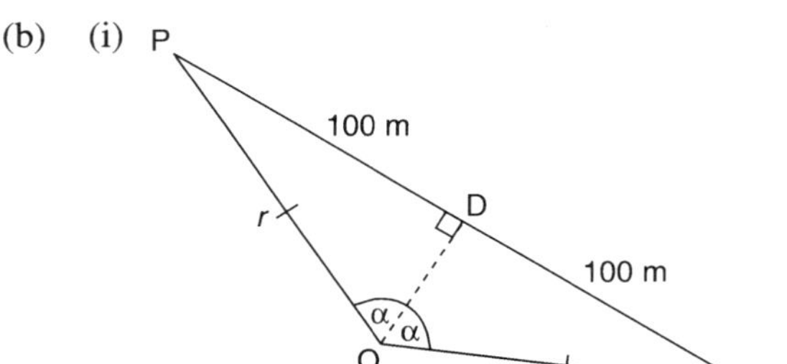

Let D be the midpoint of PQ.
In $\triangle POD$,

$$\sin\alpha = \frac{100}{r}$$
$$\therefore r\sin\alpha = 100$$
$$\therefore r = \frac{100}{\sin\alpha}$$

In the circle, centre O,

$$l = r\theta$$
$$\therefore 300 = r \times 2\alpha$$
$$\therefore r = \frac{300}{2\alpha}$$

Equating each expression for r,

$$\therefore \frac{100}{\sin\alpha} = \frac{300}{2\alpha}$$
$$\frac{\sin\alpha}{100} = \frac{2\alpha}{300}$$
$$\sin\alpha = \frac{200\alpha}{300}$$
$$\sin\alpha = \frac{2\alpha}{3}$$

(ii) $\alpha \doteqdot 1.5$

(iii) $\angle POQ = 2\alpha$
$= 2 \times 1.5$
$= 3$ radians.

and $r = \dfrac{300}{2\alpha}$
$= \dfrac{300}{3}$
$= 100$ m.

(c) From (i),

$$r = \frac{100}{\sin\alpha}$$

Now $l = r\theta$

$$\therefore r = \frac{l}{\theta}$$

Equating each expression for r,

$$\frac{l}{\theta} = \frac{100}{\sin\alpha}$$

$$l\sin\alpha = 100\theta$$
$$\frac{\sin\alpha}{\theta} = \frac{100}{l}$$

Now $\theta = 2\alpha$ (from diagram)

$$\therefore \frac{\sin\alpha}{2\alpha} = \frac{100}{l}$$

$$\therefore \frac{\sin\alpha}{\alpha} = \frac{200}{l} \quad \ldots (1)$$

Now if $\sin\alpha = m\alpha$

i.e. $\dfrac{\sin\alpha}{\alpha} = m$

then $0 < m < 1$ (from (a))

Now $\dfrac{\sin\alpha}{\alpha} = \dfrac{200}{l} = m$ from (1)

$$\therefore 0 < \frac{200}{l} < 1$$

$\therefore 0 < 200 < l$
$\therefore l > 200.$

HIGHER SCHOOL CERTIFICATE EXAMINATION

1997

MATHEMATICS

2/3 UNIT (COMMON)

Time allowed—Three hours
(Plus 5 minutes reading time)

DIRECTIONS TO CANDIDATES

- Attempt ALL questions.
- ALL questions are of equal value.
- All necessary working should be shown in every question. Marks may be deducted for careless or badly arranged work.
- Standard integrals are printed on page 12.
- Board-approved calculators may be used.
- Answer each question in a *separate* Writing Booklet.
- You may ask for extra Writing Booklets if you need them.

QUESTION 1. Use a *separate* Writing Booklet. **Marks**

(a) Find the value of $\dfrac{1}{7+5\times 3}$ correct to three significant figures. **2**

(b) Simplify $(2-3x)-(5-4x)$. **2**

(c) Write down the exact value of $135°$ in radians. **1**

(d) **1**

In the diagram, PQ is an arc of a circle with centre O. The radius $OP = 3$ cm and the angle POQ is $\dfrac{5\pi}{6}$ radians.

Find the length of the arc PQ.

(e) Using the table of standard integrals, find $\int \sec 3x \tan 3x \, dx$. **1**

(f) Forty-five balls, numbered 1 to 45, are placed in a barrel, and one ball is drawn at random. What is the probability that the number on the ball drawn is even? **2**

(g) By rationalising the denominator, express $\dfrac{8}{3-\sqrt{5}}$ in the form $a+b\sqrt{5}$. **3**

QUESTION 2. Use a *separate* Writing Booklet.

Marks

(a) **6**

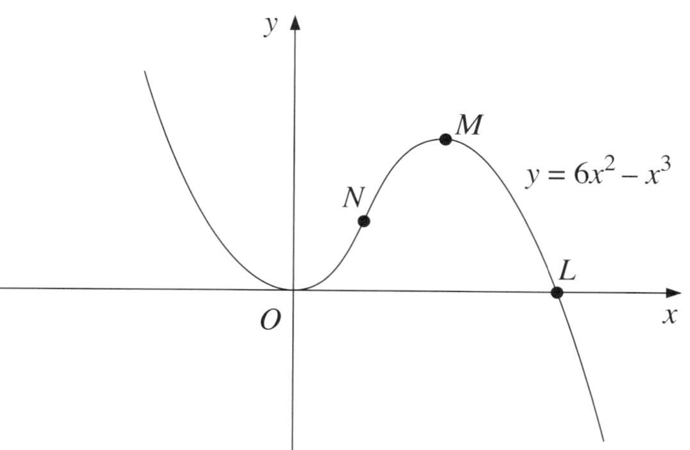

The diagram shows a sketch of the curve $y = 6x^2 - x^3$. The curve cuts the x axis at L, and has a local maximum at M and a point of inflection at N.

(i) Find the coordinates of L.

(ii) Find the coordinates of M.

(iii) Find the coordinates of N.

(b) The graph of $y = f(x)$ passes through the point $(1, 4)$ and $f'(x) = 2x + 7$. Find $f(x)$. **2**

(c) Find a primitive of $\dfrac{2x}{x^2 + 1}$. **1**

(d) Consider the parabola with equation $x^2 = 4(y - 1)$. **3**

(i) Find the coordinates of the vertex of the parabola.

(ii) Find the coordinates of the focus of the parabola.

QUESTION 3. Use a *separate* Writing Booklet. **Marks**

(a) Differentiate the following functions: **6**

(i) $\left(x^2+5\right)^3$

(ii) $\dfrac{\cos x}{x}$

(iii) $x^2 \ln x$.

(b) Let A and B be the points $(0,1)$ and $(2,3)$ respectively. **6**

(i) Find the coordinates of the midpoint of AB.

(ii) Find the slope of the line AB.

(iii) Find the equation of the perpendicular bisector of AB.

(iv) The point P lies on the line $y = 2x - 9$ and is equidistant from A and B. Find the coordinates of P.

QUESTION 4. Use a *separate* Writing Booklet.

Marks

(a) The table shows the values of a function $f(x)$ for five values of x. **3**

x	1	1·5	2	2·5	3
$f(x)$	5	1	–2	3	7

Use Simpson's rule with these five values to estimate $\int_1^3 f(x)\,dx$.

(b) (i) Sketch the graph of $y = x^2 - 6$, and label all intercepts with the axes. **6**

(ii) On the same set of axes, carefully sketch the graph of $y = |x|$.

(iii) Find the x coordinates of the two points where the graphs intersect.

(iv) Hence solve the inequality $x^2 - 6 \le |x|$.

(c) **3**

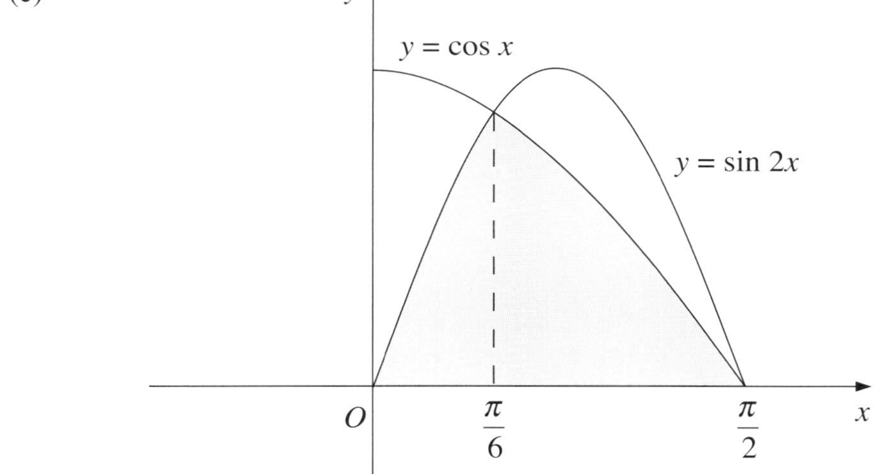

The diagram shows the graphs of the functions $y = \cos x$ and $y = \sin 2x$ between $x = 0$ and $x = \frac{\pi}{2}$. The two graphs intersect at $x = \frac{\pi}{6}$ and $x = \frac{\pi}{2}$.

Calculate the area of the shaded region.

QUESTION 5. Use a *separate* Writing Booklet. **Marks**

(a) Evaluate $\int_0^{\ln 7} e^{-x}\,dx$. **2**

(b) **4**

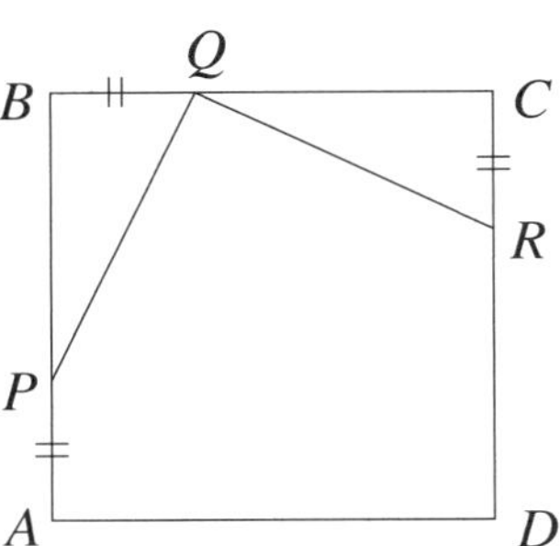

In the diagram, $ABCD$ is a square. The points P, Q, and R lie on AB, BC, and CD respectively, such that $AP = BQ = CR$.

(i) Prove that triangles PBQ and QCR are congruent.

(ii) Prove that $\angle PQR$ is a right angle.

(c) The rate of inflation measures the rate of change in prices. Between January 1996 and December 1996, prices were rising but the rate of inflation was falling. Draw a graph of prices as a function of time that fits this description. **2**

(d) A ball is dropped from a height of 2 metres onto a hard floor and bounces. After each bounce, the maximum height reached by the ball is 75% of the previous maximum height. Thus, after it first hits the floor, it reaches a height of 1·5 metres before falling again, and after the second bounce, it reaches a height of 1·125 metres before falling again. **4**

(i) What is the maximum height reached after the third bounce?

(ii) What kind of sequence is formed by the successive maximum heights?

(iii) What is the total distance travelled by the ball from the time it was first dropped until it eventually comes to rest on the floor?

QUESTION 6. Use a *separate* Writing Booklet.

Marks

(a) **6**

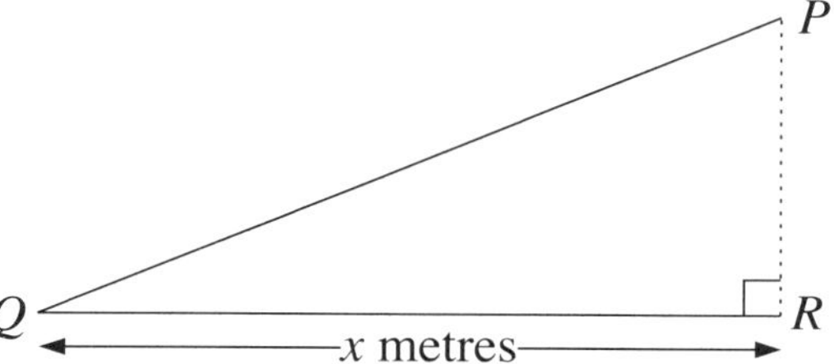

A wire of length 5 metres is to be bent to form the hypotenuse and base of a right-angled triangle PQR, as shown in the diagram. Let the length of the base QR be x metres.

(i) What is the length of the hypotenuse PQ in terms of x?

(ii) Show that the area of the triangle PQR is $\frac{1}{2}x\sqrt{25-10x}$ square metres.

(iii) What is the maximum possible area of the triangle?

(b) **6**

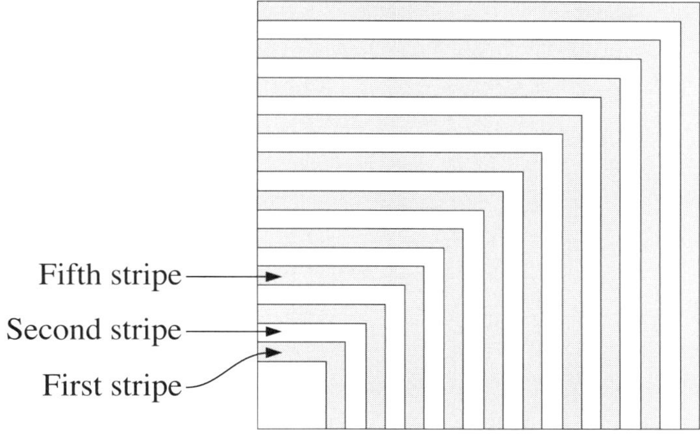

A logo is made of 20 squares with a common corner, as shown in the diagram. The odd-numbered 'stripes' between successive squares are shaded in the diagram. The shaded stripes are painted in gold paint, which costs \$9 per square centimetre.

The side length of the nth square is $(2n+4)$ cm. The nth stripe lies between the nth square and the $(n+1)$th square.

(i) Show that the area of the nth stripe is $(8n+20)$ cm^2.

(ii) Hence find the areas of the first and last stripes.

(iii) Hence find the total cost of the gold paint for the logo.

QUESTION 7. Use a *separate* Writing Booklet.

Marks

(a) By expressing $\sec\theta$ and $\tan\theta$ in terms of $\sin\theta$ and $\cos\theta$, show that **2**

$$\sec^2\theta - \tan^2\theta = 1.$$

(b) **5**

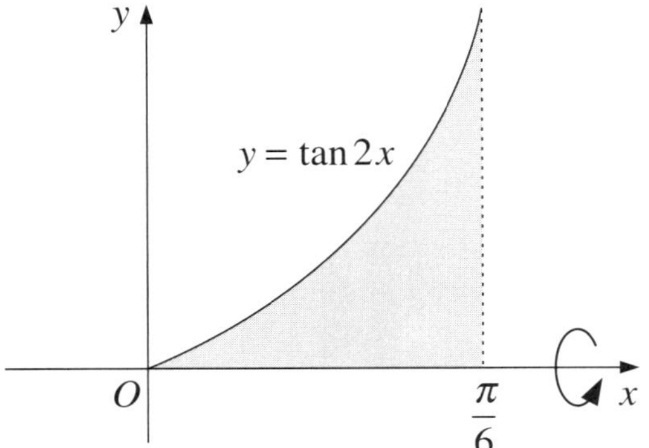

The diagram shows part of the graph of the function $y = \tan 2x$. The shaded region is bounded by the curve, the x axis, and the line $x = \frac{\pi}{6}$. The region is rotated about the x axis to form a solid.

(i) Show that the volume of the solid is given by

$$V = \pi \int_0^{\frac{\pi}{6}} \left(\sec^2 2x - 1\right) dx.$$

You may use the result of part (a).

(ii) Find the exact volume of the solid.

(c) A ball is dropped into a long vertical tube filled with honey. The rate at which the ball decelerates is proportional to its velocity. Thus **5**

$$\frac{dv}{dt} = -kv,$$

where v is the velocity in centimetres per second, t is the time in seconds, and k is a constant.

When the ball first enters the honey, at $t = 0$, $v = 100$. When $t = 0{\cdot}25$, $v = 85$.

(i) Show that $v = Ce^{-kt}$ satisfies the equation $\frac{dv}{dt} = -kv$.

(ii) Find the value of the constant C.

(iii) Find the value of the constant k.

(iv) Find the velocity when $t = 2$.

QUESTION 8. Use a *separate* Writing Booklet.

Marks

(a) **4**

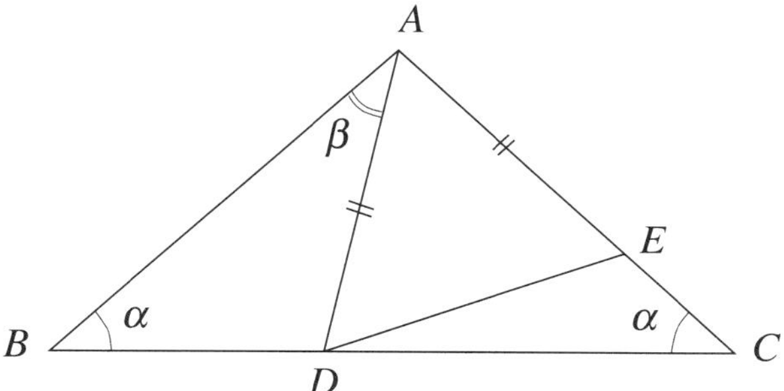

In the isosceles triangle ABC, $\angle ABC = \angle ACB = \alpha$. The points D and E lie on BC and AC, so that $AD = AE$, as shown in the diagram. Let $\angle BAD = \beta$.

(i) Explain why $\angle ADC = \alpha + \beta$.

(ii) Find $\angle DAC$ in terms of α and β.

(iii) Hence, or otherwise, find $\angle EDC$ in terms of β.

(b) A particle is moving along the x axis. Its position at time t is given by **8**

$$x = t + \sin t.$$

(i) At what times during the period $0 < t < 3\pi$ is the particle stationary?

(ii) At what times during the period $0 < t < 3\pi$ is the acceleration equal to 0?

(iii) Carefully sketch the graph of $x = t + \sin t$ for $0 < t < 3\pi$.

Clearly label any stationary points and any points of inflection.

QUESTION 9. Use a *separate* Writing Booklet. **Marks**

(a) A bag contains two red balls, one black ball, and one white ball. Andrew selects one ball from the bag and keeps it hidden. He then selects a second ball, also keeping it hidden. **5**

(i) Draw a tree diagram to show all the possible outcomes.

(ii) Find the probability that both the selected balls are red.

(iii) Find the probability that at least one of the selected balls is red.

(iv) Andrew drops one of the selected balls and we can see that it is red. What is the probability that the ball that is still hidden is also red?

(b) **7**

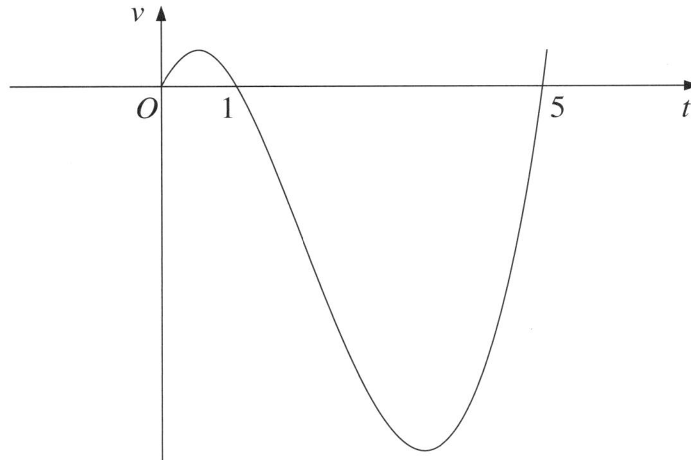

A pen moves along the x axis, ruling a line. The diagram shows the graph of the velocity of the tip of the pen as a function of time.

The velocity, in centimetres per second, is given by the equation

$$v = 4t^3 - 24t^2 + 20t,$$

where t is the time in seconds. When $t = 0$, the tip of the pen is at $x = 3$. That is, the tip is initially 3 centimetres to the right of the origin.

(i) Find an expression for x, the position of the tip of the pen, as a function of time.

(ii) What feature will the graph of x as a function of t have at $t = 1$?

(iii) The pen uses 0·05 milligrams of ink per centimetre travelled.

How much ink is used between $t = 0$ and $t = 2$?

QUESTION 10. Use a *separate* Writing Booklet.

Marks

(a) Graph the solution of $4x \le 15 \le -9x$ on a number line. **2**

(b) **10**

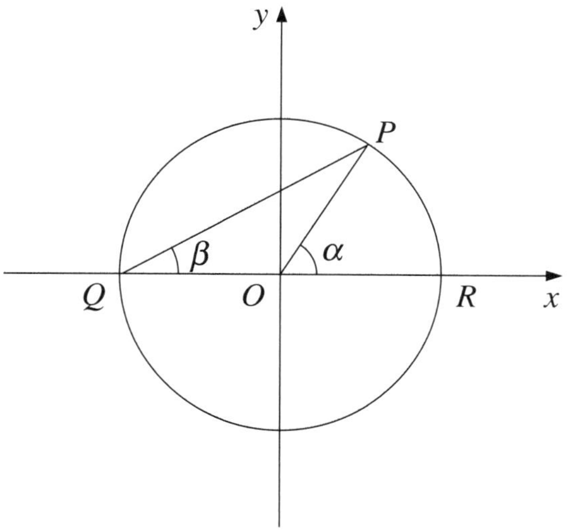

In the diagram, Q is the point $(\pm 1, 0)$, R is the point $(1, 0)$, and P is another point on the circle with centre O and radius 1. Let $\angle POR = \alpha$ and $\angle PQR = \beta$, and let $\tan \beta = m$.

(i) Explain why ΔOPQ is isosceles, and hence deduce that $\alpha = 2\beta$.

(ii) Find the equation of the line PQ.

(iii) Show that the x coordinates of P and Q are solutions of the equation

$$\left(1 + m^2\right)x^2 + 2m^2 x + m^2 - 1 = 0 .$$

(iv) Using this equation, find the coordinates of P in terms of m.

(v) Hence deduce that $\tan 2\beta = \dfrac{2\tan\beta}{1 - \tan^2\beta}$.

1997 Higher School Certificate Worked Answers

QUESTION 1

(a) $\dfrac{1}{7 + 5 \times 3} = \dfrac{1}{7 + 15}$

$= \dfrac{1}{22}$

$= 0.04545\ldots$ (by calc.)

$= 0.0455$ to 3 significant figures.

(b) $(2 - 3x) - (5 - 4x) = 2 - 3x - 5 + 4x$

$= x - 3.$

(c) $135° = \left(135 \times \dfrac{\pi}{180}\right)$ radians

$= \dfrac{3\pi}{4}$ radians.

(d) $l = r \times \theta$

$= 3 \times \dfrac{5\pi}{6}$

$= \dfrac{5\pi}{2}$ cm.

(e) $\int \sec 3x \tan 3x \, dx$

$= \frac{1}{3} \sec 3x + C.$

(f) $P(\text{Even}) = \dfrac{22}{45}.$

(g) $\dfrac{8}{3 - \sqrt{5}} = \dfrac{8}{3 - \sqrt{5}} \times \dfrac{3 + \sqrt{5}}{3 + \sqrt{5}}$

$= \dfrac{8(3 + \sqrt{5})}{9 - 5}$

$= \dfrac{8(3 + \sqrt{5})}{4}$

$= 2(3 + \sqrt{5})$

$= 6 + 2\sqrt{5}$

$= a + b\sqrt{5}$

where $a = 6$ and $b = 2$.

QUESTION 2

(a) (i) $y = 6x^2 - x^3$

$= x^2(6 - x)$

At L, $y = 0$

$\therefore 0 = x^2(6 - x)$

$\therefore x = 0$ or 6

$\therefore$ Coordinates of L are $(6, 0)$.

(ii) $\dfrac{dy}{dx} = 12x - 3x^2$

$= 3x(4 - x)$

At M, $\dfrac{dy}{dx} = 0$

$\therefore 3x(4 - x) = 0$

$x = 0$ or 4

When $x = 4$, $y = 6(4)^2 - (4)^3$

$= 96 - 64$

$= 32$

$\therefore$ Coordinates of M are $(4, 32)$.

(iii) $\dfrac{d^2y}{dx^2} = 12 - 6x$

$= 6(2 - x)$

At N, $\dfrac{d^2y}{dx^2} = 0$

$\therefore 6(2 - x) = 0$

$\therefore x = 2$

When $x = 2$, $y = 6(2)^2 - (2)^3$

$= 24 - 8$

$= 16$

$\therefore$ Coordinates of N are $(2, 16)$.

(b) $f'(x) = 2x + 7$

$f(x) = \dfrac{2x^2}{2} + 7x + C$

$= x^2 + 7x + C$

Now $f(1) = 4$

$\therefore 4 = (1)^2 + 7(1) + C$

$4 = 1 + 7 + C$

$\therefore c = -4$

$\therefore f(x) = x^2 + 7x - 4$

(c) $\int \dfrac{2x}{x^2 + 1} \, dx = \ln(x^2 + 1) + C$

(d)

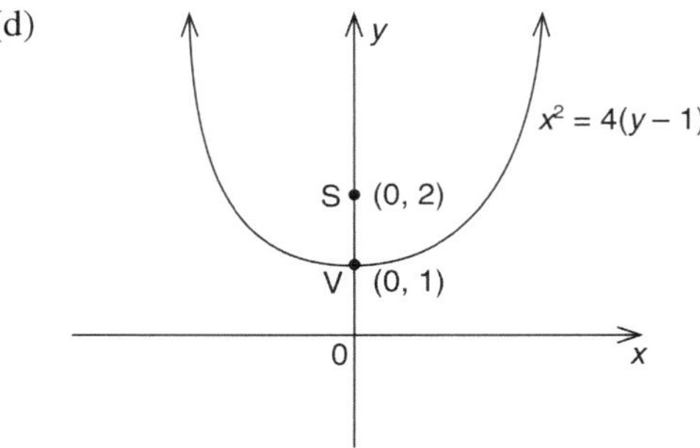

(i) Vertex $(0, 1)$

(ii) From the equation,

$4a = 4$

$\therefore a = 1$

$\therefore$ Coordinates of focus are $(0, 2)$.

QUESTION 3

(a) (i) $\frac{d}{dx}(x^2 + 5)^3 = 3(x^2 + 5)^2.\frac{d}{dx}(x^2 + 5)$
$= 3(x^2 + 5)^2.2x$
$= 6x(x^2 + 5)^2.$

(ii) $\frac{d}{dx}\left(\frac{\cos x}{x}\right) = \frac{x.\frac{d}{dx}(\cos x) - \cos x.\frac{d}{dx}(x)}{x^2}$
$= \frac{x(-\sin x) - \cos x\,(1)}{x^2}$
$= \frac{-x\sin x - \cos x}{x^2}$

(iii) $\frac{d}{dx}(x^2 \ln x) = x^2.\frac{d}{dx}(\ln x) + \ln x.\frac{d}{dx}(x^2)$
$= x^2\left(\frac{1}{x}\right) + \ln x\,(2x)$
$= x + 2x\ln x.$

(b) (i) Coordinates of midpoint of AB are
are $\left(\frac{0+2}{2}, \frac{1+3}{2}\right)$
i.e. (1, 2).

(ii) Slope of line AB, $m_1 = \frac{3-1}{2-0}$
$= \frac{2}{2}$
$= 1.$

(iii) If slope of perpendicular bisector is m_2,
$\therefore\ m_2 = -\frac{1}{m_1}$ (since $m_1 m_2 = -1$)
$\therefore\ m_2 = -\frac{1}{1}$
$= -1.$

The equation of the line passing through (1, 2) with gradient $= -1$,
is $y - 2 = -1(x - 1)$
$y - 2 = -x + 1$
$y = -x + 3.$

(iv) If P is equidistant from A and B, it will lie on the perpendicular bisector.
$\therefore$ P is the point of intersection
of $y = -x + 3$
and $y = 2x - 9$

$\therefore$ Solving simultaneously:
$-x + 3 = 2x - 9$
$12 = 3x$
$\therefore x = 4$
$\therefore y = -4 + 3$
$= -1$

$\therefore$ Coordinates of P are (4, –1).

QUESTION 4

(a)

x	1	1.5	2	2.5	3
$f(x)$	5	1	–2	3	7
	y_0	y_1	y_2	y_3	y_4

$\int_1^3 f(x)dx \doteqdot \frac{h}{3}[y_0 + y_4 + 2y_2 + 4(y_1 + y_3)]$
$= \frac{\frac{1}{2}}{3}[5 + 7 + 2(-2) + 4(1 + 3)]$
$= \frac{1}{6}[12 - 4 + 16]$
$= \frac{1}{6} \times 24$
$= 4.$

(b) (i) (ii) $y = x^2 - 6$
$= (x + \sqrt{6})(x - \sqrt{6})$
$\therefore$ x intercepts are $\pm\sqrt{6}$
y intercept occurs when $x = 0$
$\therefore y = 0 - 6$
$= -6$

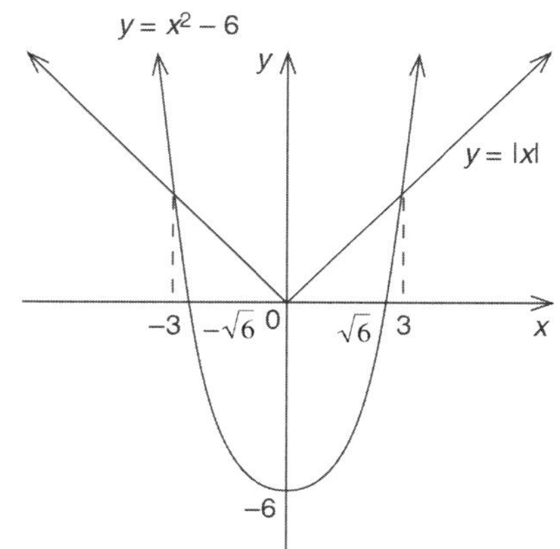

(iii) Graphs intersect when
$x^2 - 6 = |x|$
$(x^2 - 6)^2 = x^2$
$x^4 - 12x^2 + 36 = x^2$
$x^4 - 13x^2 + 36 = 0$
$(x^2 - 9)(x^2 - 4) = 0$
$(x + 3)(x - 3)(x + 2)(x - 2) = 0$
$\therefore x = \pm 3$ or ± 2

From the graph $x = \pm 3$.

(iv) $x^2 - 6 \leqslant |x|$
when $-3 \leqslant x \leqslant 3$ (from graph).

(c) Area of shaded region
$= \int_0^{\frac{\pi}{6}} \sin 2x\,dx + \int_{\frac{\pi}{6}}^{\frac{\pi}{2}} \cos x\,dx$
$= \left[-\frac{\cos 2x}{2}\right]_0^{\frac{\pi}{6}} + \left[\sin x\right]_{\frac{\pi}{6}}^{\frac{\pi}{2}}$

$$= \left[-\frac{\cos\frac{\pi}{3}}{2} - \left(-\frac{\cos 0}{2}\right)\right] + \left(\sin\frac{\pi}{2} - \sin\frac{\pi}{6}\right)$$

$$= -\tfrac{1}{4} - (-\tfrac{1}{2}) + 1 - \tfrac{1}{2}$$

$$= -\tfrac{1}{4} + \tfrac{1}{2} + 1 - \tfrac{1}{2}$$

$$= \tfrac{3}{4}\text{ units}^2.$$

QUESTION 5

(a) $\int_0^{\ln 7} e^{-x}\,dx$

$$= \left[-e^{-x}\right]_0^{\ln 7}$$

$$= -e^{-\ln 7} - (-e^0)$$

$$= -\frac{1}{e^{\ln 7}} - (-1)$$

$$= -\frac{1}{7} + 1$$

$$= \frac{6}{7}.$$

(b) (i)

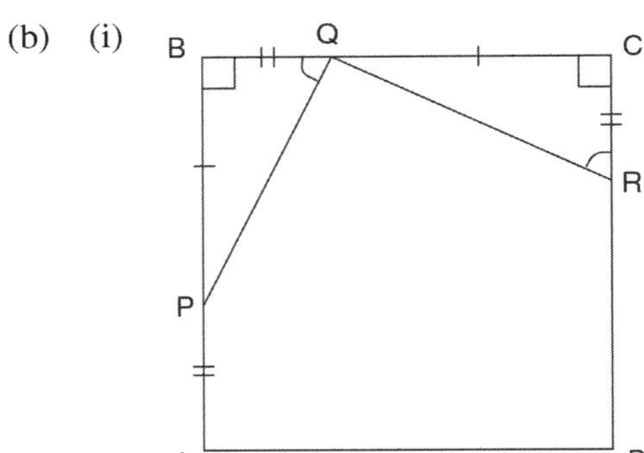

Now $AB = BC$ (sides of square)
and $AP = BQ$ (given)
$\therefore\ AB - AP = BC - BQ$
$\therefore\ BP = QC$

In the $\triangle$'s PBQ and QCR,

$BQ = CR$ (given)
$\angle PBQ = \angle QCR$ (90°, $ABCD$ a square)
$BP = QC$ (from above)

$\therefore\ \triangle PBQ \equiv \triangle QCR$ (SAS)

(ii) Now $\angle BQP = \angle QRC$
(angles of congruent triangles)
and $\angle CQR + \angle QRC = 90°$
(angle sum of $\triangle QRC$)
$\therefore\ \angle CQR + \angle BQP = 90°$
$\therefore\ \angle PQR = 180° - 90°$
(BC is a straight line)
$= 90°$

(c)

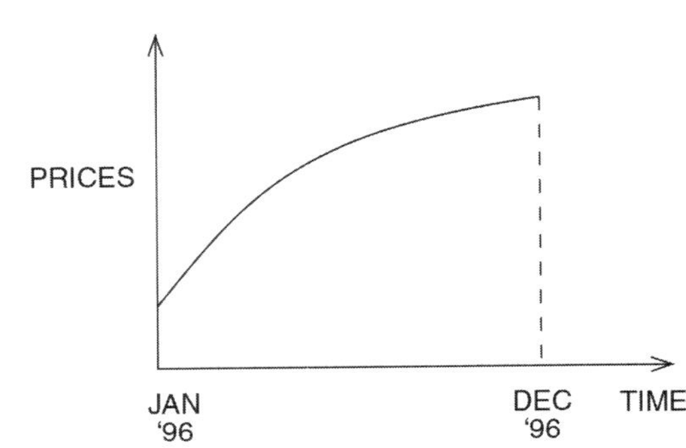

(d) (i)

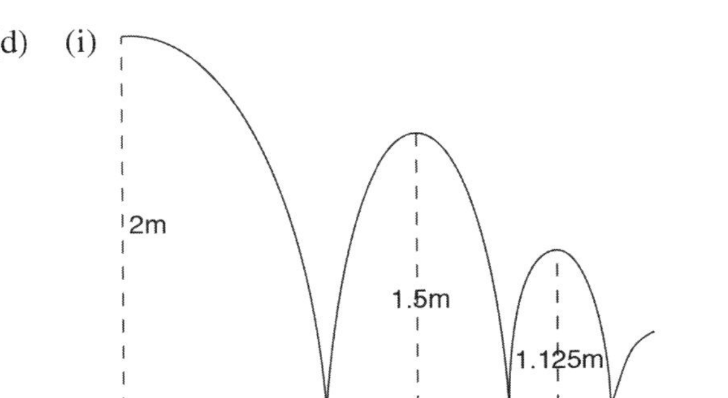

$h = 2 \times 0.75 \times 0.75 \times 0.75$
$h = 0.84375$ m

(ii) Geometric sequence or progression (infinite).

(iii) The falls constitute one series:
$2 + 1.5 + 1.25 + \ldots$

$$S_\infty = \frac{a}{1-r}$$

$a = 2,\ r = \tfrac{3}{4}$

$$\therefore S_\infty = \frac{2}{1-\frac{3}{4}} = \frac{2}{\frac{1}{4}} = 8\text{ m}$$

The rises constitute another series:
$1.5 + 1.25 + 0.84375 + \ldots$
$a = 1.5,\ r = \tfrac{3}{4}$

$$\therefore S_\infty = \frac{1.5}{1-\frac{3}{4}} = \frac{1.5}{\frac{1}{4}} = 6\text{ m}$$

$\therefore$ Total distance travelled $= 8 + 6 = 14$ m.

QUESTION 6

(a) (i)

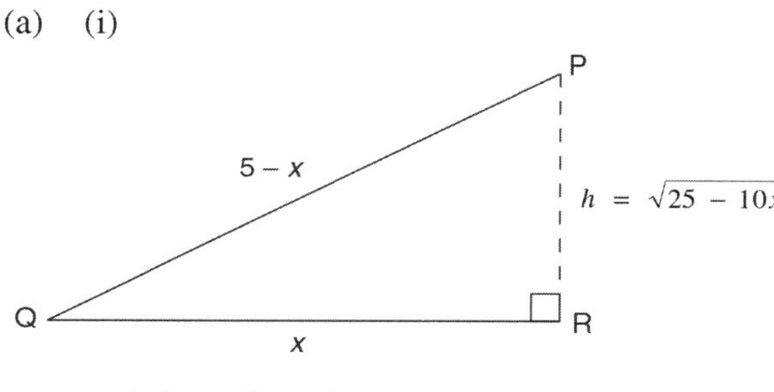

$PQ = (5 - x)$ m

(ii) Now

$$(5 - x)^2 = x^2 + h^2$$
$$25 - 10x + x^2 = x^2 + h^2$$
$$h^2 = 25 - 10x$$
$$h = \sqrt{25 - 10x} \text{ m}$$

($h > 0$ as it is a height)

$\therefore$ Area of $\triangle PQR$

$= \frac{1}{2} \times b \times h$

$= \frac{1}{2} \times x \times \sqrt{25 - 10x}$

$= \frac{1}{2}x\sqrt{25 - 10x}$ m^2

(iii) $A = \frac{1}{2}x(25 - 10x)^{\frac{1}{2}}$

$$\frac{dA}{dx} = \tfrac{1}{2}\left[x.\tfrac{1}{2}(25 - 10x)^{-\frac{1}{2}}.-10 + (25 - 10x)^{\frac{1}{2}}.1\right]$$

$$= \frac{1}{2}\left[\frac{-10x(25 - 10x)^{-\frac{1}{2}}}{2} + (25 - 10x)^{\frac{1}{2}}\right]$$

$$= \frac{1}{2}\left[-\frac{5x}{\sqrt{25 - 10x}} + \sqrt{25 - 10x}\right]$$

Stationary points occur when

$$\frac{dA}{dx} = 0$$

$$\frac{1}{2}\left[-\frac{5x}{\sqrt{25 - 10x}} + \sqrt{25 - 10x}\right] = 0$$

$$-\frac{5x}{\sqrt{25 - 10x}} + \sqrt{25 - 10x} = 0$$

$$\sqrt{25 - 10x} = \frac{5x}{\sqrt{25 - 10x}}$$

$$\left(\sqrt{25 - 10x}\right)^2 = 5x$$

$$25 - 10x = 5x$$

$$15x = 25$$

$$x = \frac{25}{15}$$

$$= \frac{5}{3}$$

Now when $x < \frac{5}{3}$, say $x = 1$,

$$\frac{dA}{dx} = \frac{1}{2}\left[-\frac{5}{\sqrt{15}} + \sqrt{15}\right]$$
$$\doteqdot 1.29 > 0$$

When $x > \frac{5}{3}$, say $x = 2$,

$$\frac{dA}{dx} = \frac{1}{2}\left[-\frac{10}{\sqrt{5}} + \sqrt{5}\right]$$
$$\doteqdot -1.12 < 0$$

$\therefore$ A maximum value occurs at $x = \frac{5}{3}$.

At $x = \frac{5}{3}$, $A = \frac{1}{2}\left(\frac{5}{3}\right)\sqrt{25 - 10\left(\frac{5}{3}\right)}$

$$= \frac{5}{6}\sqrt{25 - \frac{50}{3}}$$
$$= \frac{5}{6}\sqrt{\frac{25}{3}}$$
$$= \frac{5}{6} \times 5 \times \frac{1}{\sqrt{3}}$$
$$= \frac{25}{6} \times \frac{1}{\sqrt{3}} \times \frac{\sqrt{3}}{\sqrt{3}}$$
$$= \frac{25\sqrt{3}}{18} \text{ m}^2.$$

(b) (i) Side of nth square $= (2n + 4)$ cm

Side of $(n + 1)$th square $= 2(n + 1) + 4$

$= 2n + 2 + 4$

$= (2n + 6)$ cm

$\therefore$ Area of nth stripe

$= (2n + 6)^2 - (2n + 4)^2$

$= 4n^2 + 24n + 36 - (4n^2 + 16n + 16)$

$= 4n^2 + 24n + 36 - 4n^2 - 16n - 16$

$= (8n + 20)$ cm^2

(ii) Area of first stripe

$= 8(1) + 20$

$= 28$ cm^2

Area of 19th stripe

$= 8(19) + 20$

$= 172$ cm^2

(iii) $S = \frac{n}{2}[a + l]$

$= \frac{10}{2}[28 + 172]$

$= 5 \times 200$

$= 1000$ cm^2

Total cost $= 9 \times 1000$

$= \$9000$.

QUESTION 7

(a) Now $\sec\theta = \dfrac{1}{\cos\theta}$

and $\tan\theta = \dfrac{\sin\theta}{\cos\theta}$

$$\begin{aligned}\text{LHS} &= \sec^2\theta - \tan^2\theta \\ &= \frac{1}{\cos^2\theta} - \frac{\sin^2\theta}{\cos^2\theta} \\ &= \frac{1-\sin^2\theta}{\cos^2\theta} \\ &= \frac{\cos^2\theta}{\cos^2\theta} \\ &= 1 = \text{RHS}\end{aligned}$$

$\therefore \sec^2\theta - \tan^2\theta = 1.$

(b) (i) $V = \pi\int_0^{\frac{\pi}{6}} y^2\,dx$

$$\begin{aligned}&= \pi\int_0^{\frac{\pi}{6}} (\tan 2x)^2\,dx \\ &= \pi\int_0^{\frac{\pi}{6}} \tan^2 2x\,dx \\ &= \pi\int_0^{\frac{\pi}{6}} (\sec^2 2x - 1)\,dx \qquad \text{(from (a))}\end{aligned}$$

(ii) $V = \pi\left[\frac{\tan 2x}{2} - x\right]_0^{\frac{\pi}{6}}$

$$\begin{aligned}&= \pi\left[\left(\frac{\tan\frac{\pi}{3}}{2} - \frac{\pi}{6}\right) - (0-0)\right] \\ &= \pi\left[\frac{\sqrt{3}}{2} - \frac{\pi}{6}\right] \\ &= \frac{\pi(3\sqrt{3}-\pi)}{6}\ \text{units}^3.\end{aligned}$$

(c) (i) $v = Ce^{-kt}$

$$\begin{aligned}\frac{dv}{dt} &= -kCe^{-kt} \\ &= -k(v) \\ &= -kv\end{aligned}$$

(ii) $t = 0, v = 100$

$\therefore 100 = Ce^0$

$\therefore 100 = C(1)$

$\therefore C = 100$

(iii) $t = 0.25, v = 85$

$$\begin{aligned}\therefore 85 &= 100e^{-0.25k} \\ \tfrac{85}{100} &= e^{-0.25k} \\ \ln\tfrac{85}{100} &= \ln e^{-0.25k} \\ \ln 0.85 &= -0.25k\ln e \\ \ln 0.85 &= -0.25k \\ k &= \frac{\ln 0.85}{-0.25} \\ &\doteqdot 0.650\,075\ldots \quad \text{(by calc.)} \\ &= 0.6501 \text{ to 4 decimal places.}\end{aligned}$$

(iv) When $t = 2$,

$$\begin{aligned}V &= 100e^{-0.650075\ldots\times 2} \\ &= 27.2490\ldots\ \text{cm/s} \quad \text{(by calc.)} \\ &= 27.2 \text{ cm/s to one decimal place}\end{aligned}$$

QUESTION 8

(a) (i)

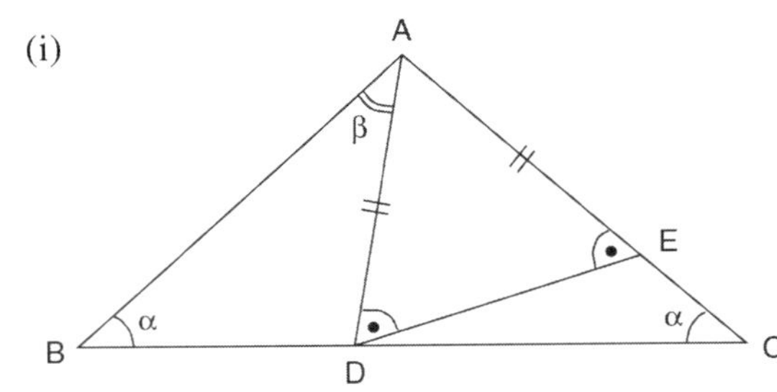

$\angle ADC$ is the exterior angle of $\triangle ABD$ and the exterior angle of a triangle is equal to the sum of the opposite 2 interior angles, α and β.

i.e. $\angle ADC = \angle ABD + \angle BAD$
$= \alpha + \beta$

(ii) $\angle DAC = 180° - (\alpha + \beta + \alpha)$ (angle sum $\triangle ADC$)
$= 180° - (2\alpha + \beta)$
$= 180° - 2\alpha - \beta$

(iii) Now $\angle ADE = \angle AED$ (angles opp. equal sides)

and $\angle ADE + \angle AED + \angle DAC = 180°$ (angle sum $\triangle ADE$)

$\therefore 2 \times \angle ADE = 180° - \angle DAC$ (since $\angle ADE = \angle AED$)

$$\begin{aligned}\angle ADE &= \frac{180° - (180° - 2\alpha - \beta)}{2} \\ &= \frac{2\alpha+\beta}{2} \\ &= \alpha + \frac{\beta}{2}\end{aligned}$$

Now $\angle ADC = \angle ADE + \angle EDC$

$\therefore \alpha + \beta = \alpha + \frac{\beta}{2} + \angle EDC$

$\therefore \angle EDC = \alpha + \beta - \alpha - \frac{\beta}{2}$

$= \frac{\beta}{2}.$

(b) (i) $x = t + \sin t$

$v = \frac{dx}{dt} = 1 + \cos t$

The particle is stationary when

$v = 0$

$\therefore 1 + \cos t = 0$

$\cos t = -1$

$\therefore t = \pi$ for $0 < t < 3\pi$.

(ii) $a = \frac{dv}{dt} = 0 - \sin t$

$= -\sin t$

Now $a = 0$ when

$0 = -\sin t$

$\therefore \sin t = 0$

$\therefore t = \pi, 2\pi$ for $0 < t < 3\pi$.

(iii) Possible points of inflection occur when

$$\frac{dv}{dt} = \frac{d^2x}{dt^2} = 0$$

i.e. when $t = \pi, 2\pi$

When $t = \pi, x = \pi + \sin \pi = \pi$.
When $t = 2\pi, x = 2\pi + \sin 2\pi = 2\pi$.

Consider the point (π, π).

t	3	π	3.2
$\frac{d^2x}{dt^2} = -\sin t$	–0.14	0	0.06

$\therefore$ A change of concavity.
$\therefore$ There is a horizontal point of inflection at (π, π). (Since at $t = \pi$, $\frac{dx}{dt} = 0$ also.)

Consider the point $(2\pi, 2\pi)$:

t	6.2	$2\pi \doteqdot (6.28)$	6.4
$\frac{dx}{dt} = -\sin t$	0.08	0	–0.12

$\therefore$ A change of concavity.
$\therefore$ There is a point of inflection at $(2\pi, 2\pi)$.

End points:
When $t = 0,\ x = 0 + \sin 0 = 0$
When $t = 3\pi,\ x = 3\pi + \sin 3\pi = 3\pi$

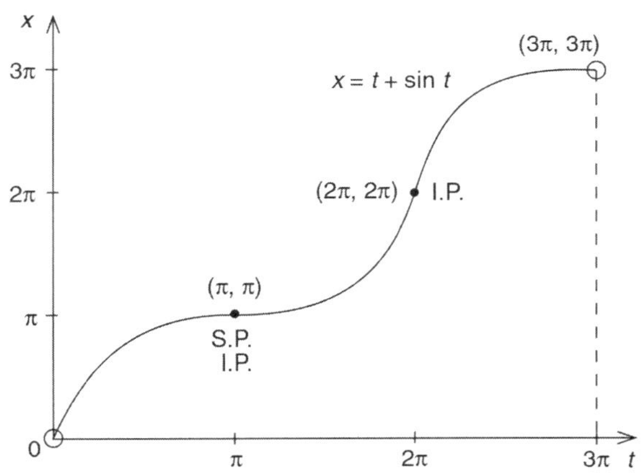

QUESTION 9

(a) (i)

First		Second		Outcome	Probability
2/4	R	1/3	R	RR	$\frac{2}{12}$
		1/3	B	RB	$\frac{2}{12}$
		1/3	W	RW	$\frac{2}{12}$
1/4	B	2/3	R	BR	$\frac{2}{12}$
		1/3	W	BW	$\frac{1}{12}$
1/4	W	2/3	R	WR	$\frac{2}{12}$
		1/3	B	WB	$\frac{1}{12}$
					$\frac{12}{12}$

(ii) $P(\text{RR}) = \frac{2}{12}$
$= \frac{1}{6}$.

(iii) $P(\text{At least 1 red}) = 1 - [P(\text{BW} + P(\text{WB})]$
$= 1 - (\frac{1}{12} + \frac{1}{12})$
$= 1 - \frac{2}{12}$
$= \frac{10}{12}$
$= \frac{5}{6}$.

(iv) If one of the selected balls is red, there are 5 possible outcomes only one of which will yield another red:

<u>RR</u>, RB, RW, BR and WR

$\therefore P(\text{Hidden Red}) = \frac{1}{5}$.

(b) (i) $v = \frac{dx}{dt} = 4t^3 - 24t^2 + 20t$

$$\therefore x = \frac{4t^4}{4} - \frac{24t^3}{3} + \frac{20t^2}{2} + C$$
$$= t^4 - 8t^3 + 10t^2 + C$$

When $t = 0, x = 3 \therefore C = 3$
$\therefore x = t^4 - 8t^3 + 10t^2 + 3$

(ii) When $t = 1, v = \frac{dx}{dt} = 4 - 24 + 20$
$= 0$

$\therefore$ There is a stationary point at $t = 1$.

Now $\frac{d^2x}{dt^2} = 12t^2 - 48t + 20$

When $t = 1$, $\frac{d^2x}{dt^2} = 12 - 48 + 20$
$= -16 < 0$

$\therefore$ At $t = 1$, there will be a maximum turning point.

(iii) $x = t^4 - 8t^3 + 10t^2 + 3$

t	0	1	2
x	3	6	–5

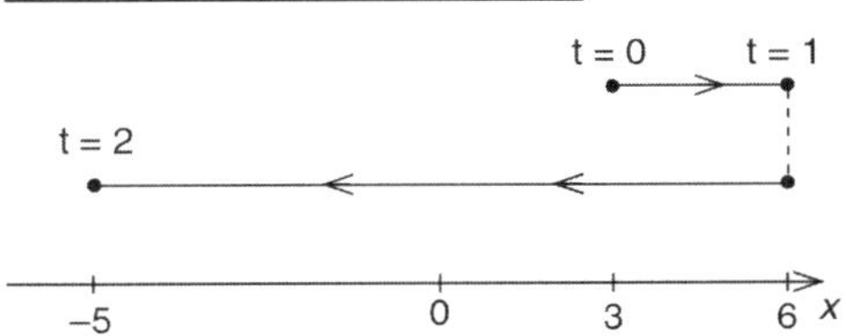

The pen changes direction at $t = 1$,
$\therefore$ Total distance travelled
$= 3 + 11$
$= 14$ cm

The amount of ink used
$= 14 \times 0.05$
$= 0.7$ milligrams of ink.

QUESTION 10

(a) $4x \leqslant 15 \leqslant -9x$

$\therefore 4x \leqslant 15 \quad \text{and} \quad -9x \geqslant 15$

$\therefore x \leqslant \frac{15}{4} \quad \text{and} \quad x \leqslant -\frac{15}{9}$

$\therefore x \leqslant -\frac{15}{9}$

$-\frac{15}{9}$ -1 0 x

(b) (i)

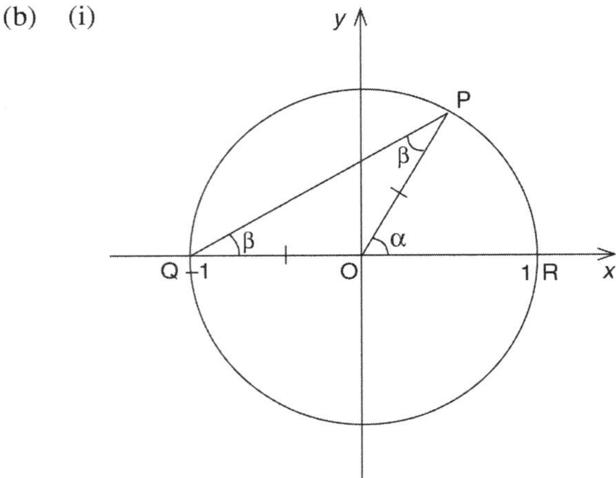

$\triangle OPQ$ is isosceles because 2 sides are equal:

$OP = OQ$ (radii of circle)

$\therefore \angle PQO = \angle QPO = \beta$
(opposite equal angles)

Now $\angle POR = \angle PQO + \angle QPO$
(exterior angle property)

$\therefore \alpha = \beta + \beta$
$= 2\beta$

(ii) Equation of line passing through $Q(-1, 0)$ with gradient $= m$ is

$$y - 0 = m(x - (-1))$$
$$y = m(x + 1)$$
$$y = mx + m$$

(iii) Equation of circle, centre $(0, 0)$, $r = 1$:

$$x^2 + y^2 = 1 \quad \ldots (1)$$

Equation of PQ:

$$y = mx + m \quad \ldots (2)$$

Solving simultaneously:

Substituting (2) into (1):

$$x^2 + (mx + m)^2 = 1$$
$$x^2 + m^2x^2 + 2m^2x + m^2 = 1$$
$$x^2(1 + m^2) + 2m^2x + m^2 - 1 = 0$$
$$(1 + m^2)x^2 + 2m^2x + m^2 - 1 = 0$$

(iv) One root of the equation will be at Q where $x = -1$.

If -1 and γ are roots of the equation,

$$\gamma(-1) = \frac{c}{a}$$

i.e. $$(-1)\gamma = \frac{m^2 - 1}{1 + m^2}$$

$$\therefore \gamma = \frac{1 - m^2}{1 + m^2}$$

$\therefore x$ coordinate of P is $\frac{1 - m^2}{1 + m^2}$.

(v) Now since $\alpha = 2\beta$
$\tan 2\beta = \tan \alpha$

Consider the $\triangle OPT$:

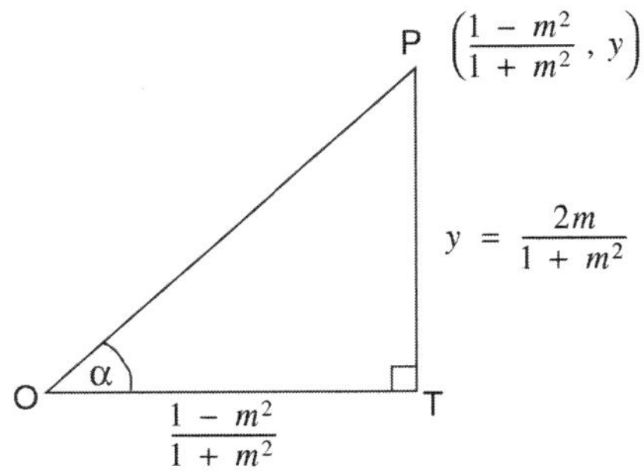

Since P lies on the line $y = mx + m$,

$$y = m\left(\frac{1 - m^2}{1 + m^2}\right) + m$$
$$= \frac{m - m^3}{1 + m^2} + m$$
$$= \frac{m - m^3 + m(1 + m^2)}{1 + m^2}$$
$$= \frac{m - m^3 + m + m^3}{1 + m^2}$$
$$= \frac{2m}{1 + m^2}$$

Now $$\tan \alpha = \frac{PT}{OT}$$
$$= \frac{2m}{1 + m^2} \div \frac{1 - m^2}{1 + m^2}$$
$$= \frac{2m}{1 + m^2} \times \frac{1 + m^2}{1 - m^2}$$
$$= \frac{2m}{1 - m^2}$$

Since $\tan 2\beta = \tan \alpha$
and $m = \tan \beta$ (given)

$$\therefore \tan 2\beta = \frac{2 \tan \beta}{1 - \tan^2 \beta}.$$

HIGHER SCHOOL CERTIFICATE EXAMINATION

1998
MATHEMATICS
2/3 UNIT (COMMON)

Time allowed—Three hours
(Plus 5 minutes reading time)

DIRECTIONS TO CANDIDATES

- Attempt ALL questions.
- ALL questions are of equal value.
- All necessary working should be shown in every question. Marks may be deducted for careless or badly arranged work.
- Standard integrals are printed on page 12.
- Board-approved calculators may be used.
- Answer each question in a SEPARATE Writing Booklet.
- You may ask for extra Writing Booklets if you need them.

QUESTION 1. Use a SEPARATE Writing Booklet. **Marks**

(a) Express $\frac{3}{11}$ as a recurring decimal. **1**

(b) Simplify $|-5|-|8|$. **1**

(c) A coin is tossed three times. What is the probability that 'heads' appears every time? **2**

(d) Find a primitive of x^2+7. **2**

(e) Find the exact value of $\sin\left(\frac{\pi}{4}\right)+\sin\left(\frac{2\pi}{3}\right)$. **2**

(f) By rationalising the denominators, express $\frac{1}{3-\sqrt{2}}+\frac{1}{3+\sqrt{2}}$ in simplest form. **2**

(g) A merchant buys tea from a wholesaler and then sells it at a profit of 37·5%. If the merchant sells a packet of tea for \$3·08, what price does he pay to the wholesaler per packet of tea? **2**

QUESTION 2. Use a SEPARATE Writing Booklet. **Marks**

(a) Differentiate the following functions: **6**

(i) $\left(3x^2+4\right)^5$

(ii) $x\ \sin(x+1)$

(iii) $\dfrac{\tan x}{x}$.

(b) Evaluate the following integrals: **4**

(i) $\displaystyle\int_1^2 \frac{1}{x^2}\,dx$

(ii) $\displaystyle\int_0^3 e^{4x}\,dx$.

(c) Find $\displaystyle\int \frac{x}{x^2+3}\,dx$. **2**

QUESTION 3. Use a SEPARATE Writing Booklet. **Marks**

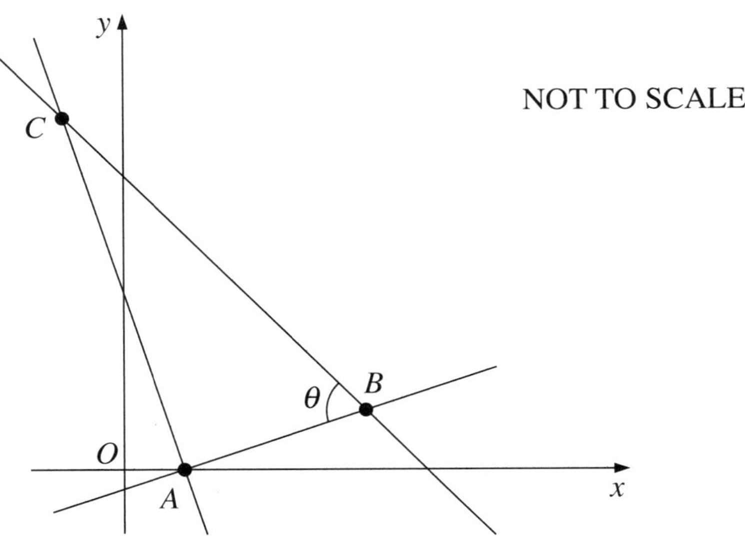

The diagram shows points $A(1,0)$, $B(4,1)$ and $C(-1,6)$ in the Cartesian plane. Angle ABC is θ.

Copy or trace this diagram into your Writing Booklet.

(a) Show that A and C lie on the line $3x + y = 3$. **2**

(b) Show that the gradient of AB is $\frac{1}{3}$. **1**

(c) Show that the length of AB is $\sqrt{10}$ units. **1**

(d) Show that AB and AC are perpendicular. **1**

(e) Find $\tan\theta$. **2**

(f) Find the equation of the circle with centre A that passes through B. **2**

(g) The point D is not shown on the diagram. The point D lies on the line $3x + y = 3$ between A and C, and $AD = AB$. Find the coordinates of D. **2**

(h) On your diagram, shade the region satisfying the inequality $3x + y \le 3$. **1**

QUESTION 4. Use a SEPARATE Writing Booklet. **Marks**

(a) The following table lists the values of a function for three values of x. **4**

x	1·0	2·0	3·0
$f(x)$	1·7	9·0	4·3

Use these three function values to estimate $\int_1^3 f(x)\,dx$ by:

(i) Simpson's rule

(ii) the trapezoidal rule.

(b) The third term of an arithmetic series is 32 and the sixth term is 17. **3**

(i) Find the common difference.

(ii) Find the sum of the first ten terms.

(c) The first term of a geometric series is 16 and the fourth term is $\frac{1}{4}$. **2**

(i) Find the common ratio.

(ii) Find the limiting sum of the series.

(d) The equation of a parabola is $x^2 = 8(y+3)$. **3**

(i) Find the coordinates of the vertex of the parabola.

(ii) Find the equation of the directrix of the parabola.

QUESTION 5. Use a SEPARATE Writing Booklet. **Marks**

(a) **8**

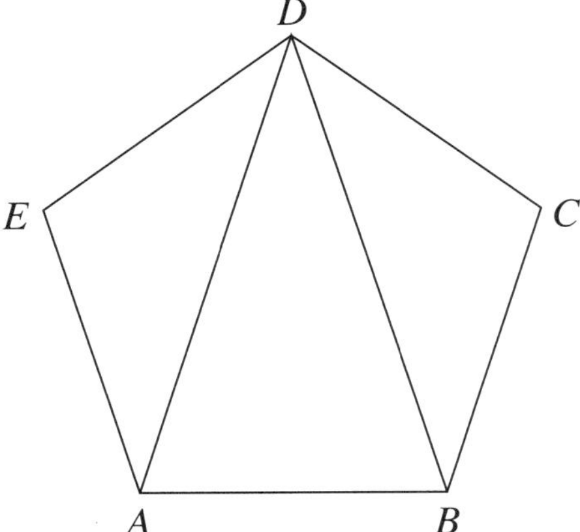

The diagram shows a regular pentagon $ABCDE$. Each of the sides AB, BC, CD, DE and EA is of length x metres. Each of the angles $\angle ABC$, $\angle BCD$, $\angle CDE$, $\angle DEA$ and $\angle EAB$ is $108°$. Two diagonals, AD and BD, have been drawn.

Copy or trace the diagram into your Writing Booklet.

(i) State why triangle BCD is isosceles, and hence find $\angle CBD$.

(ii) Show that triangles BCD and DEA are congruent.

(iii) Find the size of $\angle ADB$.

(iv) Find an expression for the area of the pentagon in terms of x and trigonometric ratios.

(b) The population P of a city is growing at a rate that is proportional to the current population. The population at time t years is given by **4**

$$P = Ae^{kt},$$

where A and k are constants.

The population at time $t = 0$ was 1 000 000 and at time $t = 2$ was 1 072 500.

(i) Find the value of A.

(ii) Find the value of k.

(iii) At what time will the population reach 2 000 000?

QUESTION 6. Use a SEPARATE Writing Booklet. **Marks**

(a) A particle P moves along a straight line for 8 seconds, starting at the fixed point S at time $t = 0$. At time t seconds, P is $x(t)$ metres to the right of S. The graph of $x(t)$ is shown in the diagram. **5**

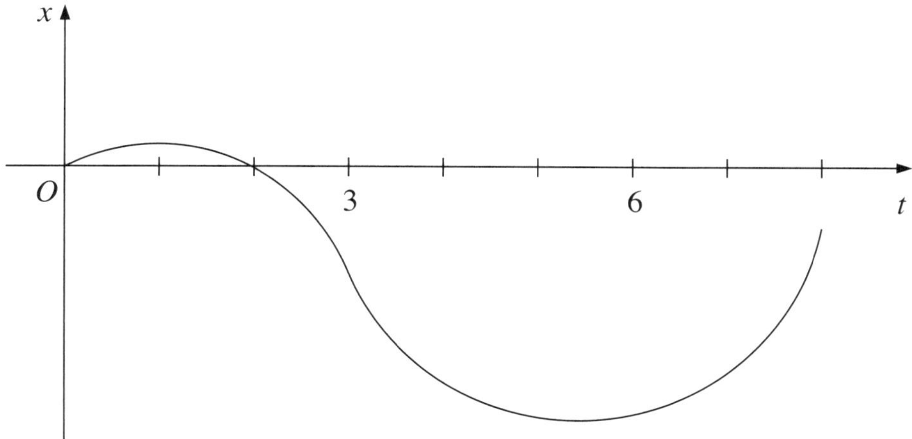

(i) At approximately what times is the velocity of the particle equal to 0?

(ii) At approximately what time is the acceleration of the particle equal to 0?

(iii) At approximately what time is the distance from S greatest?

(iv) At approximately what time is the particle moving with the greatest velocity?

(b) The function $f(x) = xe^{-2x} + 1$ has first derivative $f'(x) = e^{-2x} - 2xe^{-2x}$ and second derivative $f''(x) = 4xe^{-2x} - 4e^{-2x}$. **7**

(i) Find the value of x for which $y = f(x)$ has a stationary point.

(ii) Find the values of x for which $f(x)$ is increasing.

(iii) Find the value of x for which $y = f(x)$ has a point of inflection and determine where the graph of $y = f(x)$ is concave upwards.

(iv) Sketch the curve $y = f(x)$ for $-\frac{1}{2} \leq x \leq 4$.

(v) Describe the behaviour of the graph for very large positive values of x.

QUESTION 7. Use a SEPARATE Writing Booklet. **Marks**

(a) (i) Write down the discriminant of $3x^2 + 2x + k$. **2**

(ii) For what values of k does $3x^2 + 2x + k = 0$ have real roots?

(b) Consider the function $y = 1 + \sqrt{3}\sin x + \cos x$. **5**

(i) Find the equation of the tangent to the graph of the function at $x = \dfrac{5\pi}{6}$.

(ii) Find the maximum and minimum values of $1 + \sqrt{3}\sin x + \cos x$ in the interval $0 \le x \le 2\pi$.

(c) **5**

The diagram shows a sector of a circle with radius r cm. The angle at the centre is θ radians and the perimeter of the sector is 8 cm.

(i) Find an expression for r in terms of θ.

(ii) Show that A, the area of the sector in cm^2, is given by

$$A = \frac{32\theta}{(\theta + 2)^2}.$$

(iii) If $0 \le \theta \le \dfrac{\pi}{2}$, find the maximum area and the value of θ for which this occurs.

QUESTION 8. Use a SEPARATE Writing Booklet. **Marks**

(a) Sand is tipped from a truck onto a pile. The rate, R kg/s, at which the sand is flowing is given by the expression $R = 100t - t^3$, for $0 \le t \le T$, where t is the time in seconds after the sand begins to flow. **7**

(i) Find the rate of flow at time $t = 8$.

(ii) What is the largest value of T for which the expression for R is physically reasonable?

(iii) Find the maximum rate of flow of sand.

(iv) When the sand starts to flow, the pile already contains 300 kg of sand. Find an expression for the amount of sand in the pile at time t.

(v) Calculate the total weight of sand that was tipped from the truck in the first 8 seconds.

(b) **5**

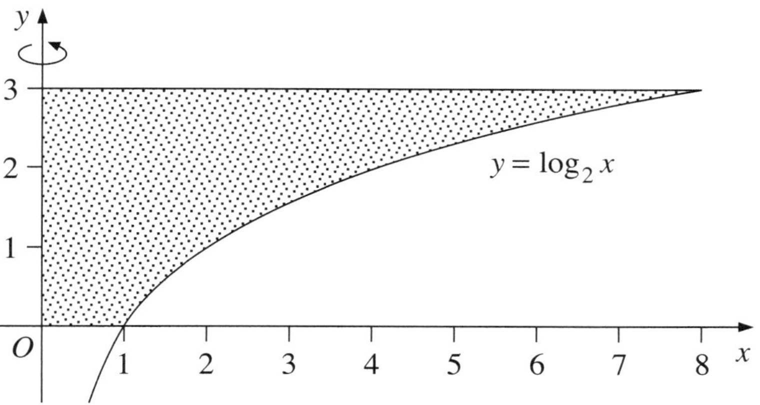

The diagram shows the graph of $y = \log_2 x$ between $x = 1$ and $x = 8$. The shaded region, bounded by $y = \log_2 x$, the line $y = 3$, and the x and y axes, is rotated about the y axis to form a solid.

(i) Show that the volume of the solid is given by

$$V = \pi \int_0^3 e^{y \ln 4}\, dy.$$

(ii) Hence find the volume of the solid.

QUESTION 9. Use a SEPARATE Writing Booklet. **Marks**

(a) Solve $\ln(7x - 12) = 2\ln x$. **2**

(b) **4**

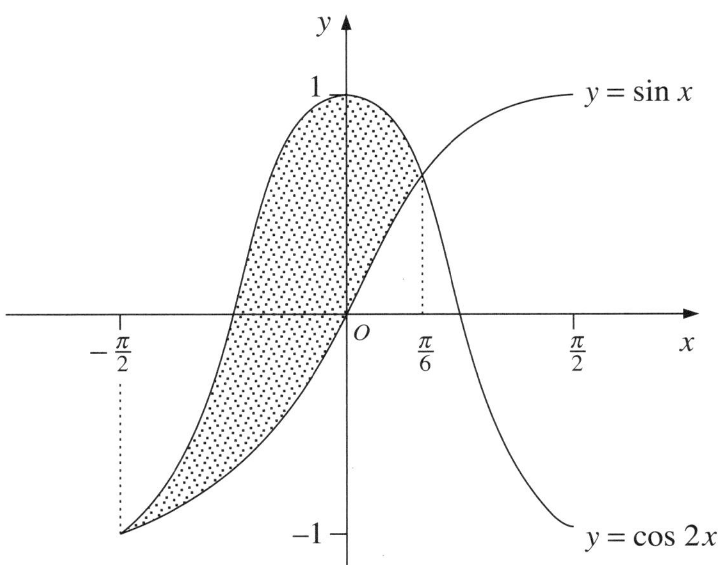

The diagram shows the graphs of the functions $y = \cos 2x$ and $y = \sin x$ between $x = -\frac{\pi}{2}$ and $x = \frac{\pi}{2}$. The two graphs intersect at $x = \frac{\pi}{6}$ and $x = -\frac{\pi}{2}$. Calculate the area of the shaded region.

(c) **6**

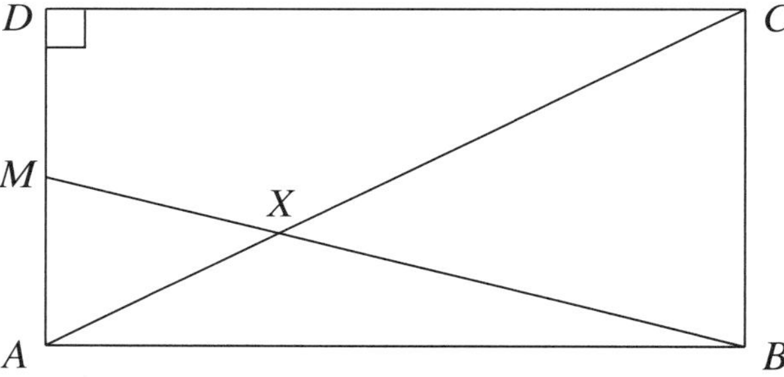

In the diagram, $ABCD$ is a rectangle and $AB = 2AD$. The point M is the midpoint of AD. The line BM meets AC at X.

(i) Show that the triangles AXM and BXC are similar.

(ii) Show that $3CX = 2AC$.

(iii) Show that $9(CX)^2 = 5(AB)^2$.

QUESTION 10. Use a SEPARATE Writing Booklet. **Marks**

(a) A game is played in which two coloured dice are thrown once. The six faces of the blue die are numbered 4, 6, 8, 9, 10 and 12. The six faces of the pink die are numbered 2, 3, 5, 7, 11 and 13. The player wins if the number on the pink die is larger than the number on the blue die. **5**

(i) By drawing up a table of possible outcomes, or otherwise, calculate the probability of the player winning a game.

(ii) Calculate the probability that the player wins at least once in two successive games.

(b) A fish farmer began business on 1 January 1998 with a stock of 100 000 fish. He had a contract to supply 15 400 fish at a price of \$10 per fish to a retailer in December each year. In the period between January and the harvest in December each year, the number of fish increases by 10%. **7**

(i) Find the number of fish just after the second harvest in December 1999.

(ii) Show that F_n, the number of fish just after the nth harvest, is given by

$$F_n = 154\,000 - 54\,000\,(1{\cdot}1)^n.$$

(iii) When will the farmer have sold all his fish, and what will his total income be?

(iv) Each December the retailer offers to buy the farmer's business by paying \$15 per fish for his entire stock. When should the farmer sell to maximise his total income?

End of paper

1998 Higher School Certificate Worked Answers

QUESTION 1

(a) $\frac{3}{11} = 0.2727\ldots$ (by calc.)
$= 0.\dot{2}\dot{7}.$

(b) $|-5| - |8| = 5 - 8$
$= -3.$

(c)

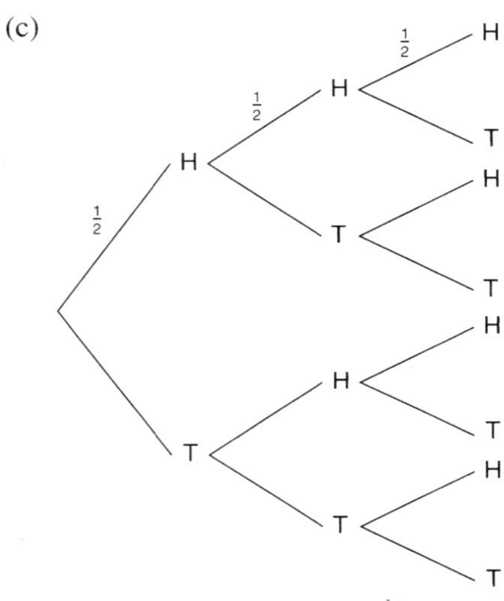

$P(\text{HHH}) = \frac{1}{2} \times \frac{1}{2} \times \frac{1}{2}$
$= \frac{1}{8}.$

(d) $\int (x^2 + 7)\, dx = \frac{x^3}{3} + 7x + c.$

(e) $\sin\left(\frac{\pi}{4}\right) + \sin\left(\frac{2\pi}{3}\right) = \frac{1}{\sqrt{2}} + \frac{\sqrt{3}}{2}$
$= \frac{\sqrt{2} + \sqrt{3}}{2}.$

(f) $\frac{1}{3 - \sqrt{2}} + \frac{1}{3 + \sqrt{2}}$

$= \frac{1}{3 - \sqrt{2}} \times \frac{3 + \sqrt{2}}{3 + \sqrt{2}} + \frac{1}{3 + \sqrt{2}} \times \frac{3 - \sqrt{2}}{3 - \sqrt{2}}$

$= \frac{3 + \sqrt{2}}{9 - 2} + \frac{3 - \sqrt{2}}{9 - 2}$

$= \frac{3 + \sqrt{2}}{7} + \frac{3 - \sqrt{2}}{7}$

$= \frac{3 + \sqrt{2} + 3 - \sqrt{2}}{7}$

$= \frac{6}{7}.$

(g) old price + 37.5% of old price = new price

$x + \frac{37.5}{100}x = \$3.08$
$x + 0.375x = \$3.08$
$1.375x = \$3.08$
$x = \frac{\$3.08}{1.375}$
$= \$2.24.$ (by calc.)

QUESTION 2

(a) (i) $\frac{d}{dx}[(3x^2 + 4)^5]$
$= 5(3x^2 + 4)^4 . \frac{d}{dx}(3x^2 + 4)$
$= 5(3x^2 + 4)^4(6x)$
$= 30x(3x^2 + 4)^4.$

(ii) $\frac{d}{dx}[x \sin(x + 1)]$
$= x . \frac{d}{dx}\sin(x + 1) + \sin(x + 1) . \frac{d}{dx}(x)$
$= x\cos(x + 1) . 1 + \sin(x + 1) . 1$
$= x\cos(x + 1) + \sin(x + 1).$

(iii) $\frac{d}{dx}\left(\frac{\tan x}{x}\right) = \frac{x . \frac{d}{dx}(\tan x) - \tan x . \frac{d}{dx}(x)}{x^2}$
$= \frac{x\sec^2 x - \tan x . 1}{x^2}$
$= \frac{x\sec^2 x - \tan x}{x^2}.$

(b) (i) $\int_1^2 \frac{1}{x^2}\, dx = \int_1^2 x^{-2}\, dx$
$= \left[\frac{x^{-1}}{-1}\right]_1^2$
$= \left[-\frac{1}{x}\right]_1^2$
$= \left(-\frac{1}{2}\right) - \left(-\frac{1}{1}\right)$
$= -\frac{1}{2} + 1$
$= \frac{1}{2}.$

(ii) $\int_0^3 e^{4x}\,dx = \left[\frac{e^{4x}}{4}\right]_0^3$

$= \left(\frac{e^{12}}{4}\right) - \left(\frac{e^0}{4}\right)$

$= \frac{e^{12}}{4} - \frac{1}{4}$

$= \frac{e^{12} - 1}{4}.$

(c) $\int \frac{x}{x^2 + 3}\,dx = \frac{1}{2}\int \frac{2x}{x^2 + 3}\,dx$

$= \frac{1}{2}\ln(x^2 + 3) + c.$

QUESTION 3

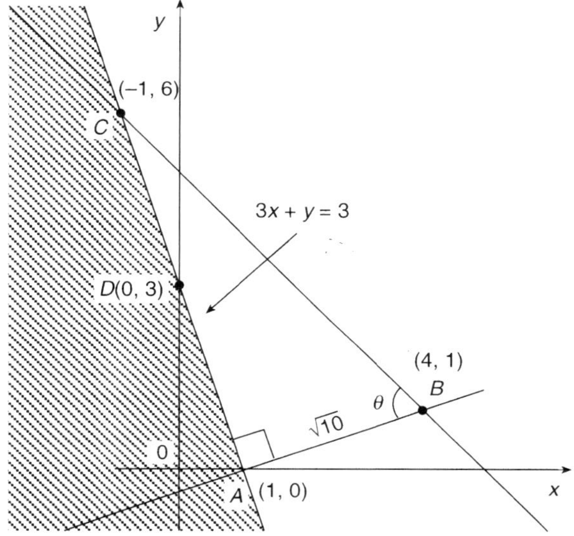

(a) Substituting (1, 0) and (−1, 6) into equation:

RHS = 3

LHS = 3(1) + 0

= 3

= RHS

∴ A lies on line

and RHS = 3

LHS = 3(−1) + 6

= −3 + 6

= 3

= RHS

∴ C lies on line.

(b) Gradient of AB: $m_1 = \frac{1 - 0}{4 - 1}$

$= \frac{1}{3}.$

(c) $AB = \sqrt{(4 - 1)^2 + (1 - 0)^2}$

$= \sqrt{9 + 1}$

$= \sqrt{10}$ units.

(d) Gradient of AC: $m_2 = \frac{6 - 0}{-1 - 1}$

$= \frac{6}{-2}$

$= -3$

Gradient of AB: $m_1 = \frac{1}{3}$ (from (b))

Now $m_1 \times m_2 = \frac{1}{3} \times -3$

$= -1$

∴ $AB \perp AC$.

(e) $AC = \sqrt{(-1 - 1)^2 + (6 - 0)^2}$

$= \sqrt{4 + 36}$

$= \sqrt{40}$

$= 2\sqrt{10}.$

Now $\tan\theta = \frac{AC}{AB}$

$= \frac{2\sqrt{10}}{\sqrt{10}}$

$= 2.$

(f) Equation of circle with centre (1, 0) and radius $AB = \sqrt{10}$ units is given by:

$(x - 1)^2 + (y - 0)^2 = (\sqrt{10})^2$

$x^2 - 2x + 1 + y^2 = 10$

$x^2 + y^2 - 2x - 9 = 0.$

(g) Now length $AB = \sqrt{10}$ (from (c))

∴ length $AD = \sqrt{10}$ (since $AD = AB$)

and length $AC = 2\sqrt{10}$ (from (e))

∴ $DC = \sqrt{10}$

∴ $AD = DC$

∴ D is the mid point of AC.

∴ Coordinates of D are

$\left(\frac{-1 + 1}{2}, \frac{0 + 6}{2}\right)$

i.e. $\left(\frac{0}{2}, \frac{6}{2}\right)$

i.e. (0, 3)

(h) See diagram.

QUESTION 4

(a)

x	1.0	2.0	3.0
$f(x)$	1.7	9.0	4.3
	y_0	y_1	y_2

(i) $\int_1^3 f(x)\,dx \doteqdot \frac{h}{3}[y_0 + y_2 + 4y_1]$

$= \frac{1}{3}[1.7 + 4.3 + 4(9.0)]$

$= \frac{1}{3}[42]$

$= 14.$

(ii) $\int_1^3 f(x)\,dx \doteqdot \frac{h}{2}[y_0 + y_2 + 2(y_1)]$

$= \frac{1}{2}[1.7 + 4.3 + 2(9.0)]$

$= \frac{1}{2}[24]$

$= 12.$

(b) (i) $u_n = a + (n-1)d$

$u_3 = a + 2d = 32 \quad \ldots (1)$

$u_6 = a + 5d = 17 \quad \ldots (2)$

(1) – (2): $-3d = 15$

$d = -5$

(Substituting $d = -5$ into (1):

$a + 2(-5) = 32$

$a - 10 = 32$

$\therefore a = 42.$)

(ii) $S_n = \frac{n}{2}[2a + (n-1)d]$

$S_{10} = \frac{10}{2}[2(42) + 9(-5)]$

$= 5[84 - 45]$

$= 5[39]$

$= 195.$

(c) (i) $U_n = ar^{n-1}$

$U_4 = ar^3$

$\frac{1}{4} = 16r^3$

$r^3 = \frac{1}{64}$

$\therefore r = \frac{1}{4}.$

(ii) $S = \frac{a}{1-r}$

$= \frac{16}{1 - \frac{1}{4}}$

$= \frac{16}{\frac{3}{4}}$

$= \frac{64}{3}.$

(d) (i) $x^2 = 8(y + 3)$

This parabola is concave up and may be expressed in the form

$(x - 0)^2 = 4.2(y + 3)$

$\therefore$ Vertex = $(0, -3)$ and focal length = 2

(ii) Equation of directrix is

i.e. $y = -3 - 2$

i.e. $y = -5$

QUESTION 5

(a) (i)

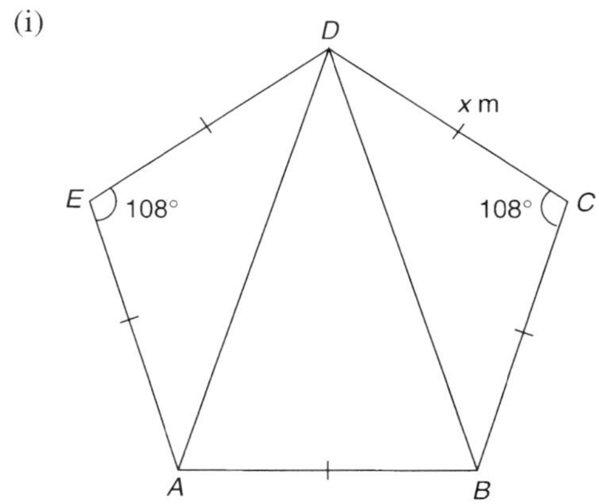

$CD = BC$ (equal sides of regular pentagon)

$\therefore \triangle BCD$ is isosceles

Now $\angle BCD = 108°$ (given)

and $\angle CDB = \angle CBD$

(equal angles of isosceles $\triangle$)

$\therefore \angle CDB + \angle CBD = 180° - 108°$

(angle sum of triangle)

$= 72°$

$\therefore \angle CBD = 36°$

(and $\angle CDB = 36°$)

(ii) In the $\triangle$'s BCD and DEA,

$CD = DE$ (given)

$BC = EA$ (given)

$\angle BCD = \angle DEA$ (given)

$\therefore \triangle BCD \equiv \triangle DEA$ (SAS).

(iii) $\angle CDE = \angle EDA + \angle ADB + \angle CDB$

Now $\angle EDA = 36°$

(corr. angle of congruent triangle)

and $\angle CDE = 108°$ (given)

$\therefore 108° = 36° + \angle ADB + 36°$

$108° = 72° + \angle ADB$

$\therefore \angle ADB = 36°$

(iv) Area $ABCDE$
= Area $\triangle BCD$ + Area $\triangle AED$ + Area $\triangle BDA$

Now
Area $\triangle BCD = \frac{1}{2} \times DC \times BC \times \sin \angle DCB$

$= \frac{1}{2}(x)(x)(\sin 108°)$

$= \frac{1}{2}x^2 \sin 108°$

and

Area $\triangle AED$ = Area $\triangle BCD$ (congruent triangles)

$= \frac{1}{2}x^2 \sin 108°.$

and
Area $\triangle BDA = \frac{1}{2} \times AD \times BD \times \sin \angle ADB$

$$BD^2 = CD^2 + BC^2 - 2(CD)(BC)\cos 108°$$
$$= x^2 + x^2 - 2(x)(x)\cos 108°$$
$$= 2x^2 - 2x^2 \cos 108°$$

$\therefore$ Area $\triangle BDA = \frac{1}{2} \times BD^2 \times \sin 36°$ (since $AD = BD$)

$= \frac{1}{2}(2x^2 - 2x^2 \cos 108°)\sin 36°$

$= (x^2 - x^2 \cos 108°)\sin 36°$

$\therefore$ Area $ABCDE$

$= \frac{1}{2}x^2 \sin 108° + \frac{1}{2}x^2 \sin 108° + (x^2 - x^2 \cos 108°)\sin 36°$

$= x^2 \sin 108° + (x^2 - x^2 \cos 108°)\sin 36°$

$= x^2 \sin 108° + x^2 \sin 36° - x^2 \cos 108° \sin 36°$

$= x^2(\sin 108° + \sin 36° - \cos 108° \sin 36°)$

(Another method yields an equivalent answer of $x^2\left(\sin 108° + \frac{1}{4}\tan 72°\right)$

(b) (i) $P = Ae^{kt}$

At $t = 0$, $P = 1\,000\,000$,
$\therefore 1\,000\,000 = Ae^0$
$1\,000\,000 = A(1)$
$\therefore A = 1\,000\,000$

(ii) $P = 1\,000\,000e^{kt}$

At $t = 2$, $P = 1\,072\,500$

$\therefore 1\,072\,500 = 1\,000\,000e^{2k}$

$\frac{1\,072\,500}{1\,000\,000} = e^{2k}$

$\ln \frac{1\,072\,500}{1\,000\,000} = \ln e^{2k}$

$\ln 1.072\,500 = 2k \ln e$

$\ln 1.072\,500 = 2k$

$k = \frac{\ln 1.072\,500}{2}$

$= 0.03499\ldots$ (by calc.)

$= 0.035$ to 3 decimal places

(iii) $2\,000\,000 = 1\,000\,000e^{kt}$ (where $k = 0.03499\ldots$)

$\frac{2\,000\,000}{1\,000\,000} = e^{kt}$

$2 = e^{kt}$

$\ln 2 = \ln e^{kt}$

$\ln 2 = kt \ln e$

$= kt$

$\therefore t = \frac{\ln 2}{k}$

$= \frac{\ln 2}{(0.03499\ldots)}$ (using calc. memory for k)

$= 19.806\ldots$ (by calc.)

$\doteqdot 19.8$ years

QUESTION 6

(a) (i) 1 and $5\frac{1}{2}$ seconds.

(ii) 3 seconds

(iii) $5\frac{1}{2}$ seconds

(iv) 8 seconds

(b) (i) Stationary points occur when

$f'(x) = 0$

$\therefore e^{-2x} - 2xe^{-2x} = 0$

$e^{-2x}(1 - 2x) = 0$

$\therefore 1 - 2x = 0$ (since $e^{-2x} > 0$)

$2x = 1$

$x = \frac{1}{2}.$

(ii) $f(x)$ is increasing when

$f'(x) > 0$

$\therefore e^{-2x}(1 - 2x) > 0$

Now $e^{-2x} > 0$ for all x

but $1 - 2x > 0$

when $1 > 2x$

$\therefore 2x < 1$

$\therefore x < \frac{1}{2}.$

$\therefore f(x)$ is increasing when $x < \frac{1}{2}.$

(iii) Possible points of inflection occur when

$f''(x) = 0$

$4xe^{-2x} - 4e^{-2x} = 0$

$4e^{-2x}(x - 1) = 0$

$e^{-2x} > 0 \quad \therefore x - 1 = 0$

$x = 1$

Checking for change of concavity:

x	$f''(x)$
0.9	$(+)(-) < 0$
1.0	0
1.1	$(+)(+) > 0$

$\therefore$ There is a change of concavity and a point of inflection at $x = 1$.

The graph is concave up
when $f''(x) > 0$
$4e^{-2x}(x - 1) > 0$
Now $e^{-2x} > 0$ and $x - 1 > 0$
when $x > 1$

$\therefore$ Graph is concave upwards when $x > 1$.

(iv) Now $f(x) = xe^{-2x} + 1$

$f\left(-\frac{1}{2}\right) = -0.4$

$f(0) = 1.0$

$f(4) = 1.0013$

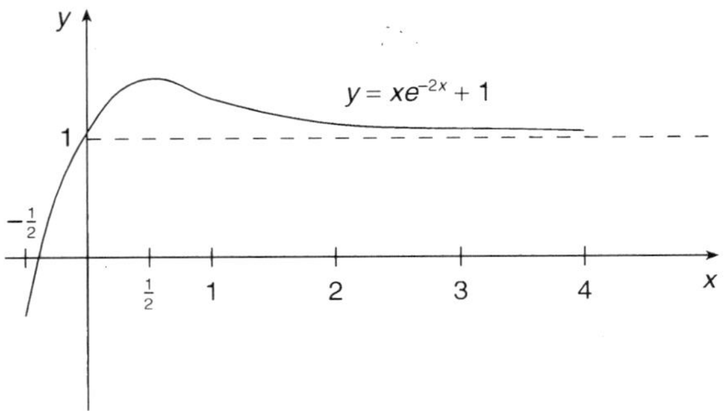

(v) $f(x) = xe^{-2x} + 1$

$= \dfrac{x}{e^{2x}} + 1$

As x gets very large, e^{2x} gets infinitesimally larger and $\dfrac{x}{e^{2x}}$ becomes very small and approaches zero.

$\therefore$ $f(x)$ will approach 1.

QUESTION 7

(a) (i) $\Delta = b^2 - 4ac$

$= (2)^2 - 4(3)(k)$

$= 4 - 12k$

(ii) Real roots occur when

$\Delta \geqslant 0$

$\therefore 4 - 12k \geqslant 0$

$4 \geqslant 12k$

$k \leqslant \dfrac{1}{3}$.

(b) (i) $y = 1 + \sqrt{3}\sin x + \cos x$

$\dfrac{dy}{dx} = \sqrt{3}\cos x - \sin x$

At $x = \frac{5\pi}{6}$, $\dfrac{dy}{dx} = \sqrt{3}\cos\left(\frac{5\pi}{6}\right) - \sin\left(\frac{5\pi}{6}\right)$

$= \sqrt{3}\left(-\frac{\sqrt{3}}{2}\right) - \left(\frac{1}{2}\right)$

$= -\frac{3}{2} - \frac{1}{2}$

$= -\frac{4}{2}$

$= -2$

and $y = 1 + \sqrt{3}\sin\left(\frac{5\pi}{6}\right) + \cos\left(\frac{5\pi}{6}\right)$

$= 1 + \sqrt{3}\left(\frac{1}{2}\right) + \left(-\frac{\sqrt{3}}{2}\right)$

$= 1 + \frac{\sqrt{3}}{2} - \frac{\sqrt{3}}{2}$

$= 1$

Equation of tangent through $\left(\frac{5\pi}{6}, 1\right)$ with gradient $= -2$ is:

$y - 1 = -2\left(x - \frac{5\pi}{6}\right)$

$y - 1 = -2x + \frac{5\pi}{3}$

$y = -2x + \frac{5\pi}{3} + 1$

(ii) $y = 1 + \sqrt{3}\sin x + \cos x$

$\dfrac{dy}{dx} = \sqrt{3}\cos x - \sin x$

$\dfrac{d^2y}{dx^2} = -\sqrt{3}\sin x - \cos x$

Possible maximum and minimum values occur when

$\dfrac{dy}{dx} = 0$

$\sqrt{3}\cos x - \sin x = 0$

$\sqrt{3}\cos x = \sin x$

$\dfrac{\sqrt{3}\cos x}{\cos x} = \dfrac{\sin x}{\cos x}$

$\sqrt{3} = \tan x$

$\therefore x = \dfrac{\pi}{3}$ and $\dfrac{4\pi}{3}$, $\quad 0 \leqslant x \leqslant 2\pi$

At $x = \frac{\pi}{3}$, $y = 1 + \sqrt{3}\sin\left(\frac{\pi}{3}\right) + \cos\left(\frac{\pi}{3}\right)$

$= 1 + \sqrt{3}\left(\frac{\sqrt{3}}{2}\right) + \frac{1}{2}$

$= 1 + \frac{3}{2} + \frac{1}{2}$

$= 3$

and $\frac{d^2y}{dx^2} = -\sqrt{3}\sin\frac{\pi}{3} - \cos\frac{\pi}{3}$

$= -\sqrt{3}\left(\frac{\sqrt{3}}{2}\right) - \frac{1}{2}$

$= -\frac{3}{2} - \frac{1}{2}$

$= -\frac{4}{2}$

$= -2$

$\therefore$ A maximum turning point at $\left(\frac{\pi}{3}, 3\right)$

At $x = \frac{4\pi}{3}$, $y = 1 + \sqrt{3}\sin\left(\frac{4\pi}{3}\right) + \cos\left(\frac{4\pi}{3}\right)$

$= 1 + \sqrt{3}\left(-\frac{\sqrt{3}}{2}\right) + \left(-\frac{1}{2}\right)$

$= 1 - \frac{3}{2} - \frac{1}{2}$

$= 1 - \frac{4}{2}$

$= -1$

and $\frac{d^2y}{dx^2} = -\sqrt{3}\sin\frac{4\pi}{3} - \cos\frac{4\pi}{3}$

$= -\sqrt{3}\left(-\frac{\sqrt{3}}{2}\right) - \left(-\frac{1}{2}\right)$

$= \frac{3}{2} + \frac{1}{2}$

$= 2$

$\therefore$ A minimum turning point at $\left(\frac{4\pi}{3}, -1\right)$

$\therefore$ Maximum and minimum values are 3 and -1 respectively.

(c) (i)

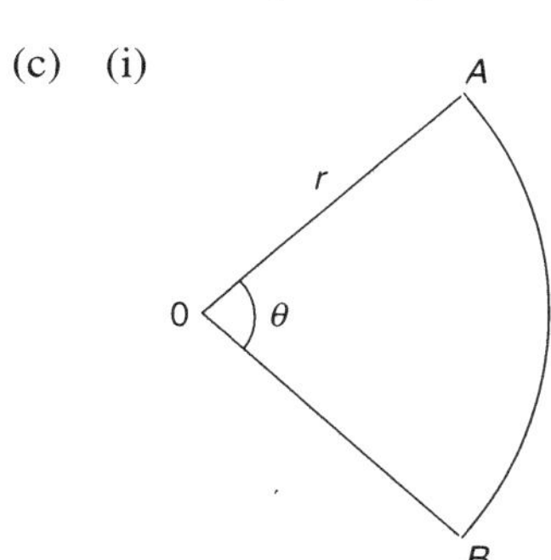

Perimeter $= OA + OB + \text{arc } AB$

$8 = r + r + r\theta$

$8 = 2r + r\theta$

$8 = r(2 + \theta)$

$r = \frac{8}{2 + \theta}$

(ii) Area of sector

$= \frac{1}{2}r^2\theta$

$= \frac{1}{2}\left(\frac{8}{2+\theta}\right)^2.\theta$

$= \frac{1}{2}.\frac{64}{(2+\theta)^2}.\theta$

$= \frac{32\theta}{(2+\theta)^2}$

(iii) $A = \frac{32\theta}{(2+\theta)^2}$ for $0 < \theta < \frac{\pi}{2}$

$\frac{dA}{d\theta} = \frac{(2+\theta)^2.\frac{d}{d\theta}(32\theta) - 32\theta.\frac{d}{d\theta}(2+\theta)^2}{[(2+\theta)^2]^2}$

$= \frac{(2+\theta)^2.32 - 32\theta.2(2+\theta)}{(2+\theta)^4}$

$= \frac{32(2+\theta)^2 - 64\theta(2+\theta)}{(2+\theta)^4}$

$= \frac{32(2+\theta)[(2+\theta) - 2\theta]}{(2+\theta)^4}$

$= \frac{32(2+\theta)(2-\theta)}{(2+\theta)^4}$

$= \frac{32(2-\theta)}{(2+\theta)^3}$

Stationery points occur when

$\frac{dA}{d\theta} = 0$

$\frac{32(2-\theta)}{(2+\theta)^3} = 0$

$\therefore \theta = 2$

Since $0 \leqslant \theta \leqslant \frac{\pi}{2}$, this solution lies outside of the range of possible values of θ.

$\therefore$ the maximum will lie on one of the endpoints of the interval.

Consider points on the graph of

$A = \frac{32\theta}{(\theta+2)^2}$ for $0 < \theta < \frac{\pi}{2}$:

At $\theta = 0$, $A = 0$

At $\theta = \frac{\pi}{2}$, $A = \frac{32\left(\frac{\pi}{2}\right)}{\left(\frac{\pi}{2}+2\right)^2}$

$= \frac{16\pi}{\left(\frac{\pi}{2}+2\right)^2}$

$\doteqdot 3.94$

There are no turning points in this interval and the curve must be rising in this interval, therefore the maximum area will occur when $\theta = \frac{\pi}{2}$ and will be approximately

$\frac{16\pi}{\left(\frac{\pi}{2}+2\right)^2} \doteqdot 3.94$ cm.

QUESTION 8

(a) (i) $R = 100t - t^3$

At $t = 8$, $R = 100(8) - (8)^3$

$= 800 - 512$

$= 288$

$\therefore$ Flow rate is 288 kg/s.

(ii) The expression is reasonable until $R = 0$

$\therefore 0 = 100t - t^3$

$= t(100 - t^2)$

$\therefore t = 0, 10, \text{ or } -10$ (not possible)

$\therefore$ The largest value of T is 10.

(iii) $R = 100t - t^3$

$\dfrac{dR}{dt} = 100 - 3t^2 \quad \dfrac{d^2R}{dt^2} = -6t$

Stationary points occur when $\dfrac{dR}{dt} = 0$

$$100 - 3t^2 = 0$$
$$3t^2 = 100$$
$$t^2 = \frac{100}{3}$$
$$t = \pm\sqrt{\frac{100}{3}}$$
$$= \frac{10}{\sqrt{3}} \quad (\text{since } t > 0)$$

When $t = \dfrac{10}{\sqrt{3}}$, $\dfrac{d^2R}{dt^2} = -6\left(\dfrac{10}{\sqrt{3}}\right)$

$$= -\frac{60}{\sqrt{3}} < 0$$

$\therefore$ A maximum value occurs at $t = \dfrac{10}{\sqrt{3}}$.

When $t = \dfrac{10}{\sqrt{3}}$, $R = 100\left(\dfrac{10}{\sqrt{3}}\right) - \left(\dfrac{10}{\sqrt{3}}\right)^3$

$$= \frac{1000}{\sqrt{3}} - \frac{1000}{3\sqrt{3}}$$
$$= \frac{3000 - 1000}{3\sqrt{3}}$$
$$= \frac{2000}{3\sqrt{3}}$$
$$= \frac{2000\sqrt{3}}{9}$$
$$\doteqdot 384.9 \quad (\text{by calc.})$$

$\therefore$ The maximum flow rate is approximately 384.9 kg/s.

(iv) $R = 100t - t^3$

If M is the amount in the pile at time t:

$\dfrac{dM}{dt} = 100t - t^3$

$\therefore M = 50t^2 - \dfrac{t^4}{4} + C$

When $t = 0$, $M = 300$ $\therefore C = 300$

$\therefore M = 50t - \dfrac{t^4}{4} + 300.$

(v) $\displaystyle\int_0^8 (100t - t^3)\, dt$

$$= \left[\frac{100t^2}{2} - \frac{t^4}{4}\right]_0^8$$
$$= \left[50t^2 - \frac{t^4}{4}\right]_0^8$$
$$= \left(50(64) - \frac{4096}{4}\right) - 0$$
$$= 3200 - 1024$$
$$= 2176$$

$\therefore$ Total weight of sand tipped from truck in first 8 seconds was 2176 kg.

(b) (i) $y = \log_2 x$

$\therefore y = \dfrac{\log_e x}{\log_e 2}$ (change of base)

$\log_e x = y \log_e 2$

$x = e^{y \log_e 2}$

$= e^{y \ln 2}$

$$V = \pi \int_0^3 x^2\, dy$$
$$= \pi \int_0^3 (e^{y \ln 2})^2\, dy$$
$$= \pi \int_0^3 e^{2y \ln 2}\, dy$$
$$= \pi \int_0^3 e^{y \ln 2^2}\, dy$$
$$= \pi \int_0^3 e^{y \ln 4}\, dy$$

(ii) $$V = \pi \int_0^3 e^{y \ln 4}\, dy$$
$$= \pi \left[\frac{e^{y \ln 4}}{\ln 4}\right]_0^3$$
$$= \pi \left(\frac{e^{3 \ln 4}}{\ln 4} - \frac{e^0}{\ln 4}\right)$$
$$= \pi \left(\frac{e^{\ln 64}}{\ln 4} - \frac{1}{\ln 4}\right)$$
$$= \pi \left(\frac{64}{\ln 4} - \frac{1}{\ln 4}\right)$$
$$= \pi \times \frac{63}{\ln 4}$$
$$\doteqdot 142.8$$

$\therefore$ Volume of the solid is approximately 142.8 units^3.

QUESTION 9

(a) $\ln(7x - 12) = 2\ln x$

$\ln(7x - 12) = \ln x^2$

$\therefore 7x - 12 = x^2$

$x^2 - 7x + 12 = 0$

$(x - 3)(x - 4) = 0$

$\therefore x = 3 \text{ or } 4.$

(b) The area of the shaded region is given by the integral of the upper function less the lower function as the curves are continuous and do not intersect in the interval $\left[-\frac{\pi}{2}, \frac{\pi}{6}\right]$

$$A = \int_{-\frac{\pi}{2}}^{\frac{\pi}{6}} (\cos 2x - \sin x)\,dx$$

$$= \left[\frac{\sin 2x}{2} + \cos x\right]_{-\frac{\pi}{2}}^{\frac{\pi}{6}}$$

$$= \left[\left(\frac{\sin\frac{\pi}{3}}{2} + \cos\frac{\pi}{6}\right) - \left(\frac{\sin(-\pi)}{2} + \cos\left(-\frac{\pi}{2}\right)\right)\right]$$

$$= \left(\frac{\sqrt{3}/2}{2} + \frac{\sqrt{3}}{2}\right) - \left(\frac{0}{2} + 0\right)$$

$$= \frac{\sqrt{3}}{4} + \frac{\sqrt{3}}{2}$$

$$= \frac{\sqrt{3} + 2\sqrt{3}}{4}$$

$$= \frac{3\sqrt{3}}{4} \text{ units}^2$$

(c) (i)

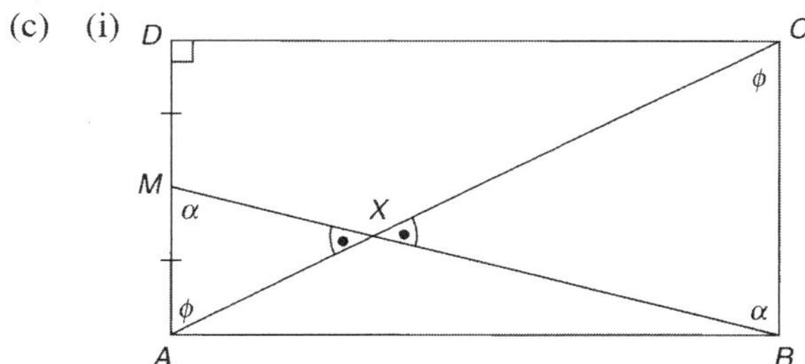

In the $\triangle$'s AXM and BXC,

$\angle MXA = \angle CXB$ (vertically opposite)
$\angle MAX = \angle BCX$ (alternate, $BC \parallel AD$)

$\therefore \triangle AXM \parallel\!| \triangle BXC$

(ii) (To show the relationship between CX and AC, it is necessary to find the relationship between CX and AX.)

Now $\dfrac{AX}{CX} = \dfrac{XM}{XB} = \dfrac{MA}{BC}$ (prop. sides of similar $\triangle$'s)

$\therefore \dfrac{AX}{CX} = \dfrac{MA}{BC}$

$\therefore \dfrac{AX}{CX} = \dfrac{MA}{DA}$ (since $BC = DA$)

$\therefore \dfrac{AX}{CX} = \dfrac{1}{2}$

i.e. $\dfrac{CX}{AX} = 2$

$\therefore \dfrac{CX}{AC} = \dfrac{2}{3}$

i.e. $3CX = 2AC$

(iii) $AC^2 = AB^2 + BC^2$ (Pythagoras' Theorem)

and $AC = \dfrac{3CX}{2}$ (from (ii))

and $BC = \dfrac{AB}{2} = \dfrac{AD}{2}$ (given)

$$\therefore \left(\frac{3CX}{2}\right)^2 = AB^2 + \left(\frac{AB}{2}\right)^2$$

$$\frac{9(CX)^2}{4} = AB^2 + \frac{AB^2}{4}$$

$$\therefore 9(CX)^2 = 4(AB)^2 + AB^2$$

$$\therefore 9(CX)^2 = 5(AB)^2$$

QUESTION 10

(a) (i)

PINK DIE

BLUE	2	3	5	7	11	13
4			✓	✓	✓	✓
6				✓	✓	✓
8					✓	✓
9					✓	✓
10					✓	✓
12						✓

$$P(\text{Win}) = \frac{14}{36} = \frac{7}{18}$$

(ii) $$P(\text{Loss}) = 1 - \frac{7}{18} = \frac{11}{18}$$

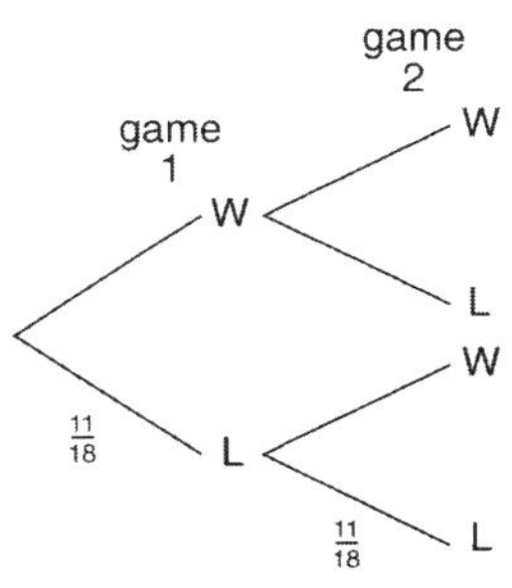

$P(\text{At least one win})$
$= 1 - P(\text{LL})$
$= 1 - \left(\frac{11}{18} \times \frac{11}{18}\right)$
$= 1 - \frac{121}{324}$
$= \frac{203}{324}.$

(b) (i) No. of fish after 1st harvest
$= 100\,000(1.1) - 15\,400$

No. of fish after 2nd harvest
$= [100\,000(1.1) - 15\,400]1.1 - 15\,400$
$= 100\,000(1.1)^2 - 15\,400(1.1) - 15\,400$
$= 88\,660$ fish

(ii) No. of fish after nth harvest, F_n
$= 100\,000(1.1)^n - 15\,400(1.1)^{n-1}$
$- 15\,400(1.1)^{n-2} \ldots - 15\,400(1.1) - 15\,400$
$= 100\,000(1.1)^n - [15\,400 + 15\,400(1.1)$
$+ \ldots + 15\,400(1.1)^{n-1}]$
$= 100\,000(1.1)^n - \left[\frac{a(r^n - 1)}{r - 1}\right]$
where $a = 15\,400$, $r = 1.1$ and $n = n$
$= 100\,000(1.1)^n - \left[\frac{15\,400[(1.1)^n - 1]}{1.1 - 1}\right]$
$= 100\,000(1.1)^n - \frac{15\,400[(1.1)^n - 1]}{0.1}$
$= 100\,000(1.1)^n - 154\,000[(1.1)^n - 1]$
$= 100\,000(1.1)^n - 154\,000(1.1)^n + 154\,000$
$= 154\,000 - 54\,000(1.1)^n$

(iii) When all fish are sold, $F_n = 0$
i.e. $154\,000 - 54\,000(1.1)^n = 0$
$54\,000(1.1)^n = 154\,000$
$(1.1)^n = \frac{154\,000}{54\,000}$
$(1.1)^n = 2.851 \ldots$ (by calc.)
$\ln(1.1)^n = \ln 2.851 \ldots$
$n \ln(1.1) = \ln 2.851 \ldots$
$n = \frac{\ln 2.851 \ldots}{\ln 1.1}$
$= 10.995 \ldots$ (by calc.)
$\doteqdot 11$
$\therefore$ The farmer will have sold all his fish after the 11th harvest.

His total income will be
$11 \times 15\,400 \times \$10 = \$1\,694\,000$

(iv) If I is the total income received after n harvests:

$I = 15 \times F_n + 15\,400 \times 10 \times n$
$= 15[154\,000 - 54\,000(1.1)^n] + 154\,000n$
$= 2\,310\,000 - 810\,000(1.1)^n + 154\,000n$

By trial and error:

When $n = 6$, $I = \$1\,799\,036$.
When $n = 7$, $I = \$1\,809\,539$.
When $n = 8$, $I = \$1\,805\,693$.

$\therefore$ The farmer should sell after the 7th harvest.

(A more advanced approach:

$\frac{dI}{dn} = -810\,000 \ln(1.1).(1.1)^n + 154\,000$

$\left(\text{NB } \frac{d}{dn}(1.1)^n = \ln(1.1).(1.1)^n\right)$

Stationary points occur when
$\frac{dI}{dn} = 0$
$\therefore 0 = -810\,000 \ln(1.1).(1.1)^n + 154\,000$
$154\,000 = 810\,000 \ln(1.1).(1.1)^n$
$\frac{154\,000}{810\,000 \ln(1.1)} = (1.1)^n$
$1.9948 = (1.1)^n$
$\ln 1.9948 = \ln(1.1)^n$
$\ln 1.9948 = n \ln(1.1)$
$n = \frac{\ln 1.9948}{\ln 1.1}$
$\doteqdot 7.25$

Now when $n = 7$, $I = \$1\,809\,539$.
and when $n = 8$, $I = \$1\,805\,693$.

$\therefore$ The farmer should sell after the 7th harvest.

BOARD OF STUDIES
NEW SOUTH WALES

HIGHER SCHOOL CERTIFICATE EXAMINATION

1999
MATHEMATICS
2/3 UNIT (COMMON)

Time allowed—Three hours
(Plus 5 minutes reading time)

DIRECTIONS TO CANDIDATES

- Attempt ALL questions.
- ALL questions are of equal value.
- All necessary working should be shown in every question. Marks may be deducted for careless or badly arranged work.
- Standard integrals are printed on page 12.
- Board-approved calculators may be used.
- Answer each question in a SEPARATE Writing Booklet.
- You may ask for extra Writing Booklets if you need them.

QUESTION 1 Use a SEPARATE Writing Booklet. **Marks**

(a) The points A and B have coordinates $(3, -4)$ and $(7, 2)$ respectively. Find the coordinates of the midpoint of AB. **2**

(b) Find the value of e^3, correct to three significant figures. **2**

(c) Solve $3 - 2x \geq 7$. **2**

(d) Solve the simultaneous equations **2**

$$x + y = 1$$

$$2x - y = 5.$$

(e) Find integers a and b such that $\left(5 - \sqrt{2}\right)^2 = a + b\sqrt{2}$. **2**

(f) **2**

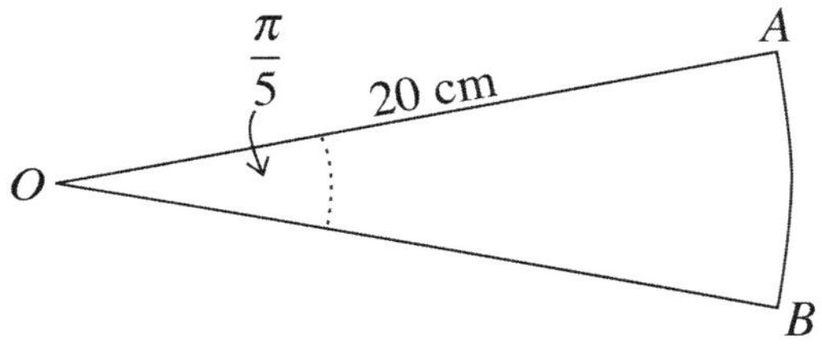

NOT TO SCALE

In the diagram, AB is an arc of a circle with centre O. The radius OA is 20 cm. The angle AOB is $\frac{\pi}{5}$ radians. Find the area of the sector AOB.

QUESTION 2 Use a SEPARATE Writing Booklet. **Marks**

(a) Find: **4**

(i) $\int \left(\frac{1}{x^2} + \frac{1}{x} \right) dx$

(ii) $\int \cos(2x+1)\, dx$.

(b) **8**

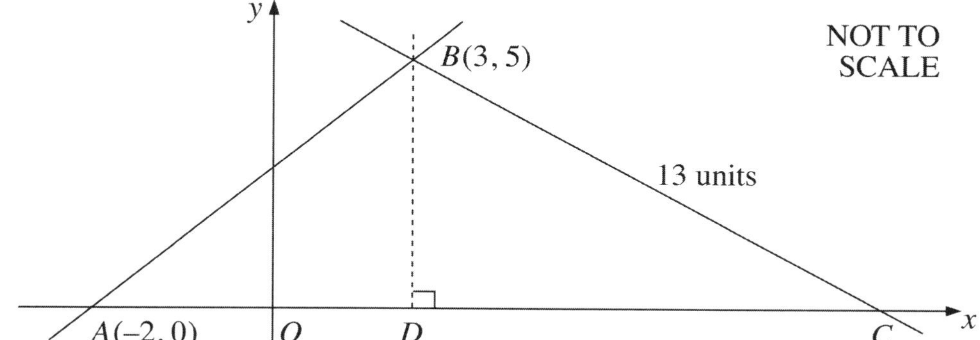

The diagram shows the points $A(-2, 0)$, $B(3, 5)$ and the point C which lies on the x axis. The point D also lies on the x axis such that BD is perpendicular to AC.

(i) Show that the gradient of AB is 1.

(ii) Find the equation of the line AB.

(iii) What is the size of $\angle BAC$?

(iv) The length of BC is 13 units. Find the length of DC.

(v) Calculate the area of ΔABC.

(vi) Calculate the size of $\angle ABC$, to the nearest degree.

QUESTION 3 Use a SEPARATE Writing Booklet. **Marks**

(a) Differentiate the following functions: **4**

(i) $x \tan x$

(ii) $\dfrac{e^x}{1+x}$.

(b) Find the equation of the normal to the curve $y = \sqrt{x+2}$ at the point $(7, 3)$. **4**

(c) **4**

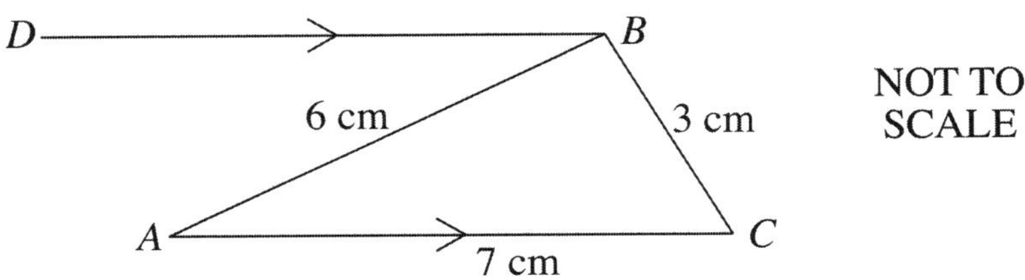

In the diagram, AC is parallel to DB, AB is 6 cm, BC is 3 cm and AC is 7 cm.

(i) Use the cosine rule to find the size of $\angle ACB$, to the nearest degree.

(ii) Hence find the size of $\angle DBC$, giving reasons for your answer.

QUESTION 4 Use a SEPARATE Writing Booklet. **Marks**

(a) An infinite geometric series has a first term of 8 and a limiting sum of 12. Calculate the common ratio. **2**

(b) A market gardener plants cabbages in rows. The first row has 35 cabbages. The second row has 39 cabbages. Each succeeding row has 4 more cabbages than the previous row. **5**

(i) Calculate the number of cabbages in the 12th row.

(ii) Which row would be the first to contain more than 200 cabbages?

(iii) The farmer plants only 945 cabbages. How many rows are needed?

(c) **5**

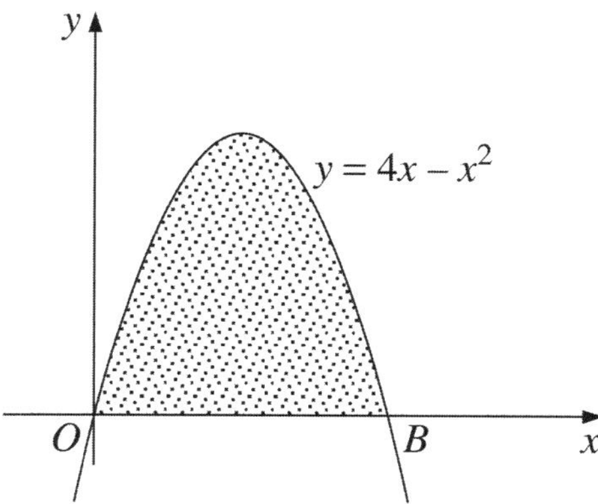

The diagram shows the graph of the function $y = 4x - x^2$.

(i) Find the x coordinate of the point B where the curve crosses the positive x axis.

(ii) Find the area of the shaded region contained by the curve $y = 4x - x^2$ and the x axis.

(iii) Write down a pair of inequalities that specify the shaded region.

QUESTION 5 Use a SEPARATE Writing Booklet. **Marks**

(a) Consider the curve given by $y = x^3 - 6x^2 + 9x + 1$. **8**

(i) Find $\frac{dy}{dx}$.

(ii) Find the coordinates of the two stationary points.

(iii) Determine the nature of the stationary points.

(iv) Sketch the curve for $x \geq 0$.

(b) Let $\log_a 2 = x$ and $\log_a 3 = y$.

Find an expression for $\log_a 12$ in terms of x and y. **2**

(c) **2**

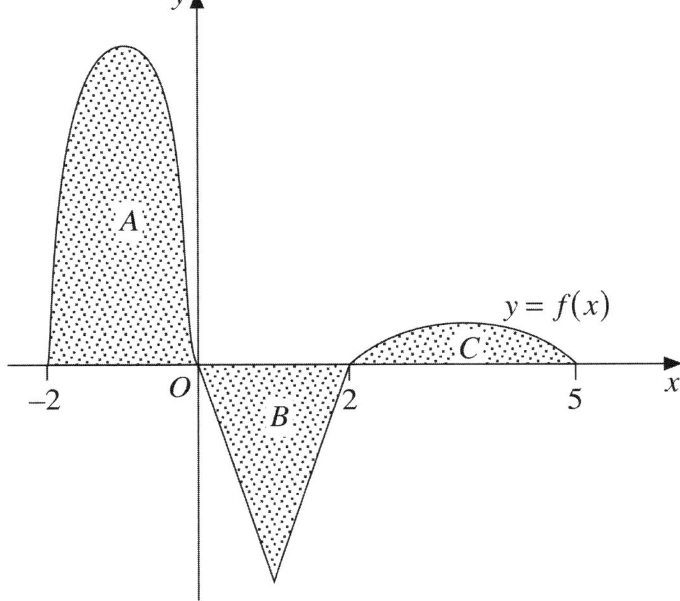

The graph of the function f is shown in the diagram. The shaded areas are bounded by $y = f(x)$ and the x axis. The shaded area A is 8 square units, the shaded area B is 3 square units and the shaded area C is 1 square unit.

Evaluate $\int_{-2}^{5} f(x)\,dx$.

QUESTION 6 Use a SEPARATE Writing Booklet. **Marks**

(a) The mass M kg of a radioactive substance present after t years is given by $M = 10e^{-kt}$, where k is a positive constant. After 100 years the mass has reduced to 5 kg. **7**

(i) What was the initial mass?

(ii) Find the value of k.

(iii) What amount of the radioactive substance would remain after a period of 1000 years?

(iv) How long would it take for the initial mass to reduce to 8 kg?

(b) **5**

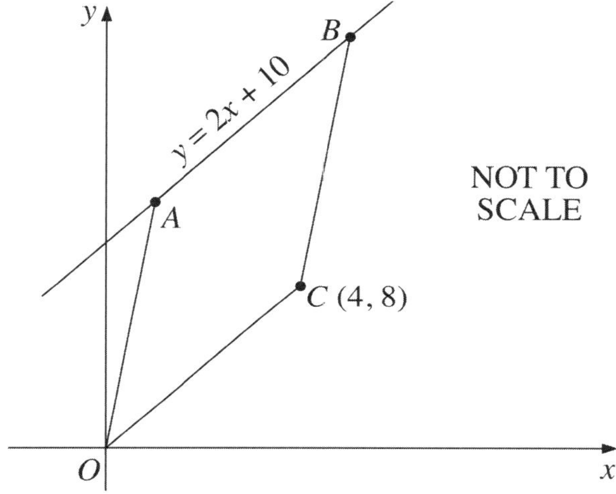

The equation of AB is $y = 2x + 10$. The point C is $(4, 8)$.

Copy or trace the diagram into your Writing Booklet.

(i) Show that OC and AB are parallel.

(ii) State why $\angle ABO = \angle BOC$.

(iii) The line OB divides the quadrilateral $OABC$ into two congruent triangles. Prove that $OABC$ is a parallelogram.

QUESTION 7 Use a SEPARATE Writing Booklet. **Marks**

(a) Isabella invests $\$P$ at 8% per annum compounded annually. She intends to withdraw \$3000 at the end of each of the next six years to cover school fees. **5**

(i) Write down an expression for the amount $\$A_1$ remaining in the account following the withdrawal of the first \$3000.

(ii) Find an expression for the amount $\$A_2$ remaining in the account after the second withdrawal.

(iii) Calculate the amount $\$P$ that Isabella needs to invest if the account balance is to be \$0 at the end of six years.

(b) A particle P is moving along the x axis. Its position at time t seconds is given by **7**

$$x = 2\sin t - t, \qquad t \geq 0.$$

(i) Find an expression for the velocity of the particle.

(ii) In what direction is the particle moving at $t = 0$?

(iii) Determine when the particle first comes to rest.

(iv) When is the acceleration negative for $0 \leq t \leq 2\pi$?

(v) Calculate the total distance travelled by the particle in the first π seconds.

QUESTION 8 Use a SEPARATE Writing Booklet. **Marks**

(a) **5**

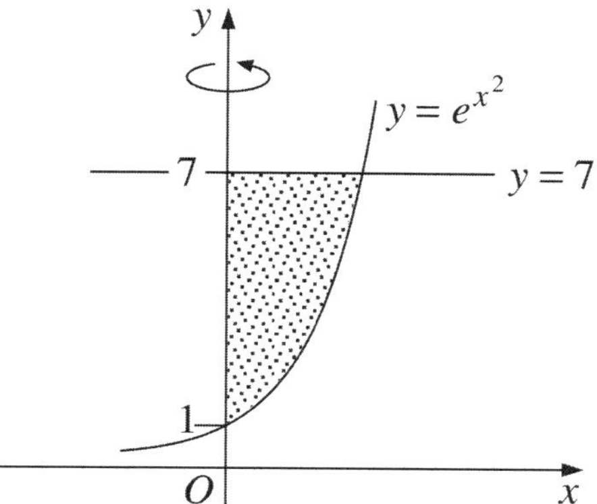

The shaded region bounded by $y = e^{x^2}$, $y = 7$ and the y axis is rotated around the y axis to form a solid.

(i) Show that the volume of the solid is given by $V = \pi \int_1^7 \log_e y \, dy$.

(ii) Copy and complete the table. Give your answers correct to 3 decimal places.

y	1	4	7
$\log_e y$			

(iii) Use the trapezoidal rule with 3 function values to approximate the volume, V.

(b) A box contains five cards. Each card is labelled with a number. The numbers on the cards are 0, 3, 3, 5, 5. **5**

Cameron draws one card at random from the box and then draws a second card at random *without* replacing the first card drawn.

(i) What is the probability that he draws a '5', then a '3'?

(ii) What is the probability that the sum of the two numbers drawn is at least 8?

(iii) What is the probability that the second card drawn is labelled '3'?

(c) **2**

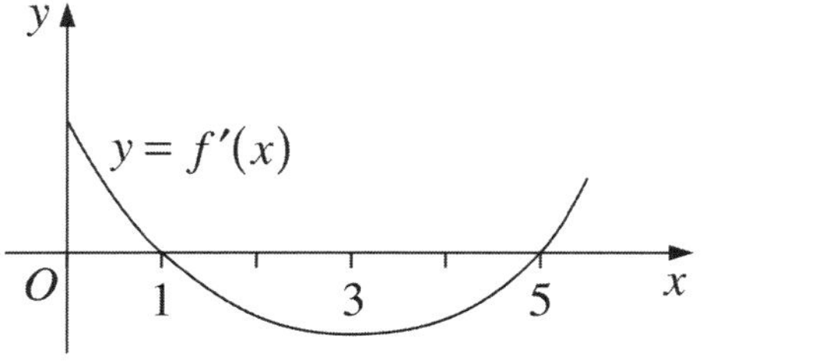

The diagram shows the graph of the gradient function of the curve $y = f(x)$.

For what value of x does $f(x)$ have a local minimum? Justify your answer.

QUESTION 9 Use a SEPARATE Writing Booklet. **Marks**

(a) 3

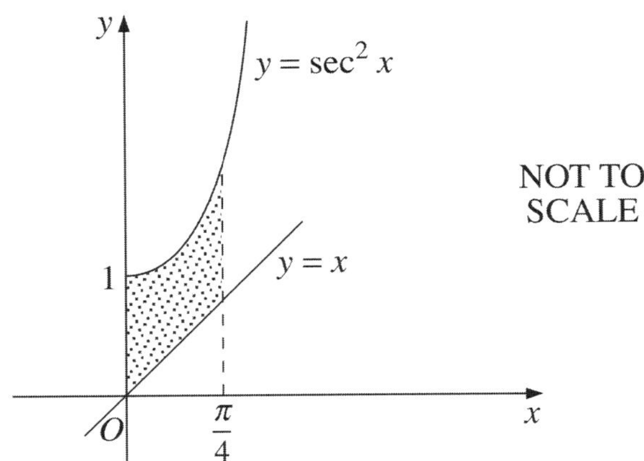

The diagram shows the graphs of the functions $y = \sec^2 x$ and $y = x$ between $x = 0$ and $x = \dfrac{\pi}{4}$.

Calculate the area of the shaded region.

(b) 9

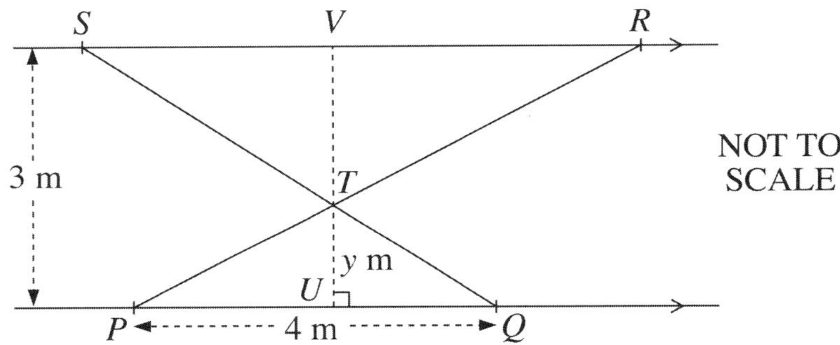

In the diagram, PQ and SR are parallel railings which are 3 metres apart. The points P and Q are fixed 4 metres apart on the lower railing. Two crossbars PR and QS intersect at T as shown in the diagram. The line through T perpendicular to PQ intersects PQ at U and SR at V. The length of UT is y metres.

(i) By using similar triangles, or otherwise, show that $\dfrac{SR}{PQ} = \dfrac{VT}{UT}$.

(ii) Show that $SR = \dfrac{12}{y} - 4$.

(iii) Hence show that the total area A of ΔPTQ and ΔRTS is $A = 4y - 12 + \dfrac{18}{y}$.

(iv) Find the value of y that minimises A. Justify your answer.

QUESTION 10 Use a SEPARATE Writing Booklet. **Marks**

(a) (i) Show that $x = \dfrac{\pi}{3}$ is a solution of $\sin x = \dfrac{1}{2}\tan x$. **6**

(ii) On the same set of axes, sketch the graphs of the functions $y = \sin x$ and $y = \dfrac{1}{2}\tan x$ for $-\pi \le x \le \pi$.

(iii) Hence find all solutions of $\sin x = \dfrac{1}{2}\tan x$ for $-\dfrac{\pi}{2} < x < \dfrac{\pi}{2}$.

(iv) Use your graphs to solve $\sin x \le \dfrac{1}{2}\tan x$ for $-\dfrac{\pi}{2} < x < \dfrac{\pi}{2}$.

(b) **6**

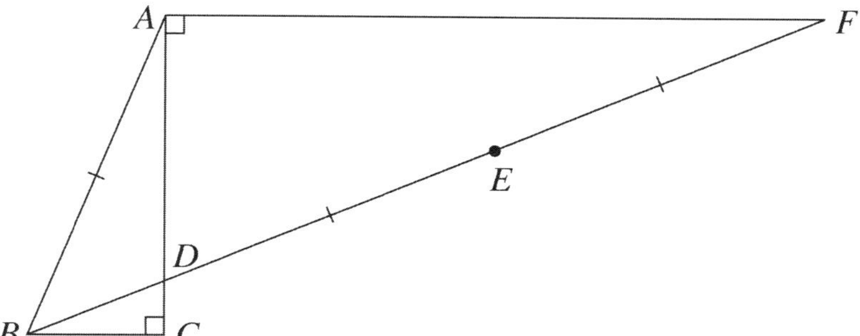

In the diagram, $AC \perp BC$, $AC \perp AF$ and $AB = DE = EF$.

Copy or trace the diagram into your Writing Booklet.

(i) Show that $\angle DBC = \angle DFA$.

(ii) On your diagram, mark the point G on the line AF such that $EG \parallel AC$.

Show that $\Delta AGE \equiv \Delta FGE$.

(iii) Prove that $\angle ABD = 2\angle DBC$.

End of paper

1999 Higher School Certificate Worked Answers

QUESTION 1

(a) Coordinates of the midpoint of AB are

$\left(\frac{3+7}{2}, \frac{-4+2}{2}\right)$

$= \left(\frac{10}{2}, \frac{-2}{2}\right)$

$= (5, -1)$.

(b) $e^3 \doteqdot 20.0855\ldots$ (by calc.)
$= 20.1$ to 3 sig. figures

(c) $3 - 2x \geqslant 7$

$\therefore -2x \geqslant 4$

$x \leqslant -\frac{4}{2}$

$\therefore x \leqslant -2$.

(d) $x + y = 1 \quad \ldots(1)$

$2x - y = 5 \quad \ldots(2)$

(1) + (2) $3x = 6$

$\therefore x = 2$

Substituting $x = 2$ into (1);

$2 + y = 1$

$\therefore y = -1$

(e) $(5 - \sqrt{2})^2 = 25 + 2 - 10\sqrt{2}$

$= 27 - 10\sqrt{2}$

$= a + b\sqrt{2}$

where $a = 27$ and $b = -10$.

(f) Aea of sector AOB

$= \frac{1}{2}r^2\theta$

$= \frac{1}{2} \times (20)^2 \times \frac{\pi}{5}$

$= \frac{1}{2} \times 400 \times \frac{\pi}{5}$

$= 40\pi \text{ cm}^2$.

QUESTION 2

(a) (i) $\int\left(\frac{1}{x^2} + \frac{1}{x}\right) dx$

$= \int\left(x^{-2} + \frac{1}{x}\right) dx$

$= \frac{x^{-1}}{-1} + \ln x + C$

$= -\frac{1}{x} + \ln x + C$.

(ii) $\int \cos(2x + 1)\, dx = \frac{\sin(2x + 1)}{2} + C$

(b) (i) Gradient of $AB = \frac{5 - 0}{3 - (-2)}$

$= \frac{5}{5}$

$= 1$.

(ii) Equation of line AB through $B(3, 5)$ with gradient = 1 is

$y - 5 = 1(x - 3)$

$y - 5 = x - 3$

$y = x + 2$

(iii) $\tan \angle BAC$ = slope of AB

$\therefore \tan \angle BAC = 1$

$\therefore \angle BAC = 45°$.

(iv) $BC^2 = BD^2 + DC^2$

$\therefore 13^2 = 5^2 + DC^2$ (Pythagoras' theorem)

$DC^2 = 169 - 25$

$= 144$

$\therefore DC = 12$ units.

(v) Area of $\triangle ABC = \frac{1}{2} \times AC \times BD$

$= \frac{1}{2} \times 17 \times 5$

$= \frac{85}{2}$

$= 42.5 \text{ units}^2$.

(vi) $\angle ABC = \angle ABD + \angle DBC$

Now $\angle ABD = 180° - 90° - 45°$ (Angle sum $\triangle ABD$)

$= 45°$

In $\triangle BDC$, $\sin \angle DBC = \frac{12}{13}$

$\therefore \angle DBC \doteqdot 67°23'$ (by calc.)

$\therefore \angle ABC = 45° + 67°23'$
$= 112°23'$
$= 112°$ to the nearest degree.

QUESTION 3

(a) (i) $y = x \tan x$

$$\frac{dy}{dx} = x \cdot \frac{d}{dx}(\tan x) + \tan x \cdot \frac{d}{dx}(x)$$
$$= x \sec^2 x + \tan x \cdot 1$$
$$= x \sec^2 x + \tan x.$$

(ii) $y = \dfrac{e^x}{1+x}$

$$\frac{dy}{dx} = \frac{(1+x) \cdot \frac{d}{dx}(e^x) - e^x \cdot \frac{d}{dx}(1+x)}{(1+x)^2}$$
$$= \frac{(1+x) \cdot e^x - e^x \cdot (1)}{(1+x)^2}$$
$$= \frac{e^x + xe^x - e^x}{(1+x)^2}$$
$$= \frac{xe^x}{(1+x)^2}.$$

(b) $y = (x+2)^{\frac{1}{2}}$

$$\frac{dy}{dx} = \frac{1}{2}(x+2)^{-\frac{1}{2}} \cdot 1$$
$$= \frac{1}{2\sqrt{x+2}}$$

At $x = 7$, $\dfrac{dy}{dx} = \dfrac{1}{2\sqrt{9}}$
$$= \frac{1}{6}$$

$\therefore$ Gradient of normal $= -\dfrac{1}{\frac{1}{6}}$
$= -6.$

$\therefore$ Equation of normal through $(7, 3)$ with gradient $= -6$ is
$y - 3 = -6(x - 7)$
$y - 3 = -6x + 42$
$6x + y - 45 = 0$

(c) (i) $\cos \angle ACB = \dfrac{7^2 + 3^2 - 6^2}{2 \times 7 \times 3}$
$$= \frac{49 + 9 - 36}{42}$$
$$= \frac{22}{42}$$

$\therefore \angle ACB \doteqdot 58°25'$ (by calc.)
$= 58°$ to the nearest degree.

(ii) $\angle DBC = 180° - 58°25'$ (co-interior angles)
$= 121°35'$
$= 122°$ to the nearest degree.

QUESTION 4

(a) $S = \dfrac{a}{1-r}$

$\therefore 12 = \dfrac{8}{1-r}$
$12 - 12r = 8$
$-12r = -4$
$r = \dfrac{-4}{-12}$
$= \dfrac{1}{3}.$

(b) 35, 39, 43 is an arithmetic sequence with $a = 35$ and $d = 4$.

(i) $U_n = a + (n-1)d$
$U_{12} = 35 + (12 - 1)4$
$= 35 + 44$
$= 79$
$\therefore$ 79 cabbages in the 12th row.

(ii) $a + (n-1)d > 200$
$35 + (n-1) \times 4 > 200$
$35 + 4n - 4 > 200$
$4n > 169$
$n > \frac{169}{4}$
$n > 42\frac{1}{4}$
$\therefore n = 43$
$\therefore$ The 43rd row would be the first to contain more than 200 cabbages.

(iii) $S_n = \dfrac{n}{2}[2a + (n-1)d]$
$S_n = 945, a = 35, d = 4$
$\therefore 945 = \dfrac{n}{2}[2 \times 35 + (n-1) \times 4]$
$945 = \dfrac{n}{2}[70 + 4n - 4]$
$945 = \dfrac{n}{2}[66 + 4n]$

$945 = 33n + 2n^2$
$0 = 2n^2 + 33n - 945$
$0 = (2n + 63)(n - 15)$
$\therefore n = 15$ (n cannot be negative)

i.e. 15 rows are needed.

(c) (i) At B, $y = 0$
$\therefore 4x - x^2 = 0$
$x(4 - x) = 0$
$\therefore x = 0$ or 4
$\therefore$ At B, $x = 4$.

(ii) $A = \int_0^4 (4x - x^2)dx$

$= \left[2x^2 - \frac{x^3}{3}\right]_0^4$

$= \left(2(4)^2 - \frac{(4)^3}{3}\right) - 0$

$= 32 - \frac{64}{3}$

$= \frac{96}{3} - \frac{64}{3}$

$= \frac{32}{3}$

$= 10\frac{2}{3}$ units^2.

(iii) $y \geqslant 0, \quad y \leqslant 4x - x^2$

QUESTION 5

(a) (i) $y = x^3 - 6x^2 + 9x + 1$

$\frac{dy}{dx} = 3x^2 - 12x + 9$

(ii) Stationary points occur when

$\frac{dy}{dx} = 0$
$3x^2 - 12x + 9 = 0$
$3(x^2 - 4x + 3) = 0$
$3(x - 3)(x - 1) = 0$
$\therefore x = 3$ or 1

When $x = 3$, $y = (3)^3 - 6(3)^2 + 9(3) + 1$
$= 27 - 54 + 27 + 1$
$= 1$

When $x = 1$, $y = (1)^3 - 6(1)^2 + 9(1) + 1$
$= 1 - 6 + 9 + 1$
$= 5$

$\therefore$ Coordinates of stationary points are $(3, 1)$ and $(1, 5)$

(iii) $\frac{d^2y}{dx^2} = 6x - 12$

When $x = 3$, $\frac{d^2y}{dx^2} = 6(3) - 12$
$= 6 > 0$
$\therefore$ A minimum turning point at $(3, 1)$

When $x = 1$, $\frac{d^2y}{dx^2} = 6(1) - 12$
$= -6 < 0$
$\therefore$ A maximum turning point at $(1, 5)$

(iv) When $x = 0, y = 1$

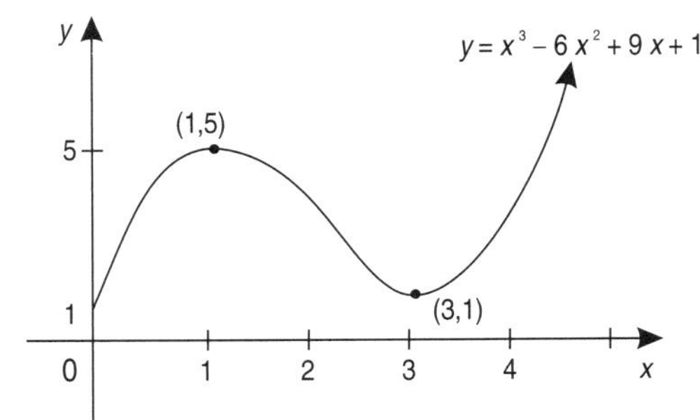

(b) $\log_a 12 = \log_a (4 \times 3)$
$= \log_a 4 + \log_a 3$
$= \log_a 2^2 + \log_a 3$
$= 2 \log_a 2 + \log_a 3$
$= 2x + y$

(c) $\int_{-2}^{5} f(x)dx =$ Area A – Area B + Area C
$= 8 - 3 + 1$
$= 6$

QUESTION 6

(a) (i) $M = 10e^{-kt}$

When $t = 0$, $M = 10e^{-k(0)}$
$= 10e^0$
$= 10$

$\therefore$ Initial mass was 10 kg.

(ii) When $t = 100$, $M = 5$
$\therefore 5 = 10e^{-k(100)}$
$\frac{5}{10} = e^{-100k}$
$\ln\left(\frac{5}{10}\right) = \ln e^{-100k}$

$$\ln 0.5 = -100k \ln e$$
$$\ln 0.5 = -100k$$
$$k = \frac{\ln 0.5}{-100}$$
$$= 0.006\,931\ldots \text{ (by calc.)}$$
$$\doteqdot 0.00693.$$

(iii) When $t = 1000$,

$$M = 10e^{-0.006\,931\ldots \times 1000}$$
$$= 0.009765\ldots \quad \text{(by calc.)}$$
$$\doteqdot 0.00977 \text{ kg}$$
$$\doteqdot 9.77 \text{ g}$$

(iv) When $M = 8$,

$$8 = 10e^{-kt}$$
$$\frac{8}{10} = e^{-kt}$$
$$\ln 0.8 = \ln e^{-kt}$$
$$\ln 0.8 = -kt$$
$$t = \frac{\ln 0.8}{-k}$$
$$= \frac{\ln 0.8}{-0.006931\ldots}$$

(using calc. memory for k)

$$\doteqdot 32.19 \text{ years} \quad \text{(by calc.)}$$

(b)

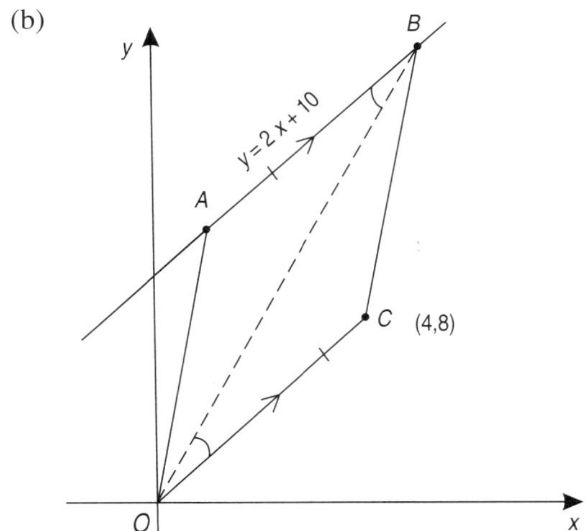

(i) Gradient of $OC = \frac{8-0}{4-0}$

$$= \frac{8}{4}$$
$$= 2.$$

Gradient of $AB = 2$ (from equation)

$\therefore OC \parallel AB$

(ii) Alternate angles ($AB \parallel OC$)

(iii) $AB = OC$ (sides of congrent $\triangle$'s)

$AB \parallel OC$ (proved above)

$\therefore OABC$ is a parallelogram

(1 pair of sides equal and parallel)

QUESTION 7

(a) (i) $\$A_1 = \$P(1.08) - \$3000$

(ii) $\$A_2 = \$A_1(1.08) - \$3000$

$$= [\$P(1.08) - \$3000]1.08 - \$3000$$
$$= \$P(1.08)^2 - \$3000(1.08) - \$3000$$

(iii) $\$A_6 = \$P(1.08)^6 - \$3000(1.08)^5 - \$3000(1.08)^4 - \ldots - \$3000$

After 6 years, $\$A_6 = 0$

$\therefore \$P(1.08)^6 - \$3000(1.08)^5 - \$3000(1.08)^4 - \ldots - \$3000 = 0$

$\therefore \$P(1.08)^6 = \$3000 + \$3000(1.08)^1 + \ldots + \$3000(1.08)^5$

$$= \$3000[1 + (1.08)^1 + \ldots + (1.08)^5]$$
$$= \$3000\left[\frac{a(r^n - 1)}{r - 1}\right]$$

where $a = 1, r = 1.08, n = 6$

$$= \$3000\left[\frac{1((1.08)^6 - 1)}{1.08 - 1}\right]$$
$$\therefore \$P = \$3000\left[\frac{1.08^6 - 1}{0.08}\right] \div (1.08)^6$$
$$= \$13\,868.64 \quad \text{(by calc.)}$$

$= \$13\,869$ to the nearest dollar.

(b) (i) $x = 2 \sin t - t \quad t \geqslant 0$

$$v = \frac{dx}{dt} = 2 \cos t - 1$$

(ii) When $t = 0$, $v = 2 \cos(0) - 1$

$$= 2(1) - 1$$
$$= +1 \text{ units/s}$$

Hence, the particle is moving away from the origin to the right (positive x direction).

(iii) Particle comes to rest when

$$v = 0$$
$$2 \cos t - 1 = 0$$
$$2 \cos t = 1$$
$$\cos t = \tfrac{1}{2}$$
$$\therefore t = \tfrac{\pi}{3}, 2\pi - \tfrac{\pi}{3}, \ldots$$

$\therefore$ Particle first comes to rest at $t = \frac{\pi}{3}$ seconds

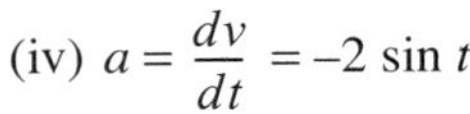

(iv) $a = \frac{dv}{dt} = -2 \sin t$

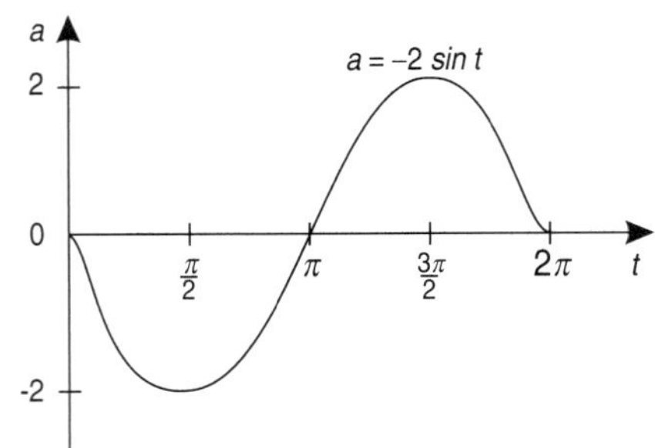

$a < 0$ when $0 < t < \pi$

(v) At $t = 0$, particle is moving to the right of 0 (from (ii)). During the first π seconds, its acceleration is always negative (from (iv)). The particle will change direction when $v = 0$, i.e. when $t = \frac{\pi}{3}$ (from (iii)).

At $t = 0$, $\quad x = 2 \sin (0) - 0 = 0$

At $t = \frac{\pi}{3}$, $\quad x = 2 \sin (\frac{\pi}{3}) - \frac{\pi}{3}$

$= 2\left(\frac{\sqrt{3}}{2}\right) - \frac{\pi}{3}$

$= \sqrt{3} - \frac{\pi}{3}$

At $t = \pi$, $\quad x = 2 \sin (\pi) - \pi$

$= 0 - \pi$

$= -\pi$

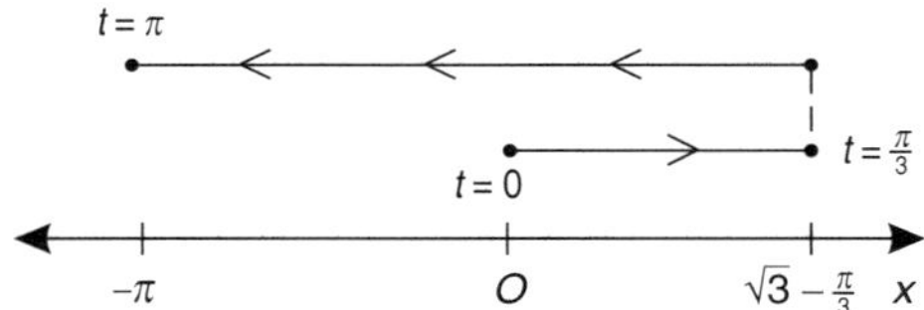

$\therefore$ Total distance travelled

$= (\sqrt{3} - \frac{\pi}{3}) + (\sqrt{3} - \frac{\pi}{3}) + \pi$

$= 2\sqrt{3} - \frac{2\pi}{3} + \pi$

$= (2\sqrt{3} + \frac{\pi}{3})$ units.

QUESTION 8

(a) (i) $V = \pi \int_1^7 x^2 \, dy$

Now $\quad y = e^{x^2}$

$\therefore \log_e y = \log_e e^{x^2}$

$\log_e y = x^2$

$\therefore V = \pi \int_1^7 \log_e y \, dy$

(ii)

y	1	4	7
$\log_e y$	0	1.386	1.946

(iii) $V = \pi \left[\frac{b-a}{2n}(y_0 + 2(y_1) + y_2)\right]$

$\doteqdot \pi \left[\frac{7-1}{2(2)}(0 + 2(1.386) + 1.946)\right]$

$\doteqdot \pi [\frac{6}{4} \times 4.718]$

$\doteqdot 22.233$ units^3.

(b)

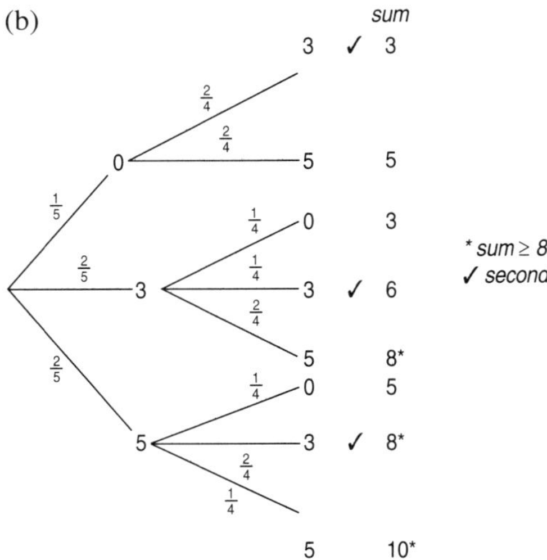

(i) $P(5 \text{ then } 3) = \frac{2}{5} \times \frac{2}{4}$

$= \frac{4}{20}$

$= \frac{1}{5}$.

(ii) $P(\text{Sum} > 8) = (\frac{2}{5} \times \frac{2}{4}) + (\frac{2}{5} \times \frac{2}{4}) + (\frac{2}{5} \times \frac{1}{4})$

$= \frac{4}{20} + \frac{4}{20} + \frac{2}{20}$

$= \frac{10}{20}$

$= \frac{1}{2}$.

(iii) P(Second card 3)

$= (\frac{1}{5} \times \frac{2}{4}) + (\frac{2}{5} \times \frac{1}{4}) + (\frac{2}{5} \times \frac{2}{4})$

$= \frac{2}{20} + \frac{2}{20} + \frac{4}{20}$

$= \frac{8}{20}$

$= \frac{2}{5}$

(c) At $x = 5$, $\quad f'(x) = 0$

and $f'(5 - \varepsilon) < 0$

and $f'(5 + \varepsilon) > 0$

$\therefore$ A local minimum at $x = 5$.

QUESTION 9

(a) $A = \int_0^{\frac{\pi}{4}} (\sec^2 x - x)\, dx$

$= \left[\tan x - \frac{x^2}{2}\right]_0^{\frac{\pi}{4}}$

$= \left(\tan\left(\frac{\pi}{4}\right) - \frac{\left(\frac{\pi}{4}\right)^2}{2}\right) - (\tan 0 - 0)$

$= 1 - \frac{\pi^2}{32} - 0$

$= \left(1 - \frac{\pi^2}{32}\right)$ units^2.

(b)

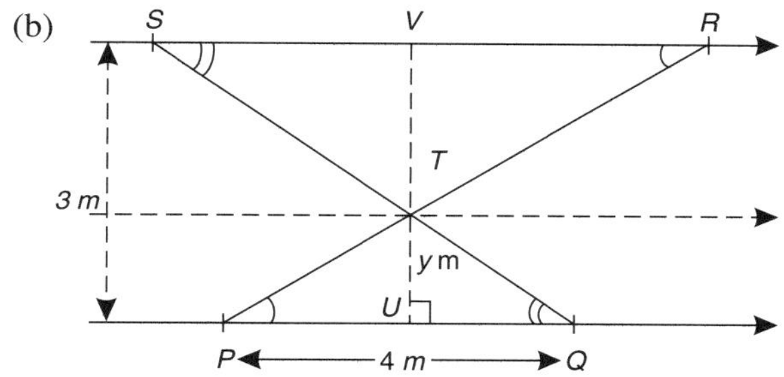

(i) In the $\triangle$'s STR and PTQ:

$\angle SRT = \angle TPQ$ (alternate, $SR \parallel PQ$)
$\angle RST = \angle TQP$ (alternate, $SR \parallel PQ$)

$\therefore \triangle STR \,|||\, \triangle PTQ$ (AAA)

$\therefore \frac{SR}{PQ} = \frac{RT}{PT}$ (corr. sides proportional)

Now $\frac{VT}{UT} = \frac{RT}{PT}$

(equal ratios on transversals by parallel lines)

$\therefore \frac{SR}{PQ} = \frac{VT}{UT}$.

(ii) From (i) $\frac{SR}{4} = \frac{3-y}{y}$

$\therefore SR = \frac{12-4y}{y}$

$= \frac{12}{y} - 4$

(iii) Total area A = Area $\triangle PTQ$ + Area $\triangle RTS$

$= \frac{1}{2} \times 4 \times y + \frac{1}{2}\left(\frac{12}{y} - 4\right)(3-y)$

$= 2y + \frac{1}{2}\left(\frac{36}{y} - 24 + 4y\right)$

$= 2y + \frac{18}{y} - 12 + 2y$

$= 4y - 12 + \frac{18}{y}$

(iv) $A = 4y - 12 + 18y^{-1}$

$\frac{dA}{dy} = 4 - 18y^{-2}$

$= 4 - \frac{18}{y^2}$

$\frac{d^2A}{dy^2} = 36y^{-3}$

$= \frac{36}{y^3}$

Possible maximum and minimum values occur when

$\frac{dA}{dy} = 0$

$\therefore 4 - \frac{18}{y^2} = 0$

$\frac{18}{y^2} = 4$

$18 = 4y^2$

$y^2 = \frac{18}{4}$

$= \frac{9}{2}$

$\therefore y = \pm\sqrt{\frac{9}{2}}$

$= \frac{3}{\sqrt{2}}$ (y must be positive)

$= \frac{3\sqrt{2}}{2}$

When $y = \frac{3\sqrt{2}}{2}$,

$\frac{d^2A}{dy^2} = \frac{36}{\left(\frac{3\sqrt{2}}{2}\right)^3} > 0$

$\therefore$ A minimum value occurs when $y = \frac{3\sqrt{2}}{2}$.

QUESTION 10

(a) (i) If $x = \frac{\pi}{3}$

LHS $= \sin\frac{\pi}{3}$

$= \frac{\sqrt{3}}{2}$

RHS $= \frac{1}{2}\tan\frac{\pi}{3}$

$= \frac{1}{2} \times \sqrt{3}$

$= \frac{\sqrt{3}}{2}$

$\therefore x = \frac{\pi}{3}$ is a solution

(ii)

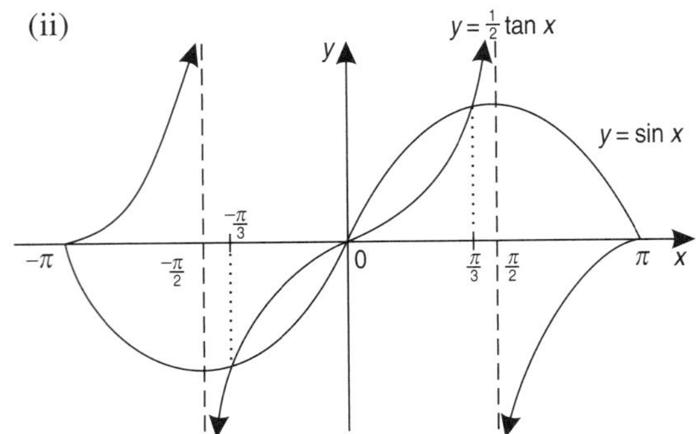

(iii) From the graph, solutions are

$x = -\frac{\pi}{3}$, 0 and $\frac{\pi}{3}$

(iv) $\sin x \leqslant \frac{1}{2}\tan x \qquad -\frac{\pi}{2} < x < \frac{\pi}{2}$

when $\frac{\pi}{3} \leqslant x < \frac{\pi}{2}$

and $-\frac{\pi}{3} \leqslant x \leqslant 0$

(b)

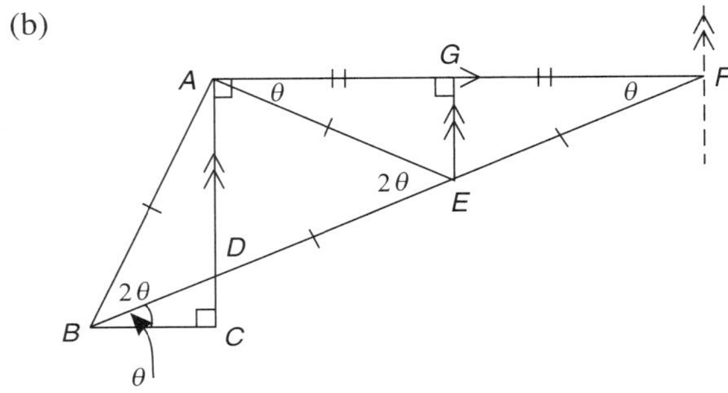

(i) $\angle FAD = \angle ACB$ (given)

$\therefore AF \parallel BC$ (alt. angles – FAD, ACB)

$\therefore \angle DBC = \angle DFA$ (alt. angles, $AF \parallel BC$)

(ii) Since $EG \parallel AC$

and $DE = EF$

$\therefore AG = GF$

(equal intercepts on transversals)

Now in $\triangle$'s AGE and FGE:

$AG = GF$ (proved)

GE is common

$\angle AGE = \angle FGE$ ($90°$, $EG \parallel AC$)

$\therefore \triangle AGE \equiv \triangle FGE$ (SAS)

(iii) Now $AE = FE$

and $\angle GAE = \angle GFE$

(corr. parts congruent $\triangle$'s)

$AE = AB$ (since $AB = EF$)

and $\angle AED = 2\angle GFE$

(exterior angle of $\triangle$)

Now $\angle AED = \angle ABD$ (opp. equal sides)

$\therefore \angle ABD = 2\angle GFE$

and $\angle GFE = \angle DBC$ (alternate: $BC \parallel AF$)

$\therefore \angle ABD = 2\angle DBC$

HIGHER SCHOOL CERTIFICATE EXAMINATION

2000
MATHEMATICS
2/3 UNIT (COMMON)

Time allowed—Three hours
(Plus 5 minutes reading time)

DIRECTIONS TO CANDIDATES

- Attempt ALL questions.
- ALL questions are of equal value.
- All necessary working should be shown in every question. Marks may be deducted for careless or badly arranged work.
- Standard integrals are printed on page 16.
- Board-approved calculators may be used.
- Answer each question in a SEPARATE Writing Booklet.
- You may ask for extra Writing Booklets if you need them.

QUESTION 1 Use a SEPARATE Writing Booklet. **Marks**

(a) Find the value of $\log_e 8$ correct to two decimal places. **2**

(b) Solve $x + 7 \geq 3$ and graph the solution on the number line. **2**

(c) What is the exact value of $\cos \dfrac{\pi}{6}$? **1**

(d) A bag contains red marbles and blue marbles in the ratio 2 : 3. **1**

A marble is selected at random.

What is the probability that the marble is blue?

(e) Solve the pair of simultaneous equations: **2**

$$x - y = 2$$
$$3x + 2y = 1 .$$

(f) Solve $|x - 5| = 3$. **2**

(g) Sketch the line $y = 2x + 3$ in the Cartesian plane. **2**

QUESTION 2 Use a SEPARATE Writing Booklet. **Marks**

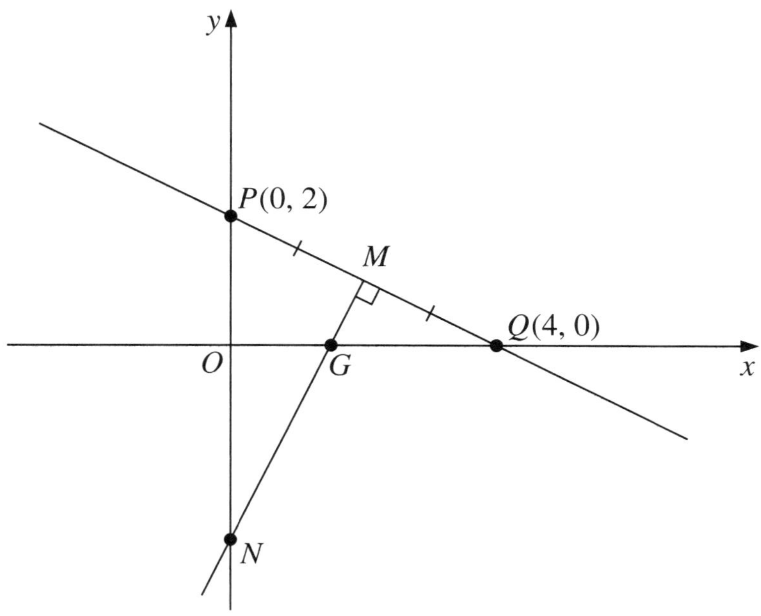

The diagram shows the points $P(0, 2)$ and $Q(4, 0)$. The point M is the midpoint of PQ. The line MN is perpendicular to PQ and meets the x axis at G and the y axis at N.

(a) Show that the gradient of PQ is $-\frac{1}{2}$. **1**

(b) Find the coordinates of M. **2**

(c) Find the equation of the line MN. **2**

(d) Show that N has coordinates $(0, -3)$. **1**

(e) Find the distance NQ. **1**

(f) Find the equation of the circle with centre N and radius NQ. **2**

(g) Hence show that the circle in part (f) passes through the point P. **1**

(h) The point R lies in the first quadrant, and $PNQR$ is a rhombus. Find the coordinates of R. **2**

QUESTION 3 Use a SEPARATE Writing Booklet. **Marks**

(a) Differentiate the following: **4**

(i) $3xe^x$

(ii) $\sin(x^2 + 1)$

(b) **2**

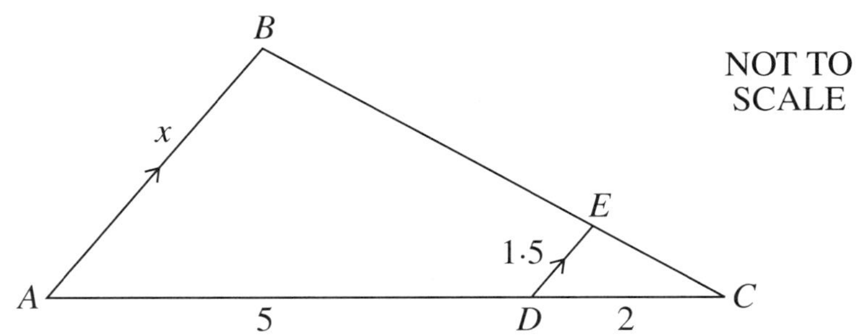

In the diagram, AB is parallel to DE, AD is 5 cm, DC is 2 cm and DE is 1·5 cm.

Find the length of AB.

(c) Find: **3**

(i) $\int \sec^2 5x \, dx$

(ii) $\int_{-2}^{1} \frac{2}{x+3} \, dx$

(d) Find the equation of the tangent to the curve $y = 2\log_e x$ at (1, 0). **3**

QUESTION 4 Use a SEPARATE Writing Booklet. **Marks**

(a) 6

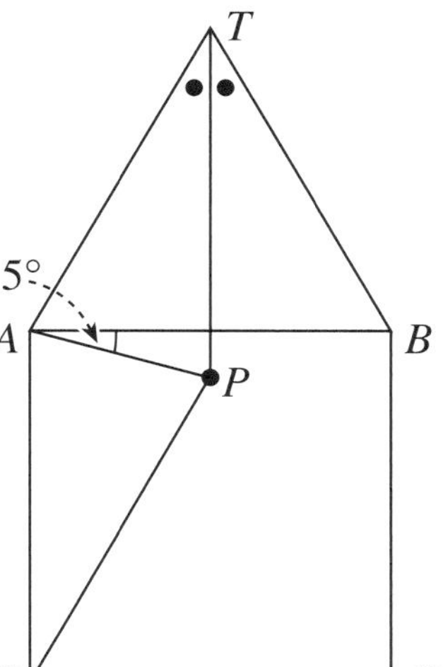

In the diagram, $ABCD$ is a square and ABT is an equilateral triangle. The line TP bisects $\angle ATB$, and $\angle PAB = 15°$.

(i) Copy the diagram into your Writing Booklet and explain why $\angle PAT = 75°$.

(ii) Prove that $\Delta TAP \equiv \Delta DAP$.

(iii) Prove that triangle DAP is isosceles.

(b) In the construction of a 5 km expressway a truck delivers materials from a base. After depositing each load, the truck returns to the base to collect the next load. The first load is deposited 200 m from the base, the second 350 m from the base, the third 500 m from the base. Each subsequent load is deposited 150 m from the previous one. 6

(i) How far is the fifteenth load deposited from the base?

(ii) How many loads are deposited along the total length of the 5 km expressway? (The last load is deposited at the end of the expressway.)

(iii) How many kilometres has the truck travelled in order to make all the deposits and then return to the base?

QUESTION 5 Use a SEPARATE Writing Booklet. **Marks**

(a) Solve $\tan x = 2$ for $0 < x < 2\pi$. **2**

Express your answer in radian measure correct to two decimal places.

(b) Four white (W) balls and two red (R) balls are placed in a bag. One ball is selected at random, removed and replaced by a ball of the other colour. The bag is then shaken and another ball is randomly selected. **5**

(i) Copy the tree diagram into your Writing Booklet. Complete the tree diagram, showing the probability on each branch.

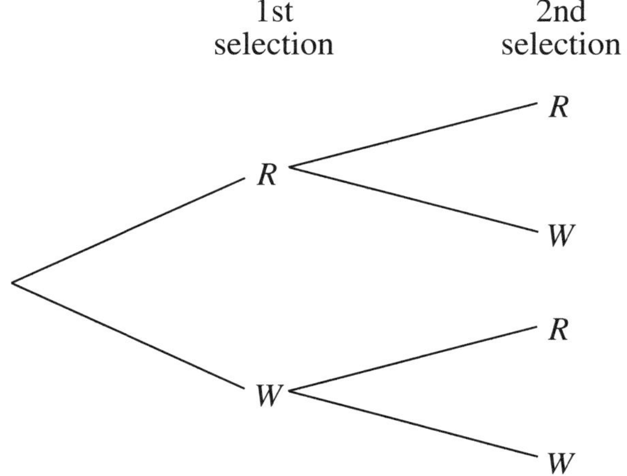

(ii) Find the probability that both balls selected are white.

(iii) Find the probability that the second ball selected is white.

(c) The population of a certain insect is growing exponentially according to $N = 200e^{kt}$, where t is the time in weeks after the insects are first counted. **5**

At the end of three weeks the insect population has doubled.

(i) Calculate the value of the constant k.

(ii) How many insects will there be after 12 weeks?

(iii) At what rate is the population increasing after three weeks?

QUESTION 6 Use a SEPARATE Writing Booklet. **Marks**

(a) Sketch the curve $y = 1 - \sin 2x$ for $0 \leq x \leq \pi$. **3**

(b) The number N of students logged onto a website at any time over a five-hour period is approximated by the formula **9**

$$N = 175 + 18t^2 - t^4, \quad 0 \leq t \leq 5.$$

(i) What was the initial number of students logged onto the website?

(ii) How many students were logged onto the website at the end of the five hours?

(iii) What was the maximum number of students logged onto the website?

(iv) When were the students logging onto the website most rapidly?

(v) Sketch the curve $N = 175 + 18t^2 - t^4$ for $0 \leq t \leq 5$.

Please turn over

QUESTION 7 Use a SEPARATE Writing Booklet. **Marks**

(a) The area under the curve $y = \dfrac{1}{\sqrt{x}}$, for $1 \le x \le e^2$, is rotated about the x axis. **4**

Find the exact volume of the solid of revolution.

(b) Estimate $\displaystyle\int_0^1 \sin\left(1 + x^2\right)dx$ by using Simpson's rule with three function values. **3**

(c) The diagram shows the graphs of $y = x^2 - 2$ and $y = x$. **5**

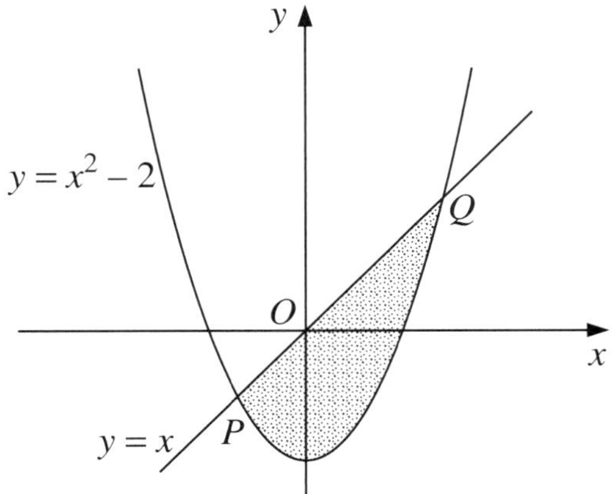

(i) Find the x values of the points of intersection, P and Q.

(ii) Calculate the area of the shaded region.

QUESTION 8 Use a SEPARATE Writing Booklet. **Marks**

(a) A particle is moving in a straight line, starting from the origin. At time t seconds the particle has a displacement of x metres from the origin and a velocity v m s^{-1}. The displacement is given by $x = 2t - 3\log_e(t+1)$. **7**

(i) Find an expression for v.

(ii) Find the initial velocity.

(iii) Find when the particle comes to rest.

(iv) Find the distance travelled by the particle in the first three seconds.

(b) An enclosure is to be built adjoining a barn, as in the diagram. The walls of the barn meet at 135°, and 117 metres of fencing is available for the enclosure, so that $x + y = 117$ where x and y are as shown in the diagram. **5**

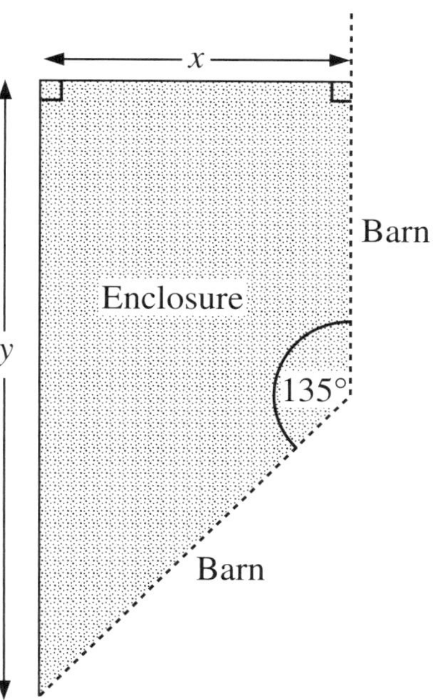

(i) Show that the shaded area of the enclosure in square metres is given by

$$A = 117x - \frac{3}{2}x^2 .$$

(ii) Show that the largest area of the enclosure occurs when $y = 2x$.

QUESTION 9 Use a SEPARATE Writing Booklet. **Marks**

(a) (i) Without using calculus, sketch $y = \log_e x$. **3**

(ii) On the same sketch, find, graphically, the number of solutions of the equation

$$\log_e x - x = -2.$$

(b) **2**

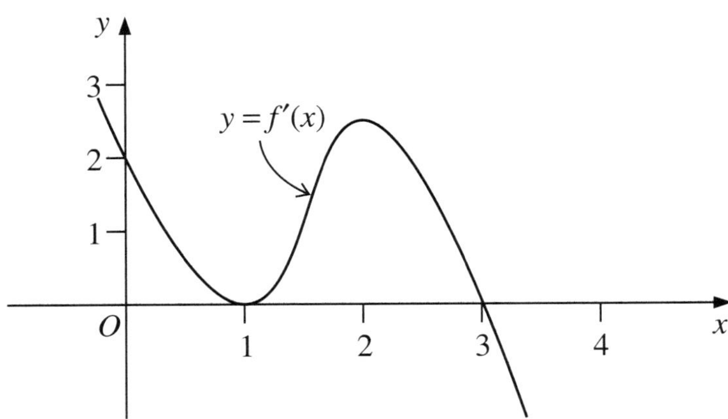

The above diagram shows a sketch of the gradient function of the curve $y = f(x)$.

In your Writing Booklet, draw a sketch of the function $y = f(x)$ given that $f(0) = 0$.

QUESTION 9 (Continued) **Marks**

(c) **7**

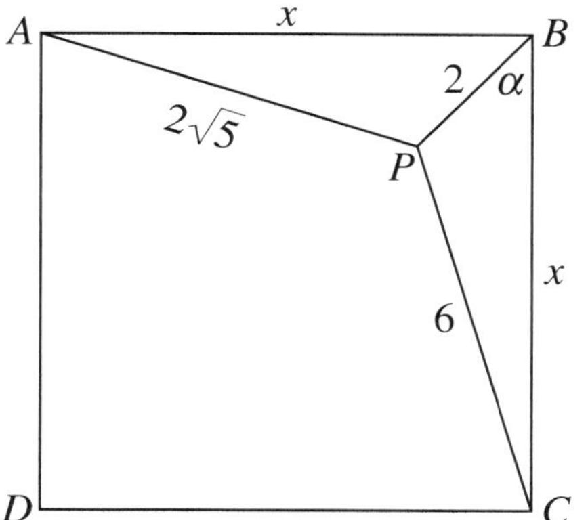

NOT TO SCALE

The diagram shows a square $ABCD$ of side x cm, with a point P within the square, such that $PC = 6$ cm, $PB = 2$ cm and $AP = 2\sqrt{5}$ cm.

Let $\angle PBC = \alpha$.

(i) Using the cosine rule in triangle PBC, show that $\cos\alpha = \dfrac{x^2 - 32}{4x}$.

(ii) By considering triangle PBA, show that $\sin\alpha = \dfrac{x^2 - 16}{4x}$.

(iii) Hence, or otherwise, show that the value of x is a solution of

$$x^4 - 56x^2 + 640 = 0.$$

(iv) Find x. Give reasons for your answer.

Please turn over

QUESTION 10 Use a SEPARATE Writing Booklet. **Marks**

(a) A store offers a loan of \$5000 on a computer for which it charges interest at the rate of 1% per month. As a special deal, the store does not charge interest for the first three months however, the first repayment is due at the end of the first month. **6**

A customer takes out the loan and agrees to repay the loan over three years by making 36 equal monthly repayments of $\$M$.

Let $\$A_n$ be the amount owing at the end of the nth repayment.

(i) Find an expression for A_3.

(ii) Show that $A_5 = (5000 - 3M)1{\cdot}01^2 - M(1 + 1{\cdot}01)$

(iii) Find an expression for A_{36}.

(iv) Find the value of M.

(b) The first snow of the season begins to fall during the night. The depth of the snow, h, increases at a constant rate through the night and the following day. At 6 am a snow plough begins to clear the road of snow. The speed, v km/h, of the snow plough is inversely proportional to the depth of snow. (This means $v = \dfrac{A}{h}$ where A is a constant.) **6**

Let x km be the distance the snow plough has cleared and let t be the time in hours from the beginning of the snowfall. Let $t = T$ correspond to 6 am.

(i) Explain carefully why, for $t \geq T$,

$$\frac{dx}{dt} = \frac{k}{t}, \text{ where } k \text{ is a constant.}$$

(ii) In the period from 6 am to 8 am the snow plough clears 1 km of road, but it takes a further 3·5 hours to clear the next kilometre.

At what time did it begin snowing?

End of paper

2000 Higher School Certificate Worked Answers

QUESTION 1

(a) $\log_e 8 = 2.0794\ldots$ (by calc.)
$= 2.08$ to 2 decimal places.

(b) $x + 7 \geqslant 3$
$\therefore x \geqslant -4$

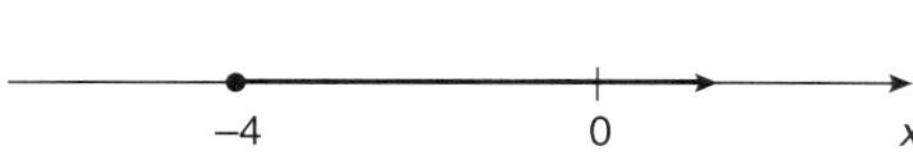

(c) $\cos \dfrac{\pi}{6} = \dfrac{\sqrt{3}}{2}$.

(d) $P(\text{Blue}) = \dfrac{3}{5}$

(e)
$$x - y = 2 \quad \ldots(1)$$
$$3x + 2y = 1 \quad \ldots(2)$$
$(1)\times(2)$ $\quad 2x - 2y = 4 \quad \ldots(3)$
$(2)+(3)$ $\quad 5x = 5$
$$\therefore x = 1$$

Substituting $x = 1$ into (1).

$$1 - y = 2$$
$$-y = 1$$
$$\therefore y = -1.$$

(f) $|x - 5| = 3$
$\therefore x - 5 = \pm 3$
$\therefore x - 5 = 3$ or $x - 5 = -3$
$\therefore x = 8$ or $x = 2$

(g) Gradient $= 2$,
y-intercept is 3,
x-intercept is $-1\frac{1}{2}$.

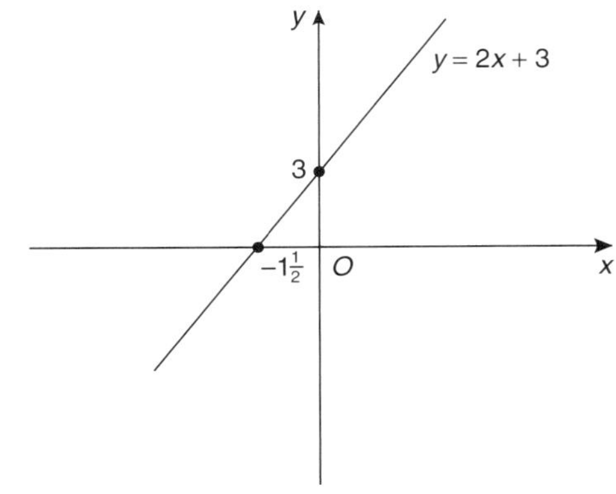

QUESTION 2

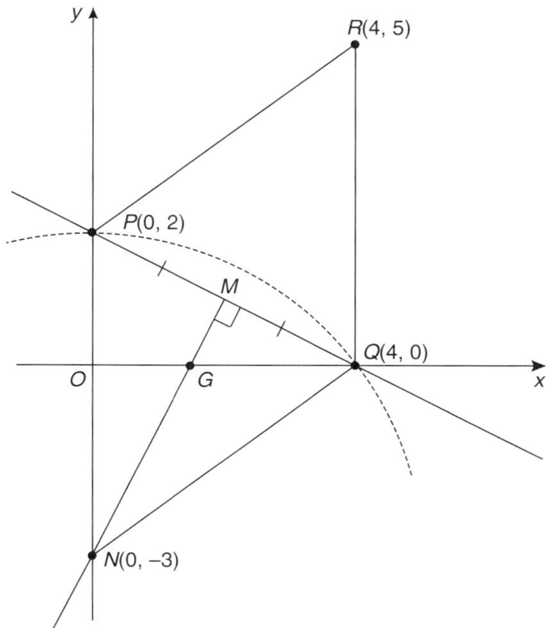

(a) Gradient of $PQ = \dfrac{y_2 - y_1}{x_2 - x_1}$
$$= \frac{0-2}{4-0}$$
$$= \frac{-2}{4}$$
$$= -\frac{1}{2}$$

(b) Coordinates of M are
$$\left(\frac{0+4}{2}, \frac{2+0}{2}\right)$$
i.e. $(2, 1)$.

(c) Gradient of $PQ = -\frac{1}{2}$
$\therefore$ Gradient of $MN = 2$ ($MN \perp PQ$)
$\therefore$ Equation of MN with gradient 2, passing through $M(2, 1)$ is given by
$y - y_1 = m(x - x_1)$
$y - 1 = 2(x - 2)$
$y - 1 = 2x - 4$
$y = 2x - 3$.

(d) Line MN cuts y-axis at N where $x = 0$.
$\therefore y = 2(0) - 3$
$= -3$
$\therefore N$ has coordinates $(0, -3)$.

(e) $NQ = \sqrt{(4-0)^2 + (0-(-3))^2}$
$= \sqrt{16+9}$
$= \sqrt{25}$
$= 5$ units.

(f) Equation of circle with centre $(0, -3)$ and radius, $NQ = 5$ units is given by:

$$(x-0)^2 + (y-(-3))^2 = 5^2$$
$$x^2 + (y+3)^2 = 25$$
$$x^2 + y^2 + 6y + 9 = 25$$
$$x^2 + y^2 + 6y - 16 = 0$$

(g) Substituting $P(0, 2)$ into $x^2 + y^2 + 6y - 16 = 0$

LHS $= 0^2 + 2^2 + 6(2) - 16$
$= 4 + 12 - 16$
$= 0$
RHS $= 0$

$\therefore$ Circle passes through P.

(h) If $PNQR$ is a rhombus, $M(2, 1)$ is the midpoint of $R(x, y)$ and $N(0, -3)$

$\therefore \dfrac{x+0}{2} = 2$ and $\dfrac{y-3}{2} = 1$
$\therefore x = 4$ and $y = 5$

$\therefore$ Coordinates of R are $(4, 5)$.

QUESTION 3

(a) (i) $\dfrac{d}{dx}(3xe^x) = 3x \cdot \dfrac{d}{dx}(e^x) + e^x \cdot \dfrac{d}{dx}(3x)$
$= 3x \cdot e^x + e^x \cdot 3$
$= 3xe^x + 3e^x$

(ii) $\dfrac{d}{dx}[\sin(x^2+1)]$
$= \cos(x^2+1) \cdot \dfrac{d}{dx}(x^2+1)$
$= \cos(x^2+1) \cdot 2x$
$= 2x\cos(x^2+1)$.

(b)

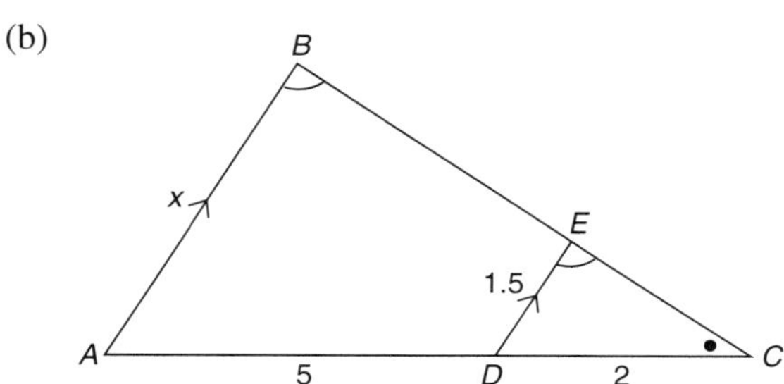

ΔABC and ΔDEC are similar
since $\angle ECD$ is common
and $\angle ABC = \angle DEC$
(corresponding angles : $AB \| DE$)

$\therefore \dfrac{AB}{DE} = \dfrac{AC}{DC}$

$\therefore \dfrac{x}{1.5} = \dfrac{7}{2}$

$\therefore x = \dfrac{7}{2} \times \dfrac{3}{2}$

$= \dfrac{21}{4}$ cm.

(c) (i) $\displaystyle\int \sec^2 5x\,dx = \frac{1}{5}\tan 5x + C$

(ii) $\displaystyle\int_{-2}^{1} \frac{2}{x+3}\,dx = 2\int_{-2}^{1} \frac{1}{x+3}\,dx$
$= 2\left[\ln(x+3)\right]_{-2}^{1}$
$= 2(\ln 4 - \ln 1)$
$= 2\ln 4$.

(d) $y = 2\log_e x$

$\dfrac{dy}{dx} = 2 \times \dfrac{1}{x}$

$= \dfrac{2}{x}$

At $x = 1$, $\dfrac{dy}{dx} = 2$

$\therefore$ Equation of tangent through $(1, 0)$ with gradient $= 2$ is:
$y - 0 = 2(x-1)$
$y = 2x - 2$.

QUESTION 4

(a) (i)

Now $\angle TAB = 60°$ (ΔABT is equilateral)

and $\angle PAT = \angle PAB + \angle TAB$
$= 15° + 60°$
$= 75°$.

(ii) Now $\angle DAP = 90° - 15°$ ($\angle DAB$ is a right angle)
$= 75°$

Also, $TA = AB$ (sides of equilateral Δ)
and $AB = DA$ (sides of square)
$\therefore TA = DA$

In ΔTAP and ΔDAP,

AP is common
$\angle PAT = \angle DAP$ (both 75°)
$TA = DA$ (from above)

$\therefore \Delta TAP \equiv \Delta DAP$ (SAS)

(iii) Since $\angle TPA = \angle APD$
(angles of contruent triangles)
and $\angle TPA = 75°$
$\therefore \angle APD = 75°$
and $\angle DAP = 75°$

$\therefore \Delta DAP$ is isosceles.

(b)

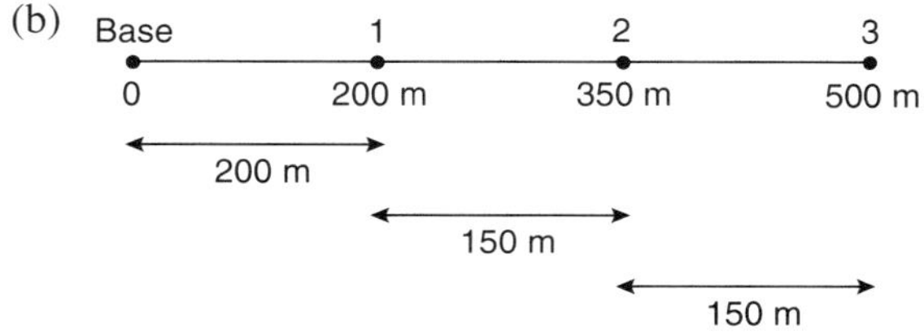

(i) 200, 350, 500, is an arithmetic sequence with $a = 200$ and $d = 150$.
$U_n = a + (n-1)d$
$U_{15} = 200 + 14(150)$
$= 200 + 2100$
$= 2300$ m.

(ii) $a = 200,\ d = 150,\ U_n = 5000$

$\therefore 5000 = 200 + (n-1)150$
$5000 = 200 + 150n - 150$
$5000 = 50 + 150n$
$150n = 4950$
$n = \dfrac{4950}{150}$
$= 33$

$\therefore$ 33 loads are deposited along the total length.

(iii) Outward distance travelled from base is given by:

$$S_n = \frac{n}{2}(2a + (n-1)d)$$
$$n = 33,\ a = 200,\ d = 150$$
$$\therefore S_{33} = \frac{33}{2}(2(200) + 32 \times 150)$$
$$= \frac{33}{2}(400 + 4800)$$
$$= \frac{33}{2} \times 5200$$
$$= 85\,800 \text{ m}$$

$\therefore$ Backward distance travelled = 85 800 m
$\therefore$ Total distance travelled
= 85 800 × 2
= 171 600 m
= 171.6 km

QUESTION 5

(a) $\tan x = 2$ for $0 < x < 2\pi$
$\therefore x = 1.1071\ldots$ or $\pi + 1.1071\ldots$
$= 1.1071\ldots$ or $4.2487\ldots$
$= 1.11$ or 4.25 radians (to 2 decimal places).

(b) (i)

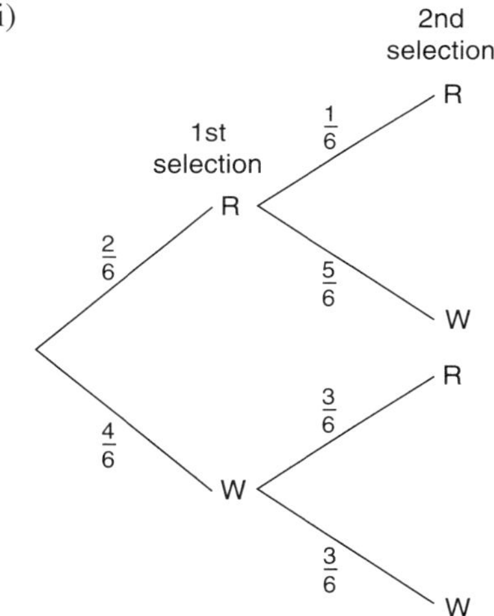

(ii) $P(WW) = \dfrac{4}{6} \times \dfrac{3}{6}$
$= \dfrac{12}{36}$
$= \dfrac{1}{3}$.

(iii) $P(WW \text{ or } RW) = \dfrac{4}{6} \times \dfrac{3}{6} + \dfrac{2}{6} \times \dfrac{5}{6}$
$= \dfrac{12}{36} + \dfrac{10}{36}$
$= \dfrac{22}{36}$
$= \dfrac{11}{18}$.

(c) (i) $N = 200e^{kt}$

When $t = 0$, $N = 200$
$\therefore$ When $t = 3$, $N = 400$

$$\therefore 400 = 200e^{3k}$$
$$2 = e^{3k}$$
$$\ln 2 = \ln e^{3k}$$
$$\ln 2 = 3k \ln e$$
$$\ln 2 = 3k$$
$$k = \frac{\ln 2}{3}$$
$$= 0.2310\ldots \quad \text{(by calc.)}$$
$$\doteqdot 0.231.$$

(ii) After 12 weeks,
$$N = 200e^{0.2310\ldots \times 12}$$
$$= 3200. \quad \text{(by calc.)}$$

(iii) $N = 200e^{kt}$
$$\frac{dN}{dt} = 200ke^{kt}$$
When $t = 3$,
$$\frac{dN}{dt} = 200(0.2310\ldots)e^{0.2310\ldots \times 3}$$
$$= 92.419\ldots \quad \text{(by calc.)}$$
$$= 92 \text{ insects per week.}$$

QUESTION 6

(a)

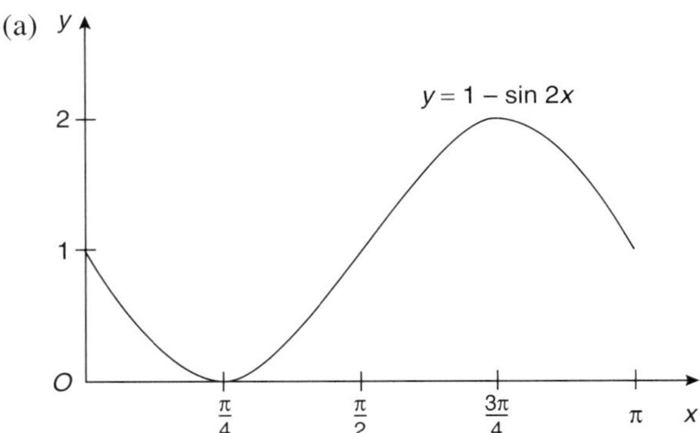

(b) (i) $N = 175 + 18t^2 - t^4, \quad 0 \leqslant t \leqslant 5$
When $t = 0$,
$$N = 175 + 0 - 0$$
$$= 175.$$

(ii) When $t = 5$,
$$N = 175 + 18(5)^2 - 5^4$$
$$= 175 + 450 - 625$$
$$= 0.$$

(iii) $$\frac{dN}{dt} = 36t - 4t^3$$
$$= 4t(9 - t^2)$$
$$= 4t(3 + t)(3 - t)$$

$$\frac{d^2N}{dt^2} = 36 - 12t^2$$
$$= 12(3 - t^2)$$

Stationary points occur when
$$\frac{dN}{dt} = 0$$
$\therefore 4t(3 + t)(3 - t) = 0$
$\therefore t = 0$ or 3 since $0 \leqslant t \leqslant 5$.

When $t = 0$, $\dfrac{d^2N}{dt^2} = 36 > 0$.
$\therefore$ A minimum turning point at $t = 0$.

When $t = 3$, $\dfrac{d^2N}{dt^2} = 12(3 - 9)$
$$= 12 \times -6$$
$$= -72 < 0.$$
$\therefore$ A maximum turning point at $t = 3$.

When $t = 3$, $N = 175 + 18(3)^2 - 3^4$
$$= 175 + 162 - 81$$
$$= 256.$$
$\therefore$ Maximum number of students was 256.

(iv) Students were logging on most rapidly
when $\dfrac{d^2N}{dt^2} = 0$
$$\therefore 12(3 - t^2) = 0$$
$$t^2 = 3$$
$$t = \pm\sqrt{3}$$
$$\therefore t = \sqrt{3} \text{ since } 0 \leqslant t \leqslant 5$$

(v) When $t = 0, N = 175$.
When $t = 5, N = 0$.

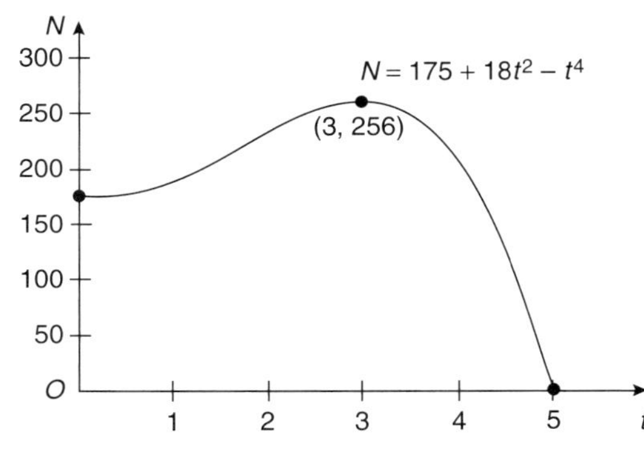

QUESTION 7

(a) $V = \pi \displaystyle\int_1^{e^2} y^2 \, dx$

$= \pi \int_1^{e^2} \left(\frac{1}{\sqrt{x}}\right)^2 dx$

$= \pi \int_1^{e^2} \frac{1}{x} dx$

$= \pi[\ln x]_1^{e^2}$

$= \pi(\ln e^2 - \ln 1)$

$= \pi(2\ln e - 0)$

$= \pi \times 2$

$= 2\pi$ units^3.

(b) $f(x) = \sin(1 + x^2)$

x	0	$\frac{1}{2}$	1
$f(x)$	sin 1 $\doteqdot 0.8415$	sin $1\frac{1}{4}$ $\doteqdot 0.9490$	sin 2 $\doteqdot 0.9093$
	y_0	y_1	y_2

$\int_0^1 \sin(1 + x^2)\ dx$

$\doteqdot \frac{h}{3}(y_0 + y_2 + 4y_1)$

$\doteqdot \frac{\frac{1}{2}}{3}[0.8415 + 0.9093 + 4(0.9490)]$

$\doteqdot \frac{1}{6} \times 5.5468$

$\doteqdot 0.9245$.

(c) (i) Solving $y = x^2 - 2$ and $y = x$ simultaneously:

$x^2 - 2 = x$

$x^2 - x - 2 = 0$

$(x - 2)(x + 1) = 0$

$\therefore x = 2$ or -1

$\therefore$ The x values are 2 and -1.

(ii) Area of shaded region

$= \int_{-1}^{2} [x - (x^2 - 2)]\ dx$

$= \int_{-1}^{2} (x - x^2 + 2)\ dx$

$= \left[\frac{x^2}{2} - \frac{x^3}{3} + 2x\right]_{-1}^{2}$

$= (\frac{4}{2} - \frac{8}{3} + 4) - (\frac{1}{2} + \frac{1}{3} - 2)$

$= 2 - \frac{8}{3} + 4 - \frac{1}{2} - \frac{1}{3} + 2$

$= 8 - \frac{8}{3} - \frac{1}{2} - \frac{1}{3}$

$= 8 - \frac{9}{3} - \frac{1}{2}$

$= 8 - 3 - \frac{1}{2}$

$= 4\frac{1}{2}$ units^2.

QUESTION 8

(a) (i) $x = 2t - 3\log_e(t + 1)$

$v = \frac{dx}{dt} = 2 - \frac{3}{t+1}$.

(ii) When $t = 0$, $v = 2 - \frac{3}{1}$

$= 2 - 3$

$= -1$ ms^{-1}.

(iii) The particle comes to rest when

$v = 0$

i.e. $2 - \frac{3}{t+1} = 0$

$2 = \frac{3}{t+1}$

$2t + 2 = 3$

$2t = 1$

$t = \frac{1}{2}$.

$\therefore$ Particle comes to rest when $t = \frac{1}{2}$ second.

(iv) At $t = 0$, $x = 0$ (given)

At $t = \frac{1}{2}$, $x = 2(\frac{1}{2}) - 3\log_e(1.5)$

$= 1 - 3\log_e 1.5$.

$(\doteqdot -0.2163\ldots)$

At $t = 3$, $x = 2(3) - 3\log_e(4)$

$= 6 - 3\log_e 4$.

$(\doteqdot 1.8411\ldots)$

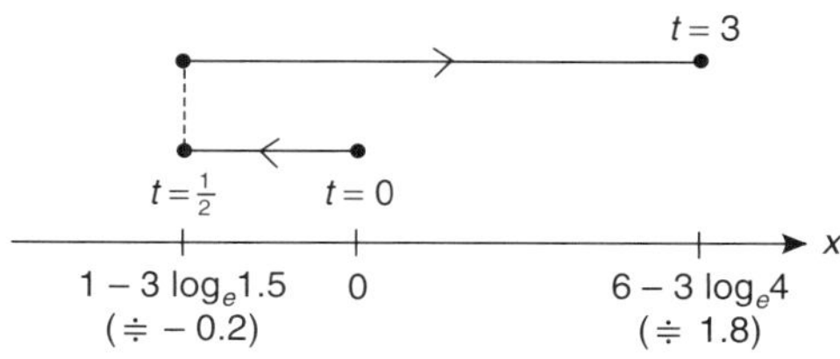

Total distance travelled

$= 2 \times -(1 - 3\log_e 1.5) + (6 - 3\log_e 4)$

$= -2 + 6\log_e 1.5 + 6 - 3\log_e 4$

$= 4 + 6\log_e 1.5 - 3\log_e 4$

$= 4 + 3(2\log_e 1.5 - \log_e 4)$

$= 4 + 3\left(\log_e \frac{1.5^2}{4}\right)$

$= 2.2739\ldots$ m (by calc.)

$= 2.27$ m to 2 decimal places.

(b)

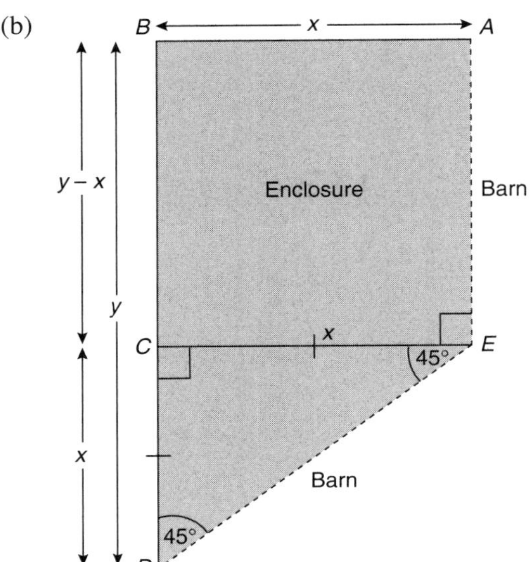

(i) Construct CE

Now $\angle DCE = 90°$
and $\angle CED = 45°$
$\therefore \angle CDE = 45°$ (angle sum ΔCDE)
$\therefore \Delta CDE$ is isosceles.

Since $AB = x$
$\therefore CE = x$ (opp. sides of rectangle)
$\therefore CD = x$ (ΔCDE isosceles)

Since $BD = y$
and $CD = x$
$\therefore BC = y - x$

Area of enclosure = Area of rectangle $ABCE$ + Area of ΔDCE

$$\therefore A = AB \times BC + \frac{1}{2}(CE \times CD)$$
$$= x(y - x) + \frac{1}{2}x^2$$
$$= xy - x^2 + \frac{1}{2}x^2$$
$$= xy - \frac{1}{2}x^2 \quad \ldots (1)$$

Now $x + y = 117$ (given)
$\therefore y = 117 - x$

Substituting into (1):

$$A = x(117 - x) - \frac{1}{2}x^2$$
$$= 117x - x^2 - \frac{1}{2}x^2$$
$$= 117x - \frac{3}{2}x^2$$

(ii) $\dfrac{dA}{dx} = 117 - 3x$

$\dfrac{d^2A}{dx^2} = -3$

Stationary points occur when

$$\frac{dA}{dx} = 0$$
$\therefore 117 - 3x = 0$
$3x = 117$
$\therefore x = 39$

When $x = 39$, $\dfrac{d^2A}{dx^2} = -3 < 0$

$\therefore$ A maximum value occurs when $x = 39$.

Since $y = 117 - x$,
$\therefore y = 117 - 39$ at $x = 39$
$= 78$.

$\therefore$ The largest area occurs when $y = 78$ and $x = 39$, i.e. when $y = 2x$.

QUESTION 9

(a) (i)

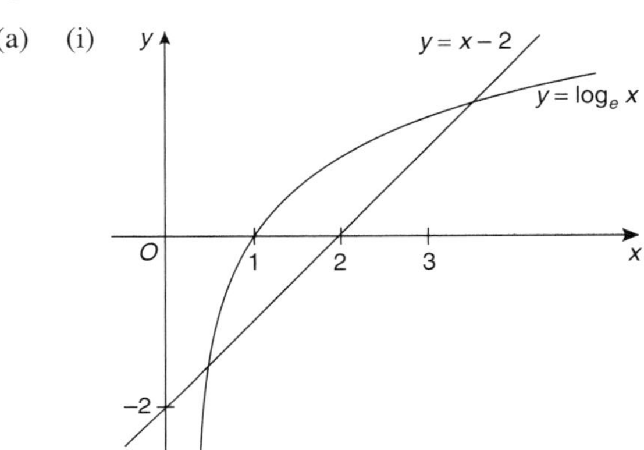

(ii) $\log_e x - x = -2$
$\therefore \log_e x = x - 2$

The solution of the equation is found by sketching the graphs of $y = \log_e x$ and $y = x - 2$.

From the sketch, there are 2 solutions.

(b)

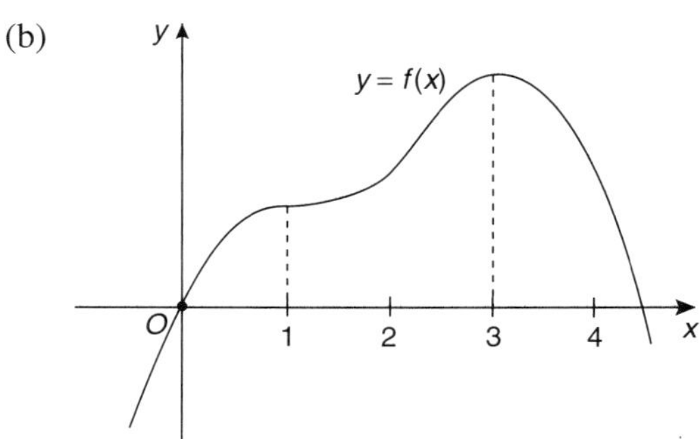

Explanation:
$f'(1) = f'(3) = 0$
$\therefore$ Stationary points occur at $x = 1$ and $x = 3$.

Now $f'(1-\varepsilon)>0$
and $f'(1+\varepsilon)>0$
$\therefore$ A horizontal point of inflexion at $x=1$.

Also, $f'(3-\varepsilon)>0$
and $f'(3+\varepsilon)<0$
$\therefore$ A maximum turning point at $x=3$.

For $x<3$, the gradient of $y=f(x)$ is always positive except at the above stationary points.

(c) (i) In $\triangle PBC$,

$$\cos\alpha=\frac{x^2+2^2-6^2}{2\times x\times 2}=\frac{x^2+4-36}{4x}=\frac{x^2-32}{4x}.$$

(ii) In $\triangle PBA$,

$$\angle ABP=90^\circ-\alpha$$

$$\cos(90^\circ-\alpha)=\frac{x^2+2^2-(2\sqrt{5})^2}{2\times x\times 2}=\frac{x^2+4-20}{4x}=\frac{x^2-16}{4x}.$$

Now $\cos(90^\circ-\alpha)=\sin\alpha$

$$\therefore\ \sin\alpha=\frac{x^2-16}{4x}.$$

(iii) Now $\cos^2\alpha+\sin^2\alpha=1$

$$\therefore\ \left(\frac{x^2-32}{4x}\right)^2+\left(\frac{x^2-16}{4x}\right)^2=1$$

$$\frac{x^4-64x^2+1024}{16x^2}+\frac{x^4-32x^2+256}{16x^2}=1$$

$$2x^4-96x^2+1280=16x^2$$

$$2x^4-112x^2+1280=0$$

$$x^4-56x^2+640=0$$

(iv) $x^4-56x^2+640=0$
Let $X=x^2$
$\therefore X^2-56X+640=0$
$(X-40)(X-16)=0$
$\therefore X=40$ or 16
$\therefore x^2=40$ or $x^2=16$
$\therefore x=\pm\sqrt{40}$ or $x=\pm4$.

Since x is a length,
$x=+\sqrt{40}$ or 4
$=2\sqrt{10}$ or 4
$\doteqdot 6.3$ or 4.

If $x=4$, $\sin\alpha=\dfrac{16-16}{16}=0$
$\therefore \alpha=0^\circ$
and $\cos\alpha=\dfrac{16-32}{16}=-1$
$\therefore \alpha=180^\circ$.

If $x=2\sqrt{10}$, $\sin\alpha=\dfrac{40-16}{8\sqrt{10}}=\dfrac{24}{8\sqrt{10}}=\dfrac{3}{\sqrt{10}}$
$\therefore \alpha=71^\circ34'$.
and $\cos\alpha=\dfrac{40-32}{8\sqrt{10}}=\dfrac{8}{8\sqrt{10}}=\dfrac{1}{\sqrt{10}}$
$\therefore \alpha=71^\circ34'$.

$\therefore x=2\sqrt{10}$, as this is the only value of x which gives the same angle α when substituted into the expressions for both $\sin\alpha$ and $\cos\alpha$.

(Alternately, the length of one side of a triangle, must be greater than the difference of the other two sides, i.e. greater than $6-2=4$.
$\therefore x=2\sqrt{10}$ ($\doteqdot 6.3$).)

QUESTION 10

(a) (i) $A_1=5000-M$
$A_2=5000-M-M$
$A_3=5000-M-M-M$
$=5000-3M$.

(ii) $A_4=(5000-3M)1.01-M$
$A_5=[(5000-3M)1.01-M]1.01-M$
$=(5000-3M)1.01^2-M(1.01)-M$
$=(5000-3M)1.01^2-M(1.01+1)$
$=(5000-3M)1.01^2-M(1+1.01)$.

(iii) $A_{36}=(5000-3M)1.01^{33}$
$-M(1+1.01+\ldots+(1.01)^{32})$.

(iv) After the 36th repayment, the loan is repaid. $\therefore A_{36} = 0$

i.e. $(5000 - 3M)1.01^{33} - M(1 + 1.01 + \ldots + (1.01)^{32}) = 0$

$$\therefore (5000 - 3M)1.01^{33} - M\left[\frac{a(r^n - 1)}{r - 1}\right] = 0$$

Where $a = 1, r = 1.01, n = 33$

$$\therefore (5000 - 3M)1.01^{33} - M\,\frac{1(1.01^{33} - 1)}{1.01 - 1} = 0$$

$$\therefore (5000 - 3M)1.01^{33} - \frac{M(1.01^{33} - 1)}{0.01} = 0$$

$$\therefore (5000 - 3M)1.01^{33} - 100M(1.01^{33} - 1) = 0$$

$$\therefore 5000(1.01)^{33} - 3M(1.01)^{33} - 100M(1.01^{33} - 1) = 0$$

$$\therefore 5000(1.01)^{33} = 3M(1.01)^{33} + 100M(1.01^{33} - 1)$$

$$\therefore 5000(1.01)^{33} = M[3(1.01)^{33} + 100(1.01^{33} - 1)]$$

$$\therefore M = \frac{5000(1.01)^{33}}{103(1.01)^{33} - 100}$$

$= \$161.34$ (by calc.)

(b) (i) If the depth of snow increases at a constant rate b, then after t hours,

$h = bt$

Now $v = \frac{A}{h}$ (given)

i.e. $\frac{dx}{dt} = \frac{A}{h}$

$$\therefore \frac{dx}{dt} = \frac{A}{bt} = \frac{A}{b} \times \frac{1}{t} = \frac{k}{t} \text{ (since } \frac{A}{b} \text{ is a constant).}$$

(ii) $\frac{dx}{dt} = \frac{k}{t}$

$\therefore x = k \ln t + C$

At 6 am, $t = T,\ x = 0$

$\therefore 0 = k \ln T + C$

$\therefore C = -k \ln T$

$$\therefore x = k \ln t - k \ln T = k(\ln t - \ln T) = k \ln\left(\frac{t}{T}\right)$$

At 8 am, $t = T + 2,\ x = 1$

$$\therefore 1 = k \ln\left(\frac{T+2}{T}\right) \quad \ldots (1)$$

At 11.30 am, $t = T + 5.5,\ x = 2$

$$\therefore 2 = k \ln\left(\frac{T+5.5}{T}\right) \quad \ldots (2)$$

Multiplying $(1) \times 2$:

$$\therefore 2 = 2k \ln\left(\frac{T+2}{T}\right) \quad \ldots (3)$$

Equating (2) and (3):

$$k \ln\left(\frac{T+5.5}{T}\right) = 2k \ln\left(\frac{T+2}{T}\right)$$

$$\ln\left(\frac{T+5.5}{T}\right) = 2 \ln\left(\frac{T+2}{T}\right)$$

$$\ln\left(\frac{T+5.5}{T}\right) = \ln\left(\frac{T+2}{T}\right)^2$$

$$\therefore \frac{T+5.5}{T} = \frac{T^2 + 4T + 4}{T^2} \quad (n.b.\ T \neq 0)$$

$$\therefore T + 5.5 = \frac{T^2 + 4T + 4}{T}$$

$$T^2 + 5.5T = T^2 + 4T + 4$$

$$1.5T = 4$$

$$T = 2.\dot{6} \text{ hours} = 2 \text{ h } 40 \text{ min.}$$

Since T is the time from the beginning of the snowfall and corresponds to 6 am, it started snowing at 3.20 am.

2001

HIGHER SCHOOL CERTIFICATE EXAMINATION

Mathematics

General Instructions

- Reading time – 5 minutes
- Working time – 3 hours
- Write using black or blue pen
- Board-approved calculators may be used
- A table of standard integrals is provided at the back of this paper
- All necessary working should be shown in every question

Total marks – 120

- Attempt Questions 1–10
- All questions are of equal value

Total marks – 120
Attempt Questions 1–10
All questions are of equal value

Answer each question in a SEPARATE writing booklet. Extra writing booklets are available.

Marks

Question 1 (12 marks) Use a SEPARATE writing booklet.

(a) Evaluate, correct to three significant figures, **2**

$$\sqrt{\frac{3^2+12^2}{231-12^2}}\ .$$

(b) Solve $|x+3|<2$. **2**

Graph your solution on a number line.

(c) Solve $x^2-2x-8=0$. **2**

(d) Find a primitive of $3+\frac{1}{x}$. **2**

(e) Simplify $\frac{x}{x^2-4}+\frac{2}{x-2}$. **2**

(f) The cost of a video recorder is \$979. This includes a 10% tax on the original price. Calculate the original price of the video recorder. **2**

Marks

Question 2 (12 marks) Use a SEPARATE writing booklet.

(a) Find the equation of the tangent to the curve $y = x^2 + 3x$ at the point $(1, 4)$. **3**

(b)

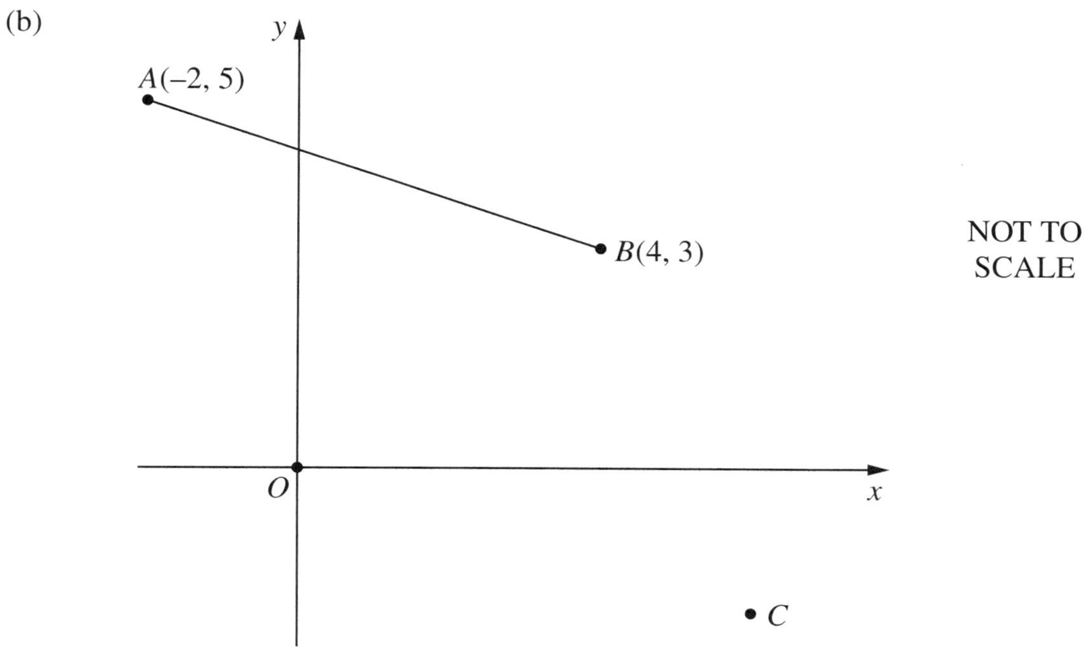

The diagram shows the points $A(-2, 5)$, $B(4, 3)$ and $O(0, 0)$. The point C is the fourth vertex of the parallelogram $OABC$.

(i) Show that the equation of AB is $x + 3y - 13 = 0$. **2**

(ii) Show that the length of AB is $2\sqrt{10}$. **1**

(iii) Calculate the perpendicular distance from O to the line AB. **2**

(iv) Calculate the area of parallelogram $OABC$. **2**

(v) Find the perpendicular distance from O to the line BC. **2**

Marks

Question 3 (12 marks) Use a SEPARATE writing booklet.

(a) Evaluate $\int_0^1 \frac{dx}{x+4}$. **2**

(b) Assume that the surface area S of a human satisfies the equation **2**

$$S = kM^{\frac{2}{3}}$$

where M is the body mass in kilograms, and k is the constant of proportionality.

A human with body mass 70 kg has surface area 18 600 cm^2.

Find the value of k, and hence find the surface area of a human with body mass 60 kg.

(c) Differentiate with respect to x:

(i) $\ln(x^2 - 9)$ **2**

(ii) $\frac{x}{e^x}$. **2**

(d) **4**

x

13

60°

7

NOT TO SCALE

The diagram shows a triangle with sides 7 cm, 13 cm and x cm, and an angle of 60° as marked.

Use the cosine rule to show that $x^2 - 7x = 120$, and hence find the exact value of x.

Marks

Question 4 (12 marks) Use a SEPARATE writing booklet.

(a) Find the values of k for which the quadratic equation $3x^2 + 2x + k = 0$ has no real roots. **2**

(b)

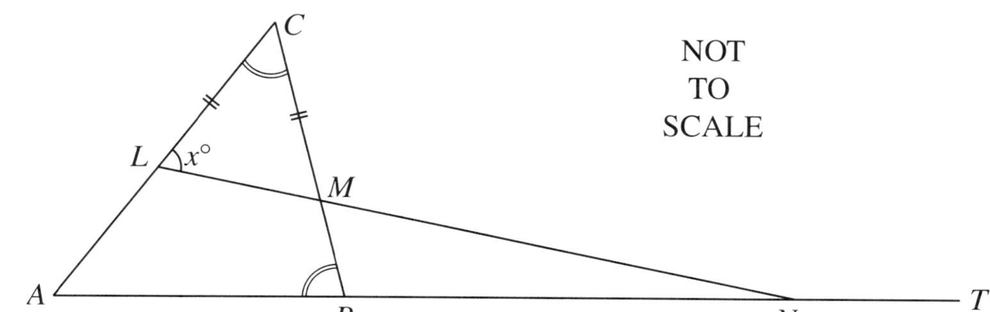

In the diagram, ABC is an isosceles triangle with $\angle ABC = \angle ACB$. The line LMN is drawn as shown so that $CL = CM$, and $\angle CLM = x°$.

Copy or trace the diagram into your Writing Booklet.

(i) Show that $\angle ABC = 180 - 2x°$. **2**

(ii) Hence show that $\angle TNL = 3x°$. **2**

(c) (i) Sketch the curve $y = 3\sin 2x$ for $-\frac{\pi}{2} \leq x \leq \frac{\pi}{2}$. **2**

(ii) On your diagram for part (i), sketch the line $y = \frac{1}{4}x$, and shade the region represented by **2**

$$\int_0^{\frac{\pi}{4}} \left(3\sin 2x - \frac{1}{4}x\right) dx .$$

(iii) Find the exact value of the integral in part (ii). **2**

Marks

Question 5 (12 marks) Use a SEPARATE writing booklet.

(a) State the domain and range of the function $y = 2\sqrt{25 - x^2}$. **3**

(b) (i) Find $\log_{10}(2^{1000})$ correct to 3 decimal places. **2**

(ii) We know that $2^{10} = 1024$, so that 2^{10} can be represented by a 4 digit numeral. How many digits are there in 2^{1000} when written as a numeral? **1**

(c) **2**

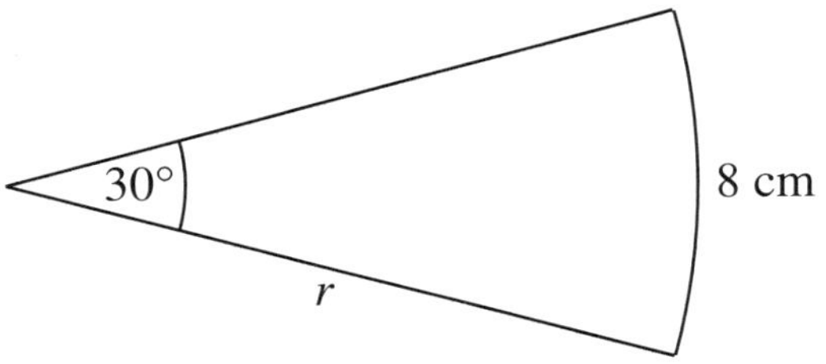

NOT TO SCALE

Find the length of the radius of the sector of the circle shown in the diagram. Give your answer correct to the nearest mm.

(d)

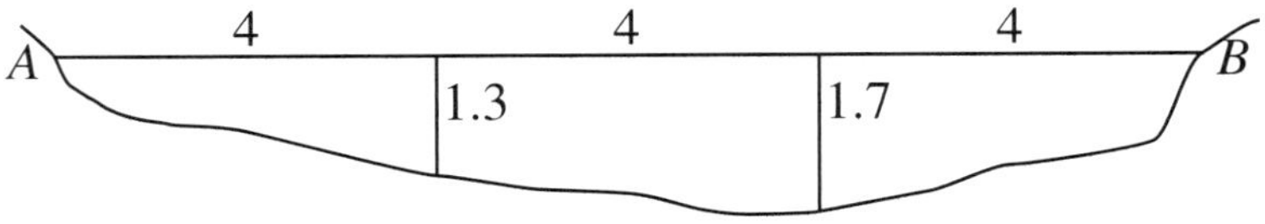

The diagram shows the cross-section of a creek, with the depths of the creek shown in metres, at 4 metre intervals. The creek is 12 metres in width.

(i) Use the trapezoidal rule to find an approximate value for the area of the cross-section. **2**

(ii) Water flows through this section of the creek at a speed of 0.5 m s^{-1}. Calculate the approximate volume of water that flows past this section in one hour. **2**

Marks

Question 6 (12 marks) Use a SEPARATE writing booklet.

(a) The first three terms of an arithmetic series are $-1+4+9+\ldots$

(i) Find the 60th term. **2**

(ii) Hence, or otherwise, find the sum of the first 60 terms of the series. **2**

(b) Find α so that the equation $P = 100(1.23)^t$ can be rewritten as $P = 100e^{\alpha t}$. Give your answer in decimal form. **2**

(c) The graph of $y = x^3 + x^2 - x + 2$ is sketched below. The points A and B are the turning points.

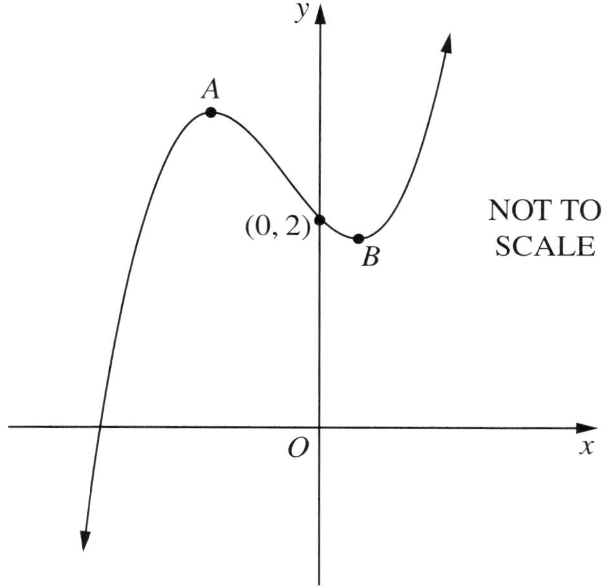

(i) Find the coordinates of A and B. **3**

(ii) For what values of x is the curve concave up? Give reasons for your answer. **2**

(iii) For what values of k has the equation $x^3 + x^2 - x + 2 = k$ three real solutions? **1**

Marks

Question 7 (12 marks) Use a SEPARATE writing booklet.

(a) **3**

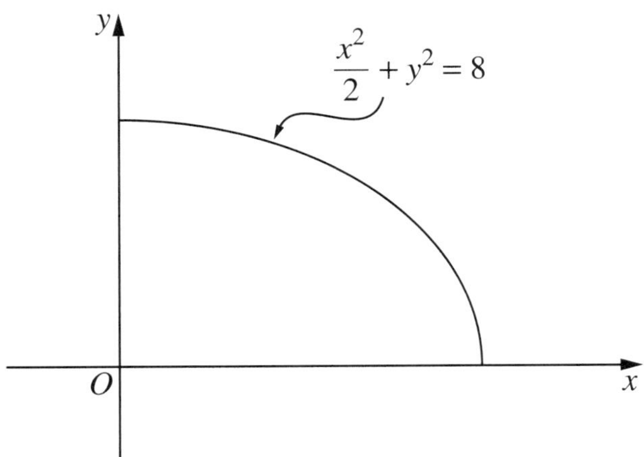

The part of the curve $\frac{x^2}{2} + y^2 = 8$ that lies in the first quadrant is rotated about the x axis.

Find the volume of the solid of revolution.

(b) Onslo tries to connect to his internet service provider. The probability that he connects on any single attempt is 0.75.

(i) What is the probability that he connects for the first time on his second attempt? **2**

(ii) What is the probability that he is still not connected after his third attempt? **1**

(c) A particle moves in a straight line so that its displacement, in metres, is given by

$$x = \frac{t-2}{t+2} \quad \text{where } t \text{ is measured in seconds.}$$

(i) What is the displacement when $t = 0$? **1**

(ii) Show that $x = 1 - \frac{4}{t+2}$. **3**

Hence find expressions for the velocity and the acceleration in terms of t.

(iii) Is the particle ever at rest? Give reasons for your answer. **1**

(iv) What is the limiting velocity of the particle as t increases indefinitely? **1**

Marks

Question 8 (12 marks) Use a SEPARATE writing booklet.

(a) In November 1923, 18 koalas were introduced on Kangaroo Island. By November 1993, the number of koalas had increased to 5000. **5**

Assume that the number N of koalas is increasing exponentially and satisfies an equation of the form $N = N_0 e^{kt}$, where N_0 and k are constants and t is measured in years from November 1923.

Find the values of N_0 and k, and predict the number of koalas that will be present on Kangaroo Island in November 2001.

(b) Five candidates, A, B, C, D and E, are standing for an election. Their names are written on pieces of cardboard that are placed in a barrel and are drawn out randomly to determine their positions on the ballot paper.

(i) What is the probability that A is drawn first? **1**

(ii) What is the probability that the order of the names on the ballot paper is that shown below? **2**

A
B
C
D
E

Question 8 continues

Marks

Question 8 (continued)

(c)

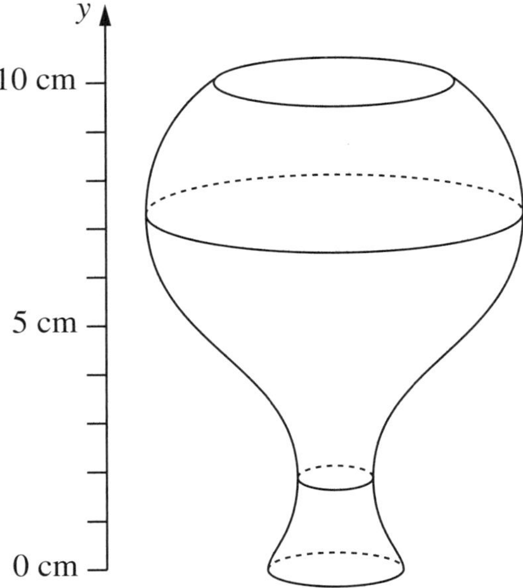

The diagram shows a ten-centimetre high glass that is being filled with water at a constant rate (by volume). Let $y = f(t)$ be the depth of water in the glass as a function of time t.

(i) Find the approximate depth y_1 at which $\dfrac{dy}{dt}$ is a maximum. **2**

Find the approximate depth y_2 at which $\dfrac{dy}{dt}$ is a minimum.

(ii) Assume that the glass takes 5 seconds to fill. **2**

Graph $y = f(t)$ and identify any points on your graph where the concavity changes.

End of Question 8

Marks

Question 9 (12 marks) Use a SEPARATE writing booklet.

(a)

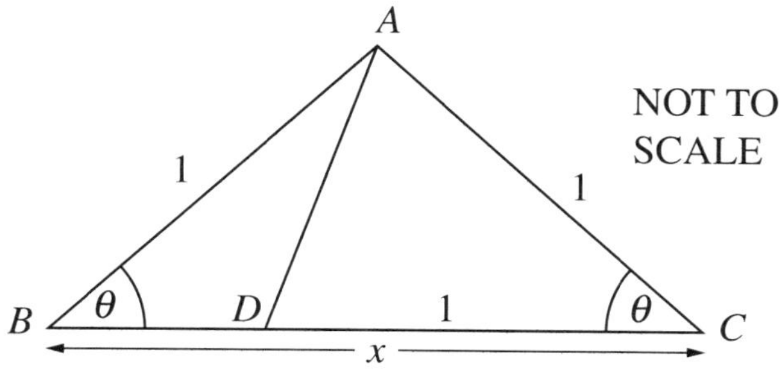

A

NOT TO SCALE

1 1

B θ D 1 θ C

x

In the diagram, ABC is an isosceles triangle where $\angle BAC = \frac{3\pi}{5}$ and $AB = AC = 1$. The point D is chosen on BC such that $CD = 1$.

Let $BC = x$, and let $\angle ABC = \theta$, and note that $\theta = \frac{\pi}{5}$.

(i) Show that $\angle ADC = 2\theta$ and hence show that triangles DBA and ABC are similar. **3**

(ii) From part (i) deduce that $x^2 - x - 1 = 0$. **1**

(iii) By using the cosine rule, deduce that **2**

$$\cos\frac{\pi}{5} = \frac{1+\sqrt{5}}{4}.$$

(b) When a valve is released, a chemical flows into a large tank that is initially empty. The volume, V litres, of chemical in the tank increases at the rate

$$\frac{dV}{dt} = 2e^{t} + 2e^{-t}$$

where t is measured in hours from the time the valve is released.

(i) At what rate does the chemical initially enter the tank? **1**

(ii) Use integration to find an expression for V in terms of t. **2**

(iii) Show that $2e^{2t} - 3e^{t} - 2 = 0$ when $V = 3$. **1**

(iv) Find t, to the nearest minute, when $V = 3$. **2**

Marks

Question 10 (12 marks) Use a SEPARATE writing booklet.

(a) Helen sets up a prize fund with a single investment of \$1000 to provide her school with an annual prize valued at \$72. The fund accrues interest at a rate of 6% per annum, compounded annually. The first prize is awarded one year after the investment is set up.

(i) Calculate the balance in the fund at the beginning of the second year. **1**

(ii) Let $\$B_n$ be the balance in the fund at the end of n years (and after the nth prize has been awarded). Show that $B_n = 1200 - 200 \times (1.06)^n$. **2**

(iii) At the end of the tenth year (and after the tenth prize has been awarded) it is decided to increase the prize value to \$90. **3**

For how many more years can the prize fund be used to award the prize?

(b)

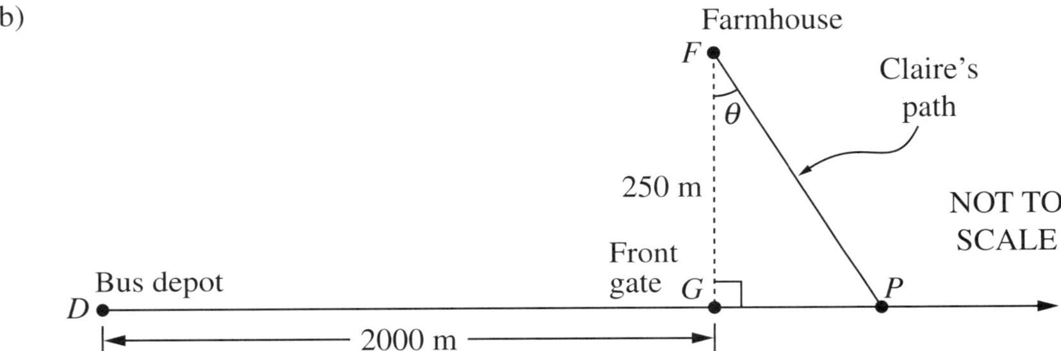

The diagram shows a farmhouse F that is located 250 m from a straight section of road. The road begins at the bus depot D, which is situated 2000 m from the front gate G of the farmhouse. The school bus leaves the depot at 8 am and travels along the road at a speed of 15 m s^{-1}. Claire lives in the farmhouse, and she can run across the open paddock between the house and the road at a speed of 4 m s^{-1}. The bus will stop for Claire anywhere on the road, but will not wait for her.

Assume that Claire catches the bus at the point P on the road where $\angle GFP = \theta$.

(i) Find two expressions in terms of θ, one expression for the time taken for the bus to travel from D to P and the other expression for the time taken by Claire to run from F to P. **2**

(ii) What is the latest time that Claire can leave home in order to catch the bus? **4**

End of paper

2001 Higher School Certificate Worked Answers

QUESTION 1

(a) $\sqrt{\dfrac{3^2+12^2}{231-12^2}} = \sqrt{\dfrac{153}{87}}$

$= \sqrt{1.7586\ldots}$ (by calc.)...

$= 1.3261\ldots$ (by calc.)

$= 1.33$ to 3 sig. figures.

(b) $|x+3|<2$

$\therefore -2 < x+3 < 2$

$\therefore -5 < x < -1.$

(number line with open circles at $^-5$ and $^-1$, x)

(c) $x^2 - 2x - 8 = 0$

$(x-4)(x+2) = 0$

$\therefore x = 4$ or -2.

(d) $\displaystyle\int \left(3+\frac{1}{x}\right) dx = 3x + \log_e x + C$

(e) $\dfrac{x}{x^2-4} + \dfrac{2}{x-2}$

$= \dfrac{x}{(x-2)(x+2)} + \dfrac{2}{x-2}$

$= \dfrac{x+2(x+2)}{(x-2)(x+2)}$

$= \dfrac{3x+4}{(x-2)(x+2)}$

(f) Original price + 10% of orig. price = Cost price

$x + 0.1x = \$979$

$1.1x = \$979$

$x = \$979 \div 1.1$

$= \$890.$ (by calc.)

QUESTION 2

(a) $y = x^2 + 3x$

$\dfrac{dy}{dx} = 2x + 3$

At $x = 1$, $\dfrac{dy}{dx} = 2(1) + 3$

$= 5$

Equation of tangent at $(1, 4)$ with gradient = 5 is

$y - 4 = 5(x-1)$

$y - 4 = 5x - 5$

$y = 5x - 1.$

(b)

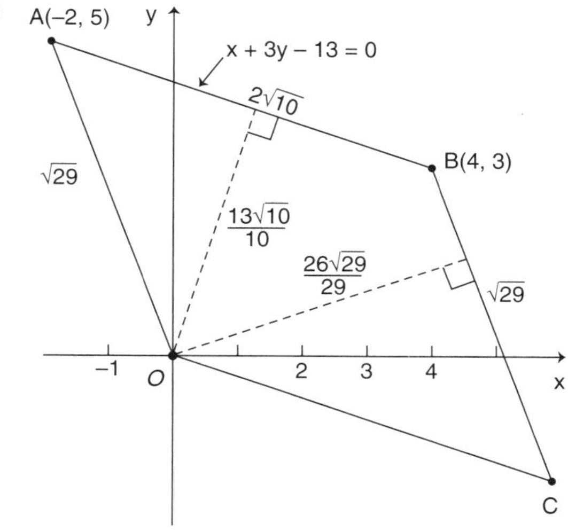

(i) Equation of line AB through $(-2, 5)$ and $(4, 3)$ is

$\dfrac{y-5}{x-(-2)} = \dfrac{3-5}{4-(-2)}$

$\dfrac{y-5}{x+2} = \dfrac{-2}{6}$

$6y - 30 = -2x - 4$

$2x + 6y - 26 = 0$

$x + 3y - 13 = 0.$

(ii) $AB = \sqrt{(4-(-2))^2 + (3-5)^2}$

$= \sqrt{6^2 + (-2)^2}$

$= \sqrt{36+4}$

$= \sqrt{40}$

$= 2\sqrt{10}$.

(iii) $O \equiv (0, 0)$

line AB: $x + 3y - 13 = 0$

$d = \left|\dfrac{Ax_1 + By_1 + C}{\sqrt{A^2 + B^2}}\right|$

$= \left|\dfrac{1(0) + 3(0) - 13}{\sqrt{1^2 + 3^2}}\right|$

$= \left|\dfrac{-13}{\sqrt{10}}\right|$

$= \dfrac{13}{\sqrt{10}} \times \dfrac{\sqrt{10}}{\sqrt{10}}$

$= \dfrac{13\sqrt{10}}{10}$.

(iv) Area = base × perpendicular height

$= 2\sqrt{10} \times \frac{13\sqrt{10}}{10}$

$= \frac{26\sqrt{100}}{10}$

$= 26$ units^2.

(v) BC = AO
(Opposite sides of parallelogram equal)

$AO = \sqrt{(-2-0)^2 + (5-0)^2}$

$= \sqrt{4+25}$

$= \sqrt{29}$

$\therefore BC = \sqrt{29}$

Now area of OABC = 26 (from (iv))

$\therefore$ BC × perp. height = 26

$\therefore$ perp. height $= \frac{26}{BC}$

$= \frac{26}{\sqrt{29}}$

$= \frac{26}{\sqrt{29}} \times \frac{\sqrt{29}}{\sqrt{29}}$

$= \frac{26\sqrt{29}}{29}$

$\therefore$ Perpendicular distance from O to BC is $\frac{26\sqrt{29}}{29}$.

QUESTION 3

(a) $\int_0^1 \frac{dx}{x+4} = \left[\ln(x+4)\right]_0^1$

$= \ln 5 - \ln 4$

$= \ln \frac{5}{4}$

$= \ln 1.25$

$\doteqdot 0.22.$ (by calc.)

(b) $S = kM^{\frac{2}{3}}$

When M = 70, S = 18 600

$\therefore 18\,600 = k(70)^{\frac{2}{3}}$

$k = \frac{18\,600}{70^{\frac{2}{3}}}$

$= 1095.0843\ldots$ (by calc.)

$\therefore S = 1095.0843\ldots \times M^{\frac{2}{3}}$

When M = 60 kg,

$S = 1095.0843\ldots \times 60^{\frac{2}{3}}$

$= 16\,783.469\ldots$ (by calc.)

$\doteqdot 16\,783$ cm^2.

(c) (i) $\frac{d}{dx}\ln(x^2-9) = \frac{1}{x^2-9} \cdot \frac{d}{dx}(x^2-9)$

$= \frac{1}{x^2-9} \cdot 2x$

$= \frac{2x}{x^2-9}.$

(ii) $\frac{d}{dx}\left(\frac{x}{e^x}\right) = \frac{e^x \cdot \frac{d}{dx}(x) - x \cdot \frac{d}{dx}(e^x)}{(e^x)^2}$

$= \frac{e^x.1 - x.e^x}{(e^x)^2}$

$= \frac{e^x - xe^x}{(e^x)^2}$

$= \frac{e^x(1-x)}{(e^x)^2}$

$= \frac{1-x}{e^x}.$

(d) $13^2 = x^2 + 7^2 - 2(x)(7)\cos 60°$

$169 = x^2 + 49 - 2(x)(7)(\tfrac{1}{2})$

$169 = x^2 + 49 - 7x$

$120 = x^2 - 7x$

i.e. $x^2 - 7x - 120 = 0$

$(x-15)(x+8) = 0$

$\therefore x = 15$ or -8

$\therefore x = 15.$

(a length must be positive)

QUESTION 4

(a) No real roots occur when

$\Delta < 0$

$b^2 - 4ac < 0$

$(2)^2 - 4(3)(k) < 0$

$4 - 12k < 0$

$-12k < -4$

$k > \frac{4}{12}$

$\therefore k > \frac{1}{3}.$

(b)

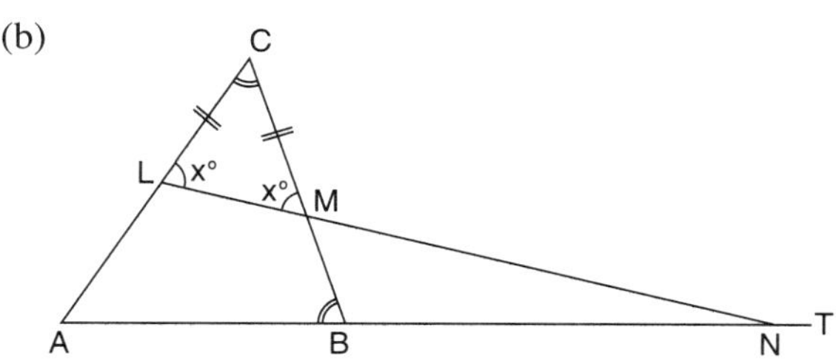

(i) Now $CL = CM$ (given)

$\therefore \angle CML = \angle CLM = x°$ (sides opp. equal angles)

$\therefore \angle LCM = 180° - \angle CML - \angle CLM$
$= 180° - x° - x°$
$= 180° - 2x°$

and $\angle LCM = \angle ABC$ (given)

$\therefore \angle ABC = 180° - 2x°$.

(ii) Now $\angle ABC = 180° - 2x°$ (from above)

$\therefore \angle CBN = 2x°$ (180° in a straight line)

and $\angle CML = \angle BMN = x°$ (vertically opposite)

Also $\angle TNL = \angle CBN + \angle BMN$ (exterior angle of triangle)
$= 2x° + x°$
$= 3x°$

(c) (i) (ii) Range: $-3 \le y \le 3$

Period: $\dfrac{2\pi}{2} = \pi$

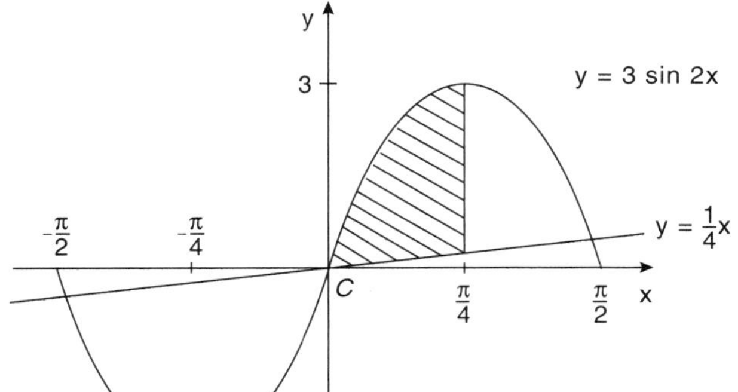

(iii) $\displaystyle\int_0^{\frac{\pi}{4}} \left(3 \sin 2x - \frac{1}{4}x\right) dx$

$$= \left[-\frac{3}{2}\cos 2x - \frac{x^2}{8}\right]_0^{\frac{\pi}{4}}$$

$$= \left[\left(-\frac{3}{2}\cos\frac{2\pi}{4} - \frac{(\frac{\pi}{4})^2}{8}\right) - \left(-\frac{3}{2}\cos 0 - 0\right)\right]$$

$$= \left[\left(-\frac{3}{2}\cos\frac{\pi}{2} - \frac{\pi^2}{128}\right) - \left(-\frac{3}{2}\right)\right]$$

$$= 0 - \frac{\pi^2}{128} + \frac{3}{2}$$

$$= \frac{3}{2} - \frac{\pi^2}{128}.$$

QUESTION 5

(a) Domain: $25 - x^2 \ge 0$
$25 \ge x^2$
$\therefore -5 \le x \le 5$

Range: $0 \le y \le 10$

(b) (i) $\log_{10}(2^{1000}) = 1000 \log_{10} 2$
$= 1000 \times 0.301\,029\,9\ldots$ (by calc.)
$= 301.0299\ldots$
$= 301.030$ to 3 decimal places.

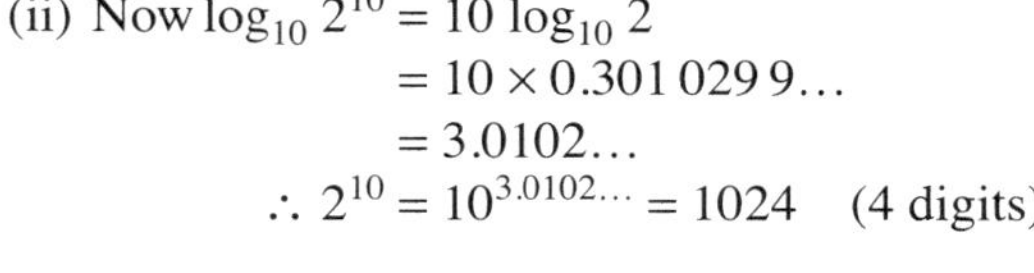

(ii) Now $\log_{10} 2^{10} = 10 \log_{10} 2$
$= 10 \times 0.301\,029\,9\ldots$
$= 3.0102\ldots$
$\therefore 2^{10} = 10^{3.0102\ldots} = 1024$ (4 digits)

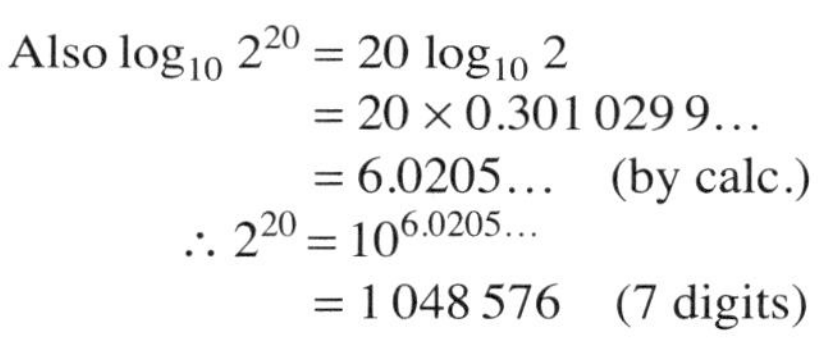

Also $\log_{10} 2^{20} = 20 \log_{10} 2$
$= 20 \times 0.301\,029\,9\ldots$
$= 6.0205\ldots$ (by calc.)
$\therefore 2^{20} = 10^{6.0205\ldots}$
$= 1\,048\,576$ (7 digits)

$\therefore 2^{1000} = 10^{301.0299\ldots}$
$\therefore$ 302 digits.

(c) $30° = \dfrac{30°}{180°} \times \pi$
$= \dfrac{\pi}{6}$.

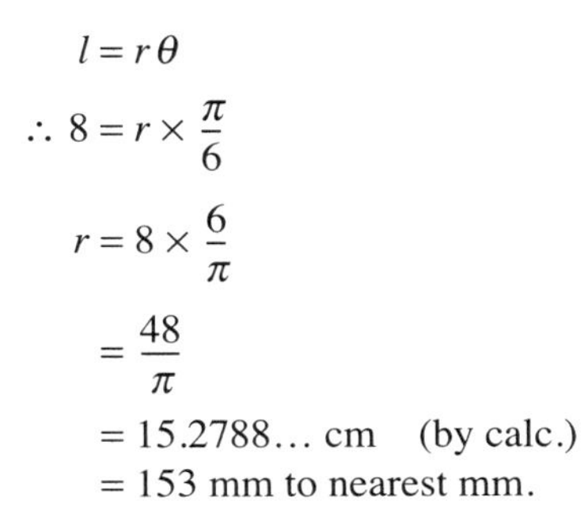

$l = r\theta$

$\therefore 8 = r \times \dfrac{\pi}{6}$

$r = 8 \times \dfrac{6}{\pi}$

$= \dfrac{48}{\pi}$

$= 15.2788\ldots$ cm (by calc.)

$= 153$ mm to nearest mm.

(d) (i)

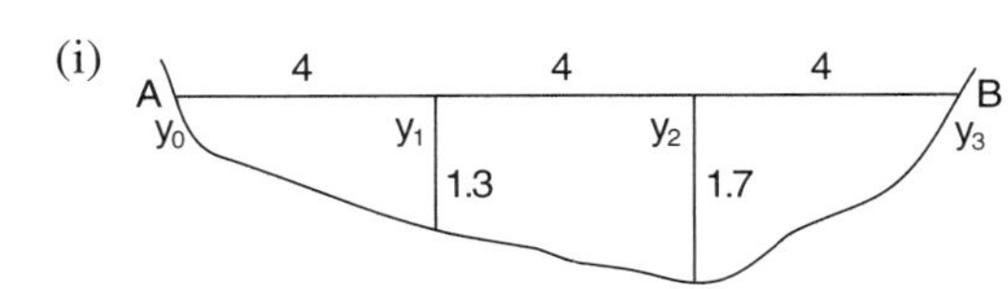

$$A \doteqdot \frac{h}{2}[y_0 + y_3 + 2(y_1 + y_2)]$$
$$= \frac{4}{2}[0 + 0 + 2(1.3 + 1.7)]$$
$$= 2 \times 6$$
$$= 12 \text{ m}^2.$$

(ii) $V = A \times \text{length}$

$$\text{length} = \text{speed} \times \text{time}$$
$$= 0.5 \times (60 \times 60)$$
$$= 1800 \text{ m} \quad \text{(by calc.)}$$

$$V = 12 \times 1800$$
$$= 21\,600 \text{ m}^3. \quad \text{(by calc.)}$$

QUESTION 6

(a) (i) $U_n = a + (n-1)d$

$$a = -1, n = 60, d = 5$$
$$U_{60} = -1 + (60-1)5$$
$$= -1 + 59(5)$$
$$= -1 + 295$$
$$= 294.$$

(ii) $S_n = \frac{n}{2}(a + l)$

$$n = 60, \quad a = -1, \quad l = 294$$
$$S_{60} = \frac{60}{2}(-1 + 294)$$
$$= 30 \times 293$$
$$= 8790. \quad \text{(by calc.)}$$

(b) $P = 100(1.23)^t$

and $P = 100e^{\alpha t}$

$$\therefore (1.23)^t = e^{\alpha t}$$
$$\therefore (1.23)^t = (e^{\alpha})^t$$
$$\therefore e^{\alpha} = 1.23$$
$$\ln e^{\alpha} = \ln 1.23$$
$$\alpha \ln e = \ln 1.23$$
$$\therefore \alpha = \ln 1.23$$
$$\doteqdot 0.21 \quad \text{(by calc.)}$$

(c) (i) $y = x^3 + x^2 - x + 2$

$$\frac{dy}{dx} = 3x^2 + 2x - 1$$
$$= (3x-1)(x+1)$$

Turning points occur when

$$\frac{dy}{dx} = 0$$
$$(3x-1)(x+1) = 0$$
$$x = \frac{1}{3} \text{ or } -1.$$

When $x = -1$, $y = (-1)^3 + (-1)^2 - (-1) + 2$

$$= -1 + 1 + 1 + 2$$
$$= 3$$

$\therefore$ Coordinates of A are $(-1, 3)$

When $x = \frac{1}{3}$, $y = \left(\frac{1}{3}\right)^3 + \left(\frac{1}{3}\right)^2 - \left(\frac{1}{3}\right) + 2$

$$= \frac{1}{27} + \frac{1}{9} - \frac{1}{3} + 2$$
$$= \frac{1}{27} + \frac{3}{27} - \frac{9}{27} + \frac{54}{27}$$
$$= \frac{49}{27}.$$

$\therefore$ Coordinates of B are $\left(\frac{1}{3}, \frac{49}{27}\right)$.

(ii) $\frac{d^2y}{dx^2} = 6x + 2$

Curve is concave up when

$$\frac{d^2y}{dx^2} > 0$$
$$6x + 2 > 0$$
$$6x > -2$$
$$x > -\frac{1}{3}.$$

(iii) If $x^3 + x^2 - x + 2 = k$ has three real solutions, then the line $y = k$ intersects the curve $y = x^3 + x^2 - x + 2$ at 3 points.

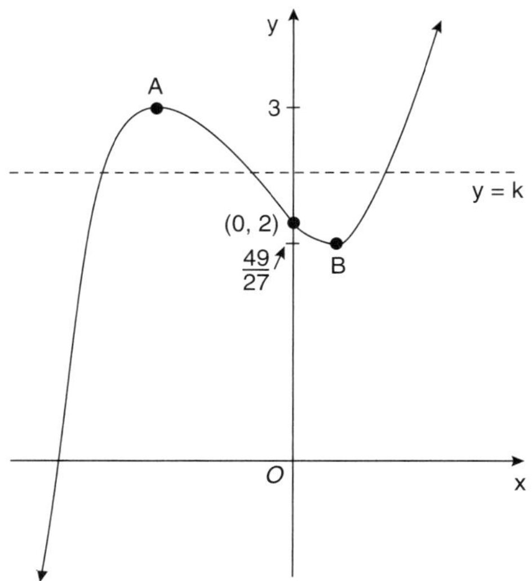

If k is between the y-coordinates of A and B, then the line $y = k$ will intersect the line at 3 points

i.e. $\frac{49}{27} < k < 3$

QUESTION 7

(a)

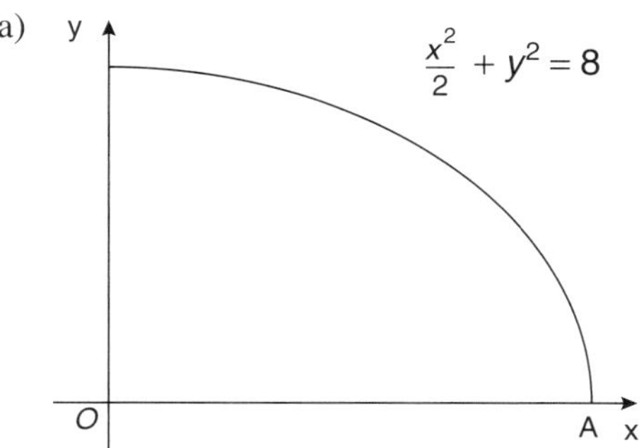

At A, $y = 0 \therefore \frac{x^2}{2} + 0 = 8$

$$x^2 = 16$$
$$\therefore x = \pm 4$$
$$\therefore \text{A} \equiv (4, 0)$$

Now $\frac{x^2}{2} + y^2 = 8$

$$\therefore y^2 = 8 - \frac{x^2}{2}$$

and $V = \pi \int_0^4 y^2 \, dx$

$$= \pi \int_0^4 \left(8 - \frac{x^2}{2}\right) dx$$
$$= \pi \left[8x - \frac{x^3}{6}\right]_0^4$$
$$= \pi\left[\left(8(4) - \frac{(4)^3}{6}\right) - 0\right]$$
$$= \pi\left(32 - \tfrac{64}{6}\right)$$
$$= \pi\left(32 - \tfrac{32}{3}\right)$$
$$= \pi\left(\tfrac{96}{3} - \tfrac{32}{3}\right)$$
$$= \frac{64\pi}{3} \text{ units}^3.$$

(b)

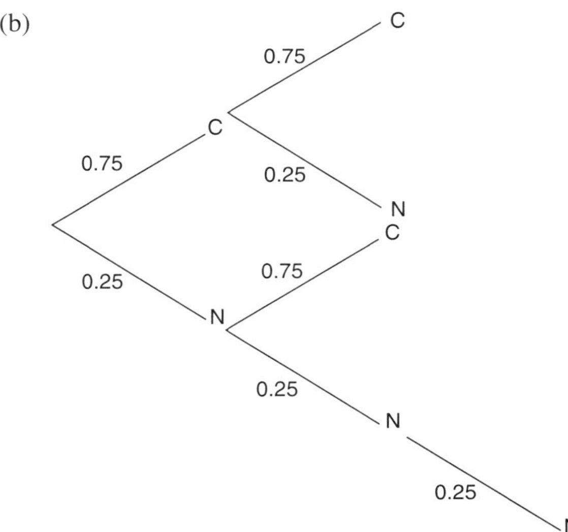

(i) $\text{P(NC)} = 0.25 \times 0.75$
$= 0.1875.$ (by calc.)

(ii) $\text{P(NNN)} = 0.25 \times 0.25 \times 0.25$
$= 0.015\,625.$ (by calc.)

(c) (i) $x = \frac{t-2}{t+2}$

When $t = 0$, $x = \frac{0-2}{0+2}$

$$= \frac{-2}{2}$$
$$= -1\text{m}.$$

(ii) $x = 1 - \frac{4}{t+2}$

$\text{RHS} = 1 - \frac{4}{t+2}$

$$= \frac{t+2}{t+2} - \frac{4}{t+2}$$
$$= \frac{t+2-4}{t+2}$$
$$= \frac{t-2}{t+2}$$
$$= x$$
$$= \text{LHS}$$

Now $x = 1 - \frac{4}{t+2}$

$$= 1 - 4(t+2)^{-1}$$

$$v = \frac{dx}{dt} = 0 + 4(t+2)^{-2}$$
$$= \frac{4}{(t+2)^2}$$
$$= 4(t+2)^{-2}$$

$$a = \frac{dv}{dt} = -8(t+2)^{-3}$$
$$= \frac{-8}{(t+2)^3}$$

(iii) The particle is at rest if

$$v = 0$$
$$\frac{4}{(t+2)^2} = 0$$

Since $t \geq 0$, $(t+2)^2 > 0$

$$\therefore \frac{4}{(t+2)^2} > 0$$

$\therefore$ Particle is never at rest.

(iv) $v = \dfrac{4}{(t+2)^2}$

As t increases indefinitely,
$(t+2)^2$ increases indefinitely

$\therefore \dfrac{4}{(t+2)^2}$ approaches zero.

$\therefore$ Limiting velocity is zero.

(ii)

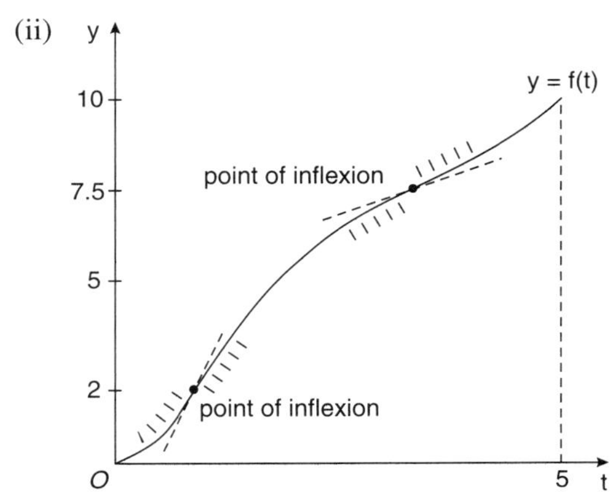

QUESTION 8

(a) $N = N_0 e^{kt}$

When $t = 0, N = 18$

$\therefore 18 = N_0 e^{k(0)}$

$\therefore 18 = N_0 e^0$

$\therefore N_0 = 18.$

Now when $t = 70, N = 5000$

$\therefore 5000 = 18e^{k(70)}$

$\dfrac{5000}{18} = e^{70k}$

$\ln\left(\dfrac{5000}{18}\right) = \ln e^{70k}$

$\ln\left(\dfrac{5000}{18}\right) = 70k \ln e$

$\ln\left(\dfrac{5000}{18}\right) = 70k$

$k = \dfrac{\ln\left(\dfrac{5000}{18}\right)}{70}$

$= 0.080\,38\ldots$ (by calc.)

$\doteqdot 0.0804$

In November 2001, $t = 78$

$\therefore N = 18e^{0.080\,38\ldots \times 78}$

$= 9511.5154\ldots$ (by calc.)

$= 9512$ koalas.

(b) (i) $P(\text{A first}) = \dfrac{1}{5}$

(ii) $P(\text{ABCDE}) = \dfrac{1}{5} \times \dfrac{1}{4} \times \dfrac{1}{3} \times \dfrac{1}{2} \times \dfrac{1}{1}$

$= \dfrac{1}{120}.$

(c) (i) $\dfrac{dy}{dx}$ will be a maximum at the depth where the glass is narrowest, i.e. at $y_1 = 2$ cm.

$\dfrac{dy}{dx}$ will be a minimum at the depth where the glass is widest, i.e. at $y_2 = 7.5$ cm.

QUESTION 9

(a) (i)

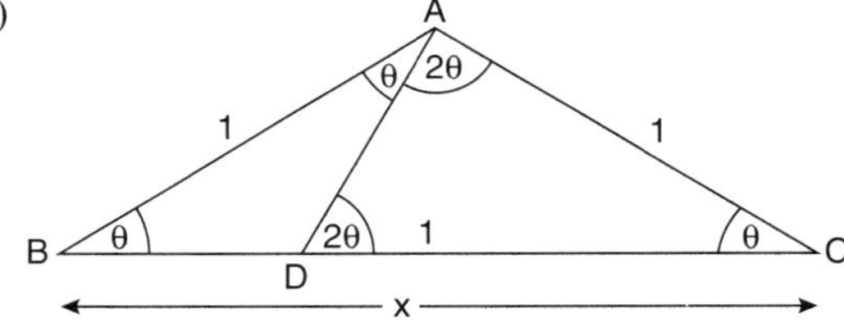

In $\triangle$ADC,

CA = CD (given)

$\therefore \angle$ADC is isosceles

$\therefore \angle CAD = \angle ADC$

Now $\angle ACD = \theta = \dfrac{\pi}{5}$ (given)

$\therefore \angle CAD + \angle ADC = \pi - \dfrac{\pi}{5}$

(180° in a triangle)

$= \dfrac{4\pi}{5}$

$\therefore \angle ADC = \dfrac{4\pi}{5} \div 2$

$= \dfrac{2\pi}{5}.$

$= 2 \times \dfrac{\pi}{5}$

$= 2\theta$

Also, $\angle ADC = \angle DBA + \angle BAD$

(exterior angle theorem)

i.e. $2\theta = \theta + \angle BAD$

$\therefore \angle BAD = \theta$

In $\triangle$'s DBA and ABC,

$\angle DBA = \angle ACB = \theta$

$\angle BAD = \angle ABC = \theta$

$\therefore \angle DBA \;|||\; \angle ABC$ (AA)

(ii) Since $\triangle$'s DBA and ABC are similar,

$$\frac{AB}{BC} = \frac{BD}{AB} = \frac{DA}{AC}$$

$$\therefore \frac{1}{x} = \frac{x-1}{1}$$

$$\therefore 1 = x^2 - x$$

$$\therefore x^2 - x - 1 = 0.$$

(iii) In $\triangle$ABC,

$$\cos\theta = \frac{x^2 + 1^2 - 1^2}{2(x)(1)}$$

$$\therefore \cos\frac{\pi}{5} = \frac{x^2}{2x}$$

$$= \frac{x}{2} \quad \ldots (1)$$

Now from (ii),
$x^2 - x - 1 = 0$

$$\therefore x = \frac{-(-1) \pm \sqrt{(-1)^2 - 4(1)(-1)}}{2(1)}$$

$$= \frac{1 \pm \sqrt{5}}{2}$$

Also, $\cos\frac{\pi}{5} = \frac{x}{2}$ (from above)

$$\therefore \cos\frac{\pi}{5} = \frac{1 \pm \sqrt{5}}{2} \times \frac{1}{2}$$

$$= \frac{1 \pm \sqrt{5}}{4}$$

Since $\frac{\pi}{5}$ is an acute angle,

$$\therefore \cos\frac{\pi}{5} = \frac{1 + \sqrt{5}}{4}.$$

(b) (i)

$$\frac{dV}{dt} = 2e^t + 2e^{-t}$$

When $t = 0$, $\frac{dV}{dt} = 2e^0 + 2e^0$

$$= 2 + 2$$
$$= 4$$

i.e. Rate = 4 litres/hour.

(ii) $V = 2e^t - 2e^{-t} + C$

When $t = 0, V = 0$

$$\therefore 0 = 2e^0 - 2e^0 + C$$
$$0 = 2 - 2 + C$$
$$\therefore C = 0$$

$$\therefore V = 2e^t - 2e^{-t}$$

(iii)

$$3 = 2e^t - 2e^{-t}$$

$$3 = 2e^t - \frac{2}{e^t}$$

$$\therefore 3e^t = 2e^{2t} - 2$$

$$\therefore 2e^{2t} - 3e^t - 2 = 0$$

(iv) Let $m = e^t$

$\therefore 2m^2 - 3m - 2 = 0$

$(2m + 1)(m - 2) = 0$

$\therefore m = -\frac{1}{2}$ or 2

but $m = e^t$ and $e^t > 0$ for $t > 0$

$\therefore m = 2$

$\therefore e^t = 2$

$\therefore \ln e^t = \ln 2$

$t \ln e = \ln 2$

$\therefore t = \ln 2$

$= 0.6931\ldots$ hours (by calc.)

$= 41.58\ldots$ min (by calc.)

$= 42$ minutes to nearest minute

QUESTION 10

(a) (i) $B_1 = 1000(1.06) - 72$
$= 988$

Balance at beginning of year 2 is \$988

(ii) $B_2 = [1000(1.06) - 72]1.06 - 72$
$= 1000(1.06)^2 - 72(1.06) - 72$
$= 1000(1.06)^2 - 72(1.06 + 1)$

$B_3 = [1000(1.06)^2 - 72(1.06 + 1)]1.06 - 72$
$= 1000(1.06)^3 - 72(1.06)^2 - 72(1.06) - 72$
$= 1000(1.06)^3 - 72(1.06^2 + 1.06 + 1)$
$= 1000(1.06)^3 - 72(1 + 1.06 + 1.06^2)$

$\therefore B_n = 1000(1.06)^n - 72[1 + (1.06) + \ldots + (1.06)^{n-1}]$

$$= 1000(1.06)^n - 72\left[\frac{a(r^n - 1)}{r - 1}\right]$$

where $a = 1, r = 1.06$

$$= 1000(1.06)^n - 72\left[\frac{1(1.06^n - 1)}{1.06 - 1}\right]$$

$$= 1000(1.06)^n - 72\left[\frac{1.06^n - 1}{0.06}\right]$$

$= 1000(1.06)^n - 1200(1.06^n - 1)$

$= 1000(1.06)^n - 1200(1.06)^n + 1200$

$\therefore B_n = 1200 - 200(1.06)^n$

(iii) Let $\$R$ be the balance in the fund at the end of n years where n is the number of years after the end of the tenth year (i.e. the number of years remaining for the fund).
After one more year,

$$R_1 = [1200 - 200(1.06)^{10}]1.06 - 90$$
$$= 1200(1.06) - 200(1.06)^{11} - 90$$

After two more years,

$$R_2 = [1200(1.06) - 200(1.06)^{11} - 90]1.06 - 90$$
$$= 1200(1.06)^2 - 200(1.06)^{12} - 90(1.06) - 90$$

After n more years,

$$R_n = 1200(1.06)^n - 200(1.06)^{n+10} - 90(1.06)^{n-1} - \dots - 90$$
$$= 1200(1.06)^n - 200(1.06)^{n+10} - 90[(1.06)^{n-1} + (1.06)^{n-2} + \dots + 1]$$
$$= 1200(1.06)^n - 200(1.06)^{n+10} - 90[1 + \dots + (1.06)^{n-2} + (1.06)^{n-1}]$$
$$= 1200(1.06)^n - 200(1.06)^{n+10} - 90\left[\frac{a(r^n - 1)}{r - 1}\right] \quad \text{where } a = 1, r = 1.06$$
$$= 1200(1.06)^n - 200(1.06)^{n+10} - 90\left[\frac{1[1.06^n - 1]}{1.06 - 1}\right]$$
$$= 1200(1.06)^n - 200(1.06)^{n+10} - 90\left[\frac{(1.06)^n - 1}{0.06}\right]$$
$$= 1200(1.06)^n - 200(1.06)^{n+10} - 1500[(1.06)^n - 1]$$
$$= 1200(1.06)^n - 200(1.06)^{n+10} - 1500(1.06)^n + 1500$$
$$= 1500 - 200(1.06)^{n+10} - 300(1.06)^n$$

When $R_n = 0$,

$$0 = 1500 - 200(1.06)^{n+10} - 300(1.06)^n$$
$$\therefore\ 1500 = 200(1.06)^{n+10} + 300(1.06)^n$$
$$15 = 2(1.06)^n(1.06)^{10} + 3(1.06)^n$$
$$15 = 3.582(1.06)^n + 3(1.06)^n$$
$$15 = 6.582(1.06)^n$$
$$\therefore\ (1.06)^n = \frac{15}{6.582}$$
$$= 2.2789$$
$$\ln (1.06)^n = \ln 2.2789$$
$$n \ln (1.06) = \ln 2.2789$$
$$n = \frac{\ln 2.2789}{\ln 1.06}$$
$$\doteqdot 14.13$$
$$\doteqdot 14$$

$\therefore$ The prize fund can be used for 14 more years.

(b) (i) Time taken by bus; T_{BUS}:

In $\triangle$GFP,

$$\tan\theta = \frac{GP}{250}$$

$\therefore$ GP = 250 tan θ

$\therefore$ DP = 2000 + 250 tan θ

$$\text{Now time} = \frac{\text{distance}}{\text{speed}}$$

$$\therefore\ T_{BUS} = \frac{2000 + 250\tan\theta}{15}$$

Time taken by Claire, T_{CLAIRE}:

In $\triangle$GFP,

$$\cos\theta = \frac{250}{FP}$$

$$\therefore \text{FP} = \frac{250}{\cos\theta}$$
$$= 250\sec\theta$$
$$\therefore T_{\text{CLAIRE}} = \frac{250\sec\theta}{4}.$$

(ii) Let $T = T_{\text{BUS}} - T_{\text{CLAIRE}}$

i.e. $T = \dfrac{2000 + 250\tan\theta}{15} - \dfrac{250(\cos\theta)^{-1}}{4}$

We ned to find the maximum value of this time difference (T).

$$\frac{dT}{d\theta} = \frac{250\sec^2\theta}{15} + \frac{250(\cos\theta)^{-2}.(-\sin\theta)}{4}$$
$$= \frac{250\sec^2\theta}{15} - \frac{250\sin\theta\sec^2\theta}{4}$$

Stationary points occur when

$$\frac{dT}{d\theta} = 0$$
$$\frac{250\sec^2\theta}{15} - \frac{250\sin\theta\sec^2\theta}{4} = 0$$
$$250\sec^2\theta\left(\frac{1}{15} - \frac{\sin\theta}{4}\right) = 0$$

Now $\sec^2\theta > 0$,

$$\therefore \frac{1}{15} - \frac{\sin\theta}{4} = 0$$
$$\frac{\sin\theta}{4} = \frac{1}{15}$$
$$\sin\theta = \frac{4}{15}$$
$$\therefore \theta = 0.2699 \text{ radians}$$

Consider $\dfrac{dT}{d\theta}$:

θ	0.2	0.2699	0.3
$\dfrac{dT}{d\theta}$	+	0	–

$\therefore$ A maximum value at $\theta = 0.2699$

At $\theta = 0.2699$,

$$T = \frac{2000 + 250\tan(0.2699)}{15} - \frac{250}{4\cos(0.2699)}$$
$= 137.944 - 64.848$ (by calc.)
$= 73.096$ (by calc.)
$= 73$ secs ($= 1$ min 13 secs)

$\therefore$ The latest Claire can leave is 1 minute 13 seconds past 8 am.

BOARD OF STUDIES
NEW SOUTH WALES

2002

HIGHER SCHOOL CERTIFICATE EXAMINATION

Mathematics

General Instructions

- Reading time – 5 minutes
- Working time – 3 hours
- Write using black or blue pen
- Board-approved calculators may be used
- A table of standard integrals is provided at the back of this paper
- All necessary working should be shown in every question

Total marks – 120

- Attempt Questions 1–10
- All questions are of equal value

Total marks – 120
Attempt Questions 1–10
All questions are of equal value

Answer each question in a SEPARATE writing booklet. Extra writing booklets are available.

Marks

Question 1 (12 marks) Use a SEPARATE writing booklet.

(a) Evaluate, correct to three significant figures, **2**

$$\frac{5.8^2 - 3.1^3}{3 \times 3.1 \times 5.8} .$$

(b) Differentiate $x^3 + 2$. **2**

(c) Solve $x^2 = 5x$. **2**

(d) Integrate $\frac{3}{x}$. **1**

(e) Solve $3x - \frac{2x - 5}{2} = 6$. **3**

(f) Solve the pair of simultaneous equations **2**

$$x - 2y = 8$$
$$2x + y = 1 .$$

Marks

Question 2 (12 marks) Use a SEPARATE writing booklet.

(a) Find the equation of the tangent to $y = e^{2x}$ at the point (0, 1). **2**

(b) Differentiate:

(i) $x \sin x$ **2**

(ii) $\dfrac{\ln x}{x^2}$. **2**

(c) **3**

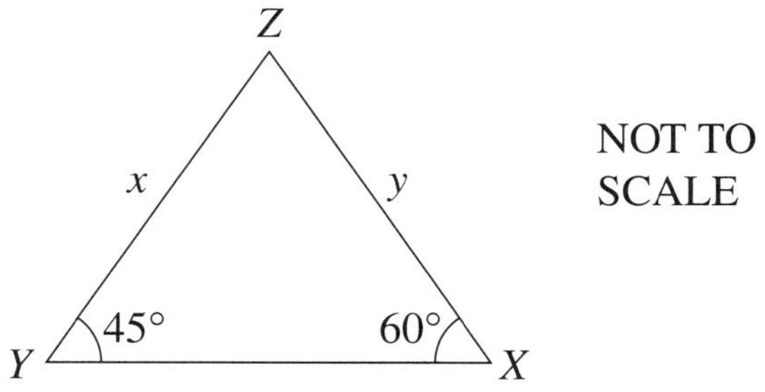

Z, x, y, 45°, 60°, Y, X

NOT TO SCALE

In the diagram, XYZ is a triangle where $\angle ZYX = 45°$ and $\angle ZXY = 60°$.

Find the exact value for the ratio $\dfrac{x}{y}$.

(d) Find:

(i) $\displaystyle\int \cos 3x \, dx$ **1**

(ii) $\displaystyle\int_0^1 \left(e^{5x} - 1\right) dx$. **2**

Marks

Question 3 (12 marks) Use a SEPARATE writing booklet.

(a) Josh invests \$1000 in a term deposit that earns 3.5% per annum compounded annually. **2**

What is the value of the investment at the end of 20 years?

(b) **3**

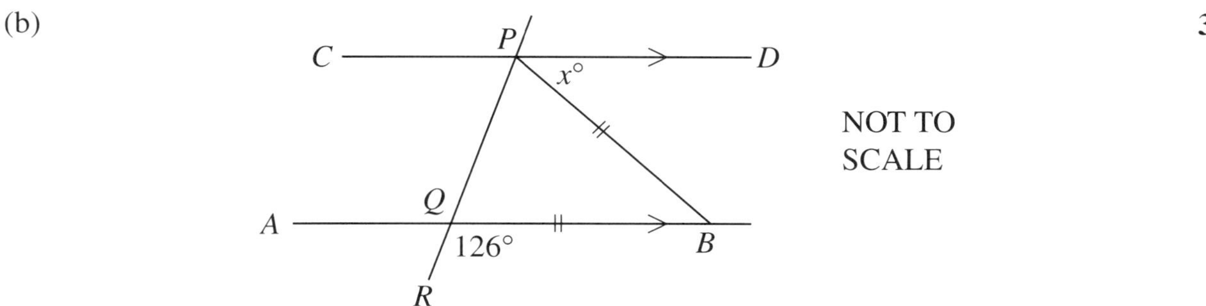

In the diagram, CD is parallel to AB, $PB = QB$, $\angle BQR = 126°$ and $\angle BPD = x°$.

Copy or trace this diagram into your writing booklet.

Find the value of x, giving complete reasons.

(c)

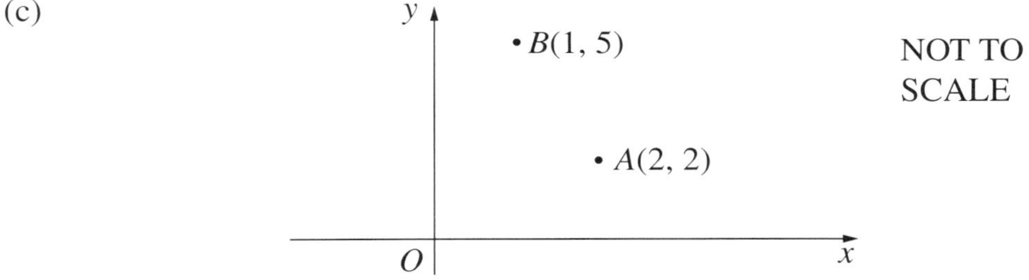

The diagram shows two points $A(2, 2)$ and $B(1, 5)$ on the number plane.

Copy the diagram into your writing booklet.

(i) Find the coordinates of M, the midpoint of AB. **1**

(ii) Show that the equation of the perpendicular bisector of AB is $x - 3y + 9 = 0$. **2**

(iii) Find the coordinates of the point C that lies on the y axis and is equidistant from A and B. **1**

(iv) The point D lies on the intersection of the line $y = 5$ and the perpendicular bisector $x - 3y + 9 = 0$. Find the coordinates of D, and mark the position of D on your diagram in your writing booklet. **1**

(v) Find the area of triangle ABD. **2**

Marks

Question 4 (12 marks) Use a SEPARATE writing booklet.

(a) Solve $|x-1| \geq 3$ and graph your solution on the number line. **2**

(b) Find all values of θ, where $0° \leq \theta \leq 360°$, that satisfy the equation **2**

$$\cos\theta - \frac{2}{5} = 0 \ .$$

Give your answer(s) to the nearest degree.

(c)

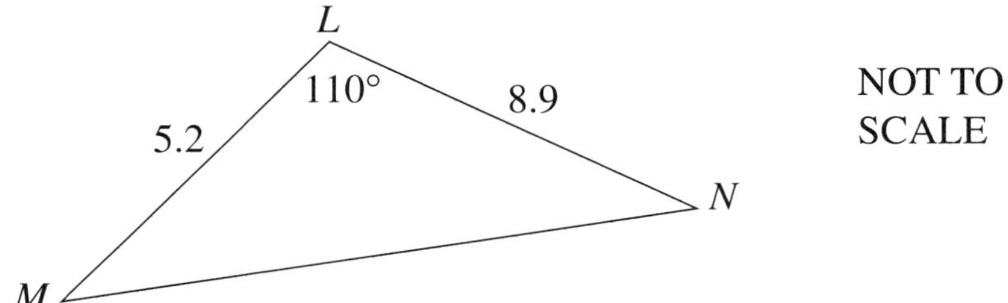

In the diagram, LMN is a triangle where $LM = 5.2$ metres, $LN = 8.9$ metres and angle $MLN = 110°$.

(i) Find the length of MN. **2**

(ii) Calculate the area of triangle LMN. **2**

(d)

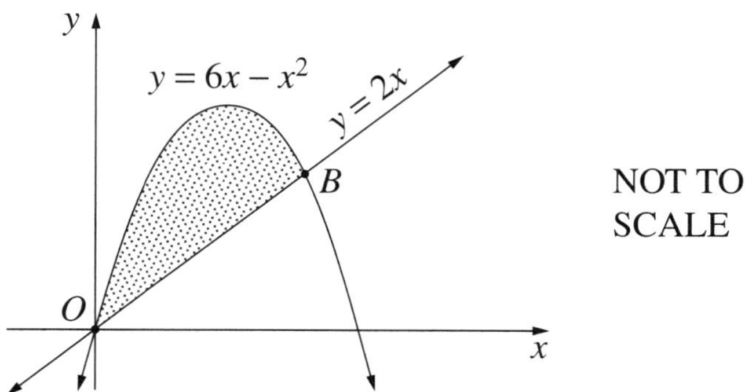

The graphs of $y = 2x$ and $y = 6x - x^2$ intersect at the origin and point B.

(i) Show that the coordinates of B are $(4, 8)$. **1**

(ii) Find the shaded area bounded by $y = 6x - x^2$ and $y = 2x$. **3**

Marks

Question 5 (12 marks) Use a SEPARATE writing booklet.

(a) Catrine is exercising her dog by throwing a stick for the dog to fetch and return. The first time, Catrine throws the stick 2 m and she continues to throw the stick after the dog has returned it. Each time, she increases the distance the stick is thrown by exactly 1.5 m. Her last throw is 32 m.

(i) How many times did Catrine throw the stick? **2**

(ii) How far did her dog run altogether in fetching and returning the stick? (Assume that the dog starts and finishes at Catrine's side.) **3**

(b) **3**

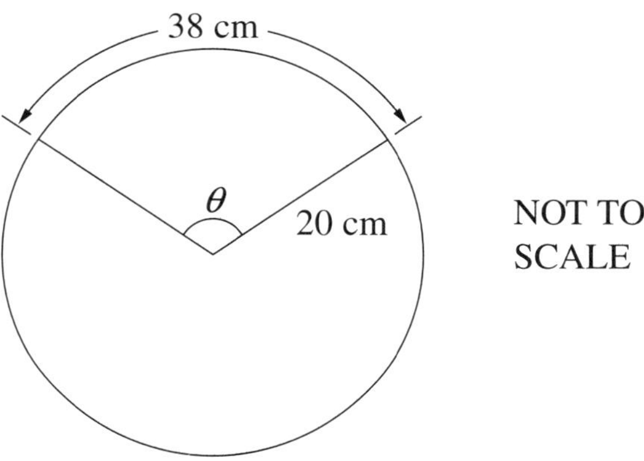

The length of the arc between two spokes on a car's steering wheel is 38 cm. Each spoke is 20 cm in length.

Calculate the angle θ between the two spokes. Give your answer correct to the nearest degree.

(c) Consider the parabola $y = x^2 - 8x + 4$. Find:

(i) the coordinates of the vertex, **2**

(ii) the coordinates of the focus. **2**

Marks

Question 6 (12 marks) Use a SEPARATE writing booklet.

(a) Sketch the graph of $y = \sqrt{4 - x^2}$, and state the range. **2**

(b) The gradient function of a curve is given by

$$f'(x) = 3(x+1)(x-3)$$

and the curve $y = f(x)$ passes through the point (0, 12).

(i) Find the equation of the curve $y = f(x)$. **2**

(ii) Sketch the curve $y = f(x)$, clearly labelling turning points and the y intercept. **3**

(iii) For what values of x is the curve concave up? **1**

(c) **4**

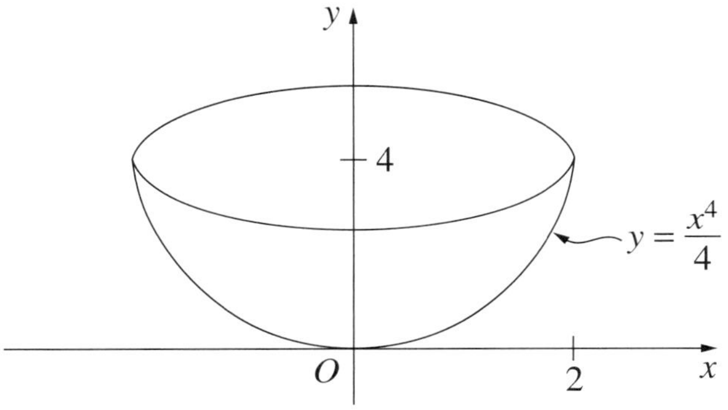

y
4
$y = \frac{x^4}{4}$
O
2
x

NOT TO SCALE

A bowl is formed by rotating the part of the curve $y = \frac{x^4}{4}$ between $x = 0$ and $x = 2$ about the y axis.

Find the volume of the bowl.

Marks

Question 7 (12 marks) Use a SEPARATE writing booklet.

(a) Consider the geometric series

$$1+\left(\sqrt{5}-2\right)+\left(\sqrt{5}-2\right)^2+\dots \quad .$$

(i) Explain why the geometric series has a limiting sum. **1**

(ii) Find the exact value of the limiting sum. Write your answer with a rational denominator. **2**

(b) A cooler, which is initially full, is drained so that at time t seconds the volume of water V, in litres, is given by

$$V=25\left(1-\frac{t}{60}\right)^2 \quad \text{for} \quad 0\le t\le 60.$$

(i) How much water was initially in the cooler? **1**

(ii) After how many seconds was the cooler one-quarter full? **2**

(iii) At what rate was the water draining out when the cooler was one-quarter full? **2**

(c) Chris has four pairs of socks in a drawer, each pair a different colour.

He selects socks one at a time and at random from the drawer.

(i) The probability that he does NOT have a matching pair after selecting the second sock is $\frac{6}{7}$. Explain why this is so. **1**

(ii) Find the probability that he does NOT have a matching pair after selecting the third sock. **2**

(iii) What is the probability that the first three socks include a matching pair? **1**

Marks

Question 8 (12 marks) Use a SEPARATE writing booklet.

(a) A drug is used to control a medical condition. It is known that the quantity Q of drug remaining in the body after t hours satisfies an equation of the form

$$Q = Q_0 e^{-kt}$$

where Q_0 and k are constants.

The initial dose is 6 milligrams and after 15 hours the amount remaining in the body is half the initial dose.

(i) Find the values of Q_0 and k. **3**

(ii) When will one-eighth of the initial dose remain? **2**

(b) A particle moves in a straight line. At time t seconds, its distance x metres from a fixed point O on the line is given by

$$x = \sin 2t + 3.$$

(i) Sketch the graph of x as a function of t for $0 \le t \le 2\pi$. **3**

(ii) Using your graph, or otherwise, find the times when the particle is at rest, and the position of the particle at those times. **2**

(iii) Describe the motion completely. **2**

Marks

Question 9 (12 marks) Use a SEPARATE writing booklet.

(a) Consider the function $y = \ln(x - 1)$ for $x > 1$.

(i) Sketch the function, showing its essential features. **2**

(ii) Use Simpson's rule with three function values to find an approximation to **2**

$$\int_2^4 \ln(x-1)\,dx.$$

(b) A superannuation fund pays an interest rate of 8.75% per annum which compounds annually. Stephanie decides to invest \$5000 in the fund at the beginning of each year, commencing on 1 January 2003. **4**

What will be the value of Stephanie's superannuation when she retires on 31 December 2023?

(c)

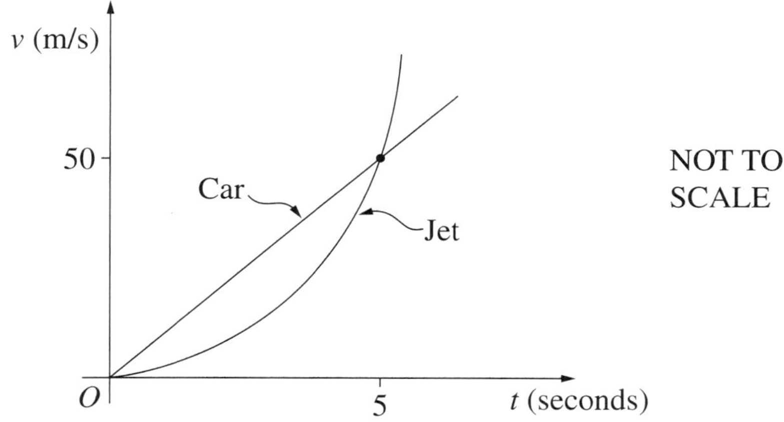

A car and a jet race one another from rest down a runway. The car increases its speed v_1 at a constant rate, while the speed of the jet is given by $v_2 = 2t^2$. After 5 seconds the car and the jet have the same speed of 50 m/s, as shown on the graph.

(i) Find an equation for the speed v_1 of the car in terms of t. **1**

(ii) How far behind the car is the jet after 5 seconds? **2**

(iii) After how many seconds does the jet catch up with the car? **1**

Marks

Question 10 (12 marks) Use a SEPARATE writing booklet.

(a) A circular pizza of radius 20 cm is cut into sectors. Each sector is to be placed on a circular plate that is just large enough to contain that sector.

(i) A sector of pizza is cut where the angle θ at its centre satisfies $0 < \theta \le \frac{\pi}{2}$. **2**

It is placed on a circular plate, of radius r cm and centre C, as shown below.

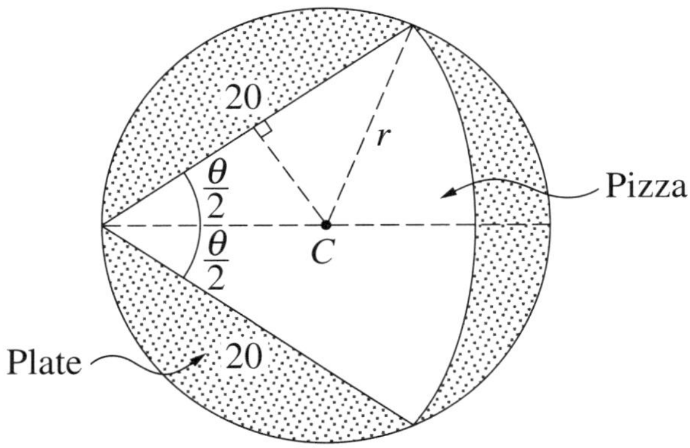

Show that $r = 10\sec\frac{\theta}{2}$ for $0 < \theta \le \frac{\pi}{2}$.

Question 10 continues

Marks

Question 10 (continued)

(ii) Another sector of pizza is cut where the angle θ at its centre satisfies **1**

$$\frac{\pi}{2} < \theta < \pi .$$

This sector of pizza is placed on a circular plate as shown below. Again, we let the radius of the plate be r cm, and we let the centre be C.

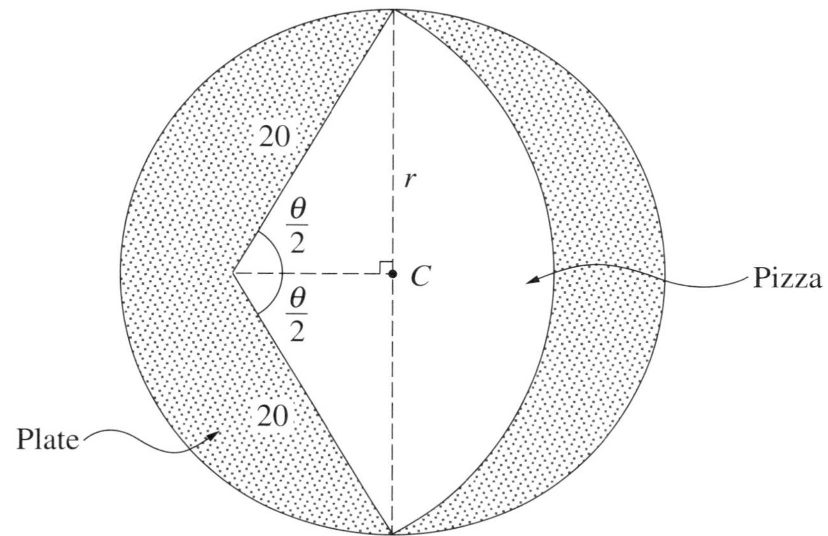

Show that $r = 20\sin\frac{\theta}{2}$ for $\frac{\pi}{2} < \theta < \pi$.

(iii) Sketch the graph of r, as defined by the equations in parts (i) and (ii), for $0 < \theta < \pi$. **3**

Question 10 continues

Marks

Question 10 (continued)

(b) On a dark night, two ships, Saga and Hero, sail parallel to a straight coastline on which there are two lights of equal brightness, 16 kilometres apart.

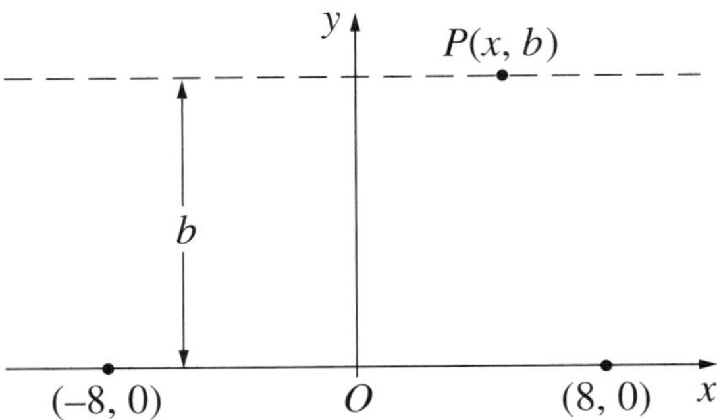

Suppose the coastline is represented by the x axis where the origin O is chosen to be the midpoint of the light sources. It is known that the (total) brightness from the lights on a ship at point $P(x, b)$ is

$$I = \frac{1}{b^2 + (x+8)^2} + \frac{1}{b^2 + (x-8)^2} \ .$$

(i) Show that $\dfrac{dI}{dx} = -\dfrac{2P}{Q}$ where **2**

$$P = \left[(x+8)\left(b^2 + (x-8)^2\right)^2 + (x-8)\left(b^2 + (x+8)^2\right)^2 \right]$$

and $Q = \left(b^2 + (x+8)^2\right)^2 \left(b^2 + (x-8)^2\right)^2$.

To answer parts (ii) and (iii), you may assume the following factorisation, given by a computer package, that

$$P = 2x\left(x^2 + 64 + b^2 + 16\sqrt{64 + b^2}\right)\left(x^2 + 64 + b^2 - 16\sqrt{64 + b^2}\right).$$

(ii) Saga sails parallel to the coast at a distance 15 km from the coast. **2**

By considering $\dfrac{dI}{dx}$, show that, as Saga sails from left to right, the brightness on Saga increases to a maximum when $x = 0$ and then decreases.

(iii) Hero sails parallel to the coast at a distance 6 km from the coast. **2**

Describe how the brightness on Hero changes as Hero sails from left to right. Give clear reasons for your answer.

End of paper

2002 Higher School Certificate Worked Answers

QUESTION 1

(a) $\dfrac{5.8^2 - 3.1^3}{3 \times 3.1 \times 5.8} = \dfrac{33.64 - 29.791}{53.94}$

$= \dfrac{3.849}{53.94}$

$= 0.07135\ldots$ (by calc.)

$= 0.0714$ to 3 sig. figures

(b) $\dfrac{d}{dx}(x^3 + 2) = 3x^2.$

(c)
$$x^2 = 5x$$
$$\therefore x^2 - 5x = 0$$
$$\therefore x(x-5) = 0$$
$$\therefore x = 0 \text{ or } 5.$$

(d) $\displaystyle\int \frac{3}{x}\,dx = 3\log_e x + C.$

(e)
$$3x - \frac{2x-5}{2} = 6$$
$$\therefore 6x - (2x-5) = 12$$
$$6x - 2x + 5 = 12$$
$$4x + 5 = 12$$
$$4x = 7$$
$$x = \frac{7}{4}.$$

(f)
$$x - 2y = 8 \quad \ldots(1)$$
$$2x + y = 1 \quad \ldots(2)$$
$(2) \times 2$ $\quad 4x + 2y = 2 \quad \ldots(3)$

$(1) + (3)$ $\quad 5x = 10$

$\therefore x = 2$

Substituting $x = 2$ into (2):
$$2(2) + y = 1$$
$$4 + y = 1$$
$$y = -3.$$

QUESTION 2

(a)
$$y = e^{2x}$$
$$\frac{dy}{dx} = 2e^{2x}$$
at $x = 0$, $\dfrac{dy}{dx} = 2e^0 = 2$

$\therefore$ Equation of tangent at (0, 1) with gradient 2

is $y - 1 = 2(x - 0)$

$y - 1 = 2x$

$y = 2x + 1.$

(b) (i) $\dfrac{d}{dx}(x\sin x) = x.\dfrac{d}{dx}(\sin x) + \sin x.\dfrac{d}{dx}(x)$

$= x\cos x + \sin x.1$

$= x\cos x + \sin x.$

(ii) $\dfrac{d}{dx}\left(\dfrac{\ln x}{x^2}\right) = \dfrac{x^2.\dfrac{d}{dx}(\ln x) - \ln x.\dfrac{d}{dx}(x^2)}{(x^2)^2}$

$= \dfrac{x^2.\dfrac{1}{x} - \ln x.\,2x}{x^4}$

$= \dfrac{x - 2x\ln x}{x^4}$

$= \dfrac{x(1 - 2\ln x)}{x^4}$

$= \dfrac{1 - 2\ln x}{x^3}.$

(c) $\dfrac{x}{\sin 60°} = \dfrac{y}{\sin 45°}$

$\therefore \dfrac{x}{y} = \dfrac{\sin 60°}{\sin 45°}$

$= \dfrac{\frac{\sqrt{3}}{2}}{\frac{1}{\sqrt{2}}}$

$= \dfrac{\sqrt{3}}{2} \times \dfrac{\sqrt{2}}{1}$

$= \dfrac{\sqrt{6}}{2}.$

(d) (i) $\displaystyle\int \cos 3x\,dx = \frac{1}{3}\sin 3x + C.$

(ii) $\displaystyle\int_0^1 (e^{5x} - 1) = \left[\frac{1}{5}e^{5x} - x\right]_0^1$

$= \left(\dfrac{1}{5}e^5 - 1\right) - \left(\dfrac{1}{5}e^0 - 0\right)$

$= \dfrac{1}{5}e^5 - 1 - \dfrac{1}{5}$

$= \dfrac{1}{5}e^5 - \dfrac{6}{5}$

$= \dfrac{1}{5}(e^5 - 6).$

QUESTION 3

(a) $A = P(1 + r)^n$

$= 1000\left(1 + \dfrac{3.5}{100}\right)^{20}$

$= 1000(1 + 0.035)^{20}$

$= 1000(1.035)^{20}$

$= \$1989.79$ to nearest cent.

(b)

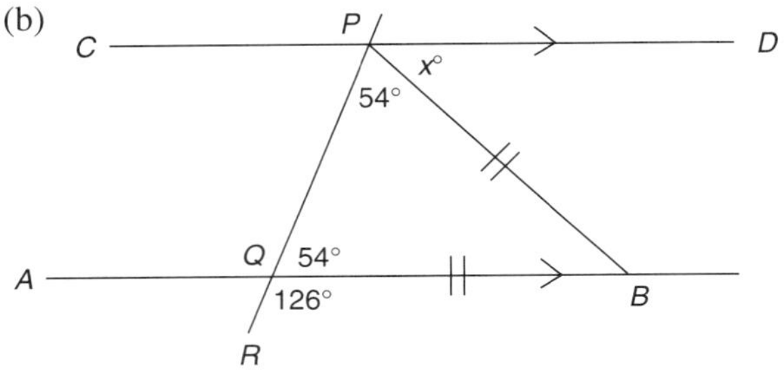

Now $\angle PQB = 180° - 126°$ (supplementary)
$= 54°$
$\therefore \angle QPB = 54°$ (ΔQPB is isosceles)
and $\angle QPD = 126°$ (corresponding; $CD \parallel AB$)

Also $\angle QPD = \angle QPB + \angle BPD$
$\therefore 126° = 54° + x°$
$\therefore x = 72.$

(c)

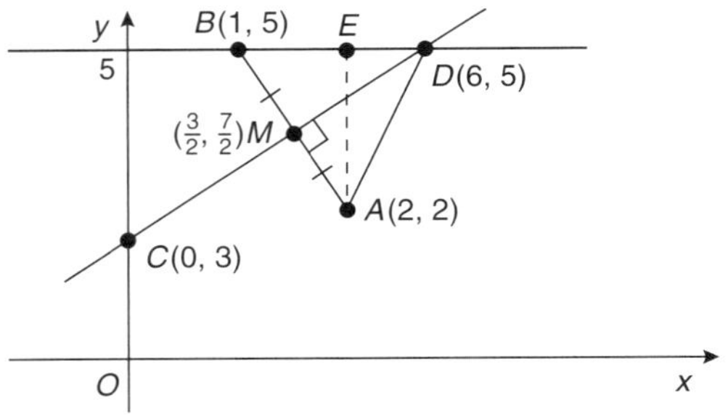

(i) Coordinates of M are,
$$\left(\frac{1+2}{2}, \frac{5+2}{2}\right)$$
i.e. $\left(\frac{3}{2}, \frac{7}{2}\right)$.

(ii) Gradient of $AB = \frac{y_2 - y_1}{x_2 - x_1}$
$$= \frac{5-2}{1-2}$$
$$= \frac{3}{-1}$$
$$= -3$$

Since for perpendicular lines $m_1 \times m_2 = -1$,
$\therefore$ Gradient of perpendicular $= \frac{1}{3}$.

$\therefore$ Equation of perpendicular bisector with gradient $= \frac{1}{3}$, passing through $M\left(\frac{3}{2}, \frac{7}{2}\right)$ is given by

$$y - \frac{7}{2} = \frac{1}{3}\left(x - \frac{3}{2}\right)$$
$$y - \frac{7}{2} = \frac{x}{3} - \frac{1}{2}$$
$$y = \frac{x}{3} - \frac{1}{2} + \frac{7}{2}$$
$$y = \frac{x}{3} + 3$$
$$\therefore 3y = x + 9$$
$$\therefore x - 3y + 9 = 0.$$

(iii) As C is equidistant from A and B, it lies on the perpendicular bisector of AB, which cuts the y axis at $x = 0$:
$\therefore 0 - 3y + 9 = 0$
$-3y = -9$
$y = 3$

$\therefore$ Coordinates of C are $(0, 3)$.

(iv) At D, $y = 5$
$\therefore x - 3(5) + 9 = 0$
$x - 15 + 9 = 0$
$x = 6$
$\therefore$ Coordinates of D are $(6, 5)$

(v) Area of $\Delta ABD = \frac{1}{2} \times b \times h$
$$= \frac{1}{2} \times BD \times AE$$
$$= \frac{1}{2} \times 5 \times 3$$
$$= \frac{15}{2}$$
$$= 7.5 \text{ units}^2.$$

QUESTION 4

(a) $|x - 1| \geq 3$
$\therefore x - 1 \geq 3$ or $-(x - 1) \geq 3$
$\therefore x \geq 4$ or $-x + 1 \geq 3$
$\therefore x \geq 4$ or $x \leq -2$

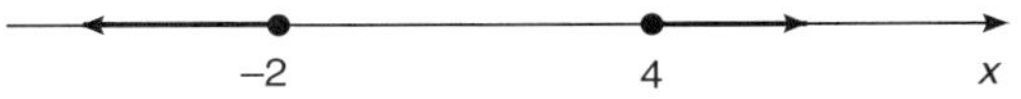

(b) $\cos\theta - \frac{2}{5} = 0$ $\quad 0° \leq \theta \leq 360°$

$\therefore \cos\theta = \frac{2}{5}$
$\therefore \theta = 66.421\ldots°$ or $360° - 66.421\ldots°$
$\therefore \theta = 66°$ or $293.578\ldots°$
$= 66°$ or $294°$ to nearest degree.

(c) (i) By the cosine rule:
$$MN^2 = LM^2 + LN^2 - 2(LM)(LN)\cos\angle MLN$$
$$= 5.2^2 + 8.9^2 - 2(5.2)(8.9)\cos 110°$$
$= 137.9073\ldots$ (by calc.)
$\therefore MN = 11.7433\ldots$ (by calc.)
$\doteqdot 11.7$ m

(ii) Area $\Delta LMN = \frac{1}{2}(LM)(LN)\sin\angle MLN$
$$= \frac{1}{2} \times 5.2 \times 8.9 \times \sin 110°$$
$= 21.7444\ldots$ (by calc.)
$\doteqdot 21.7 \text{ m}^2$.

(d) (i) Solving simultaneously:

$y = 2x \quad \ldots (1)$

$y = 6x - x^2 \quad \ldots (2)$

Equating (1) and (2):

$$6x - x^2 = 2x$$
$$4x - x^2 = 0$$
$$x(4 - x) = 0$$
$$\therefore x = 0 \text{ or } 4$$

At $x = 4$, $y = 2(4)$
$= 8$

$\therefore$ Coordinates of B are $(4, 8)$.

(ii) Required area

$$= \int_0^4 (6x - x^2)\,dx - \int_0^4 2x\,dx$$
$$= \int_0^4 (4x - x^2)\,dx$$
$$= \left[2x^2 - \frac{x^3}{3}\right]_0^4$$
$$= \left(2(4)^2 - \frac{(4)^3}{3}\right) - (0)$$
$$= 32 - \frac{64}{3}$$
$$= \frac{96 - 64}{3}$$
$$= \frac{32}{3}$$
$$= 10\tfrac{2}{3} \text{ units}^2.$$

QUESTION 5

(a) CATRINE   2 m   3.5 m   5 m

2 m   1.5 m   1.5 m

(i) The throws constitute an arithmetic sequence:
2, 3.5, 5, . . .

$$U_n = a + (n - 1)d$$
$$U_n = 32,\ a = 2,\ d = 1.5$$
$$\therefore 32 = 2 + (n - 1)1.5$$
$$32 = 2 + 1.5n - 1.5$$
$$32 = 0.5 + 1.5n$$
$$31.5 = 1.5n$$
$$n = \frac{31.5}{1.5}$$
$$= 21$$

$\therefore$ Catrine threw the stick 21 times.

(ii) Outward distance run by the dog:

$$S_n = \frac{n}{2}\,[2a + (n - 1)d]$$
$$a = 2,\ d = 1.5,\ n = 21$$
$$S_n = \frac{21}{2}\,[2(2) + (21 - 1)1.5]$$
$$= \frac{21}{2}\,[4 + (20)1.5]$$
$$= \frac{21}{2}\,[4 + 30]$$
$$= \frac{21}{2} \times 34$$
$$= 357 \text{ m.}$$

$\therefore$ Total distance run by dog
$= 2 \times 357$ m
$= 714$ m.

(b) $l = r\theta$

$$\therefore 38 = 20 \times \theta$$
$$\theta = \frac{38}{20}$$
$$= 1.9 \text{ radians}$$

Now 1.9 radians $= \left(1.9 \times \dfrac{180}{\pi}\right)^\circ$
$= 108.8619\ldots^\circ$ (by calc.)
$= 109^\circ$ to nearest degree.

(c) (i) $y = x^2 - 8x + 4$

$x^2 - 8x = y - 4$

Completing the square on LHS:

$$x^2 - 8x + 16 = y - 4 + 16$$
$$\therefore (x - 4)^2 = 1(y + 12)$$

$\therefore$ Coordinates of vertex are $(4, -12)$.

(ii) Now $4a = 1$

$$\therefore a = \frac{1}{4}.$$

i.e. Focal length $= \dfrac{1}{4}$

$\therefore$ Coordinates of focus are $\left(4, -11\frac{3}{4}\right)$.

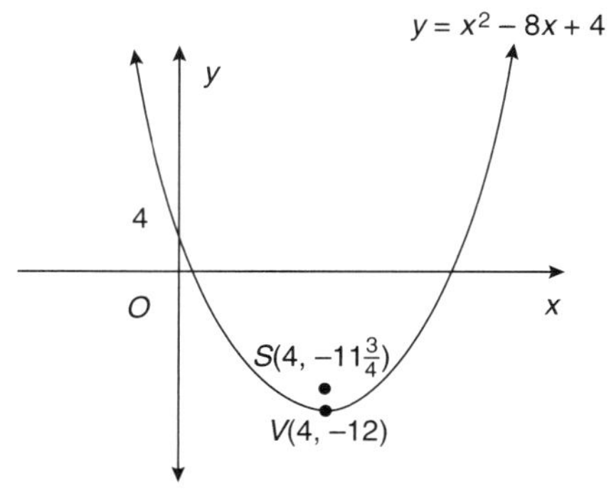

QUESTION 6

(a)

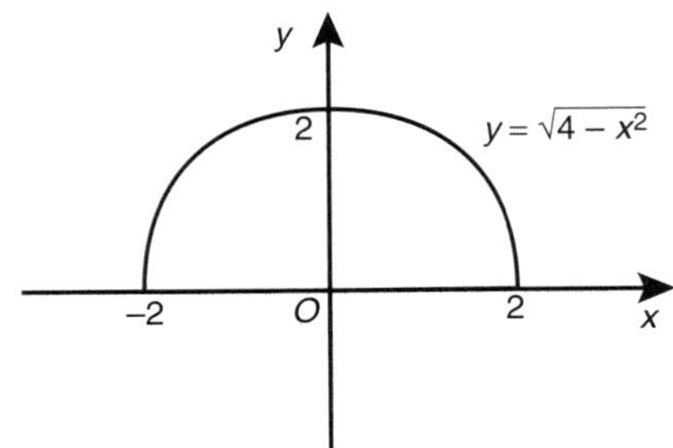

Range is $0 \leqslant y \leqslant 2$.

(b) (i) $f'(x) = 3(x + 1)(x - 3)$
$= 3(x^2 - 2x - 3)$
$= 3x^2 - 6x - 9$

$\therefore f(x) = \dfrac{3x^3}{3} - \dfrac{6x^2}{2} - 9x + C$
$= x^3 - 3x^2 - 9x + C$

The curve passes through (0, 12),

$\therefore 12 = 0 - 0 - 0 + C$
$\therefore C = 12$

$\therefore f(x) = x^3 - 3x^2 - 9x + 12$

(ii) Stationary points occur when

$f'(x) = 0$
$3(x + 1)(x - 3) = 0$
$\therefore x = 3$ or -1.

When $x = 3$, $y = 3^3 - 3(3)^2 - 9(3) + 12$
$= 27 - 27 - 27 + 12$
$= -15$

When $x = -1$, $y = (-1)^3 - 3(-1)^2 - 9(-1) + 12$
$= -1 - 3 + 9 + 12$
$= 17$

Stationary points are $(3, -15)$ and $(-1, 17)$

Now $f'(x) = 3x^2 - 6x - 9$
$\therefore f''(x) = 6x - 6$

Now $f''(3) = 6(3) - 6$
$= 12 > 0$

$\therefore$ A minimum turning point at $(3, -15)$

and $f''(-1) = 6(-1) - 6$
$= -12 < 0$

$\therefore$ A maximum turning point at $(-1, 17)$

y intercept occurs when $x = 0$,
i.e. $y = f(0) = 0^3 - 3(0)^2 - 9(0) + 12$
$= 12$

$\therefore$ y intercept is 12.

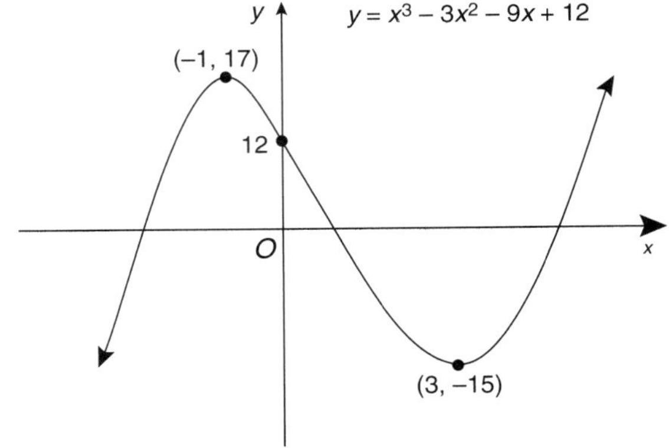

(iii) Curve is concave up when
$f''(x) > 0$
$6x - 6 > 0$
$6x > 6$
$x > 1$.

(c) $V = \pi \displaystyle\int_0^4 x^2\, dy$

Now $y = \dfrac{x^4}{4}$

$x^4 = 4y$

$x^2 = 2\sqrt{y}$

$= 2y^{\frac{1}{2}}$

$\therefore V = \pi \displaystyle\int_0^4 2y^{\frac{1}{2}}\, dy$

$= 2\pi \displaystyle\int_0^4 y^{\frac{1}{2}}\, dy$

$= 2\pi \left[\dfrac{2y^{\frac{3}{2}}}{3}\right]_0^4$

$= 2\pi \left[\dfrac{3}{2}(4)^{\frac{3}{2}} - 0\right]$

$= 2\pi \left[\dfrac{2}{3} \times 8\right]$

$= \dfrac{32\pi}{3}$ units^3.

QUESTION 7

(a) (i) A limiting sum exists if
$|r| < 1$, i.e. $-1 < r < 1$

Now $r = \sqrt{5} - 2$
$\doteqdot 0.236$ (by calc.)
and $-1 < 0.236 < 1$

$\therefore$ A limiting sum exists.

(ii) $S = \dfrac{a}{1-r}$

$= \dfrac{1}{1-(\sqrt{5}-2)}$

$= \dfrac{1}{1-\sqrt{5}+2}$

$= \dfrac{1}{3-\sqrt{5}}$

$= \dfrac{1}{3-\sqrt{5}} \times \dfrac{3+\sqrt{5}}{3+\sqrt{5}}$

$= \dfrac{3+\sqrt{5}}{9-5}$

$= \dfrac{3+\sqrt{5}}{4}.$

(b) (i) $V = 25\left(1-\dfrac{t}{60}\right)^2$ for $0 \le t \le 60$

At $t=0$, $V = 25\left(1-\dfrac{0}{60}\right)^2$

$= 25 \times 1$
$= 25$ litres.

(ii) $V = 25\left(1-\dfrac{t}{60}\right)^2$

When $V = \dfrac{25}{4}$,

$\dfrac{25}{4} = 25\left(1-\dfrac{t}{60}\right)^2$

$\dfrac{1}{4} = \left(1-\dfrac{t}{60}\right)^2$

$\pm\dfrac{1}{2} = 1-\dfrac{t}{60}$

$\dfrac{t}{60} = 1 \pm \dfrac{1}{2}$

$\dfrac{t}{60} = 1\dfrac{1}{2}$ or $\dfrac{t}{60} = \dfrac{1}{2}$

$\therefore t = 90$ or $t = 30$

$\therefore t = 30$ seconds $\quad 0 \le t \le 60$.

(ii) Now $\dfrac{dV}{dt} = 2 \times 25\left(1-\dfrac{t}{60}\right)^1 \cdot \dfrac{d}{dt}\left(1-\dfrac{t}{60}\right)$

$= 50\left(1-\dfrac{t}{60}\right) \cdot -\dfrac{1}{60}$

When $t = 30$,

$\dfrac{dV}{dt} = 50\left(1-\dfrac{30}{60}\right) \times -\dfrac{1}{60}$

$= 50\left(\dfrac{1}{2}\right)\left(-\dfrac{1}{60}\right)$

$= -\dfrac{50}{120}$

$= -\dfrac{5}{12}$ litres/second.

(Negative value indicates water is draining out.)

(c) (i) The first sock selected can be any colour. The second sock must not be the same colour as the first drawn.

$\therefore$ Probability $= \dfrac{8}{8} \times \dfrac{6}{7}$

$= \dfrac{6}{7}.$

(ii) The third sock cannot be the same colour as the first two colours drawn.

$\therefore$ Required probability $= \dfrac{8}{8} \times \dfrac{6}{7} \times \dfrac{4}{6}$

$= \dfrac{4}{7}.$

(iii) P(matching pair) $= 1 - \dfrac{4}{7}$

$= \dfrac{3}{7}.$

QUESTION 8

(a) (i) $Q = Q_0\, e^{-kt}$

When $t = 0$, $Q = 6$

$\therefore 6 = Q_0\, e^{-k(0)}$

$\therefore 6 = Q_0\, e^0$

$\therefore Q_0 = 6$

$\therefore Q = 6\, e^{-kt}$

When $t = 15$, $Q = 3$

$\therefore 3 = 6\, e^{-k(15)}$

$\therefore \dfrac{1}{2} = e^{-15k}$

$\ln\left(\dfrac{1}{2}\right) = \ln e^{-15k}$

$\ln\left(\dfrac{1}{2}\right) = -15k \ln e$

$\ln\left(\dfrac{1}{2}\right) = -15k$

$k = \dfrac{\ln\left(\dfrac{1}{2}\right)}{-15}$

$= 0.04620\ldots$ (by calc.)

$\doteqdot 0.0462.$

(ii) Now $Q = 6\, e^{-kt}$

If $Q = \dfrac{1}{8} \times 6$

$= \dfrac{3}{4}$

$$\therefore \frac{3}{4} = 6e^{-kt}$$
$$\frac{3}{24} = e^{-kt}$$
$$\frac{1}{8} = e^{-kt}$$
$$\ln\left(\frac{1}{8}\right) = \ln e^{-kt}$$
$$\ln\left(\frac{1}{8}\right) = -kt \ln e$$
$$t = \frac{\ln\left(\frac{1}{8}\right)}{-k}$$
$$= \frac{\ln\left(\frac{1}{8}\right)}{-0.04620\ldots}$$
$= 45$ hours (by calc.)

(b) (i) Range: $-1 + 3 \leqslant x \leqslant 1 + 3$
$\therefore 2 \leqslant x \leqslant 4$

Period: $\frac{2\pi}{2} = \pi$.

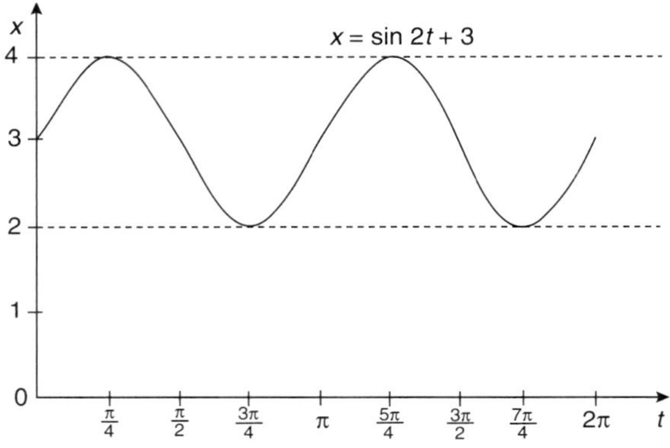

(ii) The particle is at rest when $\frac{dx}{dt} = 0$, i.e. the gradient of the tangent is 0.

Time	*Position*
$t = \frac{\pi}{4}$ s	$x = 4$ m
$t = \frac{3\pi}{4}$ s	$x = 2$ m
$t = \frac{5\pi}{4}$ s	$x = 4$ m
$t = \frac{7\pi}{4}$ s	$x = 2$ m

(iii) The particle oscillates between $x = 4$ and $x = 2$ with the centre of oscillation at $x = 3$. The particle starts at $x = 3$ and moves to the right to $x = 4$, where at $t = \frac{\pi}{4}$ s, the particle stops, changes direction and moves to the left, until at $x = 2$ and $t = \frac{3\pi}{4}$ s, it again stops, changes direction and moves to the right again, returning to the centre of oscillation at $x = 3$ ($t = \pi$ s). This motion is then repeated from $t = \pi$ s to $t = 2\pi$ s.

QUESTION 9

(a) (i)

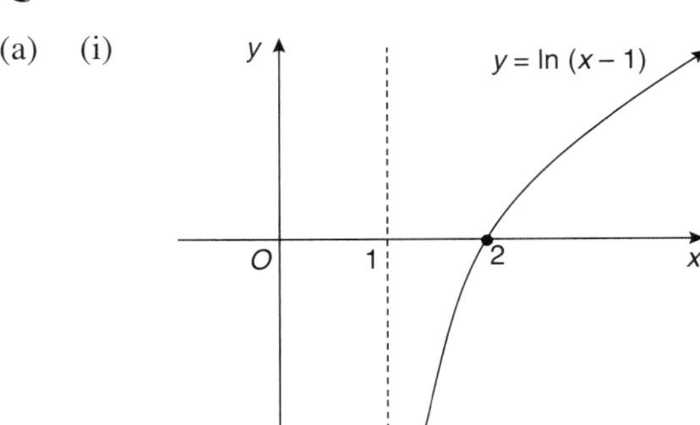

(ii) $y = \ln(x - 1)$

x	2	3	4
y	ln 1	ln 2	ln 3
	y_0	y_1	y_2

$$\int_2^4 \ln(x-1)\,dx \doteqdot \frac{h}{3}[y_0 + y_2 + 4y_1]$$
$$\doteqdot \frac{1}{3}[\ln 1 + \ln 3 + 4\ln 2]$$
$$\doteqdot \frac{1}{3}[3.8712\ldots] \quad \text{(by calc.)}$$
$$\doteqdot 1.2904\ldots$$
$= 1.29$ to 2 decimal places.

(b) $A = P\left(1 + \frac{r}{100}\right)^n$

$P = 5000, r = 8.75$

$$\therefore A = 5000\left(1 + \frac{8.75}{100}\right)^n$$
$$= 5000\,(1.0875)^n$$

$A_1 = 5000\,(1.0875)^{21}$
$A_2 = 5000\,(1.0875)^{20}$
$\vdots$
$A_{21} = 5000\,(1.0875)^1$

Total value on 31 December 2023
$= 5000\,(1.0875)^{21} + 5000\,(1.0875)^{20} + \ldots + 5000\,(1.0875)^1$
$= 5000\,[(1.0875)^{21} + (1.0875)^{20} + \ldots + (1.0875)^1]$

$= 5000\,[(1.0875)^1 + (1.0875)^2 + \ldots + (1.0875)^{21}]$

$= 5000\left[\dfrac{a(r^n - 1)}{r - 1}\right]$

where $a = 1.0875,\ r = 1.0875,\ n = 21$

$= 5000\left[\dfrac{1.0875(1.0875^{21} - 1)}{1.0875 - 1}\right]$

$= \$299\,605$ to the nearest dollar. (by calc.)

(c) (i) The equation of the line for the car is given by

$v_1 = mt$ where m is the slope of the line.

Now $m = \dfrac{50}{5} = 10$

$\therefore v_1 = 10t.$

(ii) **Car:**

$v_1 = 10t$

i.e. $\dfrac{dx_1}{dt} = 10t$

$\therefore x_1 = 5t^2 + C$

At $t = 0,\ x_1 = 0 \quad \therefore C = 0$

$\therefore x_1 = 5t^2$

At $t = 5,\ x_1 = 5 \times 5^2$

$= 125$ m

Jet:

$v_2 = 2t^2$

i.e. $\dfrac{dx_2}{dt} = 2t^2$

$\therefore x_2 = \dfrac{2t^3}{3} + C_1$

At $t = 0,\ x_2 = 0 \quad \therefore C_1 = 0$

$\therefore x_2 = \dfrac{2t^3}{3}$

At $t = 5,\ x_2 = \dfrac{2(5)^3}{3}$

$= \dfrac{250}{3}$

$= 83\frac{1}{3}$ m.

At $t = 5,\ x_1 - x_2 = 125 - 83\frac{1}{3}$

$= 41\frac{2}{3}$ m

The jet is $41\frac{2}{3}$ m behind the car.

(iii) The jet catches up with the car when $x_1 = x_2$

$\therefore 5t^2 = \dfrac{2}{3}t^3$

$5t^2 - \dfrac{2}{3}t^3 = 0$

$t^2\left(5 - \dfrac{2}{3}t\right) = 0$

$\therefore t = 0$ or $5 - \dfrac{2}{3}t = 0$

$-\dfrac{2}{3}t = -5$

$t = -5 \times -\dfrac{3}{2}$

$= \dfrac{15}{2}$

$\therefore$ The jet catches up with car after $7\frac{1}{2}$ s.

QUESTION 10

(a) (i)

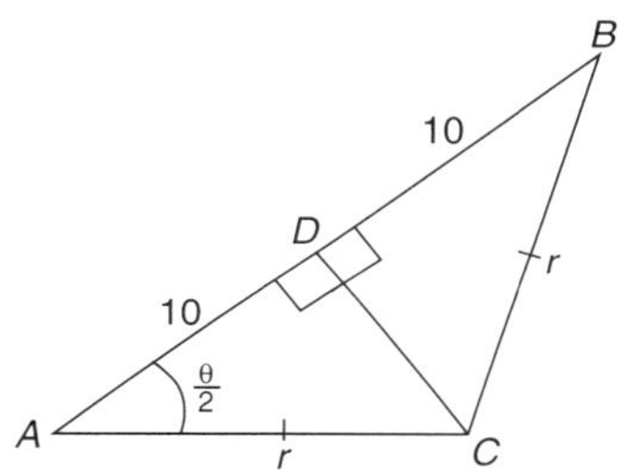

Now $AD = BD = 10$ cm
(DC bisects AB since ΔABC is isosceles and $DC \perp AB$.)

$\therefore \cos\dfrac{\theta}{2} = \dfrac{10}{r} \quad 0 < \theta \leqslant \dfrac{\pi}{2}$

$r = \dfrac{10}{\cos\frac{\theta}{2}}$

$= 10\sec\dfrac{\theta}{2}.$

(ii)

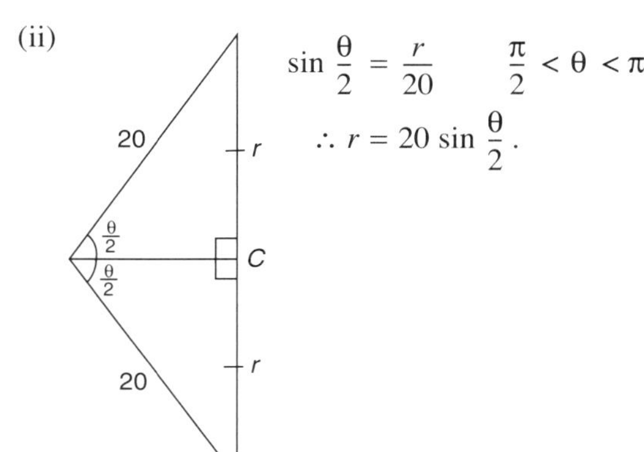

$\sin\dfrac{\theta}{2} = \dfrac{r}{20} \quad \dfrac{\pi}{2} < \theta < \pi$

$\therefore r = 20\sin\dfrac{\theta}{2}.$

(iii)

r \ θ	0	$\frac{\pi}{2}$	π
$10 \sec \frac{\theta}{2}$	10	$10\sqrt{2}$ ($\doteqdot 14.14$)	
$20 \sin \frac{\theta}{2}$		$10\sqrt{2}$ ($\doteqdot 14.14$)	20

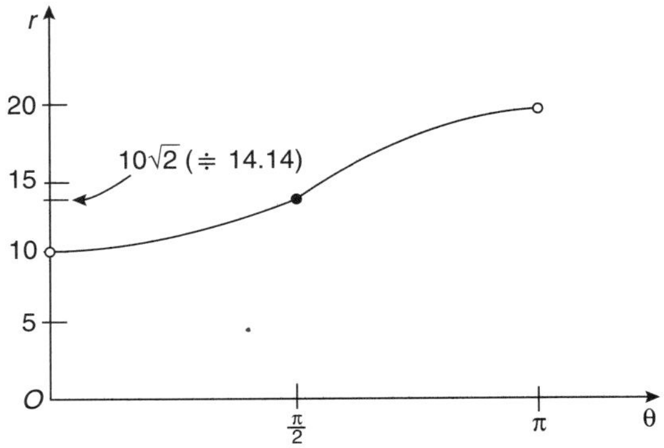

Note that for r (as defined in (i) and (ii)), $0 < \theta < \pi$.

$\therefore$ As $\theta \to 0,\ r \to 10$

and at $\theta = \frac{\pi}{2},\ r = 10\sqrt{2}$

and as $\theta \to \pi,\ r \to 20$

(b) (i)
$$I = \frac{1}{b^2 + (x+8)^2} + \frac{1}{b^2 + (x-8)^2}$$
$$= \left(b^2 + (x+8)^2\right)^{-1} + \left(b^2 + (x-8)^2\right)^{-1}$$

$$\frac{dI}{dx} = -1\left(b^2 + (x+8)^2\right)^{-2} \cdot \frac{d}{dx}(x+8)^2 + -1\left(b^2 + (x-8)^2\right)^{-2} \cdot \frac{d}{dx}(x-8)^2$$
$$= \frac{-1}{\left[b^2 + (x+8)^2\right]^2} \cdot 2(x+8)^1 + \frac{-1}{\left[b^2 + (x-8)^2\right]^2} \cdot 2(x-8)^1 \cdot 1$$
$$= \frac{-2(x+8)}{\left(b^2 + (x+8)^2\right)^2} - \frac{2(x-8)}{\left(b^2 + (x-8)^2\right)^2}$$
$$= \frac{-2(x+8)\left(b^2 + (x-8)^2\right)^2 - 2(x-8)\left(b^2 + (x+8)^2\right)^2}{\left(b^2 + (x+8)^2\right)^2\left(b^2 + (x-8)^2\right)^2}$$
$$= \frac{-2\left[(x+8)\left(b^2 + (x-8)^2\right)^2 + (x-8)\left(b^2 + (x+8)^2\right)^2\right]}{\left(b^2 + (x+8)^2\right)^2\left(b^2 + (x-8)^2\right)^2}$$
$$= -\frac{2P}{Q}.$$

(ii)
$$\frac{dI}{dx} = \frac{-2\left[2x\left(x^2 + 64 + b^2 + 16\sqrt{64 + b^2}\right)\left(x^2 + 64 + b^2 - 16\sqrt{64 + b^2}\right)\right]}{\left[b^2 + (x+8)^2\right]^2\left[b^2 + (x-8)^2\right]^2}$$

Now when $b = 15$,
$$\frac{dI}{dx} = \frac{-2\left[2x\left(x^2 + 64 + 15^2 + 16\sqrt{64 + 15^2}\right)\left(x^2 + 64 + 15^2 - 16\sqrt{64 + 15^2}\right)\right]}{\left[15^2 + (x+8)^2\right]^2\left[15^2 + (x-8)^2\right]^2}$$
$$= \frac{-2\left[2x\left(x^2 + 64 + 225 + 16\sqrt{64 + 225}\right)\left(x^2 + 64 + 225 - 16\sqrt{64 + 225}\right)\right]}{\left[225 + (x+8)^2\right]^2\left[225 + (x-8)^2\right]^2}$$
$$= \frac{-4x(x^2 + 561)(x^2 + 17)}{\left[225 + (x+8)^2\right]^2\left[225 + (x-8)^2\right]^2}$$

Possible maximum or minimum values occur when
$$\frac{dI}{dx} = 0$$

$$\therefore \quad \frac{-4x(x^2+561)(x^2+17)}{\left[225+(x+8)^2\right]^2\left[225+(x-8)^2\right]^2}=0$$

$$\therefore -4x(x^2+561)(x^2+17)=0$$

$$\therefore x=0 \quad \text{(a possible solution)}$$

$$\text{or} \quad x^2=-561 \text{ or } -17$$

(no real solutions)

To determine the nature of the stationary point at $x=0$, we only need to examine

$\frac{dI}{dx}=-4x$ because all other components of

$\frac{dI}{dx}$ are >0 for all values of x.

Hence, at $x=-0.1$, $-4x=+0.4 \therefore \frac{dI}{dx}>0$

at $x=+0.1$, $-4x=-0.4 \therefore \frac{dI}{dx}<0$

$\therefore$ A maximum value occurs at $x=0$.
i.e. The brightness increases to a maximum of $x=0$ and then decreases.

(iii) When $b=6$,

$$\frac{dI}{dx}=\frac{-2\left[2x(x^2+260)(x^2-60)\right]}{\left[36+(x+8)^2\right]^2\left[36+(x-8)^2\right]^2}$$

$$=\frac{-4x(x^2+260)(x^2-60)}{\left[36+(x+8)^2\right]^2\left[36+(x-8)^2\right]^2}$$

Possible maximum or minimum values occur when

$$\frac{dI}{dx}=0$$

$$\frac{-4x(x^2+260)(x^2-60)}{\left[36+(x+8)^2\right]^2\left[36+(x-8)^2\right]^2}=0$$

$$\therefore \quad x=0$$

$$\text{or} \quad x^2=60$$

$$\therefore \quad x=\pm\sqrt{60}$$

$$\text{or} \quad x^2=-260 \quad \text{(no real solutions)}$$

$\therefore$ Possible values of x are $x=0, \pm\sqrt{60}$ $(=\pm 7.74)$

To determine the nature of the stationary point, we only need to examine

$-4x(x^2-60)$ because all other components of $\frac{dI}{dx}$ are >0 for all values of x

x	-8	$-\sqrt{60}$	-4	0	4	$\sqrt{60}$	8
$-4x(x^2-60)$	128	0	-704	0	704	0	-128
$\frac{dI}{dx}$	>0	0	<0	0	>0	0	<0
	/	max	\	min	/	max	\

From the above table, it can be seen that as Nero sails from left to right, the brightness increases to a maximum at $x=-\sqrt{60}$ and then decreases to a minimum at $x=0$ and again increases to a maximum at $x=+\sqrt{60}$ before decreasing again.

2003

HIGHER SCHOOL CERTIFICATE EXAMINATION

Mathematics

General Instructions

- Reading time – 5 minutes
- Working time – 3 hours
- Write using black or blue pen
- Board-approved calculators may be used
- A table of standard integrals is provided at the back of this paper
- All necessary working should be shown in every question

Total marks – 120

- Attempt Questions 1–10
- All questions are of equal value

Total marks – 120
Attempt Questions 1–10
All questions are of equal value

Answer each question in a SEPARATE writing booklet. Extra writing booklets are available.

Marks

Question 1 (12 marks) Use a SEPARATE writing booklet.

(a) Evaluate $e^{-3.5}$ correct to 3 significant figures. **2**

(b) Differentiate $3x + \tan x$ with respect to x. **2**

(c) An arc of length 5 units subtends an angle θ at the centre of a circle of radius 3 units. Find the value of θ to the nearest degree. **2**

(d) Iain and Louise paid \$315.00 for a meal at a restaurant. This included a $12\frac{1}{2}\%$ tip. What was the cost of the meal without the tip? **2**

(e) Find a primitive of $3x^2 - 8$. **2**

(f) Solve $|x - 3| = 7$. **2**

Marks

Question 2 (12 marks) Use a SEPARATE writing booklet.

(a) Find the equation of the normal to the curve $y = 2\log_e x$ at the point $(e, 2)$. **3**

(b)

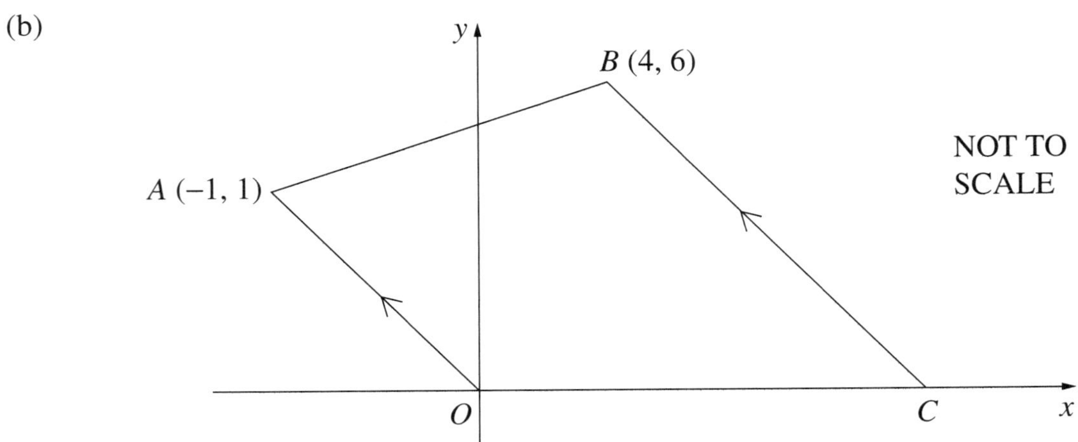

In the diagram, $OABC$ is a trapezium with $OA \parallel CB$. The coordinates of O, A and B are $(0, 0)$, $(-1, 1)$ and $(4, 6)$ respectively.

(i) Calculate the length of OA. **1**

(ii) Write down the gradient of the line OA. **1**

(iii) What is the size of $\angle AOC$? **1**

(iv) Find the equation of the line BC, and hence find the coordinates of C. **2**

(v) Show that the perpendicular distance from O to the line BC is $5\sqrt{2}$. **2**

(vi) Hence, or otherwise, calculate the area of the trapezium $OABC$. **2**

Marks

Question 3 (12 marks) Use a SEPARATE writing booklet.

(a) Differentiate with respect to x:

(i) $\left(2e^{x}-4\right)^{9}$ **2**

(ii) $x^{2}\sin x$. **2**

(b) **2**

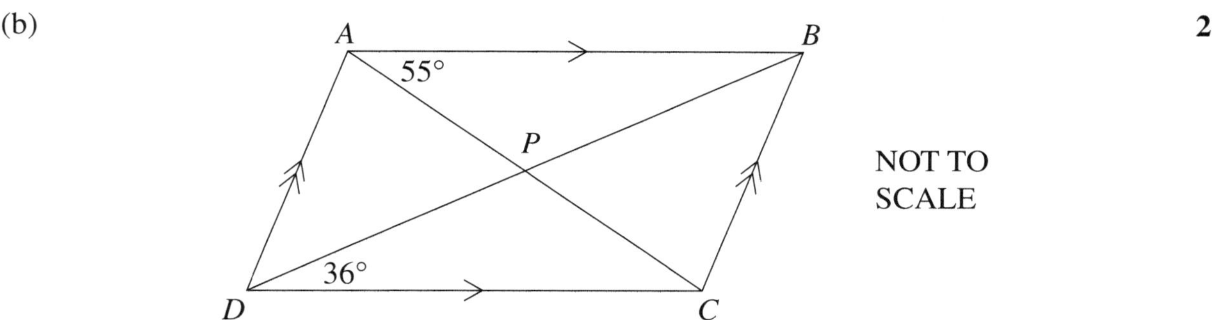

In the diagram, $ABCD$ is a parallelogram whose diagonals intersect at P. Given that $\angle BAC = 55°$ and $\angle CDB = 36°$, find the size of $\angle DPC$. Give reasons for your answer.

(c) Shade the region in the Cartesian plane for which the inequalities $y < x - 2$, $y \geq 0$ and $x \geq 6$ hold simultaneously. **3**

(d) (i) Find $\displaystyle\int \frac{2x}{x^{2}+5}\,dx$. **1**

(ii) Evaluate $\displaystyle\int_{\frac{\pi}{4}}^{\frac{\pi}{3}} \sec^{2} x\,dx$. **2**

Marks

Question 4 (12 marks) Use a SEPARATE writing booklet.

(a)

NOT TO SCALE

In the diagram, the point Q is due east of P. The point R is 38 km from P and 20 km from Q. The bearing of R from Q is 325°.

(i) What is the size of $\angle PQR$? **1**

(ii) What is the bearing of R from P? **3**

(b) The diagram shows two spinners which are spun simultaneously.

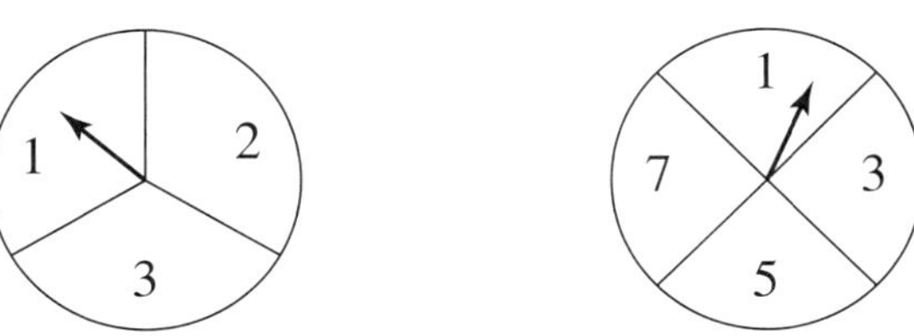

Each of the three outcomes on the first spinner are equally likely, and each of the four outcomes on the second spinner are equally likely.

(i) What is the probability that both spinners stop on the same number? **2**

(ii) What is the probability that at least one of the spinners stops on a 3? **1**

(c)

The graphs of $y = x - 4$ and $y = x^2 - 4x$ intersect at the points (4, 0) and A, as shown in the diagram.

(i) Find the coordinates of A. **2**

(ii) Find the area of the shaded region bounded by $y = x^2 - 4x$ and $y = x - 4$. **3**

Marks

Question 5 (12 marks) Use a SEPARATE writing booklet.

(a) Consider the function $f(x) = x^4 - 4x^3$.

(i) Show that $f'(x) = 4x^2(x-3)$. **1**

(ii) Find the coordinates of the stationary points of the curve $y = f(x)$, and determine their nature. **3**

(iii) Sketch the graph of the curve $y = f(x)$, showing the stationary points. **1**

(iv) Find the values of x for which the graph of $y = f(x)$ is concave down. **2**

(b) A wall is built to stop erosion when a level road is cut through a hill.

The rows of the wall, built from concrete blocks 1.5 m long, are numbered from the bottom. The bottom row (row 1) is 180 m long.

Each of the rows 2, 3, …, 20 has 3 fewer blocks than the row below it.

Above row 20, each row has 1 block fewer than the row below it. The top row has 10 blocks.

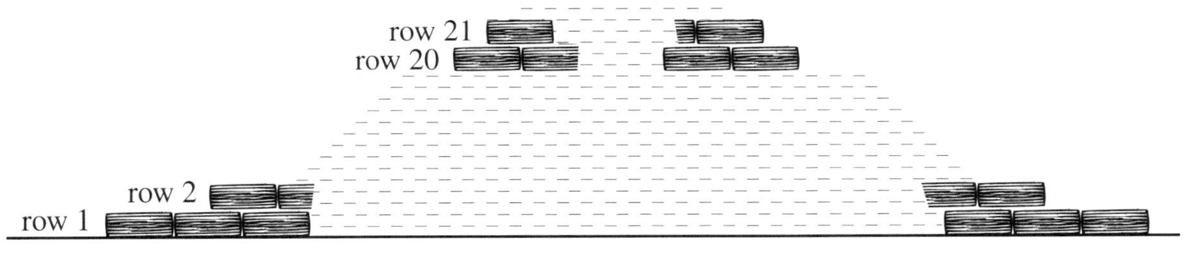

(i) How many blocks are in row 20? **2**

(ii) What is the total number of rows in the wall? **1**

(iii) How many blocks are used in the construction of the wall? **2**

Marks

Question 6 (12 marks) Use a SEPARATE writing booklet.

(a) Solve $\log_2(3x-4)=5$. **2**

(b)

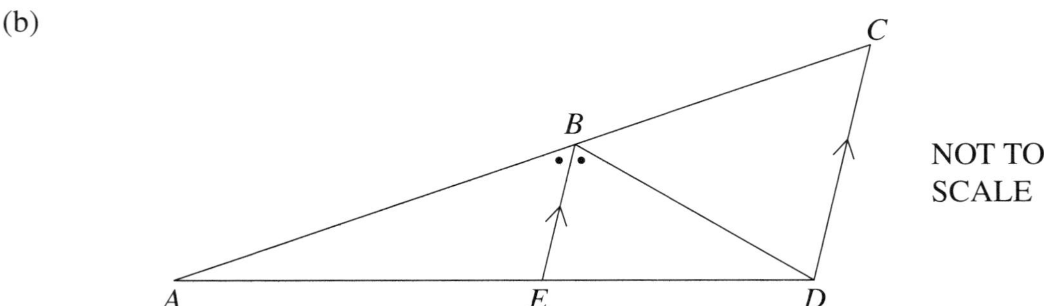

In the diagram, $BE \parallel CD$ and BE bisects $\angle ABD$.

Copy or trace the diagram into your writing booklet.

(i) Explain why $\angle EBD = \angle BDC$. **1**

(ii) Prove that ΔBCD is isosceles. **2**

(iii) Hence show that $AE:ED = AB:BD$. **2**

(c) A farmer accidentally spread a dangerous chemical on a paddock. The concentration of the chemical in the soil was initially measured at 5 kg/ha. One year later the concentration was found to be 2.8 kg/ha.

It is known that the concentration, C, is given by

$$C = C_0 e^{-kt},$$

where C_0 and k are constants, and t is measured in years.

(i) Evaluate C_0 and k. **3**

(ii) It is safe to use the paddock when the concentration is below 0.2 kg/ha. How long must the farmer wait after the accident before the paddock can be used? Give your answer in years, correct to one decimal place. **2**

Marks

Question 7 (12 marks) Use a SEPARATE writing booklet.

(a) (i) Find the limiting sum of the geometric series **2**

$$2+\frac{2}{\sqrt{2}+1}+\frac{2}{\left(\sqrt{2}+1\right)^2}+\ldots$$

(ii) Explain why the geometric series **1**

$$2+\frac{2}{\sqrt{2}-1}+\frac{2}{\left(\sqrt{2}-1\right)^2}+\ldots$$

does NOT have a limiting sum.

(b) The velocity of a particle is given by

$$v=2-4\cos t \quad \text{for} \quad 0\le t\le 2\pi,$$

where v is measured in metres per second and t is measured in seconds.

(i) At what times during this period is the particle at rest? **2**

(ii) What is the maximum velocity of the particle during this period? **2**

(iii) Sketch the graph of v as a function of t for $0\le t\le 2\pi$. **2**

(iv) Calculate the total distance travelled by the particle between $t=0$ and $t=\pi$. **3**

Marks

Question 8 (12 marks) Use a SEPARATE writing booklet.

(a) Write down the equation of the directrix of the parabola **1**

$$x^2 = -8y.$$

(b) **3**

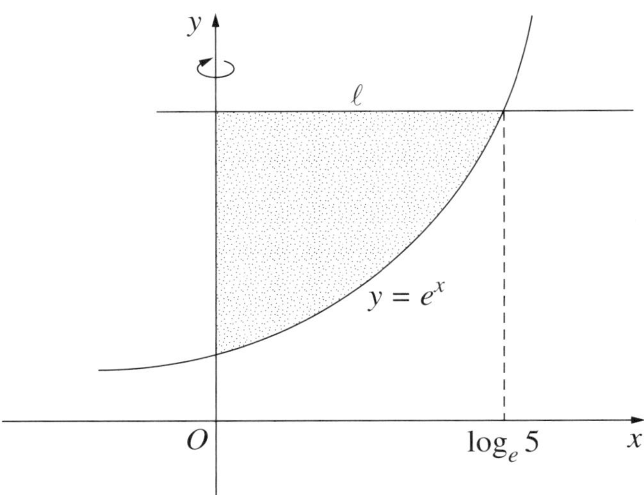

In the diagram, the shaded region is bounded by the y-axis, the curve $y = e^x$ and a horizontal line ℓ that cuts the curve at a point whose x coordinate is $\log_e 5$.

A solid is formed by rotating the shaded region about the y-axis.

Write down a definite integral whose value is the volume of the solid. (Do NOT evaluate the integral.)

(c) Use Simpson's rule with three function values to find an approximation for **3**

$$\int_2^6 \frac{x}{\log_e x}\,dx\,.$$

Give your answer correct to one decimal place.

(d) (i) Show that for all values of m, the line $y = mx - 3m^2$ touches the parabola $x^2 = 12y$. **2**

(ii) Find the values of m for which this line passes through the point (5, 2). **2**

(iii) Hence determine the equations of the two tangents to the parabola $x^2 = 12y$ from the point (5, 2). **1**

Marks

Question 9 (12 marks) Use a SEPARATE writing booklet.

(a) Solve $2\sin^2 x - 3\sin x - 2 = 0$ for $0 \le x \le 2\pi$. **2**

(b) The point A lies on the circle $\mathscr{C}_1$ with centre O and radius r. The circle $\mathscr{C}_2$ with centre A and radius $r\sqrt{3}$ intersects $\mathscr{C}_1$ at the points B and C, as shown in the diagram.

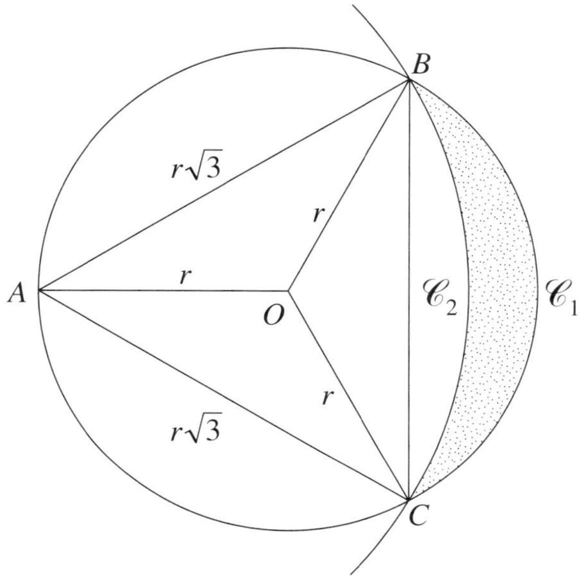

(i) Using the cosine rule (or otherwise) find $\angle BAO$. **1**

(ii) Find the area of the sector BAC of $\mathscr{C}_2$. **1**

(iii) Find the area of the sector BOC of $\mathscr{C}_1$. **1**

(iv) Hence (or otherwise) find the area of the shaded region. **2**

Question 9 continues

Marks

Question 9 (continued)

(c) A fish is swimming at a constant speed against the current. The current is moving at a constant speed of $u\ \mathrm{m\,s^{-1}}$. The speed of the fish relative to the water is $v\ \mathrm{m\,s^{-1}}$, so that the actual speed of the fish is $(v-u)\ \mathrm{m\,s^{-1}}$.

The rate at which the fish uses energy is proportional to v^3, so the amount of energy used in t seconds is given by

$$E = av^3t,$$

where a is a constant.

(i) Show that the energy used to swim L metres is given by **1**

$$E = \frac{aLv^3}{v-u}.$$

(ii) Migrating fish try to minimise the total energy used to swim a fixed distance. Find the value of v that minimises E. (You may assume $v > u > 0$.) **4**

End of Question 9

Marks

Question 10 (12 marks) Use a SEPARATE writing booklet.

(a) Barbara borrows \$120 000 to be repaid over a period of 25 years at 6% per annum reducible interest. Each year there are k regular repayments of $\$F$. Interest is calculated and charged just before each repayment.

(i) Write down an expression for the amount owing after two repayments. **1**

(ii) Show that the amount owing after n repayments is **2**

$$A_n = 120\,000\,\alpha^n - \frac{kF(\alpha^n - 1)}{0.06},$$

where $\alpha = 1 + \dfrac{0.06}{k}$.

(iii) Calculate the amount of each repayment if the repayments are made quarterly (ie. $k = 4$). **2**

(iv) How much would Barbara have saved over the term of the loan if she had chosen to make monthly rather than quarterly repayments? **2**

Question 10 continues

Marks

Question 10 (continued)

(b) A pulley P is attached to the ceiling at O by a piece of metal that can swing freely. One end of a rope is attached to the ceiling at A. The rope is passed through the pulley P and a weight is attached to the other end of the rope at M, as shown in the diagram.

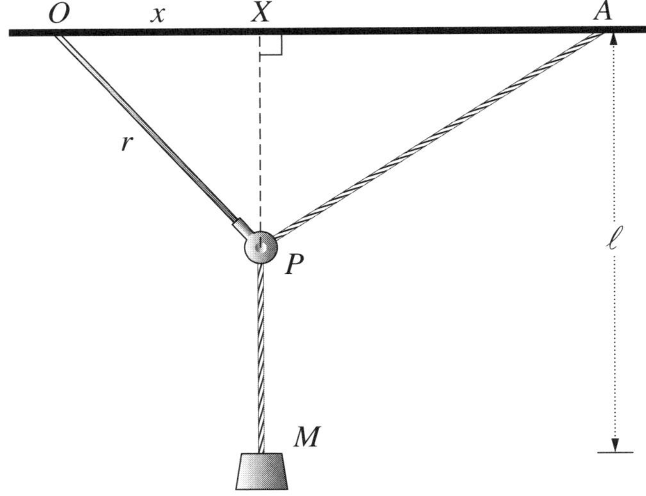

The distance OA is 1 m, the length of the rope is 2 m, and the length of the piece of metal $OP = r$ metres, where $0 < r < 1$. Let X be the point where the line MP produced meets OA. Let $OX = x$ metres and $XM = \ell$ metres.

(i) By considering triangles OXP and AXP, show that **1**

$$\ell = 2 + \sqrt{r^2 - x^2} - \sqrt{1 - 2x + r^2}\ .$$

(ii) Show that $\dfrac{d\ell}{dx} = \dfrac{\left(r^2 - x^2\right) - x^2\left(1 - 2x + r^2\right)}{\sqrt{r^2 - x^2}\sqrt{1 - 2x + r^2}\left(\sqrt{r^2 - x^2} + x\sqrt{1 - 2x + r^2}\right)}$. **2**

(iii) You are given the factorisation **2**

$$\left(r^2 - x^2\right) - x^2\left(1 - 2x + r^2\right) = (x - 1)\left(2x^2 - r^2 x - r^2\right).$$

(Do NOT prove this.)

Find the value of x for which M is closest to the floor. Justify your answer.

End of paper

2003 Higher School Certificate Worked Answers

QUESTION 1

(a) $e^{-3.5} = 0.03019...$ (by calc.)
$= 0.0302$ to 3 sig. figures.

(b) $\frac{d}{dx}(3x + \tan x) = 3 + \sec^2 x.$

(c) $l = r\theta$
$\therefore 5 = 3 \times \theta$
$\therefore \theta = \frac{5}{3}$ radians
$= \left(\frac{5}{3} \times \frac{180}{\pi}\right)^\circ$
$= 95.492...^\circ$ (by calc.)
$= 95^\circ$ to nearest degree.

(d) original cost + $12\frac{1}{2}$% of original cost = final cost
Let x represent the original price.

$$\therefore x + \frac{12\frac{1}{2}}{100}x = \$315$$
$$\therefore x + 0.125x = \$315$$
$$\therefore 1.125x = \$315$$
$$\therefore x = \frac{\$315}{1.125}$$
$$= \$280. \quad \text{(by calc.)}$$

$\therefore$ The cost of the meal without tip was \$280.

(e) $\int (3x^2 - 8)dx = \frac{3x^3}{3} - 8x + C$
$= x^3 - 8x + C.$

(f) $|x - 3| = 7$
$\therefore x - 3 = \pm 7$
$\therefore x - 3 = 7$ or $x - 3 = -7$
$\therefore x = 10$ or $x = -4.$

QUESTION 2

(a) $y = 2\log_e x$
$\frac{dy}{dx} = 2\left(\frac{1}{x}\right)$
$= \frac{2}{x}$

At $x = e$, $\frac{dy}{dx} = \frac{2}{e}$

$\therefore$ Gradient of normal
$= -\frac{1}{\frac{2}{e}}$
$= -\frac{e}{2}$

$\therefore$ Equation of normal through $(e, 2)$ with gradient $= -\frac{e}{2}$ is

$y - 2 = -\frac{e}{2}(x - e)$
$2y - 4 = -e(x - e)$
$2y - 4 = -ex + e^2$
$ex + 2y - 4 - e^2 = 0$

(b) (i) $OA = \sqrt{(-1-0)^2 + (1-0)^2}$
$= \sqrt{(-1)^2 + 1^2}$
$= \sqrt{1+1}$
$= \sqrt{2}.$

(ii) Gradient of OA
$= \frac{1-0}{-1-0}$
$= \frac{1}{-1}$
$= -1.$

(iii) $\tan \angle AOC =$ gradient of OA
$\tan \angle AOC = -1$
$\therefore \angle AOC = 180^\circ - 45^\circ$
$= 135^\circ.$

(iv) gradient of BC = gradient of OA (parallel lines)
$\therefore$ gradient of $BC = -1$

Equation of line BC, with gradient $= -1$, through the point $(4, 6)$ is given by

$y - 6 = -1(x - 4)$
$y - 6 = -x + 4$
$y = -x + 10$
$x + y = 10.$

At C, $y = 0$

$\therefore x + 0 = 10$
$\therefore x = 10$

$\therefore$ Coordinates of C are $(10, 0)$.

(v) $0 \equiv (0, 0)$
line BC: $x + y - 10 = 0$

$$d = \left|\frac{Ax_1 + By_1 + C}{\sqrt{A^2 + B^2}}\right|$$
$$= \left|\frac{1(0) + 1(0) - 10}{\sqrt{(1)^2 + (1)^2}}\right|$$
$$= \left|\frac{-10}{\sqrt{2}}\right|$$
$$= \frac{10}{\sqrt{2}} \times \frac{\sqrt{2}}{\sqrt{2}}$$
$$= \frac{10\sqrt{2}}{2}$$
$$= 5\sqrt{2}.$$

(vi) Now $BC = \sqrt{(10-4)^2 + (0-6)^2}$
$$= \sqrt{(6)^2 + (-6)^2}$$
$$= \sqrt{36 + 36}$$
$$= \sqrt{72}$$
$$= \sqrt{36 \times 2}$$
$$= 6\sqrt{2}.$$

Area of trapezium $OABC$
$$= \frac{1}{2}(OA + BC) \times h$$
$$= \frac{1}{2}(\sqrt{2} + 6\sqrt{2}) \times 5\sqrt{2}$$
$$= \frac{1}{2}(7\sqrt{2}) \times 5\sqrt{2}$$
$$= \frac{1}{2} \times 7 \times 5 \times 2$$
$$= 35 \text{ units}^2.$$

QUESTION 3

(a) (i) $\frac{d}{dx}[(2e^x - 4)^9]$
$$= 9(2e^x - 4)^8 . \frac{d}{dx}(2e^x - 4)$$
$$= 9(2e^x - 4)^8 . (2e^x)$$
$$= 18e^x(2e^x - 4)^8.$$

(ii) $\frac{d}{dx}(x^2 \sin x)$
$$= x^2 . \frac{d}{dx}(\sin x) + \sin x . \frac{d}{dx}(x^2)$$
$$= x^2 . \cos x + \sin x . 2x$$
$$= x^2 \cos x + 2x \sin x.$$

(b) $\angle BAC = \angle ACD$ (alternate angles; $AB \parallel DC$)

$\therefore \angle ACD = 55°$

In $\triangle DPC$,

$\angle CDP + \angle PCD + \angle DPC = 180°$ (angle sum of triangle DPC)

$\therefore 36° + 55° + \angle DPC = 180°$

$\therefore \angle DPC = 180° - 91°$
$= 89°.$

(c)

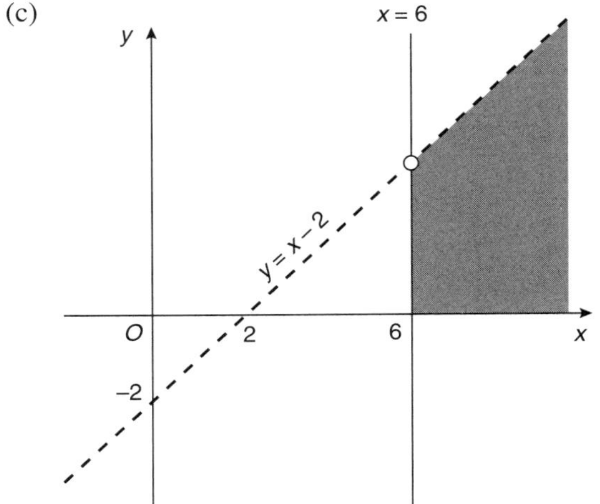

(d) (i) $\int \frac{2x}{x^2 + 5}\,dx = \ln(x^2 + 5) + C$

(ii) $\int_{\frac{\pi}{4}}^{\frac{\pi}{3}} \sec^2 x\,dx = \left[\tan x\right]_{\frac{\pi}{4}}^{\frac{\pi}{3}}$
$$= \tan\frac{\pi}{3} - \tan\frac{\pi}{4}$$
$$= \sqrt{3} - 1.$$

QUESTION 4

(a)

N, R, P, Q, 38 km, 20 km, θ, 55°, 325°

(i) $\angle PQR = 325° - 270°$
$= 55°.$

(ii) By the sine rule in $\triangle PQR$:
$$\frac{\sin\theta}{20} = \frac{\sin 55°}{38}$$

$$\sin\theta = \frac{20\times\sin 55^\circ}{38}$$
$= 0.4311...$ (by calc.)
$\therefore\ \theta = 25.539...^\circ$ (by calc.)
$= 25^\circ 32'$

$\therefore\ \angle NPR = 90^\circ - 25^\circ 32'$
$= 64^\circ 28'$

$\therefore$ Bearing of R from $P = \angle NPR$
$\doteqdot 064^\circ$.

(b)

		SPINNER 2			
		1	3	5	7
SPINNER 1	1	(11)	13*	15	17
	2	21	23*	25	27
	3	31*	(33*)	35*	37*

(i) $P(\text{Same no.}) = \frac{2}{12}$
$= \frac{1}{6}$.

(ii) $P(\text{At least one } 3) = \frac{6}{12}$
$= \frac{1}{2}$.

(c) (i) Solving simultaneously:

$y = x - 4$ ...(1)
$y = x^2 - 4x$ ...(2)

Equating (1) and (2):

$$x^2 - 4x = x - 4$$
$$x^2 - 5x + 4 = 0$$
$$(x-4)(x-1) = 0$$
$$\therefore\ x = 4 \text{ or } 1.$$

For A, $x = 1\ \therefore\ y = 1 - 4$
$= -3$

$\therefore$ Coordinates of A are $(1, -3)$.

(ii) Required Area

$$= \int_1^4 (x-4)\,dx - \int_1^4 (x^2-4x)\,dx$$
$$= \int_1^4 (x - 4 - x^2 + 4x)\,dx$$
$$= \int_1^4 (5x - x^2 - 4)\,dx$$
$$= \left[\frac{5x^2}{2} - \frac{x^3}{3} - 4x\right]_1^4$$
$$= \left[\frac{5(4)^2}{2} - \frac{(4)^3}{3} - 4(4)\right] - \left[\frac{5(1)^2}{2} - \frac{(1)^3}{3} - 4(1)\right]$$
$$= \left[40 - \frac{64}{3} - 16\right] - \left[\frac{5}{2} - \frac{1}{3} - 4\right]$$
$$= \left[\frac{120}{3} - \frac{64}{3} - \frac{48}{3}\right] - \left[\frac{15}{6} - \frac{2}{6} - \frac{24}{6}\right]$$
$$= \frac{8}{3} - \left(-\frac{11}{6}\right)$$
$$= \frac{16}{6} + \frac{11}{6}$$
$$= \frac{27}{6}$$
$$= \frac{9}{2}$$
$$= 4\tfrac{1}{2} \text{ units}^2.$$

QUESTION 5

(a) (i) $f(x) = x^4 - 4x^3$

$f'(x) = 4x^3 - 12x^2$
$= 4x^2(x-3)$

(ii) Stationary points occur when

$f'(x) = 0$
$4x^2(x-3) = 0$
$\therefore\ x = 0$ or 3

When $x = 0$, $y = 0^4 - 4(0)^3$
$= 0$

When $x = 3$, $y = 3^4 - 4(3)^3$
$= 81 - 108$
$= -27$

$\therefore$ Coordinates of stationary points are $(0, 0)$ and $(3, -27)$.

Now $f''(x) = 12x^2 - 24x$
$= 12x(x-2)$
$\therefore\ f''(3) = 12(3)(3-2)$
$= 36 > 0$

$\therefore$ A minimum turning point at $(3, -27)$

Now $f''(0) = 12(0)(-2)$
$= 0$

When $x < 0$, say $x = -0.1$

$$f''(-0.1) = 12(-0.1)(-0.1 - 2) = (+)(-)(-) > 0$$

When $x > 0$, say $x = 0.1$

$$f''(0.1) = 12(0.1)(0.1 - 2) = (+)(+)(-) < 0$$

$\therefore$ Concavity changes.
$\therefore$ A horizontal point of inflexion at $(0, 0)$.

(iii)

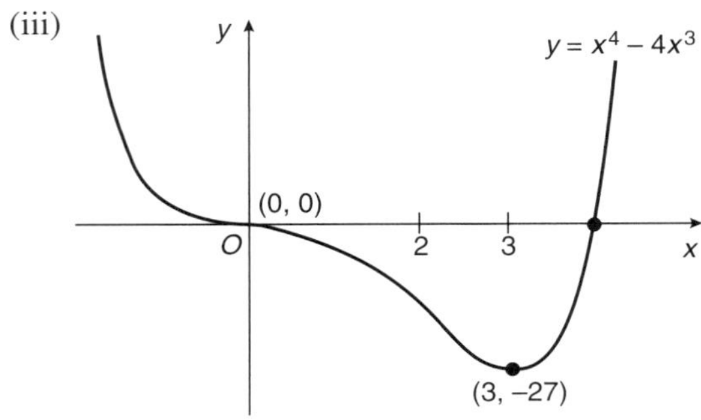

(iv) The curve is concave down when

$$f''(x) < 0$$
$$\therefore 12x^2 - 24x < 0$$
$$\therefore 12x(x - 2) < 0$$

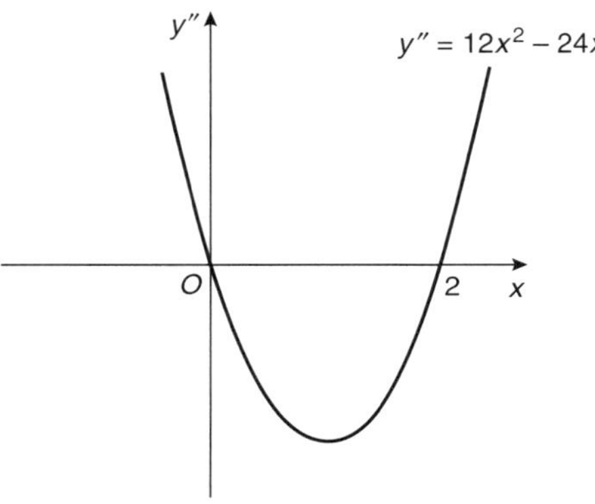

$\therefore 0 < x < 2$ (from graph)

(b) (i) Length of row 1 = 180 m
Length of each block = 1.5 m

$\therefore$ Number of blocks in row 1

$$= \frac{180}{1.5} = 120.$$

The number of blocks in each row up to row 20 are in an arithmetic sequence:

120, 117, 114 . . . (to row 20)
with $a = 120$ and $d = -3$.

Now $U_n = a + (n - 1)d$

$$\therefore U_{20} = 120 + 19(-3) = 120 - 57 = 63$$

i.e. There are 63 blocks in row 20.

(ii) Now row 21 has $63 - 1 = 62$ blocks in it.

The number of blocks in each row from row 21 to the top row are in an arithmetic sequence:
62, 61, ... 10
with $a = 62$ and $d = -1$

Now $U_n = a + (n - 1)d$

If $U_n = 10$,

$$10 = 62 + (n - 1)(-1)$$
$$10 = 62 - n + 1$$
$$10 = 63 - n$$
$$n = 63 - 10 = 53$$

$\therefore$ Total number of rows in the wall
$= 20 + 53$
$= 73.$

(iii) $S_n = \frac{n}{2}(a + l)$

For lower part of wall (rows 1–20):
$n = 20$, $a = 120$ and $l = 63$

$$\therefore S_{20} = \frac{20}{2}(120 + 63) = 10 \times 183 = 1830$$

For upper part of wall (rows 21 – 73)
$n = 53$, $a = 62$ and $l = 10$

$$\therefore S_{53} = \frac{53}{2}(62 + 10) = \frac{53}{2} \times 72 = 1908$$

$\therefore$ Total number of blocks used
$= 1830 + 1908$
$= 3738.$

QUESTION 6

(a) $\log_2(3x - 4) = 5$

$$3x - 4 = 2^5$$
$$3x - 4 = 32$$
$$3x = 36$$
$$\therefore x = 12.$$

(b)

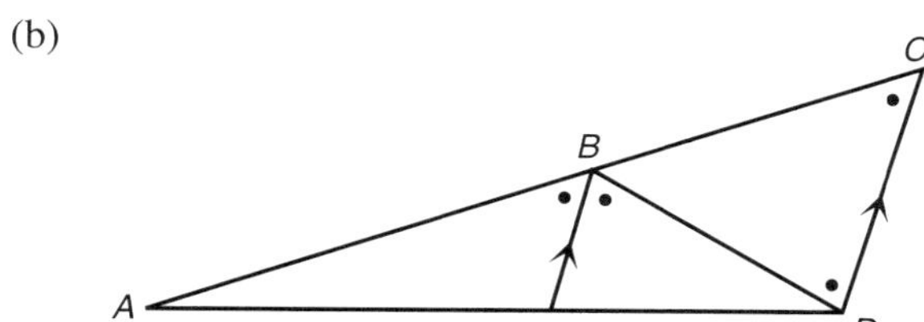

(i) $\angle EBD = \angle BDC$
(alternate angles; $BE \parallel CD$)

(ii) Now $\angle ABE = \angle BCD$
(corresponding; $BE \parallel CD$)
$\angle EBD = \angle BDC$ (proved)
$\angle EBD = \angle ABE$ (given)
$\therefore \angle BCD = \angle BDC$
$\therefore \triangle BCD$ is isosceles (two equal angles)

(iii) Now $\dfrac{AE}{ED} = \dfrac{AB}{BC}$
(equal ratios on transversals by parallel lines)
but $BC = BD$ (isosceles $\triangle BCD$)
$\therefore \dfrac{AE}{ED} = \dfrac{AB}{BD}$
i.e. $AE : ED = AB : BD$.

(c) (i) $C = C_0 e^{-kt}$
At $t = 0, C = 5$
$\therefore 5 = C_0 e^0$
$\therefore C_0 = 5$ kg/ha
$\therefore C = 5e^{-kt}$

At $t = 1, C = 2.8$
$\therefore 2.8 = 5e^{-k(1)}$
$\therefore 2.8 = 5e^{-k}$
$\therefore \dfrac{2.8}{5} = e^{-k}$
$\therefore 0.56 = e^{-k}$
$\therefore \ln(0.56) = \ln e^{-k}$
$\therefore \ln(0.56) = -k$
$\therefore k = -\ln(0.56)$
$= 0.57981\ldots$ (by calc.)
$\doteqdot 0.5798$

(ii) $C = 5e^{-kt}$
If $C = 0.2$,
$0.2 = 5e^{-kt}$
$\dfrac{0.2}{5} = e^{-kt}$
$0.04 = e^{-kt}$
$\ln(0.04) = \ln(e^{-kt})$
$\ln(0.04) = -kt \ln e$
$\ln(0.04) = -kt$
$t = \dfrac{\ln(0.04)}{-k}$
$= \dfrac{\ln(0.04)}{-(-\ln(0.56))}$
$= \dfrac{\ln(0.04)}{\ln(0.56)}$
$= 5.551\ldots$ (by calc.)
$= 5.6$ years to 1 decimal place.

QUESTION 7

(a) (i) $S = \dfrac{a}{1-r}$
Now $a = 2$,
and $r = \dfrac{\frac{2}{\sqrt{2}+1}}{2}$
$= \dfrac{1}{\sqrt{2}+1} \times \dfrac{\sqrt{2}-1}{\sqrt{2}-1}$
$= \dfrac{\sqrt{2}-1}{(\sqrt{2})^2 - 1^2}$
$= \dfrac{\sqrt{2}-1}{2-1}$
$= \sqrt{2} - 1$.

$\therefore S = \dfrac{2}{1 - (\sqrt{2}-1)}$
$= \dfrac{2}{2-\sqrt{2}} \times \dfrac{2+\sqrt{2}}{2+\sqrt{2}}$
$= \dfrac{2(2+\sqrt{2})}{2^2 - (\sqrt{2})^2}$
$= \dfrac{2(2+\sqrt{2})}{2}$
$= 2 + \sqrt{2}$.

(ii) $r = \dfrac{\frac{2}{\sqrt{2}-1}}{2}$
$= \dfrac{1}{\sqrt{2}-1} \times \dfrac{\sqrt{2}+1}{\sqrt{2}+1}$
$= \dfrac{\sqrt{2}+1}{(\sqrt{2})^2 - 1^2}$
$= \dfrac{\sqrt{2}+1}{2-1}$
$= \dfrac{\sqrt{2}+1}{1}$
$= \sqrt{2} + 1$
$\doteqdot 2.4$
For a limiting sum to exist, $-1 < r < 1$ and $\sqrt{2} + 1$ does not satisfy this inequality.

(b) (i) The particle is at rest when

$$v = 0$$

$$\therefore\ 2 - 4\cos t = 0 \quad 0 \leqslant t \leqslant 2\pi$$

$$-4\cos t = -2$$

$$\cos t = \frac{1}{2}$$

$$\therefore\ t = \frac{\pi}{3} \text{ or } \left(2\pi - \frac{\pi}{3}\right)$$

$$= \frac{\pi}{3} \text{ or } \frac{5\pi}{3} \text{ seconds.}$$

(ii) $v = 2 - 4\cos t$

$$v' = -(-4)\sin t$$
$$= 4\sin t$$

Maximum or minimum velocity occurs when

$$v' = 0 \qquad 0 \leqslant t \leqslant 2\pi$$

$$\therefore\ 4\sin t = 0$$
$$\sin t = 0$$
$$\therefore\ t = 0, \pi, 2\pi.$$

When $t = 0$, $v = 2 - 4\cos 0$
$= 2 - 4$
$= -2$

When $t = \pi$, $v = 2 - 4\cos\pi$
$= 2 - 4(-1)$
$= 2 + 4$
$= 6$

When $t = 2\pi$, $v = 2 - 4\cos(2\pi)$
$= 2 - 4(1)$
$= 2 - 4$
$= -2$

$\therefore$ Maximum velocity is 6 m/s.

(iii) $v = 2 - 4\cos t$

Period $= 2\pi$
Maximum velocity $= 6$
Minimum velocity $= -2$

t intercepts: $\frac{\pi}{3}$ and $\frac{5\pi}{3}$.

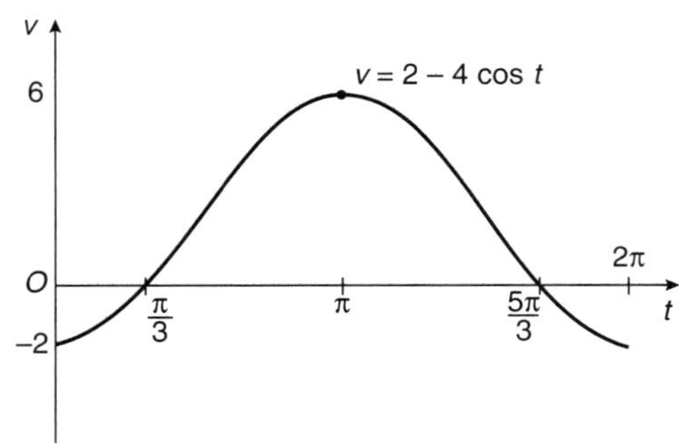

(iv) Distance travelled

$$= \left|\int_0^{\frac{\pi}{3}} (2 - 4\cos t)\,dt\right| + \int_{\frac{\pi}{3}}^{\pi} (2 - 4\cos t)\,dt$$

$$= \left|\Big[2t - 4\sin t\Big]_0^{\frac{\pi}{3}}\right| + \Big[2t - 4\sin t\Big]_{\frac{\pi}{3}}^{\pi}$$

$$= \left|\left(\frac{2\pi}{3} - 4\sin\frac{\pi}{3}\right) - (0 - 4\sin 0)\right| + (2\pi - 4\sin\pi) - \left(\frac{2\pi}{3} - 4\sin\frac{\pi}{3}\right)$$

$$= \left|\frac{2\pi}{3} - 4\left(\frac{\sqrt{3}}{2}\right) - 0\right| + 2\pi - 4(0) - \left(\frac{2\pi}{3} - 4\left(\frac{\sqrt{3}}{2}\right)\right)$$

$$= -\left(\frac{2\pi}{3} - 2\sqrt{3}\right) + 2\pi - \frac{2\pi}{3} + 2\sqrt{3}$$

$$\left(\text{since } \left(\frac{2\pi}{3} - 2\sqrt{3}\right) < 0\right)$$

$$= 2\sqrt{3} - \frac{2\pi}{3} + 2\pi - \frac{2\pi}{3} + 2\sqrt{3}$$

$$= 4\sqrt{3} + 2\pi - \frac{4\pi}{3}$$

$$= \left(4\sqrt{3} + \frac{2\pi}{3}\right) \text{metres.}$$

QUESTION 8

(a) $x^2 = -8y$
$= 4(-2)y$

Equation of directrix is

$$y = -(-2)$$
$$\therefore\ y = 2.$$

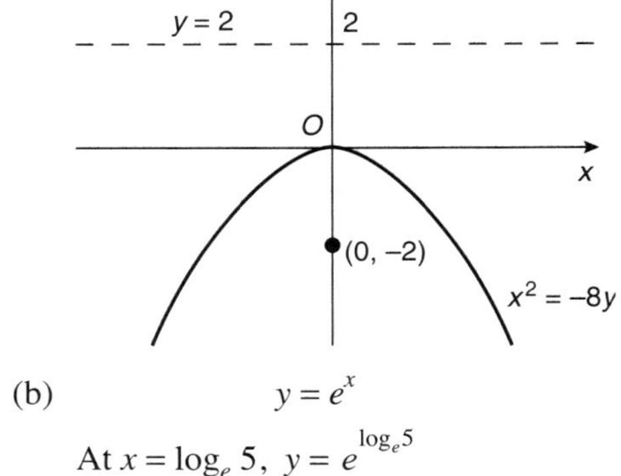

(b) $y = e^x$

At $x = \log_e 5$, $y = e^{\log_e 5}$
$= 5$

At $x = 0$, $y = e^0 = 1$

Now $V = \pi\int_1^5 x^2\,dy$

and $y = e^x$

$\therefore x = \log_e y$

$\therefore V = \pi \int_1^5 (\log_e y)^2 \, dy$

(c) $f(x) = \dfrac{x}{\log_e x}$

x	2	4	6
$f(x)$	$\dfrac{2}{\log_e 2}$	$\dfrac{4}{\log_e 4}$	$\dfrac{6}{\log_e 6}$
	y_0	y_1	y_2

$\int_2^6 \dfrac{x}{\log_e x}$

$\doteqdot \dfrac{h}{3}(y_0 + y_2 + 4y_1)$

$\doteqdot \dfrac{2}{3}\left[\dfrac{2}{\log_e 2} + \dfrac{6}{\log_e 6} + 4\left(\dfrac{4}{\log_e 4}\right)\right]$

$\doteqdot 11.8504\ldots$ (by calc.)

$= 11.9$ to one decimal place.

(d) (i) $y = mx - 3m^2$

$y = \dfrac{x^2}{12}$

Solving simultaneously:

$\dfrac{x^2}{12} = mx - 3m^2$

$x^2 = 12mx - 36m^2$

$x^2 - 12mx + 36m^2 = 0$

If the line touches the curve, then $x^2 - 12mx + 36m^2 = 0$ should have equal roots, i.e. $b^2 - 4ac = 0$.

Now $b^2 - 4ac = (-12m)^2 - 4(1)(36m^2)$
$= 144m^2 - 144m^2$
$= 0$

Hence, for all values of m, $b^2 - 4ac = 0$

$\therefore y = mx - 3m^2$ touches the parabola $x^2 = 12y$.

(ii) If (5, 2) satisfies the equation of the line,

$y = mx - 3m^2$

$\therefore 2 = m(5) - 3m^2$

$2 = 5m - 3m^2$

$3m^2 - 5m + 2 = 0$

$(3m - 2)(m - 1) = 0$

$\therefore m = \dfrac{2}{3}$ or $m = 1$.

(iii) For $m = 1$, $y = x - 3(1)^2$

i.e. $y = x - 3$

For $m = \dfrac{2}{3}$, $y = \dfrac{2}{3}x - 3\left(\dfrac{2}{3}\right)^2$

$= \dfrac{2}{3}x - 3\left(\dfrac{4}{9}\right)$

$= \dfrac{2}{3}x - \dfrac{4}{3}$

i.e. $3y = 2x - 4$.

QUESTION 9

(a) $2\sin^2 x - 3\sin x - 2 = 0 \qquad 0 \leqslant x \leqslant 2\pi$

$(2\sin x + 1)(\sin x - 2) = 0$

$\therefore 2\sin x + 1 = 0$ or $\sin x - 2 = 0$

$\therefore \sin x = -\dfrac{1}{2}$ or $\sin x = 2$

$\sin x = 2$ has no real solutions since $-1 \leqslant \sin x \leqslant 1$.

$\therefore \sin x = -\dfrac{1}{2}$

$x = \left(\pi + \dfrac{\pi}{6}\right)$ or $\left(2\pi - \dfrac{\pi}{6}\right)$

$= \dfrac{7\pi}{6}$ or $\dfrac{11\pi}{6}$.

(b) (i) $\cos\angle BAO = \dfrac{r^2 + (r\sqrt{3})^2 - r^2}{2(r)(r\sqrt{3})}$

$= \dfrac{r^2 + 3r^2 - r^2}{2\sqrt{3}r^2}$

$= \dfrac{3r^2}{2\sqrt{3}r^2}$

$= \dfrac{3}{2\sqrt{3}} \times \dfrac{\sqrt{3}}{\sqrt{3}}$

$= \dfrac{3\sqrt{3}}{6}$

i.e. $\cos\angle BAO = \dfrac{\sqrt{3}}{2}$

$\therefore \angle BAO = \dfrac{\pi}{6}$.

(ii) $A = \dfrac{1}{2}r^2\theta$

$r = r\sqrt{3}$

$\theta = \angle BAC = \dfrac{\pi}{6} + \dfrac{\pi}{6} = \dfrac{2\pi}{6} = \dfrac{\pi}{3}$

$(\angle BAC = \angle BAO + \angle CAO$

and $\angle CAO = \dfrac{\pi}{6}$ as $\triangle BAO \equiv \triangle CAO)$

$$\therefore A = \frac{1}{2}(r\sqrt{3})^2\left(\frac{\pi}{3}\right)$$
$$= \frac{1}{2}\,.\,3r^2\,.\,\frac{\pi}{3}$$
$$= \frac{\pi r^2}{2}\,.$$

(iii) $A = \frac{1}{2}r^2\theta$

$r = r$

$\theta = \angle BOC = 120° = \frac{2\pi}{3}$

(angle at centre is twice angle at circumference)

$$\therefore A = \frac{1}{2}r^2\left(\frac{2\pi}{3}\right)$$
$$= \frac{\pi r^2}{3}$$

(iv) Now triangles AOB, AOC and BOC are all congruent and hence have the same area:

$$A = \frac{1}{2}ab\sin C$$
$$= \frac{1}{2}(r)(r)\sin 120°$$
$$= \frac{1}{2}r^2\left(\frac{\sqrt{3}}{2}\right)$$
$$= \frac{\sqrt{3}}{4}r^2$$

Now,

Area of sector BAC = Area of the 3 triangles + Area of segment BC

and

Area of sector BOC = Area of 1 triangle + Area of segment BC + Area of shaded region

Therefore,

Area of sector BAC − Area of sector BOC

= Area of 2 triangles + 0 − Area of shaded region

i.e. $\frac{\pi r^2}{2} - \frac{\pi r^2}{3} = 2 \times \frac{\sqrt{3}}{4}r^2$ – Area of shaded region

$$\therefore \text{Area of shaded region} = \frac{2\sqrt{3}r^2}{4} - \frac{\pi r^2}{2} + \frac{\pi r^2}{3}$$
$$= \frac{\sqrt{3}r^2}{2} - \frac{\pi r^2}{2} + \frac{\pi r^2}{3}$$
$$= r^2\left(\frac{\sqrt{3}}{2} - \frac{\pi}{2} + \frac{\pi}{3}\right)$$
$$= r^2\left(\frac{\sqrt{3}}{2} - \frac{3\pi}{6} + \frac{2\pi}{6}\right)$$
$$= r^2\left(\frac{\sqrt{3}}{2} - \frac{\pi}{6}\right)\text{units}^2.$$

OR

OR by inspection of the diagram:

Area of shaded region
= Area of Sector BOC – (Area of Sector BAC – Area △'s AOB and AOC)

$$= \frac{\pi r^2}{3} - \left[\frac{\pi r^2}{2} - 2\left(\frac{\sqrt{3}r^2}{4}\right)\right]$$

$$= \frac{\pi r^2}{3} - \left(\frac{\pi r^2}{2} - \frac{\sqrt{3}r^2}{2}\right)$$

$$= \frac{\pi r^2}{3} - \frac{\pi r^2}{2} + \frac{\sqrt{3}r^2}{2}$$

$$= r^2\left(\frac{\pi}{3} - \frac{\pi}{2} + \frac{\sqrt{3}}{2}\right)$$

$$= r^2\left(\frac{\sqrt{3}}{2} - \frac{\pi}{6}\right)\text{units}^2.$$

(c) (i) distance = speed × time

i.e. $d = s \times t$

$$\therefore t = \frac{d}{s}$$

Now $s = v - u$

and $d = L$

$$\therefore t = \frac{L}{v-u}$$

and $E = av^3 t$

$$\therefore E = av^3\left(\frac{L}{v-u}\right)$$

$$= \frac{aLv^3}{v-u}.$$

(ii)
$$\frac{dE}{dv} = \frac{(v-u)\,.\,\frac{d}{dv}(aLv^3) - aLv^3\,.\,\frac{d}{dv}(v-u)}{(v-u)^2}$$

$$= \frac{(v-u)3aLv^2 - aLv^3\,.\,1}{(v-u)^2}$$

$$= \frac{3aLv^3 - 3aLuv^2 - aLv^3}{(v-u)^2}$$

$$= \frac{2aLv^3 - 3aLuv^2}{(v-u)^2}$$

$$= \frac{aLv^2(2v-3u)}{(v-u)^2}$$

Possible maximum or minimum values occur when

$$\frac{dE}{dv} = 0$$

$$\frac{aLv^2(2v-3u)}{(v-u)^2} = 0$$

$$\therefore v = 0 \text{ or } 2v - 3u = 0$$

$$2v = 3u$$

$$v = \frac{3u}{2}$$

Since $v > 0,\ v \neq 0$ $\quad \therefore v = \frac{3u}{2}$

To determine the nature of the stationary point:

When $v < \frac{3u}{2}\left(= \frac{6u}{4}\right)$, say $v = \frac{5u}{4}$,

$$\frac{dE}{dv} = \frac{aL\left(\frac{5u}{4}\right)^2\left(2\left(\frac{5u}{4}\right) - 3u\right)}{\left(\frac{5u}{4} - u\right)^2}$$

$$= \frac{aL\left(\frac{5u}{4}\right)^2\left(\frac{5u}{2} - 3u\right)}{\left(\frac{u}{4}\right)^2}$$

$$= \frac{(+)(+)(-)}{(+)}$$

$$< 0$$

When $v > \frac{3u}{2}$, say $v = \frac{4u}{2} = 2u$

$$\frac{dE}{dv} = \frac{aL(2u)^2[2(2u) - 3u]}{(2u-u)^2}$$

$$= \frac{aL(2u)^2(4u-3u)}{u^2}$$

$$= \frac{(+)(+)(+)}{(+)}$$

$$> 0$$

$\therefore E$ is a minimum when $v = \frac{3u}{2}$.

QUESTION 10

(a) (i) $A_1 = 120\,000\left(1 + \frac{0.06}{k}\right)^1 - F$

$$A_2 = \left[120\,000\left(1 + \frac{0.06}{k}\right)^1 - F\right]\left(1 + \frac{0.06}{k}\right) - F$$

$$= 120\,000\left(1 + \frac{0.06}{k}\right)^2 - \left(1 + \frac{0.06}{k}\right)F - F$$

(ii) $A_1 = 120\,000\,\alpha - F$

$A_2 = 120\,000\,\alpha^2 - \alpha F - F$

$A_3 = (120\,000\,\alpha^2 - \alpha F - F)(\alpha) - F$

$= 120\,000\,\alpha^3 - \alpha^2 F - \alpha F - F$

$\therefore A_n = 120\,000\alpha^n - \alpha^{n-1}F - \alpha^{n-2}F - \ldots - \alpha F - F$

$= 120\,000\,\alpha^n - F(\alpha^{n-1} + \alpha^{n-2} + \ldots + \alpha + 1)$

$= 120\,000\,\alpha^n - F(1 + \alpha + \alpha^2 + \ldots + \alpha^{n-1})$

$= 120\,000\,\alpha^n - F\left(\frac{a(r^n - 1)}{r - 1}\right)$

where $a = 1$, $r = \alpha$ and $n = n$

$= 120\,000\,\alpha^n - F\left(\frac{1(\alpha^n - 1)}{\alpha - 1}\right)$

$= 120\,000\,\alpha^n - F\left(\frac{\alpha^n - 1}{1 + \frac{0.06}{k} - 1}\right)$

$= 120\,000\,\alpha^n - F\left(\frac{\alpha^n - 1}{\frac{0.06}{k}}\right)$

$= 120\,000\,\alpha^n - \frac{kF(\alpha^n - 1)}{0.06}.$

(iii) $n = 25 \times 4$, $k = 4$, $A_n = 0$

$= 100$

$\therefore \alpha = 1 + \frac{0.06}{4}$

$= 1.015$

$\therefore 0 = 120\,000\,(1.015)^{100} - \frac{4F(1.015^{100} - 1)}{0.06}$

$= 7200\,(1.015)^{100} - 4F\,(1.015^{100} - 1)$

$4F\,(1.015^{100} - 1) = 7200\,(1.015)^{100}$

$F = \frac{7200\,(1.015)^{100}}{4\,(1.015^{100} - 1)}$

$= \$2324.47.$ (by calc.)

(iv) $n = 25 \times 12$, $k = 12$, $A_n = 0$

$= 300$

$\therefore \alpha = 1 + \frac{0.06}{12}$

$= 1.005$

$\therefore 0 = 120\,000(1.005)^{300} - \frac{12F(1.005^{300} - 1)}{0.06}$

$= 7200\,(1.005)^{300} - 12F\,(1.005^{300} - 1)$

$12F\,(1.005^{300} - 1) = 7200\,(1.005)^{300}$

$\therefore F = \frac{7200\,(1.005)^{300}}{12\,(1.005^{300} - 1)}$

$= \$773.16$ (by calc.)

Amount repaid if quarterly repayments:
$= \$2324.47 \times 100$
$= \$232\,447$

Amount repaid if monthly repayments:
$= \$773.16 \times 300$
$= \$231\,948$

Saving $= \$232\,447 - \$231\,948$
$= \$499.$

(b)

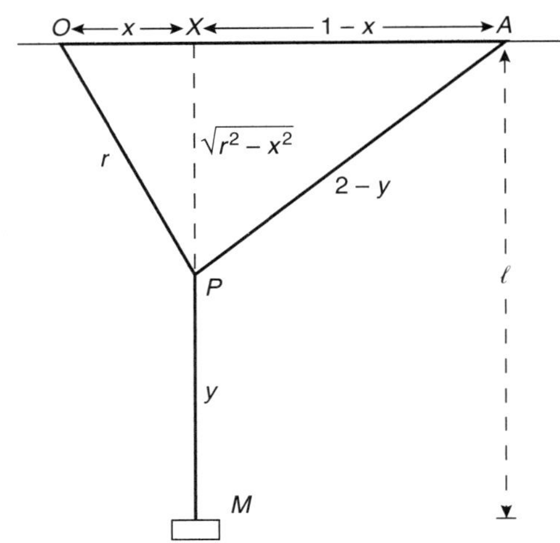

(i) Let $PM = y$ $\therefore AP = 2 - y$
Now $\ell = XP + PM$

In $\triangle OXP$,

$OX^2 + XP^2 = OP^2$ (Pythagoras' theorem)

$\therefore x^2 + XP^2 = r^2$

$XP^2 = r^2 - x^2$

$\therefore XP = \sqrt{r^2 - x^2}$

In $\triangle AXP$,

$AP^2 = AX^2 + XP^2$

$\therefore (2 - y)^2 = (1 - x)^2 + \left(\sqrt{r^2 - x^2}\right)^2$

$\therefore (2 - y)^2 = 1 - 2x + x^2 + r^2 - x^2$

$(2 - y)^2 = 1 - 2x + r^2$

$2 - y = \sqrt{1 - 2x + r^2}$

$y = 2 - \sqrt{1 - 2x + r^2}$

Now $\ell = XP + PM$

$$= \sqrt{r^2 - x^2} + y$$

$$= \sqrt{r^2 - x^2} + 2 - \sqrt{1 - 2x + r^2}$$

$$= 2 + \sqrt{r^2 - x^2} - \sqrt{1 - 2x + r^2}.$$

(ii) $\ell = 2 + \left(r^2 - x^2\right)^{\frac{1}{2}} - \left(1 - 2x + r^2\right)^{\frac{1}{2}}$

$$\frac{dl}{dx} = \frac{1}{2}\left(r^2 - x^2\right)^{-\frac{1}{2}} .-2x - \frac{1}{2}\left(1 - 2x + r^2\right)^{-\frac{1}{2}} .-2$$

$$= \frac{-x}{\sqrt{r^2 - x^2}} + \frac{1}{\sqrt{1 - 2x + r^2}}$$

$$= \frac{-x\sqrt{1 - 2x + r^2} + \sqrt{r^2 - x^2}}{\sqrt{r^2 - x^2}\,\sqrt{1 - 2x + r^2}}$$

$$= \frac{\sqrt{r^2 - x^2} - x\sqrt{1 - 2x + r^2}}{\sqrt{r^2 - x^2}\,\sqrt{1 - 2x + r^2}} \times \frac{\sqrt{r^2 - x^2} + x\sqrt{1 - 2x + r^2}}{\sqrt{r^2 - x^2} + x\sqrt{1 - 2x + r^2}}$$

(N.B. $(a - b) \times (a + b) = a^2 - b^2$)

$$= \frac{\left(r^2 - x^2\right) - x^2\left(1 - 2x + r^2\right)}{\sqrt{r^2 - x^2}\,\sqrt{1 - 2x + r^2}\left(\sqrt{r^2 - x^2} + x\sqrt{1 - 2x + r^2}\right)}.$$

(iii) For M to be closest to the floor, l must be a maximum.
Now,

$$\frac{dl}{dx} = \frac{(x - 1)\left(2x^2 - r^2x - r^2\right)}{\sqrt{r^2 - x^2}\,\sqrt{1 - 2x + r^2}\left(\sqrt{r^2 - x^2} + x\sqrt{1 - 2x + r^2}\right)}$$

(using the factorisation given)

A possible maximum value occurs when

$$\frac{dl}{dx} = 0$$

$$\therefore (x - 1)(2x^2 - r^2x - r^2) = 0$$

$$\therefore x = 1 \text{ or } 2x^2 - r^2x - r^2 = 0$$

Now if $x = 1$, $r = 1$, but $0 < r < 1$
$\therefore x \neq 1$.

Solving $2x^2 - r^2x - r^2 = 0$:

$$x = \frac{-b \pm \sqrt{b^2 - 4ac}}{2a}$$

$$= \frac{-(-r^2) \pm \sqrt{(-r^2)^2 - 4(2)(-r^2)}}{2(2)}$$

$$= \frac{r^2 \pm \sqrt{r^4 + 8r^2}}{4}$$

$$\therefore x = \frac{r^2 + \sqrt{r^4 + 8r^2}}{4} \text{ or } \frac{r^2 - \sqrt{r^4 + 8r^2}}{4}$$

Since x is a length,

$$x = \frac{r^2 + \sqrt{r^4 + 8r^2}}{4}$$

$$\left(\text{as } \frac{r^2 - \sqrt{r^4 + 8r^2}}{4} < 0\right)$$

To confirm that this value of x gives a maximum value of l, we need to sketch the curve of $\frac{dl}{dx}$, with roots

$x_1 = 1$ and

$$x_2 = \frac{r^2 - \sqrt{r^4 + 8r^2}}{4}$$

$$x_3 = \frac{r^2 + \sqrt{r^4 + 8r^2}}{4}$$

Now $x_2 < x_3$ and $x_2 < 0$
and $x_3 < 1$ (since $r < 1$)

Now the numerator of $\frac{dl}{dx}$ is a cubic equation and the denominator is always positive.

Hence as $x \to \infty$, $\frac{dl}{dx} \to +\infty$.

Hence the curve has the following shape:

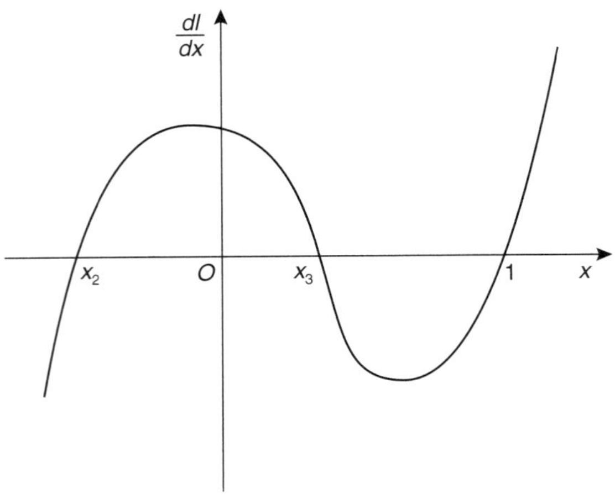

At $x_3 - \varepsilon$, $\dfrac{dl}{dx} > 0$

At $x_3 + \varepsilon$, $\dfrac{dl}{dx} < 0$

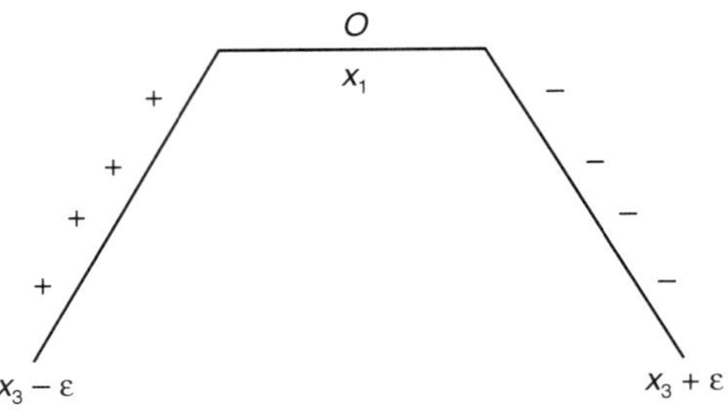

$\therefore$ At x_3, a maximum value of l occurs.

$\therefore$ The value of x for which M is closest to the floor is

$$x = \frac{r^2 + \sqrt{r^4 + 8r^2}}{4}$$

$$= \frac{r^2 + r\sqrt{r^2 + 8}}{4}$$

$$= r\left[\frac{r + \sqrt{r^2 + 8}}{4}\right].$$

BOARD OF STUDIES
NEW SOUTH WALES

2004

HIGHER SCHOOL CERTIFICATE EXAMINATION

Mathematics

General Instructions

- Reading time – 5 minutes
- Working time – 3 hours
- Write using black or blue pen
- Board-approved calculators may be used
- A table of standard integrals is provided at the back of this paper
- All necessary working should be shown in every question

Total marks – 120

- Attempt Questions 1–10
- All questions are of equal value

Total marks – 120
Attempt Questions 1–10
All questions are of equal value

Answer each question in a SEPARATE writing booklet. Extra writing booklets are available.

Marks

Question 1 (12 marks) Use a SEPARATE writing booklet.

(a) The radius of Mars is approximately 3 397 000 m. Write this number in scientific notation, correct to two significant figures. **2**

(b) Differentiate $x^4 + 5x^{-1}$ with respect to x. **2**

(c) Solve $\frac{x-5}{3} - \frac{x+1}{4} = 5$. **2**

(d) Find integers a and b such that $(3-\sqrt{2})^2 = a - b\sqrt{2}$. **2**

(e) A packet contains 12 red, 8 green, 7 yellow and 3 black jellybeans. One jellybean is selected from the packet at random. **2**

What is the probability that the selected jellybean is red or yellow?

(f) Find the values of x for which $|x+1| \leq 5$. **2**

Marks

Question 2 (12 marks) Use a SEPARATE writing booklet.

(a) The diagram shows the points $A(-1, 3)$ and $B(2, 0)$.
The line ℓ is drawn perpendicular to the x-axis through the point B.

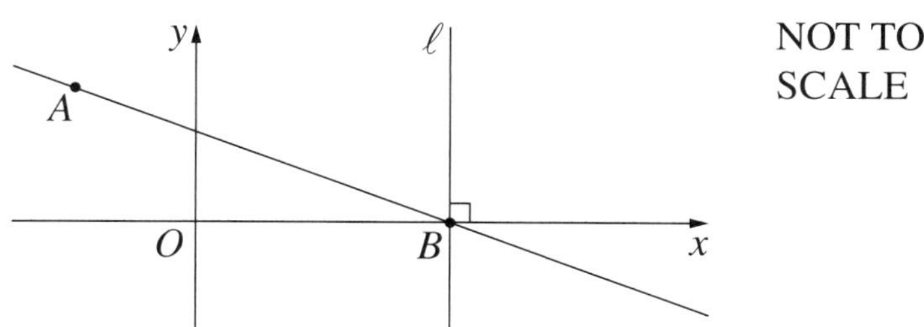

NOT TO SCALE

(i) Calculate the length of the interval AB. **1**

(ii) Find the gradient of the line AB. **1**

(iii) What is the size of the acute angle between the line AB and the line ℓ? **1**

(iv) Show that the equation of the line AB is $x + y - 2 = 0$. **1**

(v) Copy the diagram into your writing booklet and shade the region defined by $x + y - 2 \leq 0$. **1**

(vi) Write down the equation of the line ℓ. **1**

(vii) The point C is on the line ℓ such that AC is perpendicular to AB. Find the coordinates of C. **2**

(b) **2**

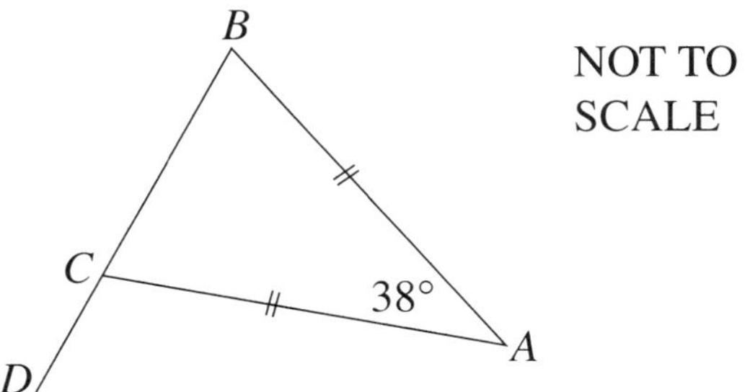

NOT TO SCALE

In the diagram, ABC is an isosceles triangle with $AB = AC$ and $\angle BAC = 38°$. The line BC is produced to D.

Copy or trace the diagram into your writing booklet.

Find the size of $\angle ACD$. Give reasons for your answer.

(c) For what values of k does $x^2 - kx + 4 = 0$ have no real roots? **2**

Marks

Question 3 (12 marks) Use a SEPARATE writing booklet.

(a) Differentiate with respect to x:

(i) $x^2 \log_e x$ **2**

(ii) $(1 + \sin x)^5$. **2**

(b) (i) Evaluate $\int_1^2 e^{3x}\,dx$. **2**

(ii) Find $\int \frac{x}{x^2 - 3}\,dx$. **2**

(c)

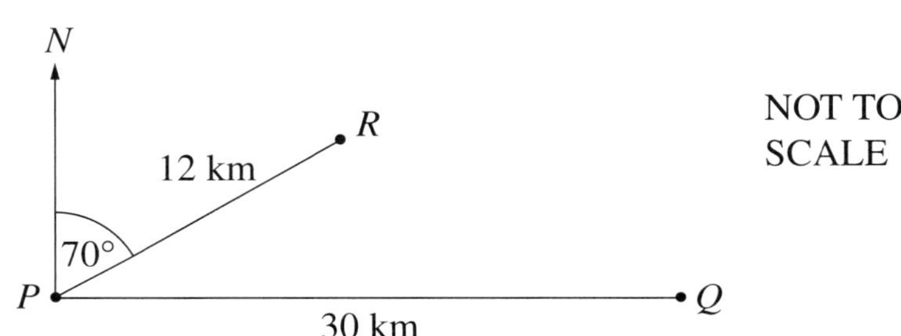

The diagram shows a point P which is 30 km due west of the point Q.
The point R is 12 km from P and has a bearing from P of 070°.

(i) Find the distance of R from Q. **2**

(ii) Find the bearing of R from Q. **2**

Marks

Question 4 (12 marks) Use a SEPARATE writing booklet.

(a) **2**

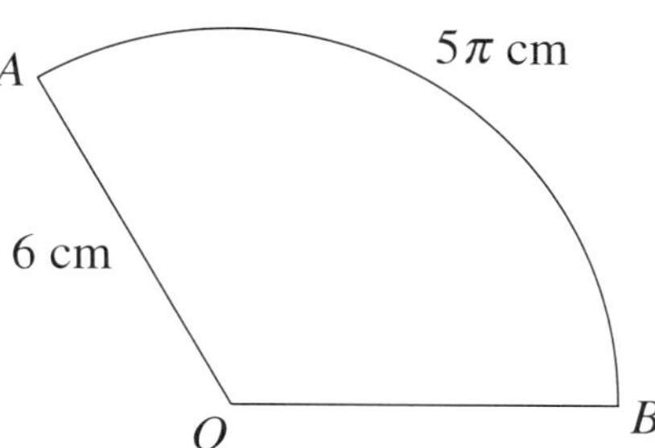

NOT TO SCALE

AOB is a sector of a circle, centre O and radius 6 cm.
The length of the arc AB is 5π cm.

Calculate the exact area of the sector AOB.

(b) Consider the function $f(x) = x^3 - 3x^2$.

(i) Find the coordinates of the stationary points of the curve $y = f(x)$ and determine their nature. **3**

(ii) Sketch the curve showing where it meets the axes. **2**

(iii) Find the values of x for which the curve $y = f(x)$ is concave up. **2**

(c) **3**

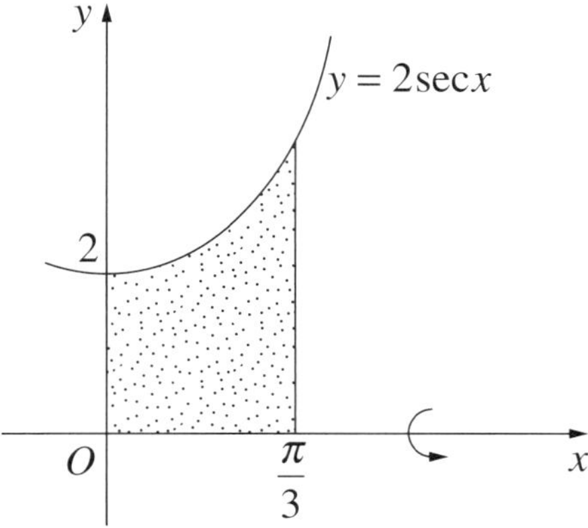

In the diagram, the shaded region is bounded by the curve $y = 2\sec x$, the coordinate axes and the line $x = \frac{\pi}{3}$. The shaded region is rotated about the x-axis.

Calculate the exact volume of the solid of revolution formed.

Marks

Question 5 (12 marks) Use a SEPARATE writing booklet.

(a) Clare is learning to drive. Her first lesson is 30 minutes long. Her second lesson is 35 minutes long. Each subsequent lesson is 5 minutes longer than the lesson before.

(i) How long will Clare's twenty-first lesson be? **1**

(ii) How many hours of lessons will Clare have completed after her twenty-first lesson? **2**

(iii) During which lesson will Clare have completed a total of 50 hours of driving lessons? **2**

(b) A particle moves along a straight line so that its displacement, x metres, from a fixed point O is given by $x = 1 + 3\cos 2t$, where t is measured in seconds.

(i) What is the initial displacement of the particle? **1**

(ii) Sketch the graph of x as a function of t for $0 \leq t \leq \pi$. **2**

(iii) Hence, or otherwise, find when AND where the particle first comes to rest after $t = 0$. **2**

(iv) Find a time when the particle reaches its maximum speed. What is this speed? **2**

Marks

Question 6 (12 marks) Use a SEPARATE writing booklet.

(a) Solve the following equation for x: **2**

$$e^{2x} + 3e^x - 10 = 0.$$

(b)

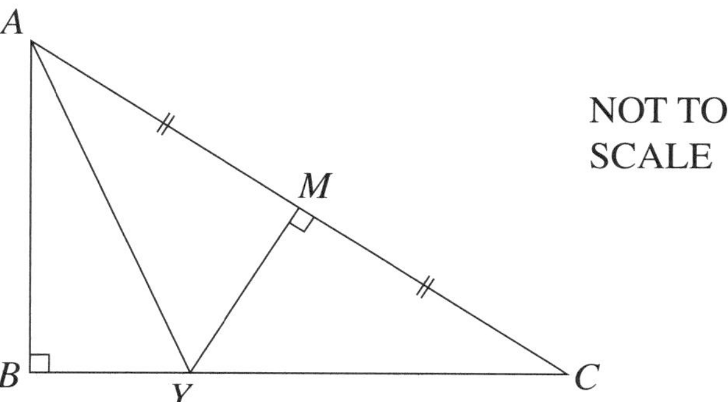

The diagram shows a right-angled triangle ABC with $\angle ABC = 90°$. The point M is the midpoint of AC, and Y is the point where the perpendicular to AC at M meets BC.

(i) Show that $\Delta AYM \equiv \Delta CYM$. **2**

(ii) Suppose that it is also given that AY bisects $\angle BAC$. Find the size of $\angle YCM$ and hence find the exact ratio $MY : AC$. **3**

(c) In a game, a turn involves rolling two dice, each with faces marked 0, 1, 2, 3, 4 and 5. The score for each turn is calculated by multiplying the two numbers uppermost on the dice.

(i) What is the probability of scoring zero on the first turn? **2**

(ii) What is the probability of scoring 16 or more on the first turn? **1**

(iii) What is the probability that the sum of the scores in the first two turns is less than 45? **2**

Marks

Question 7 (12 marks) Use a SEPARATE writing booklet.

(a) Evaluate $\sum_{n=2}^{4} n^2$. **1**

(b) At the beginning of 1991 Australia's population was 17 million. At the beginning of 2004 the population was 20 million.

Assume that the population P is increasing exponentially and satisfies an equation of the form $P = Ae^{kt}$, where A and k are constants, and t is measured in years from the beginning of 1991.

(i) Show that $P = Ae^{kt}$ satisfies $\frac{dP}{dt} = kP$. **1**

(ii) What is the value of A? **1**

(iii) Find the value of k. **2**

(iv) Predict the year during which Australia's population will reach 30 million. **2**

(c) Betty decides to set up a trust fund for her grandson, Luis. She invests \$80 at the beginning of each month. The money is invested at 6% per annum, compounded monthly.

The trust fund matures at the end of the month of her final investment, 25 years after her first investment. This means that Betty makes 300 monthly investments.

(i) After 25 years, what will be the value of the first \$80 invested? **2**

(ii) By writing a geometric series for the value of all Betty's investments, calculate the final value of Luis' trust fund. **3**

Marks

Question 8 (12 marks) Use a SEPARATE writing booklet.

(a) (i) Show that $\cos\theta\tan\theta = \sin\theta$. **1**

(ii) Hence solve $8\sin\theta\cos\theta\tan\theta = \operatorname{cosec}\theta$ for $0 \le \theta \le 2\pi$. **2**

(b) The diagram shows the graph of the parabola $x^2 = 16y$. The points $A\,(4, 1)$ and $B\,(-8, 4)$ are on the parabola, and C is the point where the tangent to the parabola at A intersects the directrix.

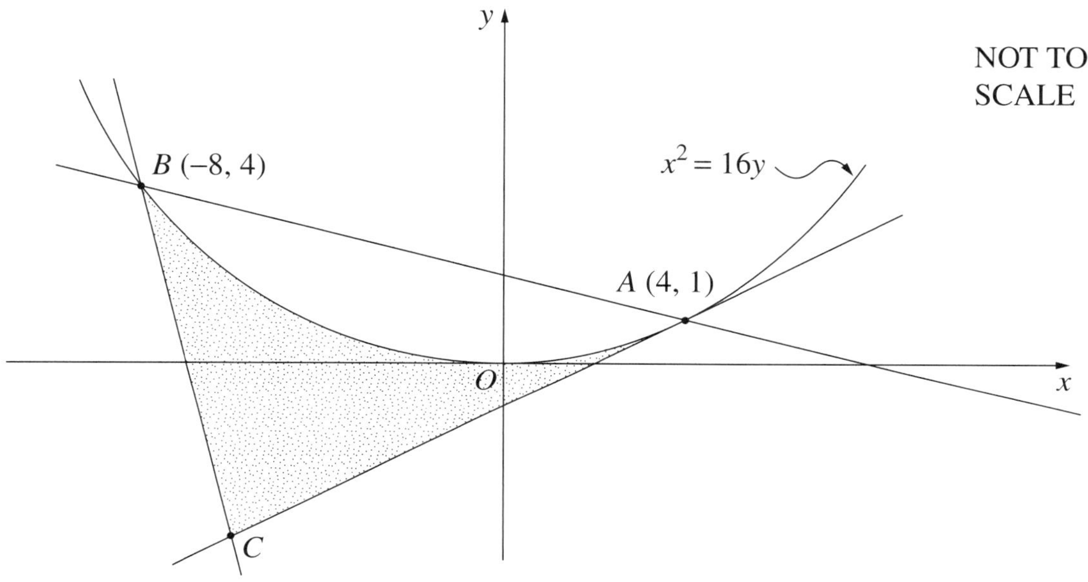

(i) Write down the equation of the directrix of the parabola $x^2 = 16y$. **1**

(ii) Find the equation of the tangent to the parabola at the point A. **2**

(iii) Show that C is the point $(-6, -4)$. **1**

(iv) Given that the equation of the line AB is $y = 2 - \frac{x}{4}$, find the area bounded by the line AB and the parabola. **2**

(v) Hence, or otherwise, find the shaded area bounded by the parabola, the tangent at A and the line BC. **3**

Marks

Question 9 (12 marks) Use a SEPARATE writing booklet.

(a) Consider the geometric series $1 - \tan^2\theta + \tan^4\theta - \ldots$

(i) When the limiting sum exists, find its value in simplest form. **2**

(ii) For what values of θ in the interval $-\frac{\pi}{2} < \theta < \frac{\pi}{2}$ does the limiting sum of the series exist? **2**

(b) A particle moves along the x-axis. Initially it is at rest at the origin. The graph shows the acceleration, a, of the particle as a function of time t for $0 \le t \le 5$.

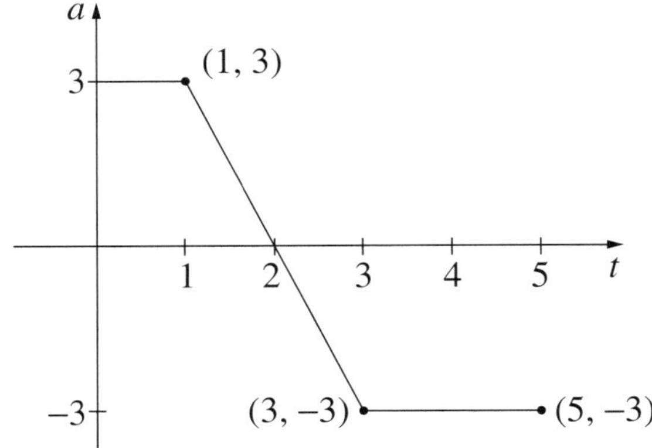

(i) Write down the time at which the velocity of the particle is a maximum. **1**

(ii) At what time during the interval $0 \le t \le 5$ is the particle furthest from the origin? Give brief reasons for your answer. **2**

(c) Consider the function $f(x) = \dfrac{\log_e x}{x}$, for $x > 0$.

(i) Show that the graph of $y = f(x)$ has a stationary point at $x = e$. **2**

(ii) By considering the gradient on either side of $x = e$, or otherwise, show that the stationary point is a maximum. **1**

(iii) Use the fact that the maximum value of $f(x)$ occurs at $x = e$ to deduce that $e^x \ge x^e$ for all $x > 0$. **2**

Marks

Question 10 (12 marks) Use a SEPARATE writing booklet.

(a) (i) Use Simpson's rule with 3 function values to find an approximation to the area under the curve $y = \frac{1}{x}$ between $x = a$ and $x = 3a$, where a is positive. **2**

(ii) Using the result in part (i), show that **1**

$$\ln 3 \doteqdot \frac{10}{9}.$$

(b)

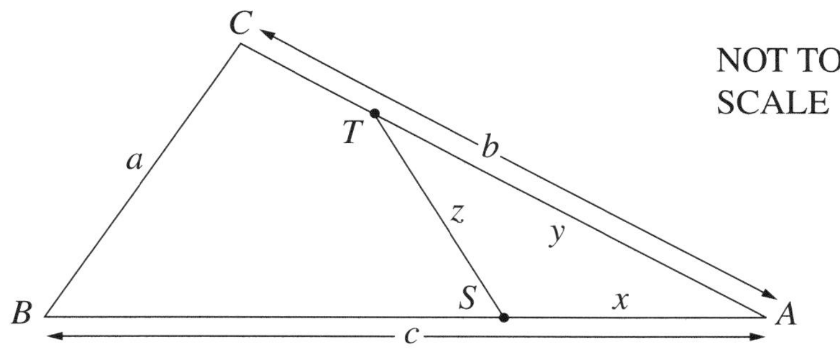

The diagram shows a triangular piece of land ABC with dimensions $AB = c$ metres, $AC = b$ metres and $BC = a$ metres, where $a \le b \le c$.

The owner of the land wants to build a straight fence to divide the land into two pieces of equal area. Let S and T be points on AB and AC respectively so that ST divides the land into two pieces of equal area.

Let $AS = x$ metres, $AT = y$ metres and $ST = z$ metres.

(i) Show that $xy = \frac{1}{2}bc$. **1**

(ii) Use the cosine rule in triangle AST to show that **2**

$$z^2 = x^2 + \frac{b^2c^2}{4x^2} - bc\cos A.$$

(iii) Show that the value of z^2 in the equation in part (ii) is a minimum when $x = \sqrt{\frac{bc}{2}}$. **4**

(iv) Show that the minimum length of the fence is $\sqrt{\frac{(P-2b)(P-2c)}{2}}$ metres, where $P = a + b + c$. **2**

(You may assume that the value of x given in part (iii) is feasible.)

End of paper

2004 Higher School Certificate
Worked Answers

QUESTION 1

(a) $3\ 397\ 000 = 3.397 \times 10^6$
$= 3.4 \times 10^6$ to 2 sig. figures.

(b) $\frac{d}{dx}(x^4 + 5x^{-1}) = 4x^3 - 5x^{-2}$.

(c)
$$\frac{x-5}{3} - \frac{x+1}{4} = 5$$
$$\therefore 12\left(\frac{x-5}{3}\right) - 12\left(\frac{x+1}{4}\right) = 5 \times 12$$
$$4(x-5) - 3(x+1) = 60$$
$$4x - 20 - 3x - 3 = 60$$
$$x - 23 = 60$$
$$\therefore x = 83.$$

(d) $(3 - \sqrt{2})^2 = a - b\sqrt{2}$

Now $(3 - \sqrt{2})^2 = 9 - 6\sqrt{2} + 2$
$= 11 - 6\sqrt{2}$
$= a - b\sqrt{2}$
where $a = 11$ and $b = 6$.

(e) P(red or yellow)
= P(red) + P(yellow)
$= \frac{12}{30} + \frac{7}{30}$
$= \frac{19}{30}$.

(f) $|x + 1| \leqslant 5$
$\therefore -5 \leqslant x + 1 \leqslant 5$
$\therefore -6 \leqslant x \leqslant 4$.

QUESTION 2

(a) (i) $AB = \sqrt{(-1-2)^2 + (3-0)^2}$
$= \sqrt{(-3)^2 + (3)^2}$
$= \sqrt{9+9}$
$= \sqrt{18}$
$= 3\sqrt{2}$.

(ii) Gradient of AB
$= \frac{3-0}{-1-2}$
$= \frac{3}{-3}$
$= -1$.

(iii) $\tan \angle ABO = -$ gradient of AB
$= 1$
$\therefore \angle ABO = 45°$

$\therefore$ Acute angle between the line AB and the line $l = 90° - 45° = 45°$.

(iv) Equation of the line AB with gradient $= -1$ through the point $(2, 0)$ is given by
$y - 0 = -1(x - 2)$
$y = -x + 2$
i.e. $x + y - 2 = 0$.

(v)

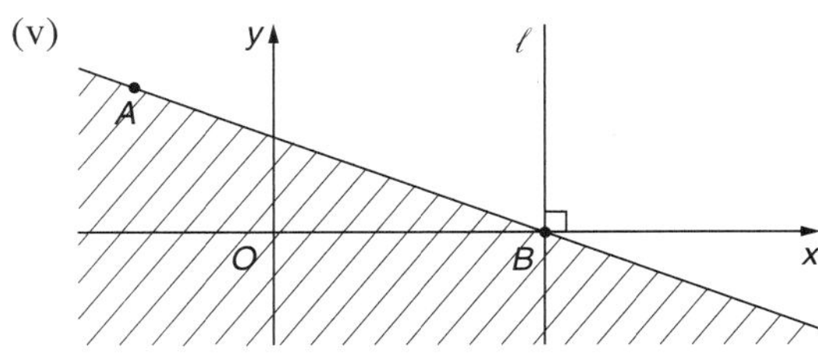

Test $(0, 0)$ in $x + y - 2 \leqslant 0$
$\therefore 0 + 0 - 2 \leqslant 0$
which is true.

$\therefore$ Shade the side of the line which contains $(0, 0)$.

(vi) $x = 2$

(vii) Gradient of $AB = -1$
$\therefore$ Gradient of $AC = 1$
(since $AB \perp AC$)

$\therefore$ Equation of AC with gradient = 1, passing through $A(-1, 3)$ is given by
$y - 3 = 1[x - (-1)]$
$y - 3 = x + 1$
$y = x + 4$

At C, $x = 2$
$\therefore y = 2 + 4$
$= 6$

$\therefore$ Coordinates of C are $(2, 6)$.

(b)

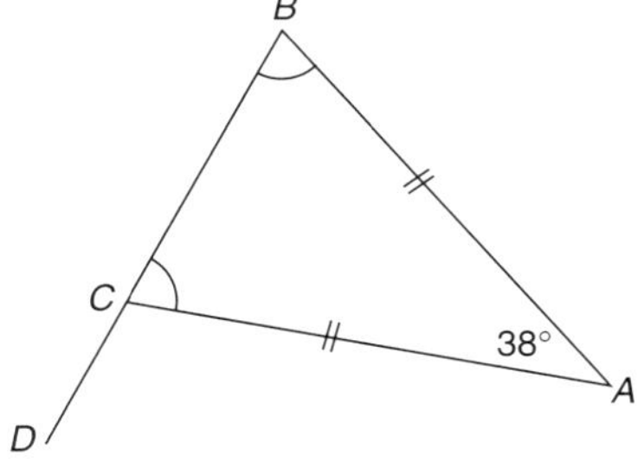

Now $\angle ABC = \angle ACB$ ($\triangle ABC$ is isosceles)

$\therefore \angle ACB = \dfrac{180° - 38°}{2}$ (angle sum $\triangle ABC = 180°$)

$= 71°$

$\angle ACD = 180° - 71°$ (BCD is a straight angle, 180°)

$= 109°$.

Alternatively:

$\angle ACD = \angle ABC + \angle BAC$ (exterior angle of triangle)

$= 71° + 38°$

$= 109°$.

(c) For $x^2 - kx + 4 = 0$,
no real roots occur when

$\Delta < 0$

$b^2 - 4ac < 0$

$(-k)^2 - 4(1)(4) < 0$

$k^2 - 16 < 0$

$(k + 4)(k - 4) < 0$

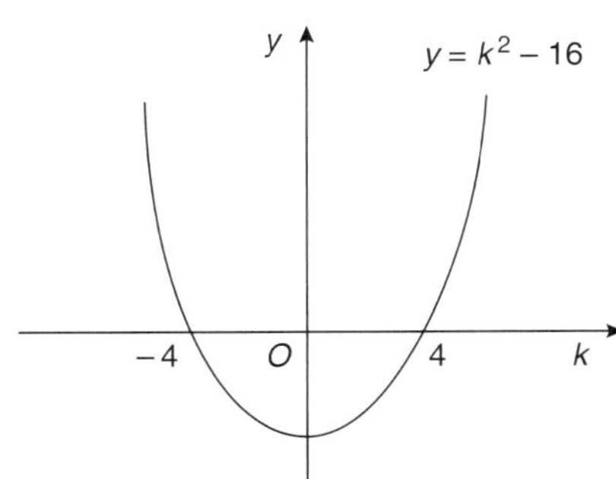

$\therefore -4 < k < 4$.

QUESTION 3

(a) (i) $\dfrac{d}{dx}(x^2 \log_e x)$

$= x^2 \cdot \dfrac{d}{dx}(\log_e x) + \log_e x \cdot \dfrac{d}{dx}(x^2)$

$= x^2 \cdot \dfrac{1}{x} + \log_e x \cdot 2x$

$= x + 2x \log_e x$

$= x(1 + 2 \log_e x)$.

(ii) $\dfrac{d}{dx}[(1 + \sin x)^5]$

$= 5(1 + \sin x)^4 \cdot \dfrac{d}{dx}(1 + \sin x)$

$= 5(1 + \sin x)^4 \cdot \cos x$

$= 5 \cos x (1 + \sin x)^4$.

(b) (i) $\displaystyle\int_1^2 e^{3x}\,dx = \left[\frac{1}{3}e^{3x}\right]_1^2$

$= \dfrac{1}{3}e^6 - \dfrac{1}{3}e^3$

$= \dfrac{1}{3}(e^6 - e^3)$.

(ii) $\displaystyle\int \frac{x}{x^2 - 3}\,dx = \frac{1}{2}\int \frac{2x}{x^2 - 3}\,dx$

$= \dfrac{1}{2}\ln(x^2 - 3) + C$.

(c)

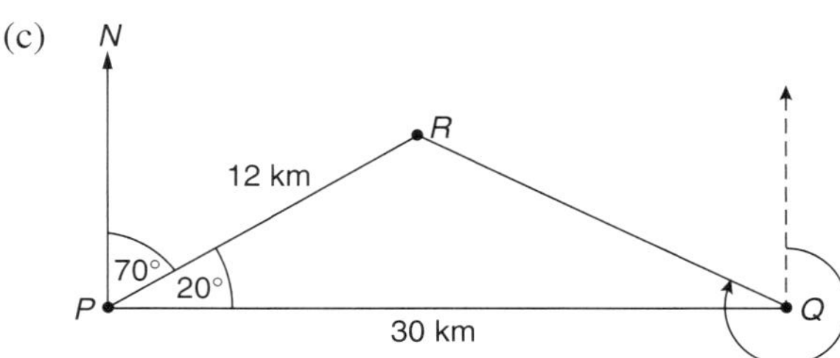

(i) By the cosine rule in $\triangle PRQ$:

$RQ^2 = PR^2 + PQ^2 - 2 \cdot PR \cdot PQ \cdot \cos \angle RPQ$

$= 12^2 + 30^2 - 2(12)(30)\cos 20°$

$= 367.4213\ldots$ (by calc.)

$\therefore RQ = 19.1682\ldots$ (by calc.)

$\doteqdot 19.2$ km to 1 decimal place.

(ii) By the sine rule in $\triangle PRQ$:

$\dfrac{\sin \angle RQP}{12} = \dfrac{\sin 20°}{RQ}$

$\sin \angle RQP = \dfrac{12 \sin 20°}{19.1682\ldots}$

$= 0.2141\ldots$ (by calc.)

$\therefore \angle RQP = 12.3637\ldots°$ (by calc.)

$\therefore$ The bearing of R from Q

$= 270° + \angle RQP$

$= 270° + 12.3637\ldots°$

$= 282.3637\ldots°$ (by calc.)

$= 282°\ 22'$ (by calc.)

$= 282°$ to the nearest degree.

QUESTION 4

(a) $l = r\theta$

$\therefore\ 5\pi = 6 \times \theta$

$\therefore\ \theta = \dfrac{5\pi}{6}$

Area of sector *AOB*

$= \dfrac{1}{2}r^2\theta$

$= \dfrac{1}{2} \times (6)^2 \times \dfrac{5\pi}{6}$

$= 15\pi\ \text{cm}^2$.

(b) (i) $f(x) = x^3 - 3x^2$

$f'(x) = 3x^2 - 6x$
$\quad = 3x(x-2)$

Stationary points occur when

$f'(x) = 0$
$3x(x-2) = 0$
$\therefore\ x = 0$ or $x = 2$

Now $f(0) = 0^3 - 3(0)^2$
$\quad = 0$
and $f(2) = 2^3 - 3(2)^2$
$\quad = 8 - 12$
$\quad = -4$

$\therefore$ Coordinates of the stationary points are $(0, 0)$ and $(2, -4)$.

Now $f''(x) = 6x - 6$

$\therefore\ f''(0) = 6(0) - 6$
$\quad = -6 < 0$

$\therefore$ A maximum turning point at $(0, 0)$

Now $f''(2) = 6(2) - 6$
$\quad = 6 > 0$

$\therefore$ A minimum turning point at $(2, -4)$.

(ii) When $y = f(x) = 0$
$x^3 - 3x^2 = 0$
$x^2(x-3) = 0$
$\therefore\ x = 0$ or 3.

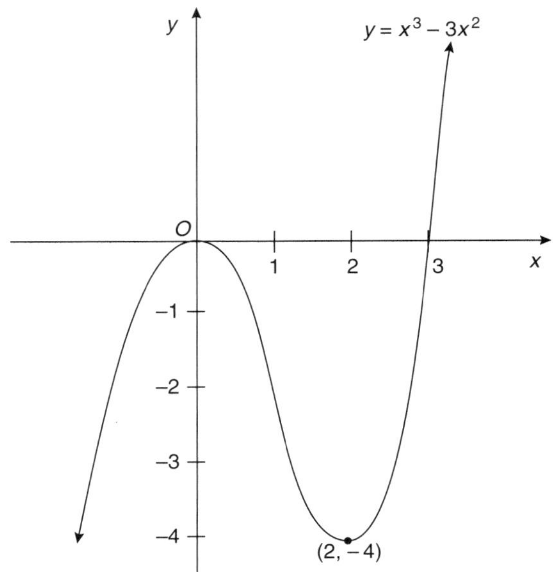

(iii) $f''(x) = 6x - 6$

The curve is concave up when

$f''(x) > 0$
$6x - 6 > 0$
$6x > 6$
$x > 1$.

(c) Now $y = 2\sec x$

$\therefore\ y^2 = 4\sec^2 x$

$$V = \pi\int_0^{\frac{\pi}{3}} y^2\,dx$$

$$= \pi\int_0^{\frac{\pi}{3}} 4\sec^2 x\,dx$$

$$= 4\pi\int_0^{\frac{\pi}{3}} \sec^2 x\,dx$$

$$= 4\pi\Big[\tan x\Big]_0^{\frac{\pi}{3}}$$

$$= 4\pi\left(\tan\frac{\pi}{3} - \tan 0\right)$$

$$= 4\pi(\sqrt{3} - 0)$$

$$= 4\sqrt{3}\,\pi\ \text{units}^3.$$

QUESTION 5

(a) (i) The length of each lesson is in an arithmetic sequence:

$30, 35, 40 \ldots$
with $a = 30$ and $d = 5$.

$T_n = a + (n-1)\,d$

$T_{21} = 30 + 20(5)$
$= 30 + 100$
$= 130$

$\therefore$ Clare's twenty-first lesson will be 130 minutes long.

(ii) $S_n = \frac{n}{2}[2a + (n-1)d]$

$n = 21,\ a = 30,\ d = 5$

$\therefore S_{21} = \frac{21}{2}[2(30) + 20 \times 5]$
$= \frac{21}{2}[60 + 100]$
$= \frac{21}{2} \times 160$
$= 1680$ minutes
$= 28$ hours.

(iii) $S_n = \frac{n}{2}[2a + (n-1)d]$

A total of 50 hours $= 50 \times 60$
$= 3000$ minutes.

$\therefore a = 30,\ d = 5,\ S_n = 3000$

$\therefore 3000 = \frac{n}{2}[2(30) + (n-1)5]$
$6000 = n[60 + 5n - 5]$
$6000 = n[55 + 5n]$
$6000 = 55n + 5n^2$

$\therefore 5n^2 + 55n - 6000 = 0$
$n^2 + 11n - 1200 = 0$

$\therefore n = \frac{-11 \pm \sqrt{(11)^2 - 4(1)(-1200)}}{2(1)}$
$= \frac{-11 \pm \sqrt{121 + 4800}}{2}$
$= \frac{-11 \pm \sqrt{4921}}{2}$
$\therefore\ n = 29.5749\ldots$ or $-40.5749\ldots$ (by calc.)
$\therefore\ n \doteqdot 29.5749$ (as $n > 0$)

$\therefore$ Clare completes 50 hours of lessons during her 30th lesson.

(b) (i) $x = 1 + 3\cos 2t$

When $t = 0$,
$x = 1 + 3\cos 0$
$= 1 + 3$
$= 4$

$\therefore$ Particle is initially 4 m to the right of O.

(ii) Period $= \frac{2\pi}{2} = \pi$

Range: $-2 \leqslant x \leqslant 4$

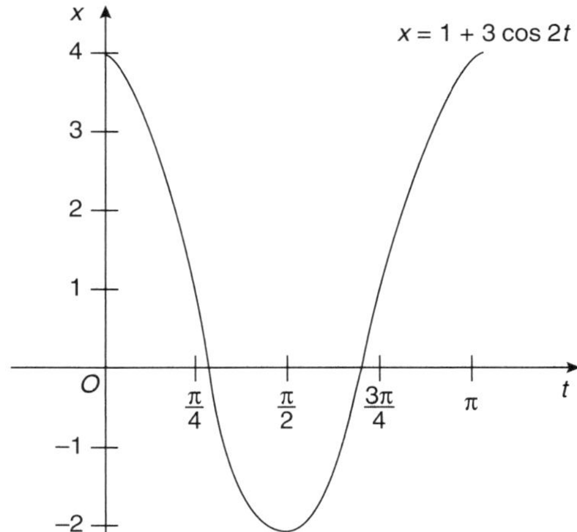

(iii) $x = 1 + 3\cos 2t$
$\therefore v = -6\sin 2t$

The particle comes to rest when
$v = 0$
i.e. $-6\sin 2t = 0$
$\sin 2t = 0$
$2t = 0, \pi, 2\pi \ldots$
$t = 0, \frac{\pi}{2}, \pi \ldots$

$\therefore$ Particle first comes to rest when $t = \frac{\pi}{2}$ seconds.

When $t = \frac{\pi}{2}$, $x = 1 + 3\cos 2\left(\frac{\pi}{2}\right)$
$= 1 + 3\cos\pi$
$= 1 + 3(-1)$
$= 1 - 3$
$= -2$

$\therefore$ Particle first comes to rest at 2 m to the left of O.

OR
The particle is at rest at the turning points of the curve when $\frac{dx}{dt} = 0$.

i.e. at $t = 0, \frac{\pi}{2}, \pi \ldots$

$\therefore$ From the graph, the particle first comes to rest when $t = \frac{\pi}{2}$ seconds, 2 m to the left of O.

(iv) $x = 1 + 3\cos 2t$

$v = \dfrac{dx}{dt} = -6\sin 2t$

$a = \dfrac{dv}{dt} = -12\cos 2t$

The particle reaches its maximum speed when

$$\frac{dv}{dt} = 0$$

$$-12\cos 2t = 0$$
$$\cos 2t = 0$$
$$2t = \frac{\pi}{2}, \frac{3\pi}{2}, \ldots$$
$$t = \frac{\pi}{4}, \frac{3\pi}{4}, \ldots$$

$\therefore$ Maximum speed is reached at $\frac{\pi}{4}$ seconds, $\frac{3\pi}{4}$ seconds, etc.

When $t = \frac{\pi}{4}$ seconds,

$$v = -6\sin 2\left(\frac{\pi}{4}\right)$$
$$= -6\sin\left(\frac{\pi}{2}\right)$$
$$= -6 \times 1$$
$$= -6$$

$\therefore$ Maximum speed is 6 m s^{-1}.

QUESTION 6

(a) $e^{2x} + 3e^x - 10 = 0$

$\therefore (e^x)^2 + 3e^x - 10 = 0$

Let $m = e^x$,

$\therefore m^2 + 3m - 10 = 0$

$(m+5)(m-2) = 0$

$\therefore m + 5 = 0$ or $m - 2 = 0$

$m = -5$ or $m = 2$

$\therefore e^x = -5$ or $\therefore e^x = 2$

no solution ($e^x > 0$) $\qquad \ln e^x = \ln 2$, $\therefore x = \ln 2$

$\therefore x = \ln 2$.

(b)

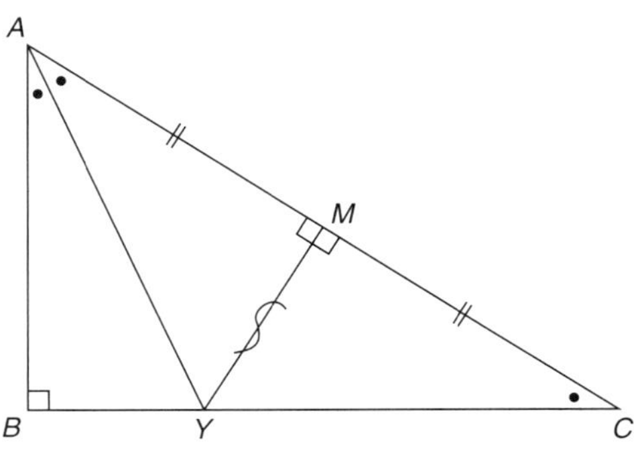

(i) In $\triangle AYM$ and $\triangle CYM$,

YM is common

$AM = CM$ (given)

$\angle AMY = \angle CMY$ ($AC \perp YM$)

$\therefore \triangle AYM \equiv \triangle CYM$ (S.A.S.)

(ii) Now $\angle YAB = \angle YAM$ (AY bisects $\angle BAC$)

and $\angle YAM = \angle YCM$ (from (i), corresponding angles in congruent triangles)

$\therefore \angle YAB = \angle YAM = \angle YCM$

In $\triangle ABC$,

$$\angle ABC + \angle BAC + \angle YCM = 180^\circ$$
$$\therefore 90^\circ + \angle YAB + \angle YAM + \angle YCM = 180^\circ$$
$$\therefore 90^\circ + 3 \times \angle YCM = 180^\circ$$
$$\therefore 3 \times \angle YCM = 90^\circ$$
$$\therefore \angle YCM = 30^\circ$$

Let $\angle YCM = \theta = 30^\circ$

$$\therefore \tan\theta = \frac{MY}{MC}$$

Since $\tan 30^\circ = \dfrac{1}{\sqrt{3}}$

$$\therefore \frac{MY}{MC} = \frac{1}{\sqrt{3}}$$

but $AC = 2MC$ (given)

$$\therefore \frac{MY}{AC} = \frac{MY}{2MC} = \frac{1}{2\sqrt{3}}$$

$\therefore MY : AC = 1 : 2\sqrt{3}$.

(c)

		Die 1					
		0	1	2	3	4	5
	0	0	0	0	0	0	0
	1	0	1	2	3	4	5
Die	2	0	2	4	6	8	10
2	3	0	3	6	9	12	15
	4	0	4	8	12	16	20
	5	0	5	10	15	20	25

(i) $P(0) = \dfrac{11}{36}$

(ii) $P(\geqslant 16) = \frac{4}{36} = \frac{1}{9}$.

(iii) $P(\text{Sum} < 45) = 1 - P(\text{Sum} \geqslant 45)$

A sum of scores $\geqslant 45$ can be obtained in the following ways:

- A score of 20 followed by a score of 25
$$P(20, 25) = \frac{2}{36} \times \frac{1}{36} = \frac{2}{1296}$$
- A score of 25 followed by a score of 20
$$P(25, 20) = \frac{1}{36} \times \frac{2}{36} = \frac{2}{1296}$$
- A score of 25 followed by a score of 25
$$P(25, 25) = \frac{1}{36} \times \frac{1}{36} = \frac{1}{1296}$$

$$\therefore P(\text{Sum} \geqslant 45) = \frac{2}{1296} + \frac{2}{1296} + \frac{1}{1296} = \frac{5}{1296}$$

$$\therefore P(\text{Sum} < 45) = 1 - \frac{5}{1296} = \frac{1291}{1296}.$$

QUESTION 7

(a) $\sum_{n=2}^{4} n^2 = 2^2 + 3^2 + 4^2$
$= 4 + 9 + 16$
$= 29$.

(b) (i) $P = Ae^{kt}$
$$\frac{dP}{dt} = kAe^{kt} = kP$$

(ii) $P = Ae^{kt}$

At $t = 0$, $P = 17\ 000\ 000 = 1.7 \times 10^7$

$\therefore 1.7 \times 10^7 = Ae^0$
$1.7 \times 10^7 = A(1)$
$\therefore A = 1.7 \times 10^7$.

(iii) $P = 1.7 \times 10^7 e^{kt}$

At $t = 13$, $P = 20\ 000\ 000 = 2 \times 10^7$

$\therefore 2 \times 10^7 = 1.7 \times 10^7 e^{13k}$

$$\frac{2 \times 10^7}{1.7 \times 10^7} = e^{13k}$$
$$\ln\left(\frac{2}{1.7}\right) = \ln e^{13k} = 13k$$
$$\therefore k = \frac{1}{13} \ln\left(\frac{2}{1.7}\right)$$
$\doteqdot 0.0125\ldots$ (by calc.)
$= 0.013$ to 3 decimal places.

(iv) $P = 1.7 \times 10^7 e^{kt}$
(where $k = 0.0125\ldots$)

If $P = 30\ 000\ 000 = 3 \times 10^7$,

$$3 \times 10^7 = 1.7 \times 10^7 e^{kt}$$
$$\frac{3 \times 10^7}{1.7 \times 10^7} = e^{kt}$$
$$\ln\left(\frac{3}{1.7}\right) = \ln e^{kt}$$
$$\ln\left(\frac{3}{1.7}\right) = kt \ln e = kt$$
$$\therefore t = \frac{\ln\left(\frac{3}{1.7}\right)}{k} = \frac{\ln\left(\frac{3}{1.7}\right)}{0.0125\ldots}$$
(using calc. memory for k)
$= 45.433\ldots$ (by calc.)
$\doteqdot 45.4$ years

$\therefore$ In the 46th year, the population will reach 30 million.

(c) (i) $A = P\left(1 + \frac{r}{100}\right)^n$

$P = \$80$
$n = 300$ months
$r = 6\%$ per annum
$= \frac{6}{12}\%$ per month
$= 0.5\%$ per month

$$\therefore A = 80\left(1 + \frac{0.5}{100}\right)^{300}$$
$$= 80(1 + 0.005)^{300}$$
$$= 80(1.005)^{300}$$
$= 357.1975\ldots$ (by calc.)
$= \$357.20$ to the nearest cent.

(ii) $A_1 = 80(1.005)^{300}$
$A_2 = 80(1.005)^{299}$
$A_3 = 80(1.005)^{298}$
$\vdots$
$A_{300} = 80(1.005)^1$

The final value of the trust fund
$= 80(1.005)^{300} + 80(1.005)^{299} + \ldots + 80(1.005)^1$
$= 80[1.005^{300} + 1.005^{299} + \ldots + 1.005^1]$
$= 80[1.005^1 + 1.005^2 + \ldots + 1.005^{300}]$
(on reversing the order of terms)
$$= 80\left[\frac{a(r^n - 1)}{r - 1}\right]$$
where $a = 1.005$, $r = 1.005$, $n = 300$
$$= 80\left[\frac{1.005(1.005^{300} - 1)}{1.005 - 1}\right]$$
$= 55\,716.714\ldots$ (by calc.)
$= \$55\,716.71$ to the nearest cent.

(b) (i) $x^2 = 16y$
If $x^2 = 4ay$
$\therefore 4a = 16$
$a = 4$

$\therefore$ Equation of the directrix is $y = -4$.

(ii) $x^2 = 16y$
$$\therefore y = \frac{x^2}{16}$$
$$\frac{dy}{dx} = \frac{2x}{16} = \frac{x}{8} = \frac{4}{8} \text{ at } x = 4 = \frac{1}{2}$$

$\therefore$ Equation of the tangent at $A(4, 1)$ with $m = \frac{1}{2}$ is
$$y - 1 = \frac{1}{2}(x - 4)$$
$$y - 1 = \frac{x}{2} - 2$$
$$y = \frac{x}{2} - 1.$$

(iii) C lies on the directrix, $y = -4$
$\therefore$ The y-coordinate of C is -4.

Now $y = \frac{x}{2} - 1$
$$\therefore -4 = \frac{x}{2} - 1$$
$$-3 = \frac{x}{2}$$
$$x = -6$$

$\therefore$ C is the point $(-6, -4)$.

(iv) Now $x^2 = 16y$
$$\therefore y = \frac{x^2}{16}$$

Required area
$$= \int_{-8}^{4}\left(2 - \frac{x}{4}\right)dx - \int_{-8}^{4}\frac{x^2}{16}\,dx$$
$$= \int_{-8}^{4}\left(2 - \frac{x}{4} - \frac{x^2}{16}\right)dx$$
$$= \left[2x - \frac{x^2}{8} - \frac{x^3}{48}\right]_{-8}^{4}$$

QUESTION 8

(a) (i) $\cos\theta\tan\theta = \sin\theta$

$$\text{LHS} = \cos\theta\tan\theta = \cos\theta\left(\frac{\sin\theta}{\cos\theta}\right) = \sin\theta = \text{RHS}$$

$\therefore \cos\theta\tan\theta = \sin\theta.$

(ii) $8\sin\theta\cos\theta\tan\theta = \operatorname{cosec}\theta \quad 0 \leqslant \theta \leqslant 2\pi$
$\therefore 8\sin\theta(\sin\theta) = \operatorname{cosec}\theta$ (from (i))
$$8\sin^2\theta = \frac{1}{\sin\theta}$$
$$8\sin^3\theta = 1$$
$$\sin^3\theta = \frac{1}{8}$$
$$\sin\theta = \frac{1}{2}$$
$$\therefore \theta = \frac{\pi}{6}, \frac{5\pi}{6}.$$

$= \left(2(4) - \frac{4^2}{8} - \frac{4^3}{48}\right) - \left(2(-8) - \frac{(-8)^2}{8} - \frac{(-8)^3}{48}\right)$
$= \left(8 - \frac{16}{8} - \frac{64}{48}\right) - \left(-16 - \frac{64}{8} - \left(\frac{-512}{48}\right)\right)$
$= \left(6 - \frac{4}{3}\right) - \left(-16 - 8 + \frac{32}{3}\right)$
$= 18$ (by calc.)
$\therefore$ Required area = 18 units^2.

(v) Shaded area = area of $\triangle ABC$ – area in (iv)

Area of $\triangle ABC = b \times h$
where h is the perpendicular distance of $C(-6, -4)$ from line AB, $y = 2 - \frac{x}{4}$

i.e. $4y = 8 - x$
i.e. $x + 4y - 8 = 0$

$$\therefore h = \left| \frac{Ax_1 + By_1 + C}{\sqrt{A^2 + B^2}} \right|$$
$$= \left| \frac{1(-6) + 4(-4) + (-8)}{\sqrt{(1)^2 + (4)^2}} \right|$$
$$= \left| \frac{-6 - 16 - 8}{\sqrt{1 + 16}} \right|$$
$$= \left| \frac{-30}{\sqrt{17}} \right|$$
$$= \frac{30}{\sqrt{17}}$$

$$AB = \sqrt{(-8-4)^2 + (4-1)^2}$$
$$= \sqrt{144 + 9}$$
$$= \sqrt{153}$$
$$= 3\sqrt{17}$$

$\therefore$ Area of $\triangle ABC$
$$= \frac{1}{2} \times b \times h$$
$$= \frac{1}{2} \times 3\sqrt{17} \times \frac{30}{\sqrt{17}}$$
$$= 45$$

$\therefore$ Area of $\triangle ABC = 45$ units^2.

$\therefore$ Shaded area = area of $\triangle ABC$ – area in (iv)
= 45 – 18
= 27 units^2.

QUESTION 9

(a) (i) $1 - \tan^2\theta + \tan^4\theta - \ldots$

$$S = \frac{a}{1-r}$$

Now $r = \frac{-\tan^2\theta}{1}$
$= -\tan^2\theta$

$$\therefore S = \frac{1}{1 - (-\tan^2\theta)}$$
$$= \frac{1}{1 + \tan^2\theta}$$
$$= \frac{1}{\sec^2\theta}$$
$$= \cos^2\theta.$$

(ii) For the limiting sum to exist,
$|r| < 1$
i.e. $|-\tan^2\theta| < 1$
$\therefore \tan^2\theta < 1$
$\therefore -1 < \tan\theta < 1$
$\therefore -\frac{\pi}{4} < \theta < \frac{\pi}{4}$.

(b) (i) The velocity of the particle is a maximum when $t = 2$.

Note: When $t = 2, a = 0$,
when $t < 2, a > 0$
and when $t > 2, a < 0$

$\therefore$ There is a maximum velocity at $t = 2$.

(ii) From $t = 0$ to $t = 2$, the acceleration of the particle is positive and velocity reaches a maximum at $t = 2$.

From $t = 2$ to $t = 4$, the acceleration is of the same magnitude as the period $t = 0$ to $t = 2$, but opposite in direction (i.e. negative).

Hence, since the velocity was zero at $t = 0$, the velocity has returned to zero by $t = 4$.

From $t = 4$, the acceleration of the particle continues to be negative and the velocity becomes negative.

$\therefore$ The particle is furthest from the origin at $t = 4$.

(c) (i) $f(x) = \frac{\log_e x}{x}$, for $x > 0$

$$f'(x) = \frac{x \cdot \frac{d}{dx}(\log_e x) - \log_e x \cdot \frac{d}{dx}(x)}{x^2}$$

$$= \frac{x\left(\frac{1}{x}\right) - \log_e x \cdot 1}{x^2}$$

$$= \frac{1 - \log_e x}{x^2}$$

Stationary points occur when

$$f'(x) = 0$$

$$\therefore \frac{1 - \log_e x}{x^2} = 0$$

$$1 - \log_e x = 0$$

$$\log_e x = 1$$

$$\therefore x = e.$$

$\therefore$ There is a stationary point at $x = e$.

(ii) Consider the graph of $y = \log_e x$:

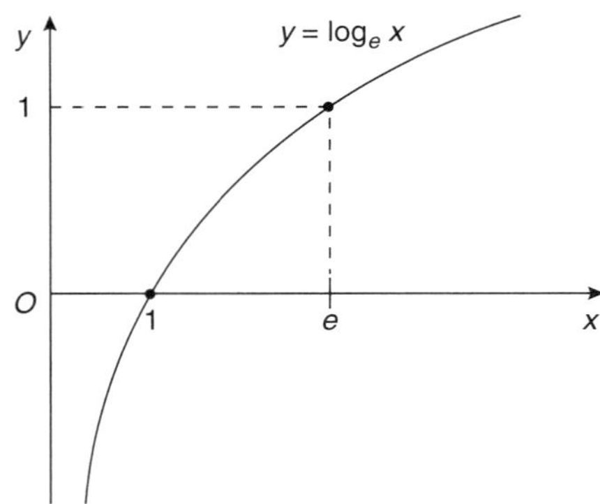

When $x < e$, $\log_e x < 1$

$$\therefore 1 - \log_e x > 0$$

$$\therefore \frac{1 - \log_e x}{x^2} > 0$$

i.e. $f'(x) > 0$

When $x > e$, $\log_e x > 1$

$$\therefore 1 - \log_e x < 0$$

$$\therefore \frac{1 - \log_e x}{x^2} < 0$$

i.e. $f'(x) < 0$

$\therefore$ The stationary point at $x = e$ is a maximum.

(iii) $f(x) = \dfrac{\log_e x}{x}$, for $x > 0$

$$\therefore f(e) = \frac{\log_e e}{e}$$

$$= \frac{1}{e}$$

Since the maximum value of

$f(x) = \dfrac{\log_e x}{x}$ is $\dfrac{1}{e}$, then

$$\frac{\log_e x}{x} \leqslant \frac{1}{e}, \quad x > 0$$

$$\log_e x \leqslant \frac{x}{e}$$

$$e \log_e x \leqslant x$$

$$\log_e x^e \leqslant x$$

$$e^{\log_e x^e} \leqslant e^x$$

$$\therefore x^e \leqslant e^x \quad (\text{Note: } e^{\log a} = a)$$

$$\therefore e^x \geqslant x^e, \text{ for all } x > 0.$$

QUESTION 10

(a) (i) $y = \dfrac{1}{x}$

x	a	$2a$	$3a$
	y $\frac{1}{a}$	$\frac{1}{2a}$	$\frac{1}{3a}$
	y_0	y_1	y_2

$$A \doteqdot \frac{h}{3}(y_0 + y_2 + 4y_1)$$

$$\doteqdot \frac{a}{3}\left(\frac{1}{a} + \frac{1}{3a} + 4\left(\frac{1}{2a}\right)\right)$$

$$\doteqdot \frac{a}{3}\left(\frac{1}{a} + \frac{1}{3a} + \frac{2}{a}\right)$$

$$\doteqdot \frac{a}{3}\left(\frac{3}{3a} + \frac{1}{3a} + \frac{6}{3a}\right)$$

$$\doteqdot \frac{a}{3} \times \frac{10}{3a}$$

$$\doteqdot \frac{10}{9}$$

$\therefore$ Area $\doteqdot \dfrac{10}{9}$ units^2.

(ii) Area in (i)

$$= \int_a^{3a} \frac{1}{x}\,dx$$

$$= \Big[\ln x\Big]_a^{3a}$$

$$= \ln 3a - \ln a$$

$$= \ln \frac{3a}{a}$$

$$= \ln 3$$

Since Simpson's rule finds the approximate area under $y = \frac{1}{x}$ in (i),

$\therefore \ln 3 \doteqdot \frac{10}{9}$.

(b) (i) Now area of $\triangle ABC$

$= \frac{1}{2} bc \sin A$

and area of $\triangle AST$

$= \frac{1}{2} xy \sin A$

Since area $\triangle AST = \frac{1}{2}$ area $\triangle ABC$ (given)

$$\frac{1}{2} xy \sin A = \frac{1}{2} \times \frac{1}{2} bc \sin A$$

$$xy \sin A = \frac{1}{2} bc \sin A$$

$$xy = \frac{1}{2} bc.$$

(ii) By the cosine rule in $\triangle AST$,

$z^2 = x^2 + y^2 - 2xy \cos A$

Now $xy = \frac{1}{2} bc$

$$\therefore y = \frac{bc}{2x}$$

$$\therefore z^2 = x^2 + \left(\frac{bc}{2x}\right)^2 - 2x\left(\frac{bc}{2x}\right)\cos A$$

$$\therefore z^2 = x^2 + \frac{b^2c^2}{4x^2} - bc \cos A$$

(iii)

$$z^2 = x^2 + \left(\frac{b^2c^2}{4}\right)x^{-2} - bc \cos A$$

$$\frac{d(z^2)}{dx} = 2x - \frac{2b^2c^2}{4}x^{-3}$$

$$= 2x - \frac{b^2c^2}{2x^3}$$

Stationary points occur when

$$\frac{d(z^2)}{dx} = 0$$

$$2x - \frac{b^2c^2}{2x^3} = 0$$

$$4x^4 - b^2c^2 = 0$$

$$4x^4 = b^2c^2$$

$$x^4 = \frac{b^2c^2}{4}$$

$$x^2 = \frac{bc}{2} \quad (x^2 \geqslant 0)$$

$$x = \sqrt{\frac{bc}{2}} \quad (x > 0)$$

Now $\frac{d(z^2)}{dx} = 2x - \left(\frac{b^2c^2}{2}\right)x^{-3}$

$$\therefore \frac{d^2(z^2)}{dx^2} = 2 + \frac{3b^2c^2}{2}x^{-4}$$

$$= 2 + \frac{3b^2c^2}{2x^4}$$

> 0 for all x

$\therefore$ A minimum value of z^2 when $x = \sqrt{\frac{bc}{2}}$.

(iv) Now $z^2 = x^2 + \frac{b^2c^2}{4x^2} - bc \cos A$

and $\cos A = \frac{b^2 + c^2 - a^2}{2bc}$

$$\therefore z^2 = x^2 + \frac{b^2c^2}{4x^2} - bc\left(\frac{b^2 + c^2 - a^2}{2bc}\right)$$

When $x = \sqrt{\frac{bc}{2}}$,

$$z^2 = \frac{bc}{2} + \frac{b^2c^2}{4\left(\frac{bc}{2}\right)} - \left(\frac{b^2 + c^2 - a^2}{2}\right)$$

$$= \frac{bc}{2} + \frac{bc}{2} - \left(\frac{b^2 + c^2 - a^2}{2}\right)$$

$$= \frac{a^2 - b^2 + 2bc - c^2}{2}$$

$$= \frac{a^2 - (b^2 - 2bc + c^2)}{2}$$

$$= \frac{1}{2}[a^2 - (b - c)^2]$$

$$= \frac{1}{2}\big[(a - (b - c))(a + (b - c))\big]$$

$$= \frac{1}{2}\big[(a - b + c)(a + b - c)\big]$$

$$= \frac{1}{2}\big[(a + b + c - 2b)(a + b + c - 2c)\big]$$

$$= \frac{(P - 2b)(P - 2c)}{2} \quad \text{where } P = a + b + c$$

$$\therefore z = \sqrt{\frac{(P - 2b)(P - 2c)}{2}} \text{ metres.}$$

2005

HIGHER SCHOOL CERTIFICATE EXAMINATION

Mathematics

General Instructions

- Reading time – 5 minutes
- Working time – 3 hours
- Write using black or blue pen
- Board-approved calculators may be used
- A table of standard integrals is provided at the back of this paper
- All necessary working should be shown in every question

Total marks – 120

- Attempt Questions 1–10
- All questions are of equal value

Total marks – 120
Attempt Questions 1–10
All questions are of equal value

Answer each question in a SEPARATE writing booklet. Extra writing booklets are available.

Marks

Question 1 (12 marks) Use a SEPARATE writing booklet.

(a) Evaluate $\sqrt{\dfrac{275.4}{5.2 \times 3.9}}$ correct to two significant figures. **2**

(b) Factorise $x^3 - 27$. **2**

(c) Find a primitive of $4 + \sec^2 x$. **2**

(d) Express $\dfrac{(2x-3)}{2} - \dfrac{(x-1)}{5}$ as a single fraction in its simplest form. **2**

(e) Find the values of x for which $|x-3| \le 1$. **2**

(f) Find the coordinates of the focus of the parabola $x^2 = 8(y-1)$. **2**

Marks

Question 2 (12 marks) Use a SEPARATE writing booklet.

(a) Solve $\cos\theta = \frac{1}{\sqrt{2}}$ for $0 \le \theta \le 2\pi$. **2**

(b) Differentiate with respect to x:

(i) $x \sin x$ **2**

(ii) $\frac{x^2}{x-1}$. **2**

(c) (i) Find $\int \frac{6x^2}{x^3+1}\,dx$. **2**

(ii) Evaluate $\int_0^{\frac{\pi}{6}} \cos 3x\,dx$. **2**

(d) Find the equation of the tangent to $y = \log_e x$ at the point $(e, 1)$. **2**

Marks

Question 3 (12 marks) Use a SEPARATE writing booklet.

(a) Evaluate $\sum_{n=3}^{5}(2n+1)$. **1**

(b) The lengths of the sides of a triangle are 7 cm, 8 cm and 13 cm.

(i) Find the size of the angle opposite the longest side. **2**

(ii) Find the area of the triangle. **1**

(c)

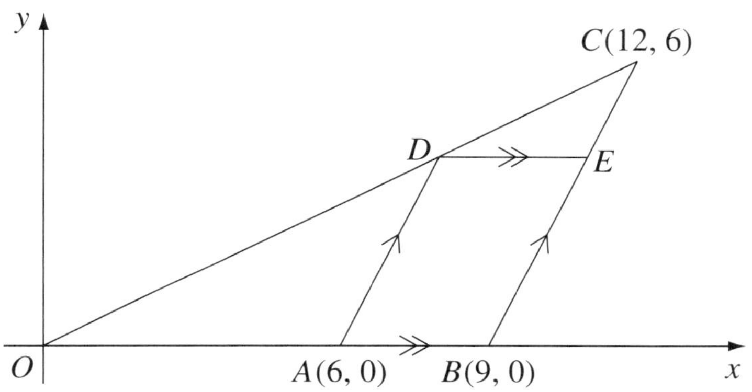

y $C(12, 6)$ D E O $A(6, 0)$ $B(9, 0)$ x

NOT
TO
SCALE

In the diagram, A, B and C are the points $(6, 0)$, $(9, 0)$ and $(12, 6)$ respectively. The equation of the line OC is $x-2y=0$. The point D on OC is chosen so that AD is parallel to BC. The point E on BC is chosen so that DE is parallel to the x-axis.

(i) Show that the equation of the line AD is $y=2x-12$. **2**

(ii) Find the coordinates of the point D. **2**

(iii) Find the coordinates of the point E. **1**

(iv) Prove that $\Delta OAD \parallel \Delta DEC$. **2**

(v) Hence, or otherwise, find the ratio of the lengths AD and EC. **1**

Marks

Question 4 (12 marks) Use a SEPARATE writing booklet.

(a)

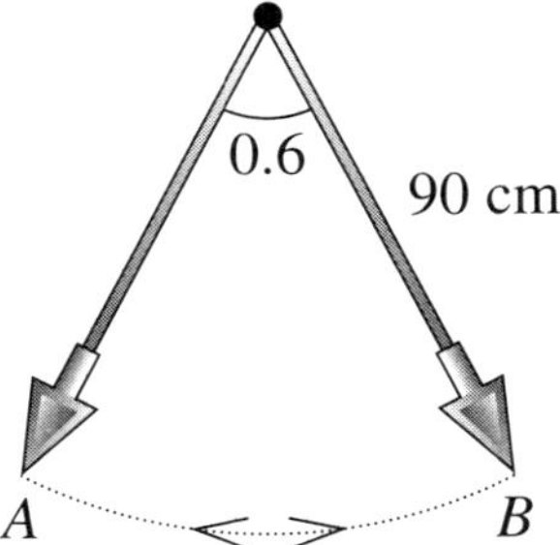

NOT TO SCALE

A pendulum is 90 cm long and swings through an angle of 0.6 radians. The extreme positions of the pendulum are indicated by the points A and B in the diagram.

(i) Find the length of the arc AB. **1**

(ii) Find the straight-line distance between the extreme positions of the pendulum. **2**

(iii) Find the area of the sector swept out by the pendulum. **1**

(b) A function $f(x)$ is defined by $f(x)=(x+3)(x^2-9)$.

(i) Find all solutions of $f(x)=0$. **2**

(ii) Find the coordinates of the turning points of the graph of $y=f(x)$, and determine their nature. **3**

(iii) Hence sketch the graph of $y=f(x)$, showing the turning points and the points where the curve meets the x-axis. **2**

(iv) For what values of x is the graph of $y=f(x)$ concave down? **1**

Marks

Question 5 (12 marks) Use a SEPARATE writing booklet.

(a) Use the change of base formula to evaluate $\log_3 7$, correct to two decimal places. **1**

(b) **3**

The diagram shows a parallelogram $ABCD$ with $\angle DAB = 120°$. The side DC is produced to E so that $AD = BE$.

Copy or trace the diagram into your writing booklet.

Prove that ΔBCE is equilateral.

(c) Find the coordinates of the point P on the curve $y = 2e^x + 3x$ at which the tangent to the curve is parallel to the line $y = 5x - 3$. **3**

(d) A total of 300 tickets are sold in a raffle which has three prizes. There are 100 red, 100 green and 100 blue tickets.

At the drawing of the raffle, winning tickets are NOT replaced before the next draw.

(i) What is the probability that each of the three winning tickets is red? **2**

(ii) What is the probability that at least one of the winning tickets is not red? **1**

(iii) What is the probability that there is one winning ticket of each colour? **2**

Marks

Question 6 (12 marks) Use a SEPARATE writing booklet.

(a) Five values of the function $f(x)$ are shown in the table. **3**

x	0	5	10	15	20
$f(x)$	15	25	22	18	10

Use Simpson's rule with the five values given in the table to estimate

$$\int_0^{20} f(x)\,dx.$$

(b) A tank initially holds 3600 litres of water. The water drains from the bottom of the tank. The tank takes 60 minutes to empty.

A mathematical model predicts that the volume, V litres, of water that will remain in the tank after t minutes is given by

$$V = 3600\left(1 - \frac{t}{60}\right)^2, \quad \text{where} \quad 0 \le t \le 60.$$

(i) What volume does the model predict will remain after ten minutes? **1**

(ii) At what rate does the model predict that the water will drain from the tank after twenty minutes? **2**

(iii) At what time does the model predict that the water will drain from the tank at its fastest rate? **2**

(c)

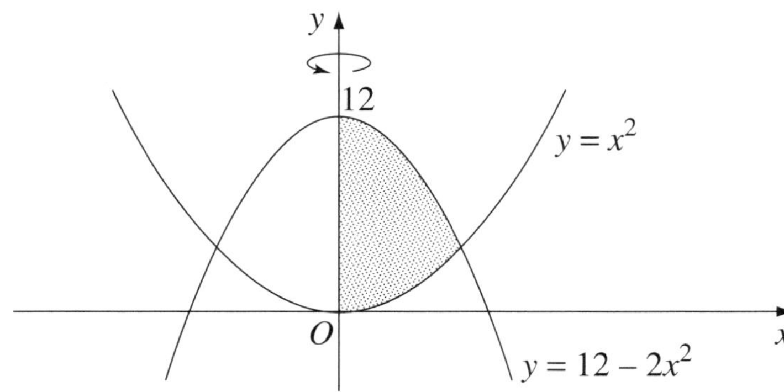

The graphs of the curves $y = x^2$ and $y = 12 - 2x^2$ are shown in the diagram.

(i) Find the points of intersection of the two curves. **1**

(ii) The shaded region between the curves and the y-axis is rotated about the y-axis. By splitting the shaded region into two parts, or otherwise, find the volume of the solid formed. **3**

Marks

Question 7 (12 marks) Use a SEPARATE writing booklet.

(a) Anne and Kay are employed by an accounting firm.

Anne accepts employment with an initial annual salary of \$50 000. In each of the following years her annual salary is increased by \$2500.

Kay accepts employment with an initial annual salary of \$50 000. In each of the following years her annual salary is increased by 4%.

(i) What is Anne's annual salary in her thirteenth year? **2**

(ii) What is Kay's annual salary in her thirteenth year? **2**

(iii) By what amount does the total amount paid to Kay in her first twenty years exceed that paid to Anne in her first twenty years? **3**

(b)

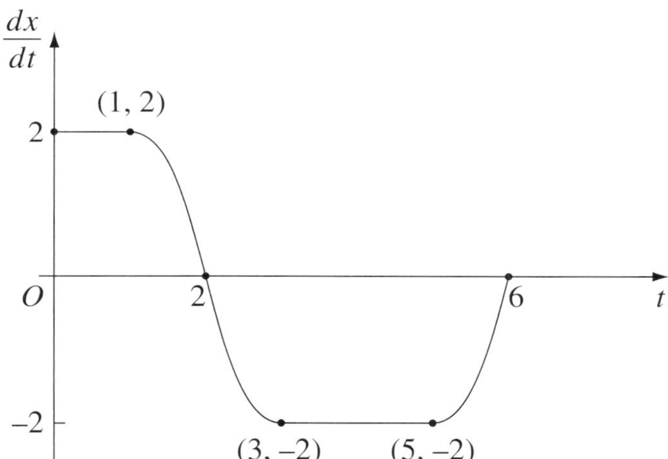

The graph shows the velocity, $\frac{dx}{dt}$, of a particle as a function of time. Initially the particle is at the origin.

(i) At what time is the displacement, x, from the origin a maximum? **1**

(ii) At what time does the particle return to the origin? Justify your answer. **2**

(iii) Draw a sketch of the acceleration, $\frac{d^2x}{dt^2}$, as a function of time for $0 \le t \le 6$. **2**

Marks

Question 8 (12 marks) Use a SEPARATE writing booklet.

(a)

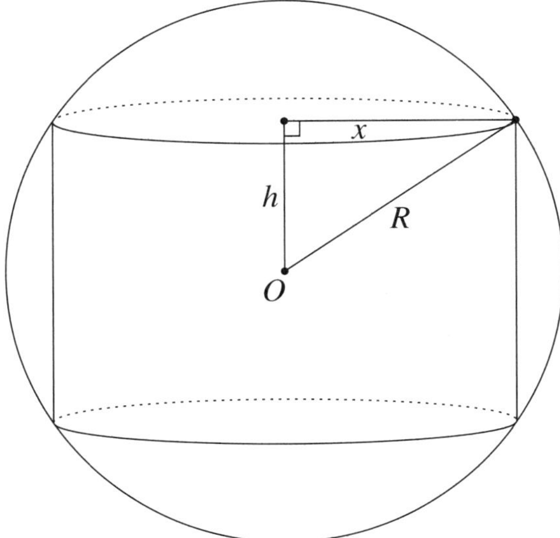

A cylinder of radius x and height $2h$ is to be inscribed in a sphere of radius R centred at O as shown.

(i) Show that the volume of the cylinder is given by **1**

$$V = 2\pi h\left(R^2 - h^2\right).$$

(ii) Hence, or otherwise, show that the cylinder has a maximum volume when $h = \dfrac{R}{\sqrt{3}}$. **3**

Question 8 continues

Marks

Question 8 (continued)

(b) **3**

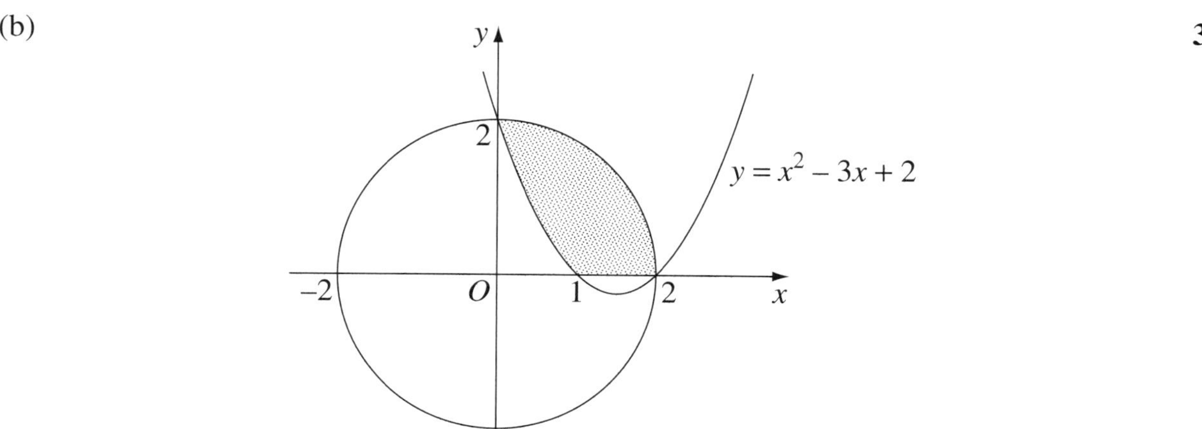

The shaded region in the diagram is bounded by the circle of radius 2 centred at the origin, the parabola $y = x^2 - 3x + 2$, and the x-axis.

By considering the difference of two areas, find the area of the shaded region.

(c) Weelabarrabak Shire Council borrowed \$3 000 000 at the beginning of 2005. The annual interest rate is 12%. Each year, interest is calculated on the balance at the beginning of the year and added to the balance owing. The debt is to be repaid by equal annual repayments of \$480 000, with the first repayment being made at the end of 2005.

Let A_n be the balance owing after the n-th repayment.

(i) Show that $A_2 = (3 \times 10^6)(1.12)^2 - (4.8 \times 10^5)(1 + 1.12)$. **1**

(ii) Show that $A_n = 10^6\left[4 - (1.12)^n\right]$. **2**

(iii) In which year will Weelabarrabak Shire Council make the final repayment? **2**

End of Question 8

Marks

Question 9 (12 marks) Use a SEPARATE writing booklet.

(a) A particle is initially at rest at the origin. Its acceleration as a function of time, t, is given by

$$\ddot{x} = 4\sin 2t.$$

(i) Show that the velocity of the particle is given by $\dot{x} = 2 - 2\cos 2t$. **2**

(ii) Sketch the graph of the velocity for $0 \le t \le 2\pi$ AND determine the time at which the particle first comes to rest after $t = 0$. **3**

(iii) Find the distance travelled by the particle between $t = 0$ and the time at which the particle first comes to rest after $t = 0$. **2**

(b)

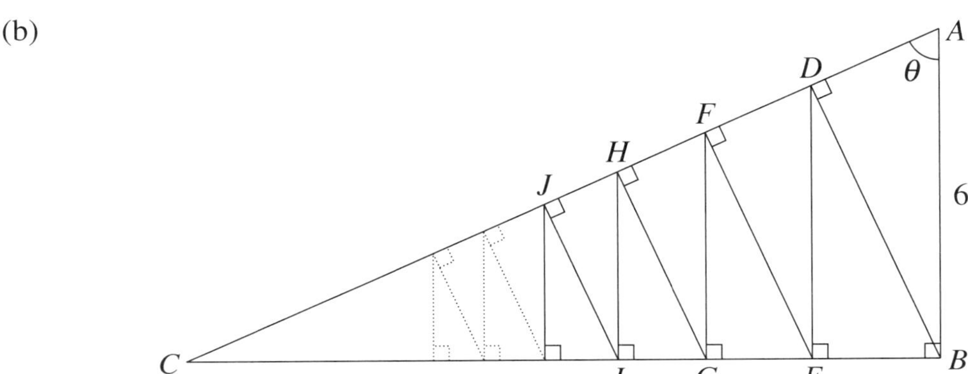

The triangle ABC has a right angle at B, $\angle BAC = \theta$ and $AB = 6$. The line BD is drawn perpendicular to AC. The line DE is then drawn perpendicular to BC. This process continues indefinitely as shown in the diagram.

(i) Find the length of the interval BD, and hence show that the length of the interval EF is $6\sin^3\theta$. **2**

(ii) Show that the limiting sum **3**

$$BD + EF + GH + \cdots$$

is given by $6\sec\theta\tan\theta$.

Marks

Question 10 (12 marks) Use a SEPARATE writing booklet.

(a)

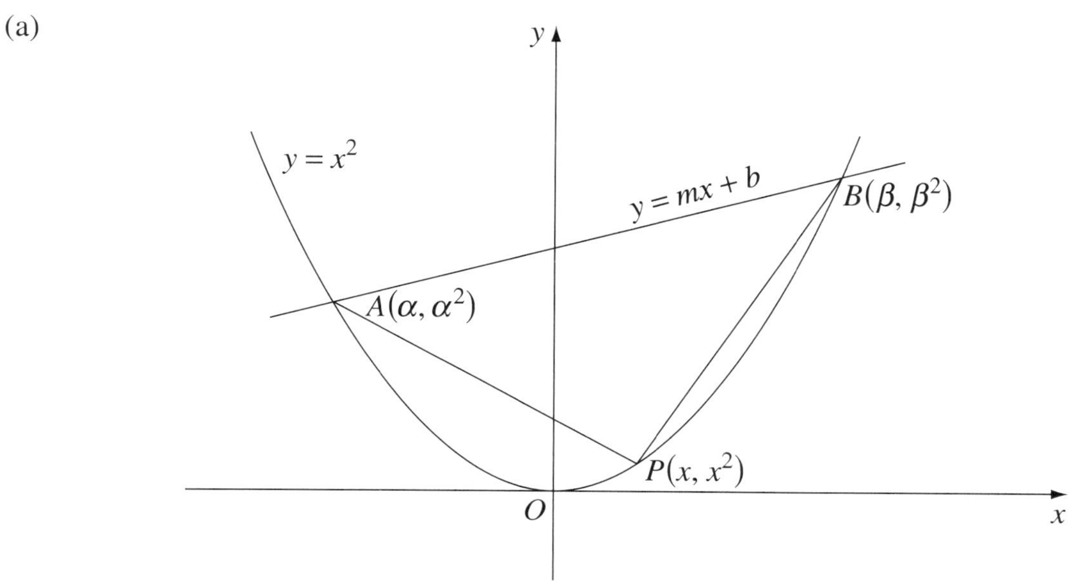

The parabola $y = x^2$ and the line $y = mx + b$ intersect at the points $A(\alpha, \alpha^2)$ and $B(\beta, \beta^2)$ as shown in the diagram.

(i) Explain why $\alpha + \beta = m$ and $\alpha\beta = -b$. **1**

(ii) Given that $(\alpha - \beta)^2 + (\alpha^2 - \beta^2)^2 = (\alpha - \beta)^2 \left[1 + (\alpha + \beta)^2\right]$, show that the distance $AB = \sqrt{(m^2 + 4b)(1 + m^2)}$. **2**

(iii) The point $P(x, x^2)$ lies on the parabola between A and B. Show that the area of the triangle ABP is given by $\frac{1}{2}(mx - x^2 + b)\sqrt{m^2 + 4b}$. **2**

(iv) The point P in part (iii) is chosen so that the area of the triangle ABP is a maximum. **2**

Find the coordinates of P in terms of m.

Question 10 continues

Marks

Question 10 (continued)

(b) Xuan and Yvette would like to meet at a cafe on Monday. They each agree to come to the cafe sometime between 12 noon and 1 pm, wait for 15 minutes, and then leave if they have not seen the other person.

Their arrival times can be represented by the point (x, y) in the Cartesian plane, where x represents the fraction of an hour after 12 noon that Xuan arrives, and y represents the fraction of an hour after 12 noon that Yvette arrives.

Thus $\left(\frac{1}{3}, \frac{2}{5}\right)$ represents Xuan arriving at 12:20 pm and Yvette arriving at 12:24 pm. Note that the point (x, y) lies somewhere in the unit square $0 \le x \le 1$ and $0 \le y \le 1$ as shown in the diagram.

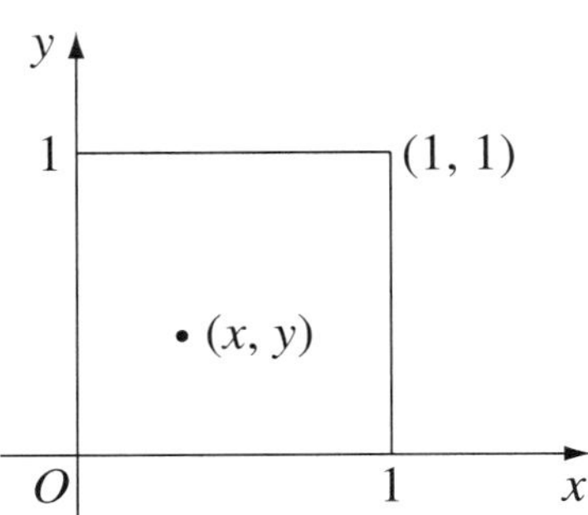

(i) Explain why Xuan and Yvette will meet if $x - y \le \frac{1}{4}$ or $y - x \le \frac{1}{4}$. **1**

(ii) The probability that they will meet is equal to the area of the part of the region given by the inequalities in part (i) that lies within the unit square $0 \le x \le 1$ and $0 \le y \le 1$. **2**

Find the probability that they will meet.

(iii) Xuan and Yvette agree to try to meet again on Tuesday. They agree to arrive between 12 noon and 1 pm, but on this occasion they agree to wait for t minutes before leaving. **2**

For what value of t do they have a 50% chance of meeting?

End of paper

2005 Higher School Certificate Worked Answers

QUESTION 1

(a) $\sqrt{\dfrac{275.4}{5.2 \times 3.9}} = \sqrt{\dfrac{275.4}{20.28}}$
$= 3.6850\ldots$ (by calc.)
$= 3.7$ to 2 sig. figures.

(b) $x^2 - 27 = x^3 - 3^3$
$= (x-3)(x^2 + 3x + 9)$

(c) $\displaystyle\int (4 + \sec^2 x)\, dx = 4x + \tan x + C$

(using standard integrals)

(d) $\dfrac{(2x-3)}{2} - \dfrac{(x-1)}{5}$

$= \dfrac{5(2x-3) - 2(x-1)}{10}$

$= \dfrac{10x - 15 - 2x + 2}{10}$

$= \dfrac{8x - 13}{10}$

(e) $|x - 3| \leqslant 1$
$\therefore -1 \leqslant x - 3 \leqslant 1$
$\therefore 2 \leqslant x \leqslant 4.$

(f) $x^2 = 8(y - 1)$

Coordinates of the vertex are $(0, 1)$.

Now $4a = 8$
$\therefore a = 2$
i.e. The focal length is 2.
$\therefore$ Coordinates of the focus are $(0, 3)$.

QUESTION 2

(a) $\cos\theta = \dfrac{1}{\sqrt{2}}$ for $0 \leqslant \theta \leqslant 2\pi$

$\therefore \theta = \dfrac{\pi}{4}$ or $2\pi - \dfrac{\pi}{4}$

$= \dfrac{\pi}{4}$ or $\dfrac{7\pi}{4}$.

(b) (i) $\dfrac{d}{dx}(x \sin x)$

$= x \cdot \dfrac{d}{dx}(\sin x) + \sin x \cdot \dfrac{d}{dx}(x)$
$= x\cos x + \sin x \cdot 1$
$= x\cos x + \sin x.$

(ii) $\dfrac{d}{dx}\left(\dfrac{x^2}{x-1}\right)$

$= \dfrac{(x-1) \cdot \dfrac{d}{dx}(x^2) - x^2 \cdot \dfrac{d}{dx}(x-1)}{(x-1)^2}$

$= \dfrac{(x-1) \cdot 2x - x^2 \cdot 1}{(x-1)^2}$

$= \dfrac{2x^2 - 2x - x^2}{(x-1)^2}$

$= \dfrac{x^2 - 2x}{(x-1)^2}$

$= \dfrac{x(x-2)}{(x-1)^2}$

(c) (i) $\displaystyle\int \frac{6x^2}{x^3+1}\, dx = 2\int \frac{3x^2}{x^3+1}\, dx$
$= 2\ln(x^3 + 1) + C.$

(ii) $\displaystyle\int_0^{\frac{\pi}{6}} \cos 3x\, dx$

$= \left[\dfrac{1}{3}\sin 3x\right]_0^{\frac{\pi}{6}}$

$= \dfrac{1}{3}\left[\sin 3\left(\dfrac{\pi}{6}\right) - \sin 3(0)\right]$

$= \dfrac{1}{3}\left(\sin\dfrac{\pi}{2} - \sin 0\right)$

$= \dfrac{1}{3}(1 - 0)$

$= \dfrac{1}{3}.$

(d) $y = \log_e x$

$\dfrac{dy}{dx} = \dfrac{1}{x}$

At $x = e$, $\dfrac{dy}{dx} = \dfrac{1}{e}$

$\therefore$ Equation of tangent at $(e, 1)$ with gradient $= \dfrac{1}{e}$ is

$y - 1 = \dfrac{1}{e}(x - e)$

$y - 1 = \dfrac{x}{e} - 1$

$\therefore y = \dfrac{x}{e}.$

QUESTION 3

(a) $\sum_{n=3}^{5}(2n+1) = 7+9+11$

$= 27.$

(b) (i)

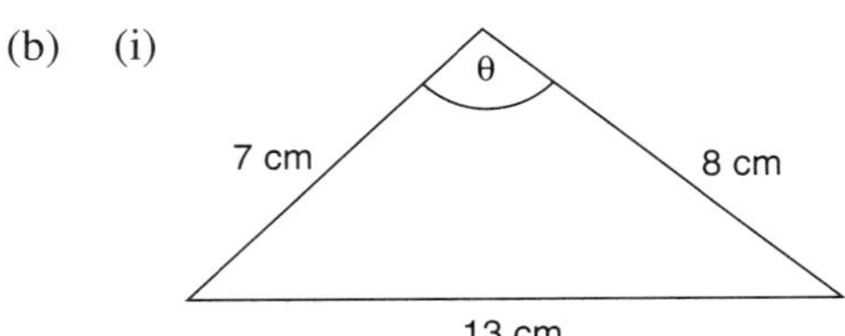

By the cosine rule:

$\cos\theta = \dfrac{7^2+8^2-13^2}{2\times7\times8}$

$= \dfrac{-56}{112}$

$= -0.5$

$\therefore \theta = 180° - 60°$

(N.B. angle is obtuse)

$= 120°.$

(ii) $A = \dfrac{1}{2}ab\sin C$

$= \dfrac{1}{2}\times7\times8\times\sin 120°$

$= 28\times\dfrac{\sqrt{3}}{2}$

$= 14\sqrt{3}$

$\therefore$ Area $= 14\sqrt{3}\ \text{cm}^2.$

(c)

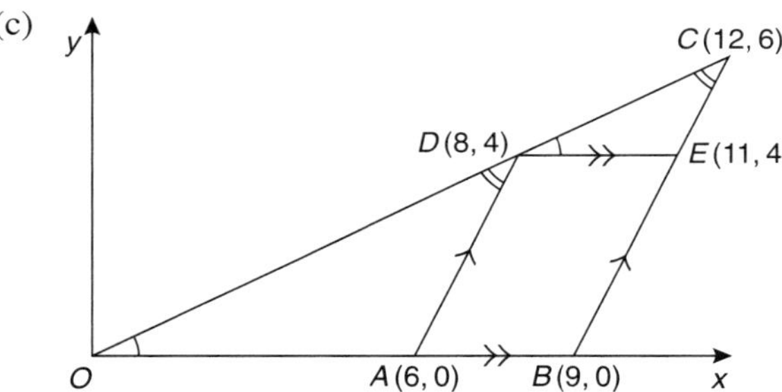

(i) Gradient of $BC = \dfrac{y_2-y_1}{x_2-x_1}$

$= \dfrac{6-0}{12-9}$

$= \dfrac{6}{3}$

$= 2.$

Now $m_{AD} = m_{BC}$

$\therefore m_{AD} = 2.$

$\therefore$ Equation of line AD with gradient $= 2$, passing through $A(6, 0)$ is given by

$y - 0 = 2(x-6)$

$\therefore y = 2x - 12.$

(ii) D is the point of intersection of

$x - 2y = 0 \quad \ldots(1)$

$y = 2x - 12 \quad \ldots(2)$

Solving simultaneously:

Substituting (2) into (1),

$x - 2(2x-12) = 0$

$x - 4x + 24 = 0$

$-3x = -24$

$\therefore x = 8$

$\therefore y = 2(8) - 12$

$= 4.$

$\therefore$ Coordinates of D are $(8, 4)$.

(iii) Coordinates of E are $(11, 4)$.
(Since $AB = 3$, $DE = 3$)

(iv) In $\triangle OAD$ and $\triangle DEC$,

$\angle DOA = \angle CDE$
(corresponding angles, $OA \parallel DE$)

$\angle ODA = \angle DCE$
(corresponding angles, $DA \parallel CE$)

$\therefore \triangle OAD\ |||\ \triangle DEC$ (equiangular)

(v) $\dfrac{OA}{DE} = \dfrac{AD}{EC}$

(corresponding sides in same ratio)

$\dfrac{6}{3} = \dfrac{AD}{EC}$

$\therefore \dfrac{AD}{EC} = \dfrac{6}{3}$

$= 2$

$\therefore AD : EC = 2 : 1.$

QUESTION 4

(a) (i) $l = r\theta$

$= 90\times0.6$

$= 54$ cm (by calc.)

(ii) By the cosine rule:

$AB^2 = 90^2 + 90^2 - 2(90)(90)\cos 0.6$

$= 2829.5630\ldots$ (by calc.)

$\therefore AB = 53.1936\ldots$ (by calc.)

$\doteqdot 53.2$ cm.

(iii) $A = \frac{1}{2}r^2\theta$

$= \frac{1}{2}(90)^2(0.6)$

$= 2430$ (by calc.)

$\therefore$ Area $= 2430$ cm^2.

(b) (i) $f(x) = 0$

$\therefore (x+3)(x^2-9) = 0$

$(x+3)(x+3)(x-3) = 0$

$\therefore x = 3$ or -3.

(ii) $f(x) = (x+3)(x^2-9)$

$= x^3 - 9x + 3x^2 - 27$

$= x^3 + 3x^2 - 9x - 27$

$f'(x) = 3x^2 + 6x - 9$

Stationary points occur when

$f'(x) = 0$

$3x^2 + 6x - 9 = 0$

$3(x^2 + 2x - 3) = 0$

$3(x+3)(x-1) = 0$

$\therefore x = -3$ or 1.

Now $f(-3) = (-3+3)\left[(-3)^2 - 9\right]$

$= 0$

and $f(1) = (1+3)(1^2 - 9)$

$= 4 \times -8$

$= -32$

$\therefore$ Coordinates of stationary points are $(-3, 0)$ and $(1, -32)$.

Now $f'(x) = 3x^2 + 6x - 9$

$\therefore f''(x) = 6x + 6$

$\therefore f''(-3) = 6(-3) + 6$

$= -18 + 6$

$= -12 < 0$.

$\therefore$ A maximum turning point occurs at $(-3, 0)$.

and $f''(1) = 6(1) + 6$

$= 12 > 0$.

$\therefore$ A minimum turning point occurs at $(1, -32)$.

The curve meets the x-axis when

$f(x) = 0$

$(x+3)(x^2-9) = 0$

$(x+3)(x+3)(x-3) = 0$

$\therefore x = \pm 3$.

(iii)

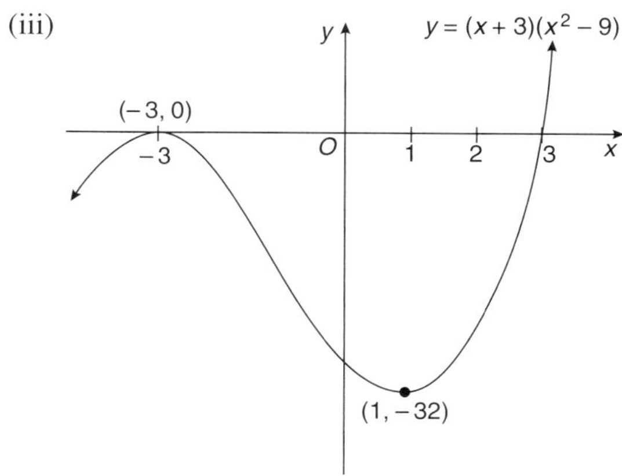

(iv) Graph of $y = f(x)$ is concave down

when $f''(x) < 0$

$6x + 6 < 0$

$6x < -6$

$x < -1$.

QUESTION 5

(a) $\log_3 7 = \frac{\log_{10} 7}{\log_{10} 3}$

$= 1.7712\ldots$ (by calc.)

$= 1.77$ to 2 decimal places.

(b)

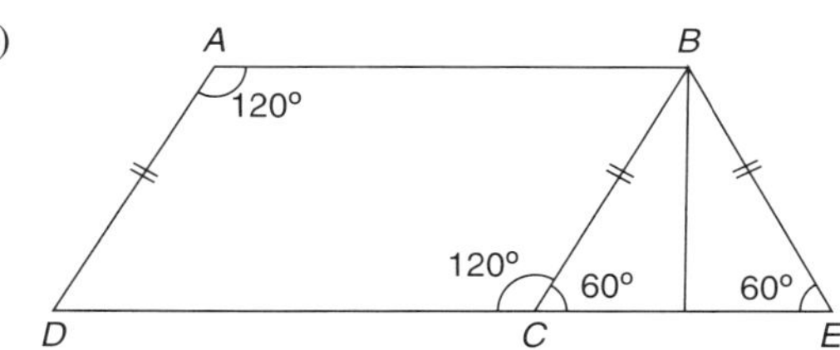

Now $\angle BCD = \angle DAB = 120°$ (opp. angles of a parallelogram)

and $BC = AD$ (opp. sides of a parallelogram)

and $\angle BCE = 180° - \angle BCD$ (DCE is a straight angle, 180°)

$= 180° - 120°$

$= 60°$

Also, $\angle BCE = \angle BEC$ (angles opp. equal sides; $BC = BE$)

$\therefore \angle BCE = \angle BEC = 60°$

Now $\angle BCE + \angle BEC + \angle CBE = 180°$ (angle sum of $\triangle BCE$ is 180°)

$\therefore 60° + 60° + \angle CBE = 180°$

$\therefore \angle CBE = 180° - 120°$

$= 60°$

$\therefore \triangle BCE$ is equilateral.

(c) Gradient of $y = 5x - 3$ is 5.
$\therefore$ Gradient of tangent must be 5.

Now $y = 2e^x + 3x$

$\therefore \frac{dy}{dx} = 2e^x + 3$

If $\frac{dy}{dx} = 5$

$\therefore 2e^x + 3 = 5$

$2e^x = 2$

$e^x = 1$

$\therefore x = 0$

When $x = 0, y = 2e^0 + 3(0)$
$= 2$

$\therefore$ Coordinates of P are $(0, 2)$.

(d) (i) P(each ticket red)
= P(RRR)

$= \frac{100}{300} \times \frac{99}{299} \times \frac{98}{298}$

$= \frac{1617}{44\,551}$ (by calc.)

$= 0.03629\ldots$ (by calc.)

$\doteqdot 0.0363$.

(ii) P(At least 1 not red)
= 1 − P(RRR)

$= 1 - \frac{1617}{44\,551}$

$= 0.96370\ldots$ (by calc.)

$\doteqdot 0.9637$.

(iii) The following outcomes contain one ticket of each colour:

R G B G R B B R G
R B G G B R B G R

$\therefore$ There are 6 favourable outcomes.

$\therefore$ P(One ticket of each colour)

$= 6 \times \frac{100}{300} \times \frac{100}{299} \times \frac{100}{298}$

$= 0.22446\ldots$ (by calc.)

$\doteqdot 0.2245$.

QUESTION 6

x	0	5	10	15	20
$f(x)$	15	25	22	18	10
	y_0	y_1	y_2	y_3	y_4

(a) $\int_0^{20} f(x)\,dx$

$\doteqdot \frac{h}{3}\left[y_0 + y_4 + 2y_2 + 4(y_1 + y_3)\right]$

$\doteqdot \frac{5}{3}\left[15 + 10 + 2(22) + 4(25 + 18)\right]$

$= \frac{5}{3}(15 + 10 + 44 + 4 \times 43)$

$= \frac{5}{3} \times 241$

$= \frac{1205}{3}$

$= 401\frac{2}{3}$.

(b) (i) $V = 3600\left(1 - \frac{t}{60}\right)^2$, where $0 \leqslant t \leqslant 60$

When $t = 10$,

$V = 3600\left(1 - \frac{10}{60}\right)^2$

$= 2500$ litres (by calc.)

(ii) $V = 3600\left(1 - \frac{t}{60}\right)^2$

$\frac{dV}{dt} = 2 \times 3600\left(1 - \frac{t}{60}\right) \cdot \frac{d}{dt}\left(1 - \frac{t}{60}\right)$

$= 7200\left(1 - \frac{t}{60}\right) \cdot -\frac{1}{60}$

$= -120\left(1 - \frac{t}{60}\right)$

When $t = 20$,

$\frac{dv}{dt} = -120\left(1 - \frac{20}{60}\right)$

$= -120 \times \frac{40}{60}$

$= -80$

$\therefore$ After 20 minutes, water is draining at 80 L/min.

(iii) Now $\frac{dV}{dt} = -120\left(1 - \frac{t}{60}\right)$

$= -120 + 2t$

To find when water will drain fastest, consider $\frac{d^2V}{dt^2}$:

Now $\frac{d^2V}{dt^2} = 2$

Since, $\frac{d^2V}{dt^2}$ is a constant, check end-points:

When $t = 0$, $\frac{dV}{dt} = -120$

When $t = 60$, $\frac{dV}{dt} = 0$

$\therefore$ The water will **drain** at the fastest rate when $t = 0$.

(c) (i) Solving simultaneously:

$$y = 12 - 2x^2 \quad \ldots (1)$$
$$y = x^2 \quad \ldots (2)$$

Equating (1) and (2):

$$x^2 = 12 - 2x^2$$
$$3x^2 - 12 = 0$$
$$3(x^2 - 4) = 0$$
$$3(x + 2)(x - 2) = 0$$
$$\therefore x = \pm 2$$

When $x = -2, y = 4$
When $x = 2, y = 4$

$\therefore$ Points of intersection are $(-2, 4)$ and $(2, 4)$.

(ii) For $y = x^2$
$\therefore x^2 = y$

For $y = 12 - 2x^2$
$2x^2 = 12 - y$
$\therefore x^2 = \frac{12 - y}{2}$

$$V = \pi\int_0^4 x^2\,dy + \pi\int_4^{12} x^2\,dy$$
$$= \pi\int_0^4 y\,dy + \pi\int_4^{12} \left(\frac{12 - y}{2}\right)dy$$
$$= \pi\int_0^4 y\,dy + \frac{\pi}{2}\int_4^{12} (12 - y)\,dy$$
$$= \pi\left[\frac{y^2}{2}\right]_0^4 + \frac{\pi}{2}\left[12y - \frac{y^2}{2}\right]_4^{12}$$
$$= \pi\left(\frac{16}{2} - 0\right) + \frac{\pi}{2}\left[\left(12(12) - \frac{12^2}{2}\right) - \left(12(4) - \frac{4^2}{2}\right)\right]$$
$$= 8\pi + \frac{\pi}{2}\left[(144 - 72) - (48 - 8)\right]$$
$$= 8\pi + \frac{\pi}{2}(72 - 40)$$
$$= 8\pi + \frac{\pi}{2}(32)$$
$$= 8\pi + 16\pi$$
$$= 24\pi \text{ units}^3.$$

QUESTION 7

(a) (i) Anne's salary represents an arithmetic sequence:
50 000, 52 500, 55 000 . . .
with $a = 50\,000,\ d = 2500,\ n = 13$.

$$T_n = a + (n - 1)d$$
$$\therefore T_{13} = 50\,000 + 12 \times 2500$$
$$= 50\,000 + 30\,000$$
$$= 80\,000$$

$\therefore$ Anne's salary in her thirteenth year is \$80 000.

(ii) Kay's salary represents a geometric sequence with $a = 50\,000,\ r = 1.04,\ n = 13$.

$$T_n = ar^{n-1}$$
$$\therefore T_{13} = 50\,000(1.04)^{12}$$
$$= 80\,051.6109\ldots \quad \text{(by calc.)}$$

$\therefore$ Kay's salary in her thirteenth year is \$80 051.61.

(iii) Total amount for Anne in first 20 years:

$$S_n = \frac{n}{2}\left(2a + (n - 1)d\right)$$

$n = 20,\ a = 50\,000,\ d = 2500$

$$\therefore S_{20} = \frac{20}{2}\left[2(50\,000) + (20 - 1)2500\right]$$
$$= 10(100\,000 + 19 \times 2500)$$
$$= 1\,475\,000 \quad \text{(by calc.)}$$

$\therefore$ Total amount paid to Anne is \$1 475 000.

Total amount paid to Kay in first 20 years is given by

$$S_n = a\frac{(r^n - 1)}{r - 1}$$

with $n = 20$, $a = 50\,000$, $r = 1.04$

$$\therefore S_{20} = \frac{50\,000(1.04^{20} - 1)}{1.04 - 1}$$
$$= 1\,488\,903.929 \quad \text{(by calc.)}$$

Total amount paid to Kay is \$1 488 903.93.

Kay's total exceeds Anne's total by
\$1 488 903.93 – \$1 475 000
= \$13 903.93.

(b) (i) The displacement x, from the origin is a maximum when $t = 2$.
(N.B. $v > 0$ when $t < 2$
$v = 0$ when $t = 2$
$v < 0$ when $t > 2$)

(ii) The particle returns to the origin at $t = 4$, by symmetry. (N.B. Area under curve from $t = 0$ to $t = 2$, is of the same magnitude but opposite in sign, as area under the curve between $t = 2$ and $t = 4$.)

The particle is initially at the origin. It moves in a positive direction away from the origin for 2 seconds. It then comes to rest at $t = 2$. It then moves back towards the origin for the next 2 seconds, returning to the origin when $t = 4$. (Note the symmetry.)

(iii)

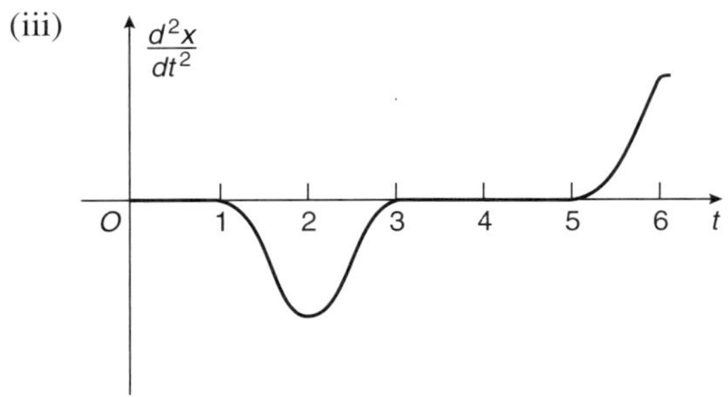

QUESTION 8

(a) (i) $V = \pi r^2 h$
where $r = x$, $h = 2h$

$$\therefore V = \pi x^2(2h)$$

Now $R^2 = x^2 + h^2$ (Pythagoras' theorem)

$$\therefore x^2 = R^2 - h^2$$

$$\therefore V = \pi(R^2 - h^2)(2h)$$
$$= 2\pi h(R^2 - h^2).$$

(ii)
$$V = 2\pi h(R^2 - h^2)$$
$$= 2\pi R^2 h - 2\pi h^3$$
$$\frac{dV}{dh} = 2\pi R^2 - 6\pi h^2$$
$$\frac{d^2V}{dh^2} = -12\pi h$$

A maximum volume may occur when

$$\frac{dV}{dh} = 0$$
$$2\pi R^2 - 6\pi h^2 = 0$$
$$6\pi h^2 = 2\pi R^2$$
$$h^2 = \frac{2\pi R^2}{6\pi}$$
$$= \frac{R^2}{3}$$
$$\therefore h = \frac{R}{\sqrt{3}}$$
$$\left(\text{since } h > 0, h \neq -\frac{R}{\sqrt{3}}\right)$$

When $h = \frac{R}{\sqrt{3}}$,

$$\frac{d^2V}{dh^2} = -12\pi h$$
$$= -12\pi\left(\frac{R}{\sqrt{3}}\right) < 0$$

$\therefore$ Cylinder has a maximum volume when $h = \frac{R}{\sqrt{3}}$.

(b) Area of shaded region

$$= \text{Area of quadrant} - \int_0^1 (x^2 - 3x + 2)\,dx$$

Now area of circle $= \pi r^2$
$= \pi(2)^2$
$= 4\pi$

$$\therefore \text{Area of quadrant} = \frac{1}{4} \times 4\pi$$
$$= \pi$$

$\therefore$ Area of shaded region

$$= \pi - \int_0^1 (x^2 - 3x + 2)\,dx$$

$$= \pi - \left[\frac{x^3}{3} - \frac{3x^2}{2} + 2x\right]_0^1$$

$$= \pi - \left[\left(\frac{1}{3} - \frac{3}{2} + 2\right) - 0\right]$$

$$= \pi - \frac{5}{6} \quad \text{(by calc.)}$$

$$\therefore \text{Area} = \left(\pi - \frac{5}{6}\right) \text{ units}^2.$$

(c) (i) $A_1 = 3\,000\,000 \times 1.12 - 480\,000$

$= (3 \times 10^6)(1.12) - (4.8 \times 10^5)$

$A_2 = A_1 \times 1.12 - (4.8 \times 10^5)$

$= \left[(3 \times 10^6)(1.12) - (4.8 \times 10^5)\right] \times 1.12 - (4.8 \times 10^5)$

$= (3 \times 10^6)(1.12)^2 - (4.8 \times 10^5)(1.12) - (4.8 \times 10^5)$

$= (3 \times 10^6)(1.12)^2 - (4.8 \times 10^5)(1.12 + 1)$

$= (3 \times 10^6)(1.12)^2 - (4.8 \times 10^5)(1 + 1.12).$

(ii) $A_3 = A_2 \times 1.12 - 4.8 \times 10^5$

$= \left[(3 \times 10^6)(1.12)^2 - (4.8 \times 10^5)(1 + 1.12)\right] \times 1.12 - (4.8 \times 10^5)$

$= (3 \times 10^6)(1.12)^3 - (4.8 \times 10^5)(1.12 + 1.12^2) - (4.8 \times 10^5)$

$= (3 \times 10^6)(1.12)^3 - (4.8 \times 10^5)(1 + 1.12 + 1.12^2)$

Continuing the pattern:

$\therefore A_n = (3 \times 10^6)(1.12)^n - (4.8 \times 10^5)(1 + 1.12 + \ldots + 1.12^{n-1})$

Now $(1 + 1.12 + 1.12^2 + \ldots + 1.12^{n-1})$ is a geometric series with $a = 1, r = 1.12$ and $n = n$.

$$\therefore A_n = (3 \times 10^6)(1.12)^n - (4.8 \times 10^5)\left(\frac{a(r^n - 1)}{r - 1}\right)$$

$$= (3 \times 10^6)(1.12)^n - (4.8 \times 10^5)\left(\frac{1(1.12^n - 1)}{1.12 - 1}\right)$$

$$= (3 \times 10^6)(1.12)^n - (4.8 \times 10^5)\left(\frac{1.12^n - 1}{0.12}\right)$$

$= (3 \times 10^6)(1.12)^n - (4 \times 10^6)(1.12^n - 1)$

$= (3 \times 10^6)(1.12)^n - (4 \times 10^6)(1.12^n) + (4 \times 10^6)$

$= (1.12)^n\left[(3 \times 10^6) - (4 \times 10^6)\right] + (4 \times 10^6)$

$= (4 \times 10^6) - (1 \times 10^6)(1.12)^n$

$= 10^6\left[4 - (1.12)^n\right].$

(iii) The loan will be repaid when

$A_n = 0$

$10^6\left[4 - (1.12)^n\right] = 0$

$4 - 1.12^n = 0$

$1.12^n = 4$

$\ln (1.12)^n = \ln 4$

$n \ln (1.12) = \ln 4$

$$n = \frac{\ln 4}{\ln 1.12}$$

$= 12.2325\ldots$ (by calc.)

$\therefore$ The loan will be repaid in the 13th year, 2017.

QUESTION 9

(a) (i) $\ddot{x} = 4 \sin 2t$

$$\dot{x} = \int 4 \sin 2t \, dt$$

$$= -\frac{1}{2}(4 \cos 2t) + C$$

$= -2 \cos 2t + C$

When $t = 0, \dot{x} = 0$

$\therefore 0 = -2 \cos 0 + C$

$\therefore 0 = -2 + C$

$\therefore C = 2$

$\therefore \dot{x} = -2 \cos 2t + 2$

$= 2 - 2 \cos 2t.$

(ii) Period $= \dfrac{2\pi}{2} = \pi$

Range: $-1 \leqslant -\cos t \leqslant 1$
$\therefore -2 \leqslant -2\cos t \leqslant 2$
$\therefore 0 \leqslant 2 - 2\cos t \leqslant 4$

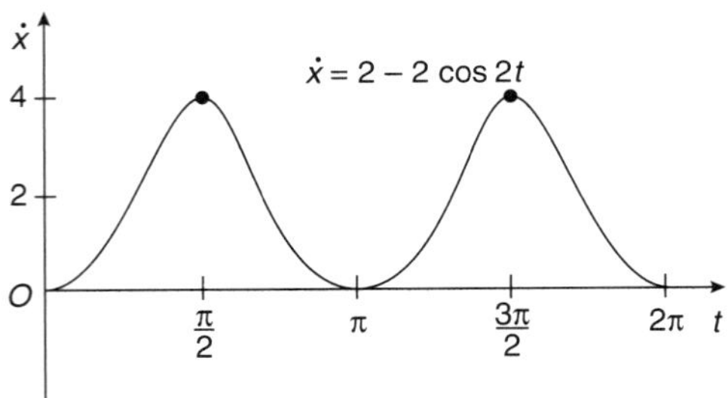

The particle is at rest when $\dot{x} = 0$.
The first time this occurs after $t = 0$ is $t = \pi$.

(iii) Distance travelled

$$= \int_0^{\pi} (2 - 2\cos 2t)\, dt$$

$$= \left[2t - \frac{2\sin 2t}{2}\right]_0^{\pi}$$

$$= \left[2t - \sin 2t\right]_0^{\pi}$$

$$= (2\pi - 0) - (0 - 0)$$
$$= 2\pi.$$

$\therefore$ Distance travelled is 2π units.

(b) (i)

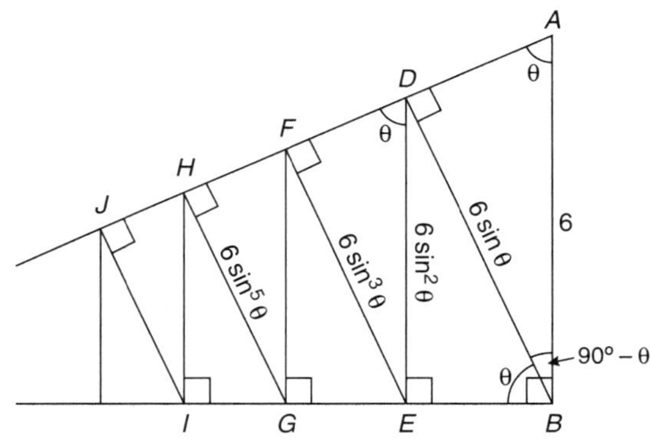

In $\triangle ABD$,

$$\sin\theta = \frac{BD}{6}$$

$\therefore BD = 6\sin\theta$

Now $\angle ABD = 90° - \theta$
(angle sum of triangle = 180°)
$\therefore \angle DBE = \theta$
($\angle ABE = 90°$, and $\angle ABE = \angle ABD + \angle DBE$)

(N.B. Each triangle has angles of θ, $90° - \theta$ and $90°$. Alternate and corresponding angles are equal; $AB \parallel DE \parallel FG$ and so on. Therefore all triangles are similar because they are equi-angular.)

In $\triangle BDE$,

$$\sin\theta = \frac{ED}{BD}$$

$$ED = BD\sin\theta$$
$$= 6\sin\theta\sin\theta$$
$$= 6\sin^2\theta$$

In $\triangle DEF$,
$\angle FDE = \angle DAB = \theta$
(corresponding angles; $DE \parallel AB$)

$$\therefore \sin\theta = \frac{EF}{ED}$$

$$EF = ED\sin\theta$$
$$= 6\sin^2\theta \,.\, \sin\theta$$
$$= 6\sin^3\theta.$$

(ii) Now $BD = 6\sin\theta$
and $EF = 6\sin^3\theta$
$\therefore GH = 6\sin^5\theta$
(on continuing the pattern)

$\therefore BD + EF + GH + \ldots$
$= 6\sin\theta + 6\sin^3\theta + 6\sin^5\theta + \ldots$

Which is a geometric series with $a = 6\sin\theta$ and $r = \sin^2\theta$.

For a limiting sum to exist:
$|r| < 1$
i.e. $|\sin^2\theta| < 1$
i.e. $-1 < \sin^2\theta < 1$

Since $0 < \theta < 90°$, $-1 < \sin^2\theta < 1$, therefore a limiting sum exists.

$$S_\infty = \frac{a}{1 - r}$$

$$= \frac{6\sin\theta}{1 - \sin^2\theta}$$

$$= \frac{6\sin\theta}{\cos^2\theta} \quad (\text{since } \sin^2\theta + \cos^2\theta = 1)$$

$$= \frac{6\sin\theta}{\cos\theta \,.\, \cos\theta}$$

$$= \frac{6\sin\theta}{\cos\theta} \times \frac{1}{\cos\theta}$$

$$= 6\tan\theta \times \sec\theta$$
$$= 6\sec\theta\tan\theta.$$

QUESTION 10

(a) (i) To find the x-coordinates, α and β, of the points of intersection, we need to solve simultaneously:

$$y = mx + b \quad \ldots(1)$$
$$y = x^2 \quad \ldots(2)$$

Substituting (2) into (1):

$$x^2 = mx + b$$
$$x^2 - mx - b = 0$$

Now sum of the roots $= \alpha + \beta$

$$= \frac{-(-m)}{1}$$
$$= m.$$

Product of the roots $= \alpha\beta$

$$= \frac{(-b)}{1}$$
$$= -b.$$

(ii) $AB = \sqrt{(\alpha - \beta)^2 + (\alpha^2 - \beta^2)^2}$

$$= \sqrt{(\alpha - \beta)^2[1 + (\alpha + \beta)^2]}$$

(using given substitution)

N.B. We know $\alpha + \beta = m$ and $\alpha\beta = -b$, $\therefore$ the above expression needs to be expressed in terms of $(\alpha + \beta)$ and $(\alpha\beta)$.

Now $\quad (\alpha - \beta)^2 = \alpha^2 + \beta^2 - 2\alpha\beta$

and $\quad (\alpha + \beta)^2 = \alpha^2 + \beta^2 + 2\alpha\beta$

$$\therefore (\alpha + \beta)^2 - 4\alpha\beta = \alpha^2 + \beta^2 + 2\alpha\beta - 4\alpha\beta$$
$$= \alpha^2 + \beta^2 - 2\alpha\beta$$
$$= (\alpha - \beta)^2$$

$$\therefore AB = \sqrt{[(\alpha + \beta)^2 - 4\alpha\beta][1 + (\alpha + \beta)^2]}$$
$$= \sqrt{[m^2 - 4(-b)](1 + m^2)}$$

(on substitution)

$$= \sqrt{(m^2 + 4b)(1 + m^2)}.$$

(iii) Area of triangle ABP

$$= \frac{1}{2} \times \text{base} \times \text{perpendicular height}$$

where base $= AB$ and perpendicular height is the perpendicular distance of the point $P(x, x^2)$ from the line $y = mx + b$, i.e. $mx - y + b = 0$.

Now $d = \left|\dfrac{Ax_1 + By_1 + C}{\sqrt{A^2 + B^2}}\right|$

$$= \left|\frac{mx - 1x^2 + b}{\sqrt{m^2 + (-1)^2}}\right|$$

$$= \frac{mx - x^2 + b}{\sqrt{m^2 + 1}}$$

$\therefore$ Area of triangle ABP

$$= \frac{1}{2} \times \sqrt{(m^2 + 4b)(1 + m^2)} \times \frac{mx - x^2 + b}{\sqrt{m^2 + 1}}$$

$$= \frac{1}{2}\sqrt{m^2 + 4b}\,.\sqrt{1 + m^2} \times \frac{mx - x^2 + b}{\sqrt{m^2 + 1}}$$

$$= \frac{1}{2}\sqrt{m^2 + 4b} \times (mx - x^2 + b)$$

$$= \frac{1}{2}(mx - x^2 + b)\sqrt{m^2 + 4b}.$$

(iv) $A = \frac{1}{2}\sqrt{m^2 + 4b}\,(mx - x^2 + b)$

$$\frac{dA}{dx} = \frac{1}{2}\sqrt{m^2 + 4b}\,(m - 2x)$$

A possible maximum may occur when

$$\frac{dA}{dx} = 0$$

$$\frac{1}{2}\sqrt{m^2 + 4b}\,(m - 2x) = 0$$
$$m - 2x = 0$$
$$2x = m$$
$$x = \frac{m}{2}$$

Now $\dfrac{d^2A}{dx^2} = \dfrac{1}{2}\sqrt{m^2 + 4b} \times (-2)$

$$< 0$$

$\therefore$ A maximum area occurs when $x = \dfrac{m}{2}$.

When $x = \dfrac{m}{2}$, $y = \left(\dfrac{m}{2}\right)^2 = \dfrac{m^2}{4}$.

$\therefore$ Coordinates of P are $\left(\dfrac{m}{2}, \dfrac{m^2}{4}\right)$.

(b) (i) Both Xuan and Yvette wait for a quarter of an hour, so if the difference between their arrival times ($x - y$ or $y - x$) is less than or equal to a quarter of an hour, then they will both be there at the same time and hence meet.

(ii) *Sketch* $x - y = \dfrac{1}{4}$

i.e. $\quad y = x - \dfrac{1}{4}$

When $y = 0, x = \dfrac{1}{4}$

$\therefore$ x intercept of $x - y = \frac{1}{4}$ is $\left(\frac{1}{4}, 0\right)$.

Also when $y = 1, x = \frac{3}{4}$.

Sketch $y - x = \frac{1}{4}$

i.e. $\quad y = \frac{1}{4} + x$

When $x = 0, y = \frac{1}{4}$

$\therefore$ y intercept of $y - x = \frac{1}{4}$ is $\left(0, \frac{1}{4}\right)$.

Also when $x = 1, y = \frac{3}{4}$.

Determine the region where each inequality holds:

Test $(0, 0)$ in $x - y \leqslant \frac{1}{4}$:

$$0 - 0 \leqslant \frac{1}{4}$$

$\therefore$ Region is above the line $x - y = 4$.

Test $(0, 0)$ in $y - x \leqslant \frac{1}{4}$:

$$0 - 0 \leqslant \frac{1}{4}$$

$\therefore$ Region is below the line $y - x = \frac{1}{4}$.

Also, $0 \leqslant x \leqslant 1$ and $0 \leqslant y \leqslant 1$.

$\therefore$ Shaded region is region where all 4 inequalities hold.

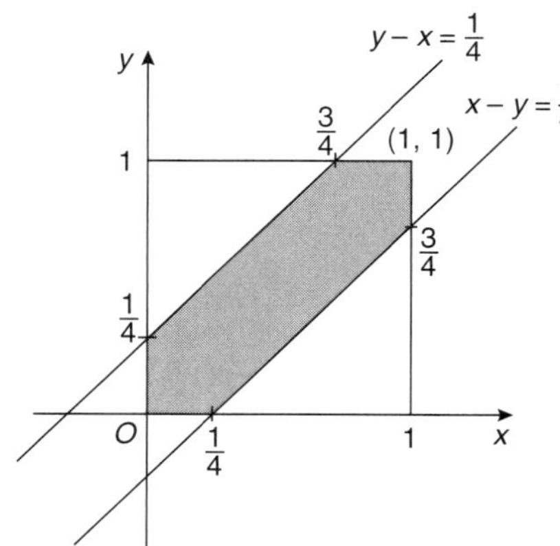

Area shaded = Area of square − 2 × unshaded triangles

$$= 1 - 2 \times \frac{1}{2} \times \frac{3}{4} \times \frac{3}{4}$$

$$= 1 - \frac{9}{16}$$

$$= \frac{7}{16} \text{ units}^2$$

$\therefore$ P(They will meet) $= \frac{7}{16}$.

(iii) For the probability of meeting to be 50%, half of the square must be shaded. Therefore, half of the square must be unshaded, i.e. unshaded area $= \frac{1}{2}$.

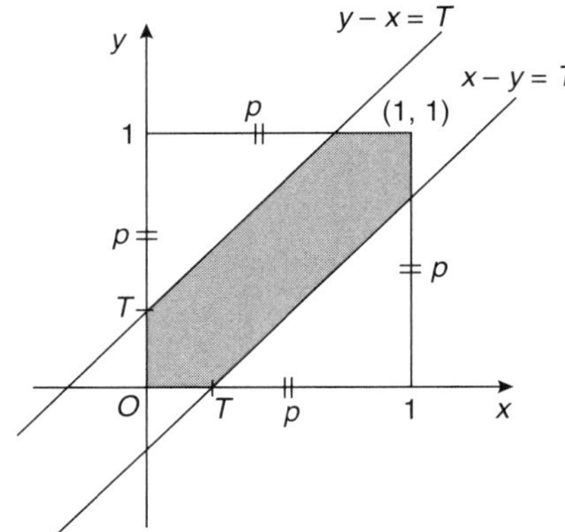

Let p be the distance as shown on the diagram.

Let $T = \frac{t}{60}$ represent the fraction of an hour they wait before leaving.

$$\therefore \text{ Unshaded area} = 2 \times \frac{b \times h}{2} = \frac{1}{2}$$

$$= 2 \times \frac{p^2}{2} = \frac{1}{2}$$

$$\therefore p^2 = \frac{1}{2}$$

$$\therefore p = \pm\frac{1}{\sqrt{2}}$$

But $p > 0$

$$\therefore p = \frac{1}{\sqrt{2}}$$

Now $T = 1 - p$

$$= 1 - \frac{1}{\sqrt{2}}$$

$$\therefore \frac{t}{60} = 1 - \frac{1}{\sqrt{2}}$$

$$t = 60\left(1 - \frac{1}{\sqrt{2}}\right)$$

$= 17.5735\ldots$ (by calc.)

$\doteqdot 18$ minutes

$\therefore$ They have a 50% chance of meeting if $t \doteqdot 18$ minutes.

BOARD OF STUDIES
NEW SOUTH WALES

2006

HIGHER SCHOOL CERTIFICATE EXAMINATION

Mathematics

General Instructions

- Reading time – 5 minutes
- Working time – 3 hours
- Write using black or blue pen
- Board-approved calculators may be used
- A table of standard integrals is provided at the back of this paper
- All necessary working should be shown in every question

Total marks – 120

- Attempt Questions 1–10
- All questions are of equal value

Total marks – 120
Attempt Questions 1–10
All questions are of equal value

Answer each question in a SEPARATE writing booklet. Extra writing booklets are available.

Marks

Question 1 (12 marks) Use a SEPARATE writing booklet.

(a) Evaluate $e^{-0.5}$ correct to three decimal places. **2**

(b) Factorise $2x^2 + 5x - 3$. **2**

(c) Sketch the graph of $y = |x + 4|$. **2**

(d) **2**

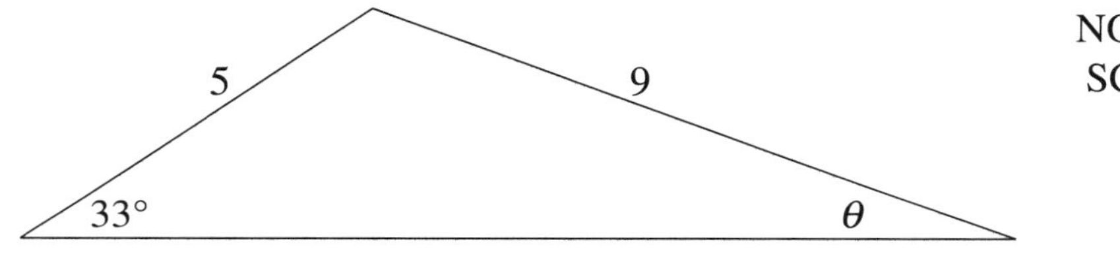

Find the value of θ in the diagram. Give your answer to the nearest degree.

(e) Solve $3 - 5x \leq 2$. **2**

(f) Find the limiting sum of the geometric series $\frac{13}{5} + \frac{13}{25} + \frac{13}{125} + \cdots$. **2**

Marks

Question 2 (12 marks) Use a SEPARATE writing booklet.

(a) Differentiate with respect to x:

(i) $x \tan x$ **2**

(ii) $\dfrac{\sin x}{x+1}$. **2**

(b) (i) Find $\int 1 + e^{7x}\,dx$. **2**

(ii) Evaluate $\displaystyle\int_0^3 \frac{8x}{1+x^2}\,dx$. **3**

(c) Find the equation of the tangent to the curve $y = \cos 2x$ at the point whose x-coordinate is $\dfrac{\pi}{6}$. **3**

Marks

Question 3 (12 marks) Use a SEPARATE writing booklet.

(a)

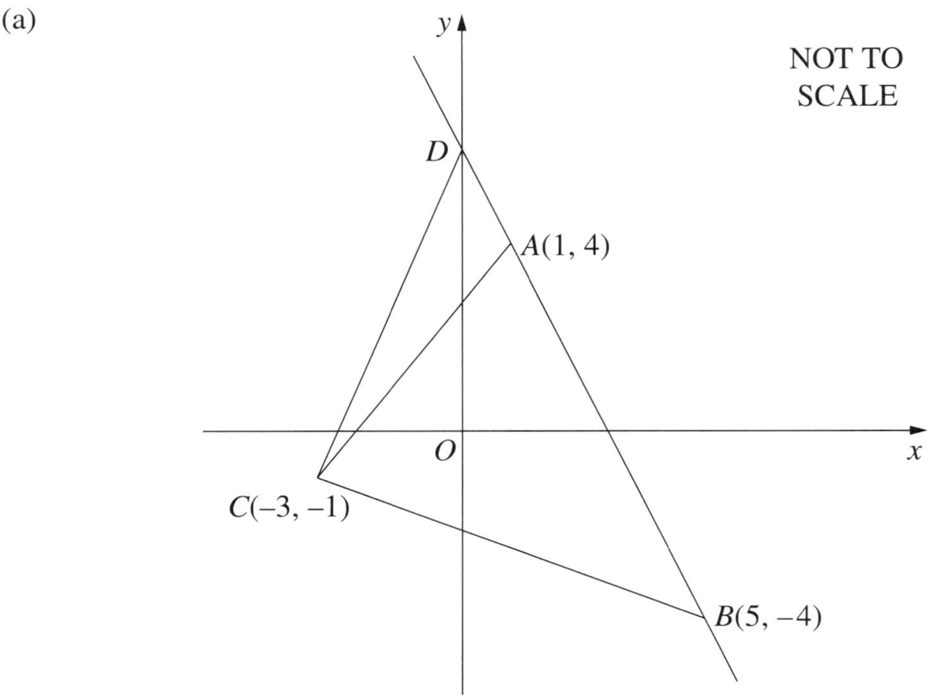

In the diagram, A, B and C are the points (1, 4), (5, –4) and (–3, –1) respectively. The line AB meets the y-axis at D.

(i) Show that the equation of the line AB is $2x + y - 6 = 0$. **2**

(ii) Find the coordinates of the point D. **1**

(iii) Find the perpendicular distance of the point C from the line AB. **1**

(iv) Hence, or otherwise, find the area of the triangle ADC. **2**

Question 3 continues

Marks

Question 3 (continued)

(b) Evaluate $\sum_{r=2}^{4} \frac{1}{r}$. **1**

(c) On the first day of the harvest, an orchard produces 560 kg of fruit. On the next day, the orchard produces 543 kg, and the amount produced continues to decrease by the same amount each day.

(i) How much fruit is produced on the fourteenth day of the harvest? **2**

(ii) What is the total amount of fruit that is produced in the first 14 days of the harvest? **1**

(iii) On what day does the daily production first fall below 60 kg? **2**

End of Question 3

Marks

Question 4 (12 marks) Use a SEPARATE writing booklet.

(a)

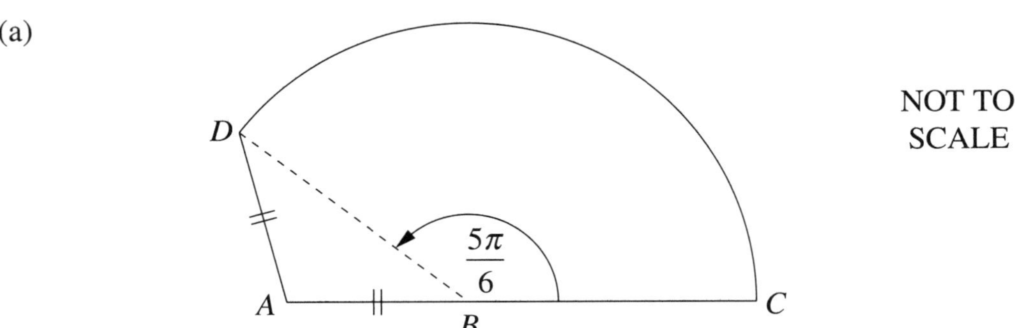

In the diagram, $ABCD$ represents a garden. The sector BCD has centre B and $\angle DBC = \frac{5\pi}{6}$.

The points A, B and C lie on a straight line and $AB = AD = 3$ metres.

Copy or trace the diagram into your writing booklet.

(i) Show that $\angle DAB = \frac{2\pi}{3}$. **1**

(ii) Find the length of BD. **2**

(iii) Find the area of the garden $ABCD$. **2**

Question 4 continues

Marks

Question 4 (continued)

(b) **3**

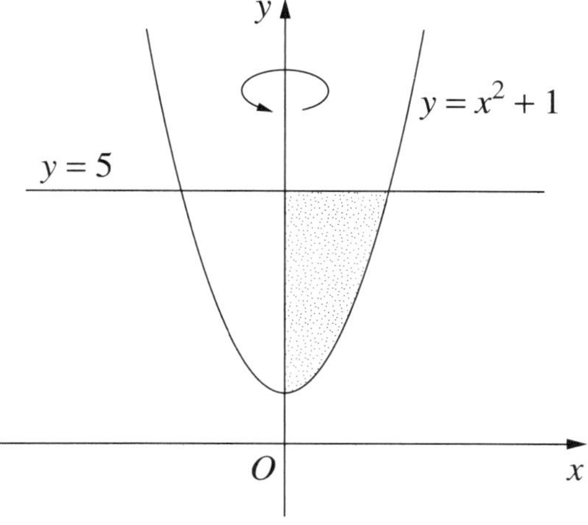

In the diagram, the shaded region is bounded by the parabola $y = x^2 + 1$, the y-axis and the line $y = 5$.

Find the volume of the solid formed when the shaded region is rotated about the y-axis.

(c) A chessboard has 32 black squares and 32 white squares. Tanya chooses three different squares at random.

(i) What is the probability that Tanya chooses three white squares? **2**

(ii) What is the probability that the three squares Tanya chooses are the same colour? **1**

(iii) What is the probability that the three squares Tanya chooses are not the same colour? **1**

End of Question 4

Marks

Question 5 (12 marks) Use a SEPARATE writing booklet.

(a) A function $f(x)$ is defined by $f(x) = 2x^2(3 - x)$.

(i) Find the coordinates of the turning points of $y = f(x)$ and determine their nature. **3**

(ii) Find the coordinates of the point of inflexion. **1**

(iii) Hence sketch the graph of $y = f(x)$, showing the turning points, the point of inflexion and the points where the curve meets the x-axis. **3**

(iv) What is the minimum value of $f(x)$ for $-1 \le x \le 4$? **1**

(b) (i) Show that $\frac{d}{dx} \log_e(\cos x) = -\tan x$. **1**

(ii) **3**

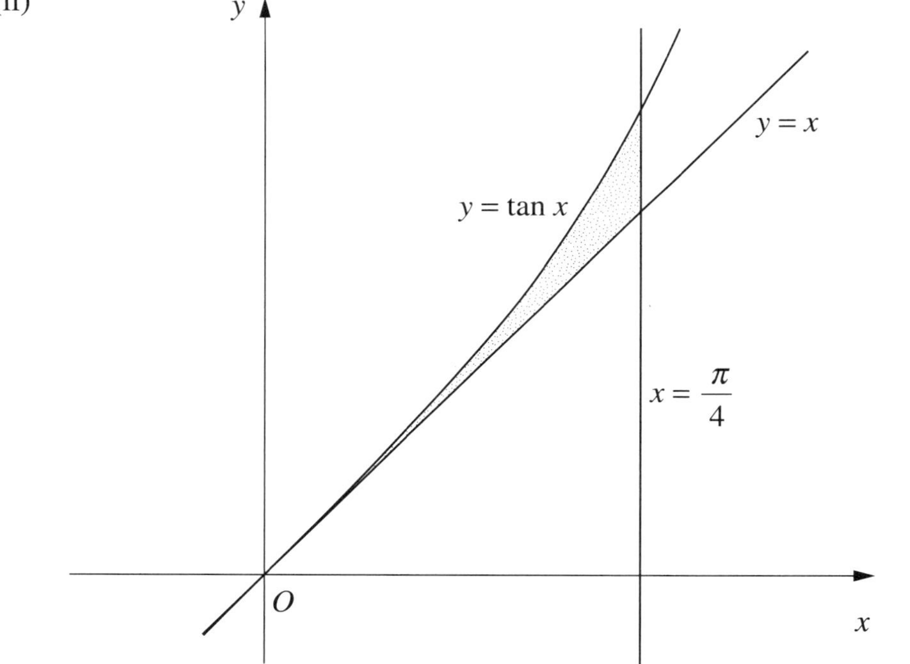

The shaded region in the diagram is bounded by the curve $y = \tan x$ and the lines $y = x$ and $x = \frac{\pi}{4}$.

Using the result of part (i), or otherwise, find the area of the shaded region.

Marks

Question 6 (12 marks) Use a SEPARATE writing booklet.

(a)

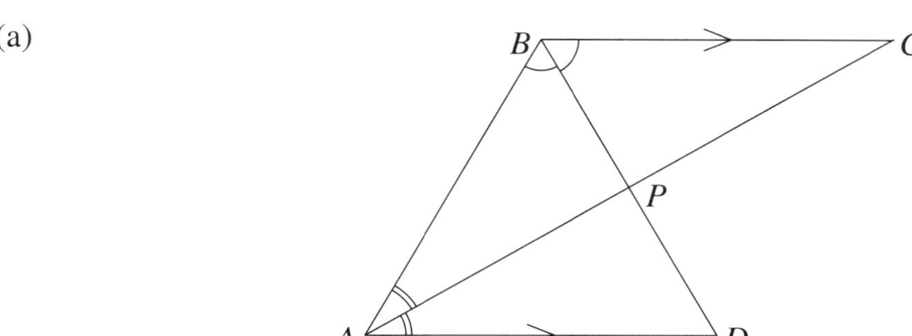

In the diagram, AD is parallel to BC, AC bisects $\angle BAD$ and BD bisects $\angle ABC$. The lines AC and BD intersect at P.

Copy or trace the diagram into your writing booklet.

(i) Prove that $\angle BAC = \angle BCA$. **1**

(ii) Prove that $\Delta ABP \equiv \Delta CBP$. **2**

(iii) Prove that $ABCD$ is a rhombus. **3**

(b) A rare species of bird lives only on a remote island. A mathematical model predicts that the bird population, P, is given by

$$P = 150 + 300e^{-0.05t}$$

where t is the number of years after observations began.

(i) According to the model, how many birds were there when observations began? **1**

(ii) According to the model, what will be the rate of change in the bird population ten years after observations began? **2**

(iii) What does the model predict will be the limiting value of the bird population? **1**

(iv) The species will become eligible for inclusion in the endangered species list when the population falls below 200. When does the model predict that this will occur? **2**

Marks

Question 7 (12 marks) Use a SEPARATE writing booklet.

(a) Let α and β be the solutions of $x^2 - 3x + 1 = 0$.

(i) Find $\alpha\beta$. **1**

(ii) Hence find $\alpha + \frac{1}{\alpha}$. **1**

(b) A function $f(x)$ is defined by $f(x) = 1 + 2\cos x$.

(i) Show that the graph of $y = f(x)$ cuts the x-axis at $x = \frac{2\pi}{3}$. **1**

(ii) Sketch the graph of $y = f(x)$ for $-\pi \le x \le \pi$ showing where the graph cuts each of the axes. **3**

(iii) Find the area under the curve $y = f(x)$ between $x = -\frac{\pi}{2}$ and $x = \frac{2\pi}{3}$. **3**

(c) (i) Write down the discriminant of $2x^2 + (k - 2)x + 8$, where k is a constant. **1**

(ii) Hence, or otherwise, find the values of k for which the parabola $y = 2x^2 + kx + 9$ does not intersect the line $y = 2x + 1$. **2**

Marks

Question 8 (12 marks) Use a SEPARATE writing booklet.

(a) A particle is moving in a straight line. Its displacement, x metres, from the origin, O, at time t seconds, where $t \geq 0$, is given by $x = 1 - \dfrac{7}{t+4}$.

(i) Find the initial displacement of the particle. **1**

(ii) Find the velocity of the particle as it passes through the origin. **3**

(iii) Show that the acceleration of the particle is always negative. **1**

(iv) Sketch the graph of the displacement of the particle as a function of time. **2**

(b) Joe borrows \$200 000 which is to be repaid in equal monthly instalments. The interest rate is 7.2% per annum reducible, calculated monthly.

It can be shown that the amount, $\$A_n$, owing after the nth repayment is given by the formula:

$$A_n = 200\,000r^n - M(1 + r + r^2 + \cdots + r^{n-1}),$$

where $r = 1.006$ and $\$M$ is the monthly repayment. (Do NOT show this.)

(i) The minimum monthly repayment is the amount required to repay the loan in 300 instalments. **3**

Find the minimum monthly repayment.

(ii) Joe decides to make repayments of \$2800 each month from the start of the loan. **2**

How many months will it take for Joe to repay the loan?

Marks

Question 9 (12 marks) Use a SEPARATE writing booklet.

(a) Find the coordinates of the focus of the parabola $12y = x^2 - 6x - 3$. **2**

(b) During a storm, water flows into a 7000-litre tank at a rate of $\frac{dV}{dt}$ litres per minute, where $\frac{dV}{dt} = 120 + 26t - t^2$ and t is the time in minutes since the storm began.

(i) At what times is the tank filling at twice the initial rate? **2**

(ii) Find the volume of water that has flowed into the tank since the start of the storm as a function of t. **1**

(iii) Initially, the tank contains 1500 litres of water. When the storm finishes, 30 minutes after it began, the tank is overflowing. **2**

How many litres of water have been lost?

(c)

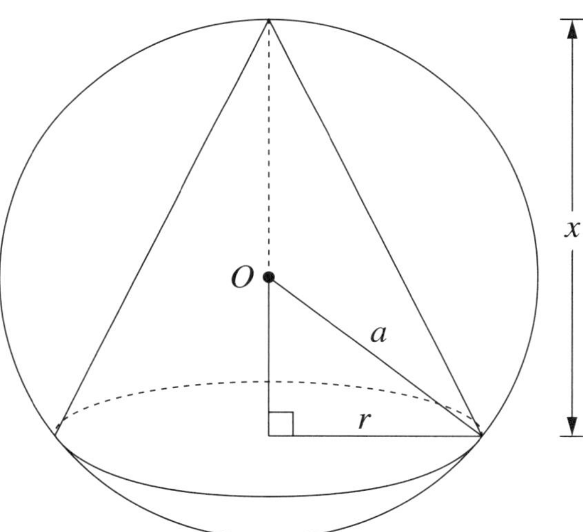

A cone is inscribed in a sphere of radius a, centred at O. The height of the cone is x and the radius of the base is r, as shown in the diagram.

(i) Show that the volume, V, of the cone is given by $V = \frac{1}{3}\pi\left(2ax^2 - x^3\right)$. **2**

(ii) Find the value of x for which the volume of the cone is a maximum. You must give reasons why your value of x gives the maximum volume. **3**

Marks

Question 10 (12 marks) Use a SEPARATE writing booklet.

(a) Use Simpson's rule with three function values to find an approximation to the value of $\displaystyle\int_{0.5}^{1.5} (\log_e x)^3\, dx$. **2**

Give your answer correct to three decimal places.

Question 10 continues

Marks

Question 10 (continued)

(b)

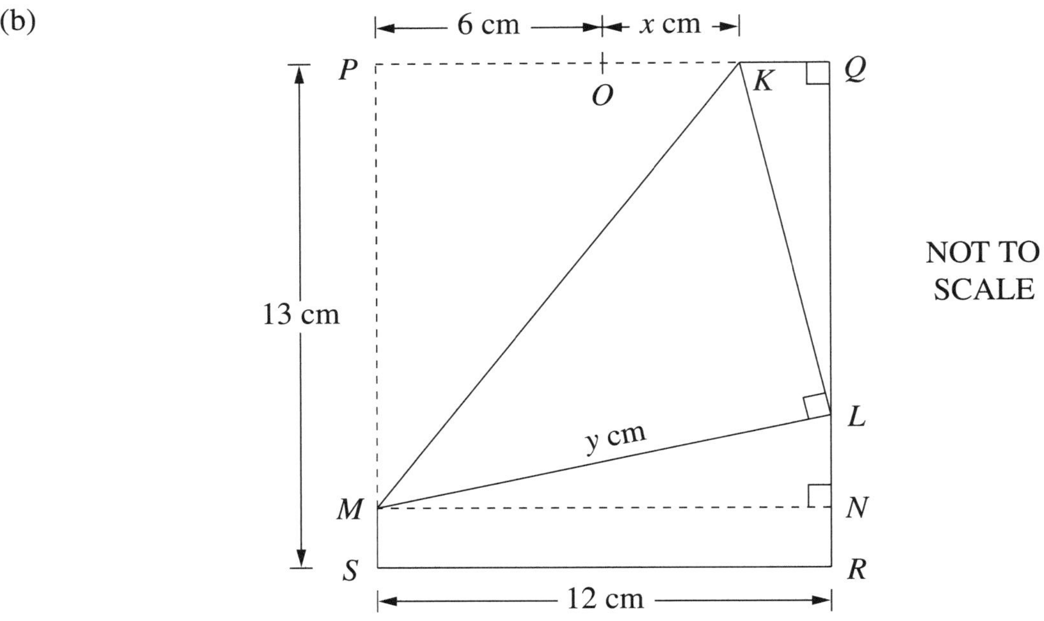

A rectangular piece of paper $PQRS$ has sides $PQ = 12$ cm and $PS = 13$ cm. The point O is the midpoint of PQ. The points K and M are to be chosen on OQ and PS respectively, so that when the paper is folded along KM, the corner that was at P lands on the edge QR at L. Let $OK = x$ cm and $LM = y$ cm.

Copy or trace the diagram into your writing booklet.

(i) Show that $QL^2 = 24x$. **1**

(ii) Let N be the point on QR for which MN is perpendicular to QR. **3**

By showing that $\Delta QKL \,|||\, \Delta NLM$, deduce that $y = \dfrac{\sqrt{6}(6+x)}{\sqrt{x}}$.

(iii) Show that the area, A, of ΔKLM is given by $A = \dfrac{\sqrt{6}(6+x)^2}{2\sqrt{x}}$. **1**

(iv) Use the fact that $12 \le y \le 13$ to find the possible values of x. **2**

(v) Find the minimum possible area of ΔKLM. **3**

End of paper

2006 Higher School Certificate Worked Answers

QUESTION 1

(a) $e^{-0.5} = 0.6065\ldots$ (by calc.)
$= 0.607$ to 3 decimal places.

(b) $2x^2 + 5x - 3 = (2x - 1)(x + 3)$.

(c)

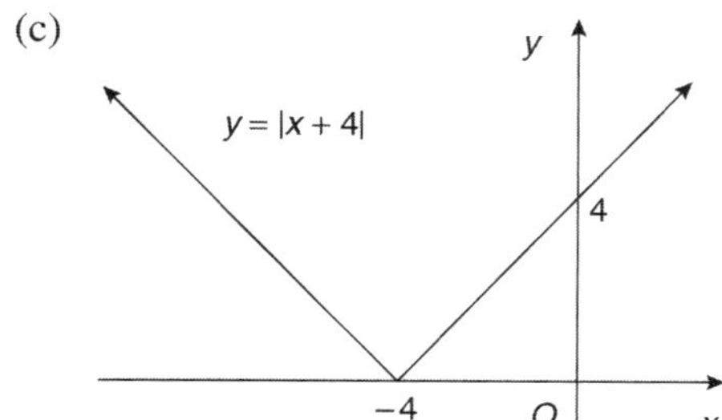

(d) By the sine rule:

$$\frac{\sin\theta}{5} = \frac{\sin 33^\circ}{9}$$

$$\therefore \sin\theta = \frac{5\sin 33^\circ}{9}$$

$= 0.3025\ldots$ (by calc.)

$\therefore \theta = 17.6124\ldots^\circ$ (by calc.)

$= 18^\circ$ to the nearest degree.

(e) $3 - 5x \leqslant 2$

$-5x \leqslant -1$

$x \geqslant \frac{1}{5}$.

(f) $S = \dfrac{a}{1-r}$

$a = \dfrac{13}{5}, r = \dfrac{1}{5}$

$$S = \frac{\frac{13}{5}}{1 - \frac{1}{5}}$$

$= 3\frac{1}{4}$.

QUESTION 2

(a) (i) $\dfrac{d}{dx}(x\tan x)$

$= x \cdot \dfrac{d}{dx}(\tan x) + \tan x \cdot \dfrac{d}{dx}(x)$

$= x \cdot \sec^2 x + \tan x \cdot 1$

$= x\sec^2 x + \tan x$.

(ii) $\dfrac{d}{dx}\left(\dfrac{\sin x}{x+1}\right)$

$$= \frac{(x+1)\cdot\frac{d}{dx}(\sin x) - \sin x \cdot \frac{d}{dx}(x+1)}{(x+1)^2}$$

$$= \frac{(x+1)\cdot\cos x - \sin x \cdot 1}{(x+1)^2}$$

$$= \frac{(x+1)\cdot\cos x - \sin x}{(x+1)^2}.$$

(b) (i) $\displaystyle\int 1 + e^{7x}\,dx = x + \frac{1}{7}e^{7x} + C$.

(ii) $\displaystyle\int_0^3 \frac{8x}{1+x^2}\,dx$

$$= 4\int_0^3 \frac{2x}{1+x^2}\,dx$$

$$= 4\Big[\log_e(1+x^2)\Big]_0^3$$

$= 4\log_e 10 - 4\log_e 1$

$= 4\log_e 10 - 0$

$= 4\log_e 10$.

(c) $y = \cos 2x$

$\dfrac{dy}{dx} = -2\sin 2x$

At $x = \dfrac{\pi}{6}$,

$\dfrac{dy}{dx} = -2\sin 2\left(\dfrac{\pi}{6}\right)$

$= -2\sin\dfrac{\pi}{3}$

$= -2\left(\dfrac{\sqrt{3}}{2}\right)$

$= -\sqrt{3}$.

At $x = \dfrac{\pi}{6}$,

$y = \cos 2x$

$= \cos 2\left(\dfrac{\pi}{6}\right)$

$= \cos\dfrac{\pi}{3}$

$= \dfrac{1}{2}$

$\therefore$ Equation of the tangent at $\left(\frac{\pi}{6}, \frac{1}{2}\right)$

with gradient of $-\sqrt{3}$ is:

$$y - \frac{1}{2} = -\sqrt{3}\left(x - \frac{\pi}{6}\right)$$

$$2y - 1 = -2\sqrt{3}x + \frac{2\sqrt{3}\pi}{6}$$

$$12y - 6 = -12\sqrt{3}x + 2\sqrt{3}\pi$$

$$12\sqrt{3}x + 12y - 6 - 2\sqrt{3}\pi = 0$$

$$6\sqrt{3}x + 6y - 3 - \sqrt{3}\pi = 0.$$

QUESTION 3

(a)

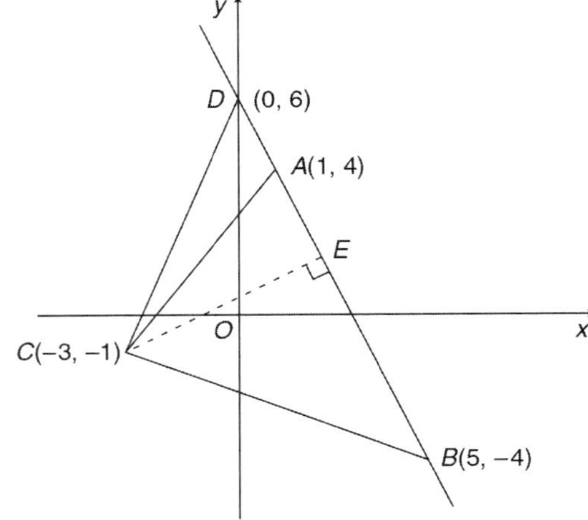

(i) Gradient of $AB = \frac{y_2 - y_1}{x_2 - x_1}$

$$= \frac{-4-4}{5-1}$$

$$= \frac{-8}{4}$$

$$= -2$$

$\therefore$ The equation of line AB with gradient $= -2$ passing through $A(1, 4)$ is given by

$$y - 4 = -2(x - 1)$$

$$y - 4 = -2x + 2$$

$$\therefore\ 2x + y - 6 = 0.$$

(ii) Line AB cuts the y-axis at D where $x = 0$.

$$\therefore\ 2(0) + y - 6 = 0$$

$$\therefore\ y = 6$$

$\therefore$ D has coordinates $(0, 6)$.

(iii) $C \equiv (-3, -1)$

line AB: $2x + y - 6 = 0$

$$d = \frac{|Ax_1 + By_1 + C|}{\sqrt{A^2 + B^2}}$$

$$= \frac{|2(-3) + 1(-1) - 6|}{\sqrt{(2)^2 + (1)^2}}$$

$$= \frac{|-13|}{\sqrt{5}}$$

$$= \frac{13}{\sqrt{5}}.$$

(iv) Let E be the foot of the perpendicular drawn from point C to the line AB.

Now $AD = \sqrt{(0-1)^2 + (6-4)^2}$

$$= \sqrt{(-1)^2 + 2^2}$$

$$= \sqrt{1+4}$$

$$= \sqrt{5}.$$

Area $\triangle ADC = \frac{1}{2} \times b \times h$

$$= \frac{1}{2} \times AD \times EC$$

$$= \frac{1}{2} \times \sqrt{5} \times \frac{13}{\sqrt{5}}$$

$$= 6\tfrac{1}{2} \text{ units}^2.$$

(b) $\sum_{r=2}^{4} \frac{1}{r} = \frac{1}{2} + \frac{1}{3} + \frac{1}{4}$

$$= 1\tfrac{1}{12}. \quad \text{(by calc.)}$$

(c) (i) The amount of fruit produced each day is an arithmetic sequence:
560, 543, 526, . . .
with $a = 560$ and $d = -17$

$$T_n = a + (n-1)d$$

$$\therefore\ T_{14} = 560 + 13(-17)$$

$$= 339$$

$\therefore$ 339 kg of fruit will be produced on the fourteenth day.

(ii) $S_n = \frac{n}{2}\left[2a + (n-1)d\right]$

$n = 14, \quad a = 560, \quad d = -17$

$$\therefore\ S_{14} = \frac{14}{2}\left[2(560) + 13(-17)\right]$$

$= \frac{14}{2}\left[1120 - 221\right]$

$= \frac{14}{2} \times 899$

$= 6293$ (by calc.)

$\therefore$ 6293 kg of fruit will be produced in the first fourteen days of harvest.

(iii)

$$a + (n-1)d < 60$$
$$560 + (n-1) \times (-17) < 60$$
$$560 - 17n + 17 < 60$$
$$-17n < -517$$
$$n > \frac{517}{17}$$
$$n > 30\frac{7}{17}$$
$$\therefore n = 31$$

$\therefore$ The daily production first falls below 60 kg on the thirty-first day.

QUESTION 4

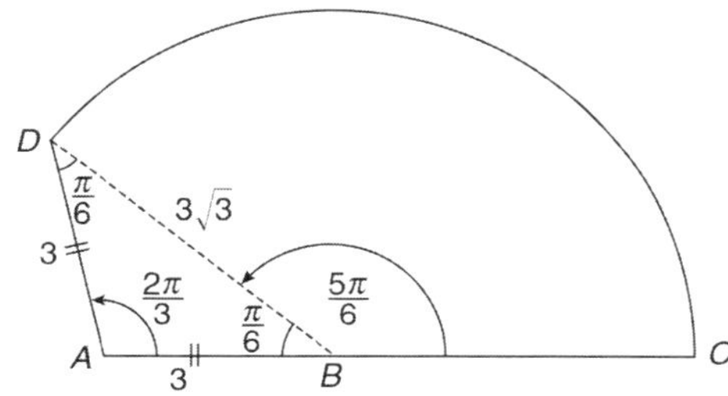

(a) (i) In $\triangle ABD$,

$\angle ABD = \pi - \angle CBD$ (ABC is a straight angle, π)

$= \pi - \frac{5\pi}{6}$

$= \frac{\pi}{6}$

Also $\angle ABD = \angle ADB$ (angles opp. equal sides; $AD = AB$)

$= \frac{\pi}{6}$

Now $\angle DAB + \angle ABD + \angle ADB = \pi$ (angle sum of $\triangle ABD$ is π)

$\therefore \angle DAB + \frac{\pi}{6} + \frac{\pi}{6} = \pi$

$\therefore \angle DAB = \pi - \frac{2\pi}{6}$

$= \frac{2\pi}{3}$.

(ii) By the sine rule:

$$\frac{BD}{\sin\frac{2\pi}{3}} = \frac{3}{\sin\frac{\pi}{6}}$$

$$\therefore BD = \frac{3 \times \sin\frac{2\pi}{3}}{\sin\frac{\pi}{6}}$$

$$= \frac{3 \times \frac{\sqrt{3}}{2}}{\frac{1}{2}}$$

$$= \frac{6 \times \sqrt{3}}{2}$$

$$= 3\sqrt{3} \text{ m.}$$

(iii) Area of Garden $ABCD$
= Area of $\triangle ABD$ + Area of Sector BCD:

$$\text{Area of } \triangle ABD = \frac{1}{2}bd \sin A$$
$$= \frac{1}{2} \times 3^2 \times \sin\frac{2\pi}{3}$$
$$= \frac{1}{2} \times 9 \times \frac{\sqrt{3}}{2}$$
$$= \frac{9\sqrt{3}}{4} \text{ m}^2$$

$$\text{Area of Sector } BCD = \frac{1}{2}r^2\theta$$
$$= \frac{1}{2} \times \left(3\sqrt{3}\right)^2 \times \frac{5\pi}{6}$$
$$= \frac{1}{2} \times 27 \times \frac{5\pi}{6}$$
$$= \frac{135\pi}{12}$$
$$= \frac{45\pi}{4} \text{ m}^2$$

Area of Garden
= Area of $\triangle ABD$ + Area of Sector BCD

$= \frac{9\sqrt{3}}{4} + \frac{45\pi}{4}$

$= \frac{9\sqrt{3} + 45\pi}{4}$ m^2.

(b) Now $y = x^2 + 1$

$\therefore x^2 = y - 1$

Now $V = \pi \int_a^b x^2\,dy$

$$\therefore V = \pi \int_1^5 (y-1)\,dy$$

$$= \pi \left[\frac{y^2}{2} - y \right]_1^5$$

$$= \pi \left[\left(\frac{5^2}{2} - 5 \right) - \left(\frac{1^2}{2} - 1 \right) \right]$$

$$= \pi \left(7\frac{1}{2} + \frac{1}{2} \right)$$

$$= 8\pi \text{ units}^3.$$

(c) (i) P(Three white squares)

$= \text{P(WWW)}$

$= \frac{32}{64} \times \frac{31}{63} \times \frac{30}{62}$

$= \frac{29\,760}{249\,984}$

$= \frac{5}{42}$

$= 0.11904\ldots$ (by calc.)

$\doteqdot 0.1190.$

(ii) P(Three squares are the same colour)

$= \text{P(WWW)} + \text{P(BBB)}$

$= \frac{29\,760}{249\,984} + \frac{29\,760}{249\,984}$

$= \frac{5}{42} + \frac{5}{42}$

$= \frac{5}{21}$

$= 0.23809\ldots$ (by calc.)

$\doteqdot 0.2381.$

(iii) P(Three squares not the same colour)

$= 1 - (\text{P(WWW)} + \text{P(BBB)})$

$= 1 - \frac{5}{21}$

$= \frac{16}{21}$

$= 0.76190\ldots$ (by calc.)

$\doteqdot 0.7619.$

QUESTION 5

(a) (i) $f(x) = 2x^2(3 - x)$

$= 6x^2 - 2x^3$

$f'(x) = 12x - 6x^2$

Stationary points occur when

$f'(x) = 0$

$12x - 6x^2 = 0$

$6x(2 - x) = 0$

$\therefore x = 0$ or 2

Now $f(0) = 2(0)^2[3 - (0)]$

$= 0$

and $f(2) = 2(2)^2[3 - (2)]$

$= 8$

$\therefore$ Coordinates of the stationary points are $(0, 0)$ and $(2, 8)$.

Now $f'(x) = 12x - 6x^2$

$\therefore f''(x) = 12 - 12x$

$\therefore f''(0) = 12 - 12(0)$

$= 12 - 0$

$= 12 > 0$

$\therefore$ A minimum turning point occurs at $(0, 0)$.

and $f''(2) = 12 - 12(2)$

$= 12 - 24$

$= -12 < 0$

$\therefore$ A maximum turning point occurs at $(2, 8)$.

(ii) Possible points of inflexion occur when

$f''(x) = 0$

$12 - 12x = 0$

$12(1 - x) = 0$

$x = 1$

Checking the change in concavity:

x	0.9	1	1.1
$f''(x)$	>0	0	<0

$\therefore$ There is change in concavity and a point of inflexion at $x = 1$.

Now $f(1) = 2(1)^2[3 - 1]$

$= 4$

$\therefore$ $(1, 4)$ are the coordinates of the point of inflexion.

(iii) Curve meets the x-axis where

$$f(x) = 0$$
$$2x^2(3 - x) = 0$$
$$\therefore x = 0 \text{ or } 3$$

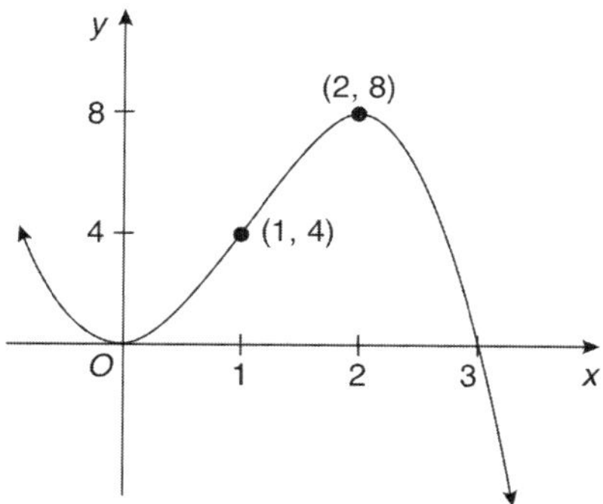

(iv) Minimum value of $f(x)$ for $-1 \leq x \leq 4$ occurs when $x = 4$,

$$f(4) = 2(4)^2[3 - (4)]$$
$$= -32.$$

(b) (i) $$\frac{d}{dx}\log_e(\cos x) = \frac{1}{\cos x} \cdot \frac{d}{dx}(\cos x)$$
$$= \frac{1}{\cos x} \cdot -\sin x$$
$$= -\frac{\sin x}{\cos x}$$
$$= -\tan x.$$

(ii) Required area

= area under curve – area of triangle

$$= \int_0^{\frac{\pi}{4}} \tan x \, dx - \frac{1}{2} \times \frac{\pi}{4} \times \frac{\pi}{4}$$
$$= -\int_0^{\frac{\pi}{4}} (-\tan x) dx - \frac{\pi^2}{32}$$
$$= -\left[\log_e(\cos x)\right]_0^{\frac{\pi}{4}} - \frac{\pi^2}{32}$$
$$= -\left[\log_e\left(\cos\frac{\pi}{4}\right) - \log_e(\cos 0)\right] - \frac{\pi^2}{32}$$
$$= -\left[\log_e\left(\frac{1}{\sqrt{2}}\right) - \log_e 1\right] - \frac{\pi^2}{32}$$
$$= -\left[\log_e 1 - \log_e \sqrt{2} - \log_e 1\right] - \frac{\pi^2}{32}$$
$$= -\left[-\log_e \sqrt{2}\right] - \frac{\pi^2}{32}$$
$$= -\left[-\log_e 2^{\frac{1}{2}}\right] - \frac{\pi^2}{32}$$
$$= -\left[-\frac{1}{2}\log_e 2\right] - \frac{\pi^2}{32}$$
$$= \frac{1}{2}\log_e 2 - \frac{\pi^2}{32}$$

$\therefore$ Area is $\left(\frac{1}{2}\log_e 2 - \frac{\pi^2}{32}\right)$ units^2.

QUESTION 6

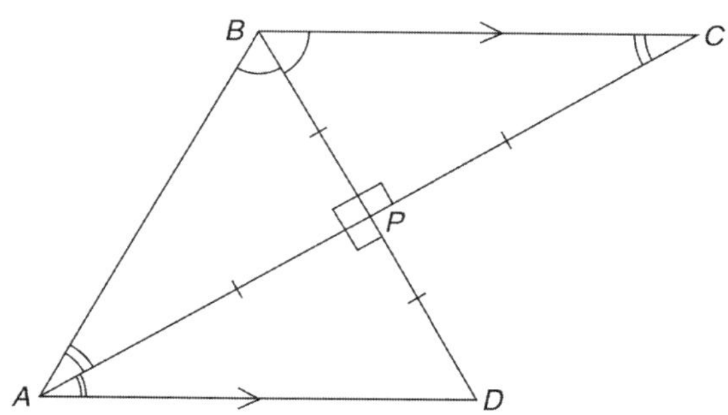

(a) (i) Now $\angle BAC = \angle CAD$ (AC bisects $\angle BAD$)

and $\angle CAD = \angle BCA$ (alternate angles, $BC \parallel AD$)

$\therefore \angle BAC = \angle BCA$.

(ii) In $\triangle ABP$ and $\triangle CBP$,

$\angle BAC = \angle BCA$ (from (i))

$\angle ABP = \angle CBP$ (BP bisects $\angle ABC$)

BP is common ($BP \perp AC$)

$\therefore \triangle ABP \equiv \triangle CBP$ (AAS)

(iii) Since $\triangle ABP \equiv \triangle CBP$,

$\therefore AP = CP$

and $\angle APB = \angle CPB$.

Also $\angle APB + \angle CPB = 180°$

(APC is a straight angle, 180°)

$\therefore \angle APB = \angle CPB$

$= 90°$

In $\triangle ABP$ and $\triangle ADP$,

$\angle BAP = \angle DAP$ (AP bisects $\angle BAD$)

$\angle APB = \angle APD$ ($BP \perp AC$)

AP is common

$\therefore \triangle ABP \equiv \triangle ADP$ (AAS)

Since $\triangle ABP \equiv \triangle ADP$

$\therefore BP = PD$

$\therefore ABCD$ is a rhombus as diagonals bisect each other at right angles.

(b) (i) $$P = 150 + 300e^{-0.05t}$$

When $t = 0$,

$$P = 150 + 300e^{-0.05 \times 0}$$
$$= 150 + 300$$
$$= 450.$$

(ii) $P = 150 + 300e^{-0.05t}$

$$\frac{dP}{dt} = -0.05 \times 300e^{-0.05t}$$
$$= -15e^{-0.05t}$$

When $t = 10$,

$$\frac{dP}{dt} = -15e^{-0.05 \times 10}$$
$$= -15e^{-0.5}$$
$$= -9.0979\ldots \quad \text{(by calc.)}$$

$\therefore$ The bird population is decreasing at a rate of 9.0979 . . . birds per year.

(iii) Now $\lim\limits_{t \to \infty} e^{-0.05t} = 0$

and $P = 150 + 300e^{-0.05t}$

$\therefore P = 150 + 300 \times 0$

$= 150.$

(iv) When $P = 200$,

$$200 = 150 + 300e^{-0.05t}$$
$$50 = 300e^{-0.05t}$$
$$\frac{50}{300} = e^{-0.05t}$$
$$\log_e \frac{50}{300} = -0.05t$$
$$t = \log_e \frac{50}{300} \div -0.05$$
$$= 35.8351\ldots \quad \text{(by calc.)}$$

$\therefore$ The bird population will be eligible for inclusion in the endangered species list during the 36th year.

QUESTION 7

(a) (i) Product of the roots $= \alpha\beta$

$$= \frac{1}{1}$$
$$= 1.$$

(ii) Sum of the roots $= \alpha + \beta$

$$= \alpha + \frac{1}{\alpha} \quad \text{(using (i))}$$
$$= \frac{3}{1}$$
$$= 3.$$

(b) (i) The curve cuts the x-axis when

$$f(x) = 0$$
$$1 + 2\cos x = 0$$
$$2\cos x = -1$$
$$\cos x = -\frac{1}{2}$$
$$x = -\frac{2\pi}{3} \text{ and } \frac{2\pi}{3}.$$
$$(\text{for } -\pi \leqslant x \leqslant \pi)$$

(ii) Period $= 2\pi$

Range: $-1 \leqslant \cos x \leqslant 1$

$\therefore -2 \leqslant 2\cos x \leqslant 2$

$\therefore -1 \leqslant 1 + 2\cos x \leqslant 3$

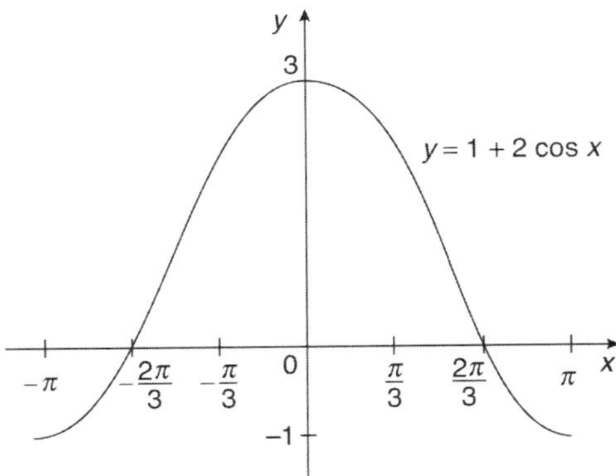

(iii) Area $= \displaystyle\int_{-\frac{\pi}{2}}^{\frac{2\pi}{3}} (1 + 2\cos x)\,dx$

$$= \Big[x + 2\sin x\Big]_{-\frac{\pi}{2}}^{\frac{2\pi}{3}}$$
$$= \left[\frac{2\pi}{3} + 2\sin\frac{2\pi}{3}\right] - \left[-\frac{\pi}{2} + 2\sin\left(-\frac{\pi}{2}\right)\right]$$
$$= \left[\frac{2\pi}{3} + 2\left(\frac{\sqrt{3}}{2}\right)\right] - \left[-\frac{\pi}{2} + 2(-1)\right]$$
$$= \frac{2\pi}{3} + \sqrt{3} + \frac{\pi}{2} + 2$$
$$= \left(\frac{7\pi}{6} + 2 + \sqrt{3}\right) \text{units}^2.$$

(c) (i) $\Delta = b^2 - 4ac$

$$= (k-2)^2 - 4(2)(8)$$
$$= (k-2)^2 - 64.$$

(ii) $y = 2x^2 + kx + 9$

$y = 2x + 1$

Solving simultaneously:

$$2x^2 + kx + 9 = 2x + 1$$
$$2x^2 + kx - 2x + 8 = 0$$
$$2x^2 + (k-2)x + 8 = 0$$

If the parabola does not intersect the line, then $2x^2 + (k-2)x + 8 = 0$ has no real roots,

i.e. $\Delta < 0$

$$(k-2)^2 - 4(2)(8) < 0$$
$$k^2 - 4k + 4 - 64 < 0$$

$$k^2 - 4k - 60 < 0$$
$$(k - 10)(k + 6) < 0$$

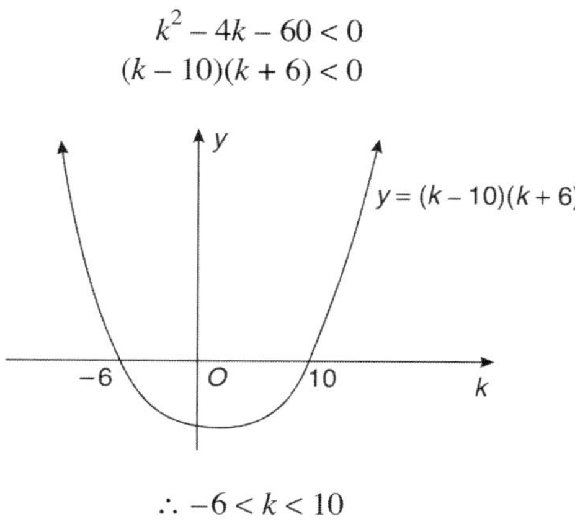

$$\therefore -6 < k < 10$$

Now $v = \dfrac{7}{(t+4)^2}$

When $t = 3$,

$$v = \frac{7}{(3+4)^2}$$
$$= \frac{1}{7}$$
$$\doteqdot 0.1428\ldots \quad \text{(by calc.)}$$

$\therefore$ The velocity of the particle as it passes through the origin is $\frac{1}{7}$ ms^{-1}.

QUESTION 8

(a) (i) $x = 1 - \dfrac{7}{t+4}$

When $t = 0$,

$$x = 1 - \frac{7}{4}$$
$$= -\frac{3}{4}$$

$\therefore$ The initial displacement of the particle is $\frac{3}{4}$ m to the left of the origin.

(ii) $v = \dfrac{dx}{dt}$

$$= \frac{d}{dt}\left(1 - \frac{7}{t+4}\right)$$
$$= \frac{d}{dt}\left(1 - 7(t+4)^{-1}\right)$$
$$= 7(t+4)^{-2}$$
$$= \frac{7}{(t+4)^2}$$

Now $x = 1 - \dfrac{7}{t+4}$

When the particle is at the origin, $x = 0$.

$$\therefore 0 = 1 - \frac{7}{t+4}$$
$$-1 = -\frac{7}{t+4}$$
$$-1(t+4) = -7$$
$$-t - 4 = -7$$
$$\therefore t = 3$$

$\therefore$ The particle is at the origin after 3 seconds.

(iii) $a = \dfrac{dv}{dt}$

$$= \frac{d}{dt}\left(\frac{7}{(t+4)^2}\right)$$
$$= \frac{d}{dt}\left(7(t+4)^{-2}\right)$$
$$= -14(t+4)^{-3}$$
$$= -\frac{14}{(t+4)^3}$$

Now for $t \geqslant 0$, $(t+4)^3 > 0$

$$\therefore -\frac{14}{(t+4)^3} < 0$$

$\therefore$ Acceleration of the particle is always negative.

(iv)

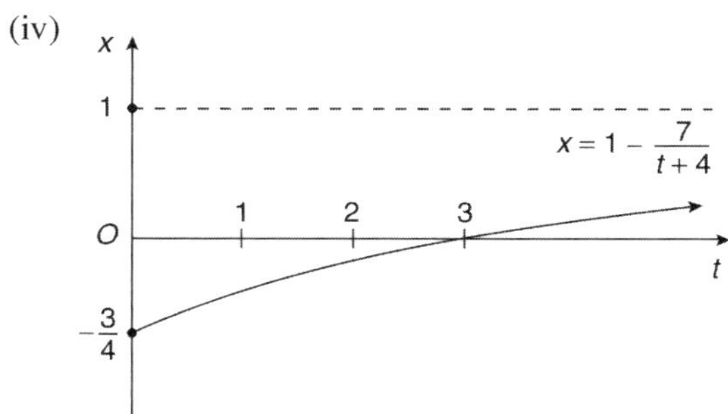

N.B.

From (i), when $t = 0, x = -\frac{3}{4}$

From (ii), when $t = 3, x = 0, v = \frac{1}{4}$

From (ii), $a < 0$

Also as $t \to \infty, x \to 1$

(b) (i) $$A_n = 200\,000r^n - M\left(1 + r + r^2 + \ldots r^{n-1}\right)$$

where $r = 1.006,\ \ n = 300,\ \ M =$ monthly repayment

$$\therefore A_{300} = 200\,000(1.006)^{300} - M\left(1 + 1.006 + 1.006^2 + \ldots + 1.006^{299}\right)$$

Now $\left(1 + 1.006 + 1.006^2 + \ldots + 1.006^{299}\right)$ is a geometric series,
where $a = 1,\ \ r = 1.006,\ \ n = 300.$

$$\text{Now } S_n = \frac{a(r^n - 1)}{r - 1}$$

$$\therefore S_{300} = \frac{1 \times (1.006^{300} - 1)}{1.006 - 1}$$

$$= \left[\frac{1.006^{300} - 1}{0.006}\right]$$

$$\therefore A_{300} = 200\,000(1.006)^{300} - M\left[\frac{1.006^{300} - 1}{0.006}\right]$$

The loan will be repaid when $A_{300} = 0$.

$$\therefore 0 = 200\,000(1.006)^{300} - M\left[\frac{1.006^{300} - 1}{0.006}\right]$$

$$M\left[\frac{1.006^{300} - 1}{0.006}\right] = 200\,000(1.006)^{300}$$

$$M = 200\,000(1.006)^{300} \div \left[\frac{1.006^{300} - 1}{0.006}\right]$$

$$M = 1\,203\,439.36 \div 836.1994\ldots$$
$$= 1439.18 \quad \text{(by calc.)}$$

$\therefore$ The minimum monthly repayment is \$1439 to the nearest dollar.

(ii) $$A_n = 200\,000(1.006)^n - M\left[\frac{1.006^n - 1}{0.006}\right]$$

When $M = 2800$,

$$\therefore A_n = 200\,000(1.006)^n - 2800\left[\frac{1.006^n - 1}{0.006}\right]$$

The loan will be repaid when $A_n = 0$.

$$\therefore 0 = 200\,000(1.006)^n - 2800\left(\frac{1.006^n - 1}{0.006}\right)$$

$$0 = 200\,000(1.006)^n - \left(\frac{2800 \times 1.006^n - 2800 \times 1}{0.006}\right)$$

$$0 = 200\,000(1.006)^n \times 0.006 - (2800 \times 1.006^n - 2800)$$
$$0 = 1200(1.006)^n - 2800(1.006)^n + 2800$$
$$0 = 1.006^n \times (1200 - 2800) + 2800$$

$$0 = -1600(1.006)^n + 2800$$
$$1600(1.006)^n = 2800$$
$$1.006^n = \frac{2800}{1600}$$
$$n \log_e 1.006 = \log_e \frac{2800}{1600}$$
$$n = \log_e \frac{2800}{1600} \div \log_e 1.006$$
$$\doteqdot 93.5488\ldots \quad \text{(by calc.)}$$

$\therefore$ The loan will be repaid in 94 months.

QUESTION 9

(a) $12y = x^2 - 6x - 3$

$\therefore x^2 - 6x = 12y + 3$

Completing the square on LHS:

$$x^2 - 6x + 9 = 12y + 3 + 9$$
$$(x-3)^2 = 12(y+1)$$

$\therefore$ Coordinates of vertex are $(3, -1)$.

Now $4a = 12$

$\therefore a = 3$

i.e. Focal length is 3 units.

$\therefore$ Coordinates of focus are $(3, 2)$.

(b) (i) $\dfrac{dv}{dt} = 120 + 26t - t^2$

At $t = 0$, $\dfrac{dv}{dt} = 120 + 26(0) - (0)^2$
$= 120$

When $\dfrac{dv}{dt} = 240$,

$$240 = 120 + 26t - t^2$$
$$t^2 - 26t + 120 = 0$$
$$(t-20)(t-6) = 0$$
$$\therefore t = 6 \text{ or } 20$$

$\therefore$ The tank is filling at twice the initial rate at 6 minutes and 20 minutes.

(ii) $\dfrac{dv}{dt} = 120 + 26t - t^2$

$$\therefore v = \int (120 + 26t - t^2)\,dt$$
$$= 120t + \frac{26}{2}t^2 - \frac{t^3}{3} + C$$
$$= 120t + 13t^2 - \frac{1}{3}t^3 + C$$

If $t = 0, v = 0 \;\therefore C = 0$

$$\therefore v = 120t + 13t^2 - \frac{1}{3}t^3.$$

(iii) When $t = 30$,

$$V = 120(30) + 13(30)^2 - \frac{1}{3}(30)^3$$
$$= 3600 + 11\,700 - 9000$$
$$= 6300.$$

$\therefore$ Total water in the tank
$= 6300 + 1500$
$= 7800$ litres.

$\therefore$ Amount of water lost
$= 7800 - 7000$
$= 800$ litres.

(c) (i) $V = \frac{1}{3}\pi r^2 h$ (where $h = x$)

$$\therefore V = \frac{1}{3}\pi r^2 x$$

Now $a^2 = (x-a)^2 + r^2$
(Pythagoras' Theorem)

$$a^2 = x^2 - 2ax + a^2 + r^2$$
$$\therefore r^2 = 2ax - x^2$$

$$\therefore V = \frac{1}{3}\pi(2ax - x^2)(x)$$
$$= \frac{1}{3}\pi(2ax^2 - x^3)$$

(ii) $V = \frac{1}{3}\pi(2ax^2 - x^3)$

$$= \frac{2}{3}\pi a x^2 - \frac{1}{3}\pi x^3$$
$$\frac{dV}{dx} = \frac{4}{3}\pi a x - \pi x^2$$
$$\frac{d^2V}{dx^2} = \frac{4}{3}\pi a - 2\pi x$$

A maximum volume occurs when

$$\frac{dV}{dx} = 0$$

$$\frac{4}{3}\pi ax - \pi x^2 = 0$$

$$\pi x\left(\frac{4}{3}a - x\right) = 0$$

$$x = 0 \text{ or } \frac{4}{3}a$$

$$\therefore x = \frac{4}{3}a$$

When $x = \frac{4}{3}a$,

$$\frac{d^2v}{dx^2} = \frac{4}{3}\pi a - 2\pi\left(\frac{4}{3}a\right)$$

$$= \frac{4}{3}\pi a - \frac{8}{3}\pi a$$

$$= -\frac{4}{3}\pi a < 0$$

$\therefore$ Cone has a maximum value when $x = \frac{4}{3}a$.

QUESTION 10

(a) $f(x) = (\log_e x)^3$

x	0.5	1	1.5
$f(x)$	$(\log_e 0.5)^3$	$(\log_e 1)^3$	$(\log_e 1.5)^3$
	y_0	y_1	y_2

$$\int_{0.5}^{1.5} (\log_e x)^3\, dx$$

$$\doteqdot \frac{h}{3}(y_0 + y_2 + 4y_1)$$

$$\doteqdot \frac{0.5}{3}\left[(\log_e 0.5)^3 + (\log_e 1.5)^3 + 4(\log_e 1)^3\right]$$

$$\doteqdot \frac{0.5}{3}(-0.3330\ldots + 0.0666\ldots + 0)$$

$\doteqdot -0.04439\ldots$ (by calc.)
$= -0.044$ to three decimal places.

(b)

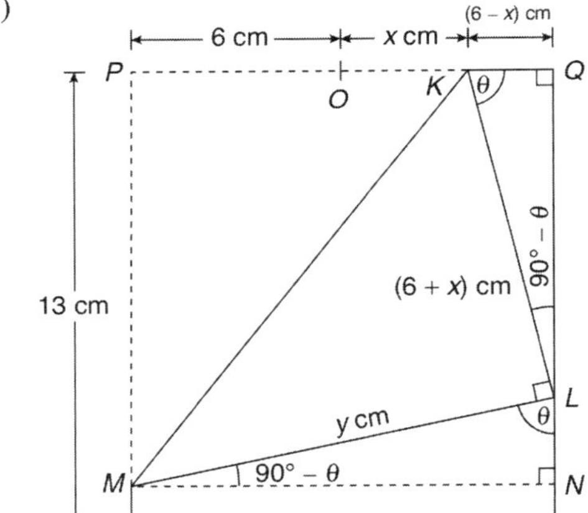

(i)

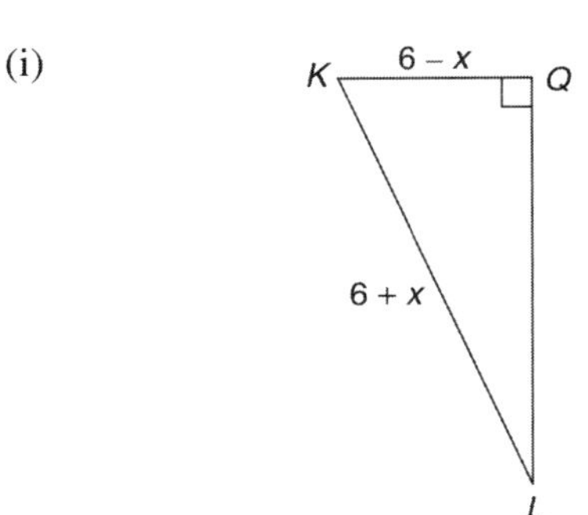

In $\triangle QKL$,

$KQ = 6 - x$ (since $OQ = 6$)
$KL = 6 + x$ (since $KL = PK$)

By Pythagoras' Theorem,

$$KL^2 = KQ^2 + QL^2$$

$$QL^2 = KL^2 - KQ^2$$

$$= (6 + x)^2 - (6 - x)^2$$

$$= 36 + 12x + x^2 - (36 - 12x + x^2)$$

$$= 36 + 12x + x^2 - 36 + 12x - x^2$$

$$\therefore QL^2 = 24x.$$

(ii) In $\triangle QKL$,

Let $\angle QKL = \theta$
$\therefore \angle QLK = 90° - \theta$ (angle sum of $\triangle QKL$)

Now $\angle QLK + \angle KLM + \angle MLN = 180°$
(QLN is a straight angle, 180°)
$90° - \theta + 90° + \angle MLN = 180°$
$\therefore \angle MLN = \theta$
and $\angle LMN = 90° - \theta$
(angle sum of $\triangle NLM$)

In $\triangle QKL$ and $\triangle NLM$,

$\angle QKL = \angle NLM = \theta$
$\angle QLK = \angle NML = 90° - \theta$

$\angle KQL = \angle LNM$ (given)

$\therefore \triangle QKL \parallel\!| \triangle NLM$ (equiangular)

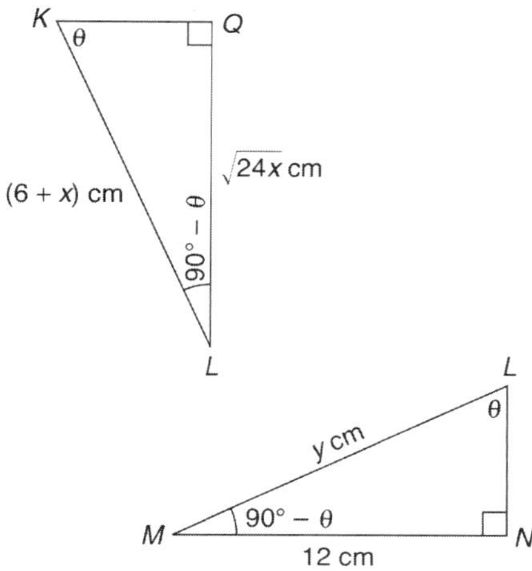

Since $\triangle QKL \parallel\!| \triangle NLM$,

$$\frac{LM}{KL} = \frac{NM}{QL}$$

$$\therefore \frac{y}{6+x} = \frac{12}{\sqrt{24x}}$$

$$y = \frac{12(6+x)}{\sqrt{24x}}$$

$$= \frac{12(6+x)}{2\sqrt{6x}}$$

$$= \frac{6(6+x)}{\sqrt{6}\sqrt{x}}$$

$$= \frac{6(6+x)}{\sqrt{6}\sqrt{x}} \times \frac{\sqrt{6}}{\sqrt{6}}$$

$$= \frac{6\sqrt{6}(6+x)}{6\sqrt{x}}$$

$$= \frac{\sqrt{6}(6+x)}{\sqrt{x}}.$$

(iii) Area of $\triangle KLM = \frac{1}{2} \times ML \times KL$

$$= \frac{1}{2} \times y \times (6+x)$$

$$= \frac{1}{2} \times \frac{\sqrt{6}(6+x)}{\sqrt{x}} \times (6+x)$$

$$= \frac{\sqrt{6}(6+x)^2}{2\sqrt{x}}.$$

(iv) Now $12 \leqslant y \leqslant 13$

$$\therefore 12 \leqslant \frac{\sqrt{6}(6+x)}{\sqrt{x}} \leqslant 13$$

$$12^2 \leqslant \left(\frac{\sqrt{6}(6+x)}{\sqrt{x}}\right)^2 \leqslant 13^2$$

$$144 \leqslant \frac{6(6+x)^2}{x} \leqslant 169$$

$$144 \leqslant \frac{6(6+x)^2}{x} \text{ and } \frac{6(6+x)^2}{x} \leqslant 169,$$

(for $x > 0$,)

Consider $144 \leqslant \frac{6(6+x)^2}{x}$:

$$\therefore 144x \leqslant 6(6+x)^2$$
$$144x \leqslant 6(36 + 12x + x^2)$$
$$24x \leqslant 36 + 12x + x^2$$
$$0 \leqslant 36 - 12x + x^2$$
$$(x-6)^2 \geqslant 0$$
$$x - 6 \geqslant 0$$
$$\therefore x \geqslant 6$$

True for all x, however, the condition that will satisfy information from the diagram must be $x = 6$.
(Since $SR = 12$ cm, and $OP = 6$ cm
$\therefore x \leqslant 6$ cm)

Consider $\frac{6(6+x)^2}{x} \leqslant 169$:

$$6(6+x)^2 \leqslant 169x$$
$$6(36 + 12x + x^2) \leqslant 169x$$
$$216 + 72x + 6x^2 - 169x \leqslant 0$$
$$6x^2 - 97x + 216 \leqslant 0$$

Using the quadratic formula:

$$x = \frac{-b \pm \sqrt{b^2 - 4ac}}{2a}$$

$$= \frac{97 \pm \sqrt{(-97)^2 - 4(6)(216)}}{2 \times 6}$$

$$= \frac{97 \pm \sqrt{4225}}{12}$$

$$= \frac{97 \pm 65}{12}$$

$$= \frac{162}{12} \text{ or } \frac{32}{12}$$

$$= 13\tfrac{1}{2} \text{ or } 2\tfrac{2}{3}$$

$\therefore$ For $6x^2 - 97x + 216 \leqslant 0$,

$$2\tfrac{2}{3} \leqslant x \leqslant 13\tfrac{1}{2}$$

True for all x, however, $6 < x \leqslant 13\frac{1}{2}$, will not satisfy the conditions in the diagram.

$\therefore$ When $12 \leqslant y \leqslant 13$, the possible values for x are $2\frac{2}{3} \leqslant x \leqslant 6$.

(v) Area of $\triangle KLM = \dfrac{\sqrt{6}(6+x)^2}{2\sqrt{x}}$

Now $\dfrac{d}{dx}\left(\sqrt{6}(6+x)^2\right) = 2\sqrt{6}\,(6+x)$

and $\dfrac{d}{dx}(2\sqrt{x}) = \dfrac{d}{dx}\left(2x^{\frac{1}{2}}\right)$

$= x^{-\frac{1}{2}}$

$= \dfrac{1}{\sqrt{x}}$

$$\therefore \frac{d}{dx}\left(\frac{\sqrt{6}(6+x)^2}{2\sqrt{x}}\right)$$

$$= \frac{(2\sqrt{x})\frac{d}{dx}\left(\sqrt{6}(6+x)^2\right) - \left(\sqrt{6}(6+x)^2\right)\frac{d}{dx}(2\sqrt{x})}{(2\sqrt{x})^2}$$

$$= \frac{2\sqrt{x}.2\sqrt{6}(6+x) - \sqrt{6}(6+x)^2.\frac{1}{\sqrt{x}}}{4x}$$

$$= \frac{2\sqrt{x}.2\sqrt{6}(6+x).\sqrt{x} - \sqrt{6}(6+x)^2}{\sqrt{x}} \div 4x$$

$$= \frac{2\sqrt{x}.2\sqrt{6}(6+x).\sqrt{x} - \sqrt{6}(6+x)^2}{4x\sqrt{x}}$$

$$= \frac{4x\sqrt{6}(6+x) - \sqrt{6}(6+x)^2}{4x\sqrt{x}}$$

$$= \frac{\sqrt{6}(6+x)(4x-(6+x))}{4x\sqrt{x}}$$

$$= \frac{\sqrt{6}(6+x)(3x-6)}{4x\sqrt{x}}$$

A possible minimum value occurs when

$$\frac{d}{dx}\left(\frac{\sqrt{6}(6+x)^2}{2\sqrt{x}}\right) = 0$$

i.e. $\dfrac{\sqrt{6}(6+x)(3x-6)}{4x\sqrt{x}} = 0$

$\sqrt{6}\,(6+x)(3x-6) = 0$

$(6+x)(3x-6) = 0$

$\therefore\ 6+x=0$ or $3x-6=0$

$\therefore\ x=-6$ or $\therefore\ x=2$

Since $x > 0,\ x \neq -6$

Since $2\frac{2}{3} \leqslant x \leqslant 6$, this solution lies outside of the range of possible values of x.

$\therefore$ The minimum will lie on one of the endpoints of the interval.

Consider the endpoints on the graph of

Area $= \dfrac{\sqrt{6}(6+x)^2}{2\sqrt{x}}$ for $2\frac{2}{3} \leqslant x \leqslant 6$:

At $x = 2\frac{2}{3}$,

$$A = \frac{\sqrt{6}\left(6+2\frac{2}{3}\right)^2}{2\sqrt{2\frac{2}{3}}}$$

$= 56\frac{1}{3}$ cm^2. (by calc.)

At $x = 6$,

$$A = \frac{\sqrt{6}(6+6)^2}{2\sqrt{6}}$$

$= 72$ cm^2. (by calc.)

There are no turning points in this interval and the curve must be rising in this interval, therefore the minimum area will occur when $x = 2\frac{2}{3}$.

$\therefore$ The minimum possible area of $\triangle KLM$ is $56\frac{1}{3}$ cm^2.

BOARD OF STUDIES
NEW SOUTH WALES

2007

HIGHER SCHOOL CERTIFICATE EXAMINATION

Mathematics

General Instructions

- Reading time – 5 minutes
- Working time – 3 hours
- Write using black or blue pen
- Board-approved calculators may be used
- A table of standard integrals is provided at the back of this paper
- All necessary working should be shown in every question

Total marks – 120

- Attempt Questions 1–10
- All questions are of equal value

Total marks – 120
Attempt Questions 1–10
All questions are of equal value

Answer each question in a SEPARATE writing booklet. Extra writing booklets are available.

Marks

Question 1 (12 marks) Use a SEPARATE writing booklet.

(a) Evaluate $\sqrt{\pi^2 + 5}$ correct to two decimal places. **2**

(b) Solve $2x - 5 > -3$ and graph the solution on a number line. **2**

(c) Rationalise the denominator of $\dfrac{1}{\sqrt{3} - 1}$. **2**

(d) Find the limiting sum of the geometric series **2**

$$\frac{3}{4} + \frac{3}{16} + \frac{3}{64} + \cdots.$$

(e) Factorise $2x^2 + 5x - 12$. **2**

(f) Find the equation of the line that passes through the point $(-1, 3)$ and is perpendicular to $2x + y + 4 = 0$. **2**

Marks

Question 2 (12 marks) Use a SEPARATE writing booklet.

(a) Differentiate with respect to x:

(i) $\dfrac{2x}{e^x + 1}$ **2**

(ii) $(1 + \tan x)^{10}$. **2**

(b) (i) Find $\int (1 + \cos 3x)\, dx$. **2**

(ii) Evaluate $\int_1^4 \dfrac{8}{x^2}\, dx$. **3**

(c) The point $P(\pi, 0)$ lies on the curve $y = x \sin x$. Find the equation of the tangent to the curve at P. **3**

Marks

Question 3 (12 marks) Use a SEPARATE writing booklet.

(a)

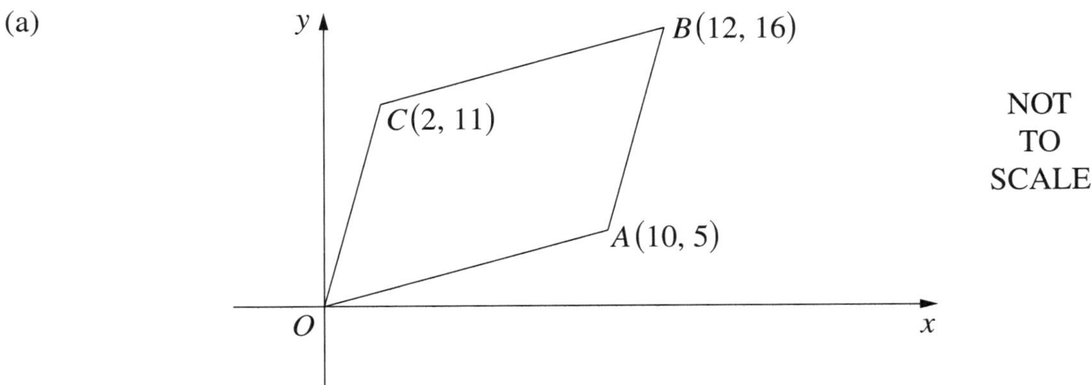

In the diagram, A, B and C are the points $(10, 5)$, $(12, 16)$ and $(2, 11)$ respectively.

Copy or trace this diagram into your writing booklet.

(i) Find the distance AC. **1**

(ii) Find the midpoint of AC. **1**

(iii) Show that $OB \perp AC$. **2**

(iv) Find the midpoint of OB and hence explain why $OABC$ is a rhombus. **2**

(v) Hence, or otherwise, find the area of $OABC$. **1**

(b) Heather decides to swim every day to improve her fitness level.

On the first day she swims 750 metres, and on each day after that she swims 100 metres more than the previous day. That is, she swims 850 metres on the second day, 950 metres on the third day and so on.

(i) Write down a formula for the distance she swims on the nth day. **1**

(ii) How far does she swim on the 10th day? **1**

(iii) What is the total distance she swims in the first 10 days? **1**

(iv) After how many days does the total distance she has swum equal the width of the English Channel, a distance of 34 kilometres? **2**

Marks

Question 4 (12 marks) Use a SEPARATE writing booklet.

(a) Solve $\sqrt{2}\sin x = 1$ for $0 \le x \le 2\pi$. **2**

(b) Two ordinary dice are rolled. The score is the sum of the numbers on the top faces.

(i) What is the probability that the score is 10? **2**

(ii) What is the probability that the score is not 10? **1**

(c)

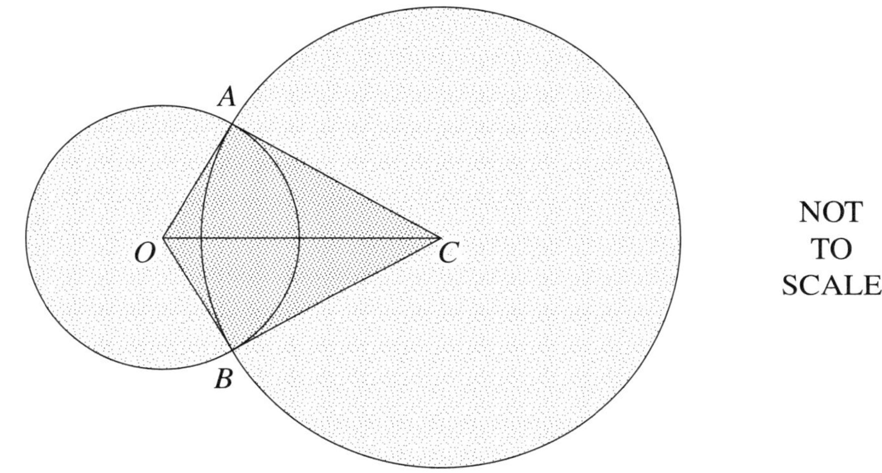

An advertising logo is formed from two circles, which intersect as shown in the diagram.

The circles intersect at A and B and have centres at O and C.

The radius of the circle centred at O is 1 metre and the radius of the circle centred at C is $\sqrt{3}$ metres. The length of OC is 2 metres.

(i) Use Pythagoras' theorem to show that $\angle OAC = \frac{\pi}{2}$. **1**

(ii) Find $\angle ACO$ and $\angle AOC$. **2**

(iii) Find the area of the quadrilateral $AOBC$. **1**

(iv) Find the area of the major sector ACB. **1**

(v) Find the total area of the logo (the sum of all the shaded areas). **2**

Marks

Question 5 (12 marks) Use a SEPARATE writing booklet.

(a)

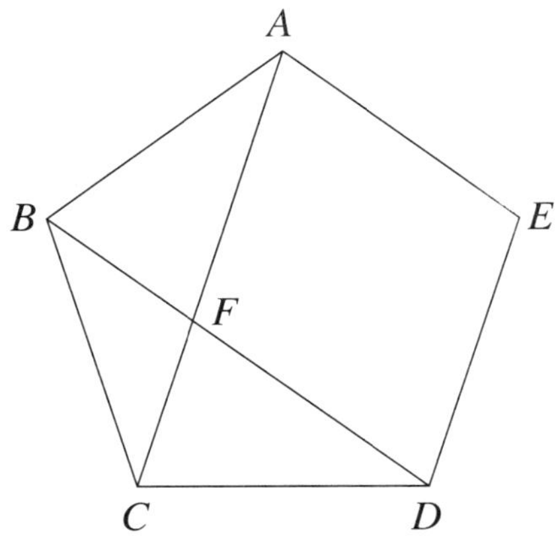

A, B, C, D, E, F

NOT TO SCALE

In the diagram, $ABCDE$ is a regular pentagon. The diagonals AC and BD intersect at F.

Copy or trace this diagram into your writing booklet.

(i) Show that the size of $\angle ABC$ is $108°$. **1**

(ii) Find the size of $\angle BAC$. Give reasons for your answer. **2**

(iii) By considering the sizes of angles, show that ΔABF is isosceles. **2**

(b) A particle is moving on the x-axis and is initially at the origin. Its velocity, v metres per second, at time t seconds is given by

$$v = \frac{2t}{16 + t^2}.$$

(i) What is the initial velocity of the particle? **1**

(ii) Find an expression for the acceleration of the particle. **2**

(iii) Find the time when the acceleration of the particle is zero. **1**

(iv) Find the position of the particle when $t = 4$. **3**

Marks

Question 6 (12 marks) Use a SEPARATE writing booklet.

(a) Solve the following equation for x: **2**

$$2e^{2x} - e^{x} = 0.$$

(b) Let $f(x) = x^4 - 4x^3$.

(i) Find the coordinates of the points where the curve crosses the axes. **2**

(ii) Find the coordinates of the stationary points and determine their nature. **4**

(iii) Find the coordinates of the points of inflexion. **1**

(iv) Sketch the graph of $y = f(x)$, indicating clearly the intercepts, stationary points and points of inflexion. **3**

End of Question 6

Marks

Question 7 (12 marks) Use a SEPARATE writing booklet.

(a) (i) Find the coordinates of the focus, S, of the parabola $y = x^2 + 4$. **2**

(ii) The graphs of $y = x^2 + 4$ and the line $y = x + k$ have only one point of intersection, P. Show that the x-coordinate of P satisfies **1**

$$x^2 - x + 4 - k = 0.$$

(iii) Using the discriminant, or otherwise, find the value of k. **1**

(iv) Find the coordinates of P. **2**

(v) Show that SP is parallel to the directrix of the parabola. **1**

(b)

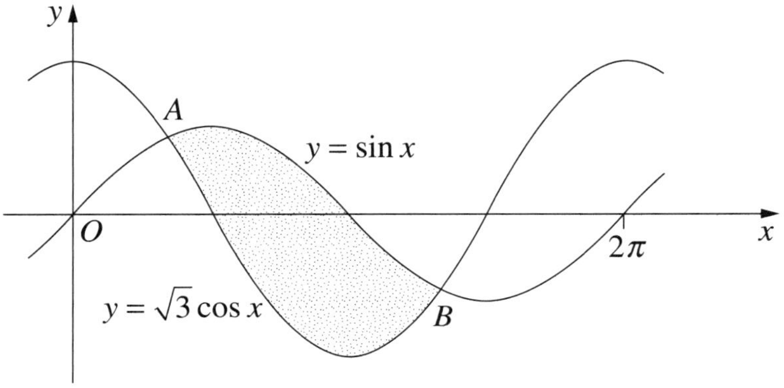

The diagram shows the graphs of $y = \sqrt{3}\cos x$ and $y = \sin x$. The first two points of intersection to the right of the y-axis are labelled A and B.

(i) Solve the equation $\sqrt{3}\cos x = \sin x$ to find the x-coordinates of A and B. **2**

(ii) Find the area of the shaded region in the diagram. **3**

Marks

Question 8 (12 marks) Use a SEPARATE writing booklet.

(a) One model for the number of mobile phones in use worldwide is the exponential growth model,

$$N = Ae^{kt},$$

where N is the estimate for the number of mobile phones in use (in millions), and t is the time in years after 1 January 2008.

(i) It is estimated that at the start of 2009, when $t = 1$, there will be 1600 million mobile phones in use, while at the start of 2010, when $t = 2$, there will be 2600 million. Find A and k. **3**

(ii) According to the model, during which month and year will the number of mobile phones in use first exceed 4000 million? **2**

(b)

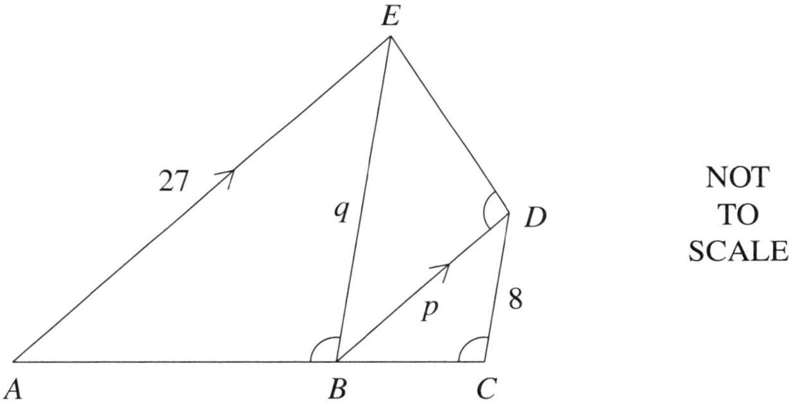

E
27
q
D
p
8
A
B
C
NOT TO SCALE

In the diagram, AE is parallel to BD, $AE = 27$, $CD = 8$, $BD = p$, $BE = q$ and $\angle ABE$, $\angle BCD$ and $\angle BDE$ are equal.

Copy or trace this diagram into your writing booklet.

(i) Prove that $\Delta ABE \ ||| \ \Delta BCD$. **2**

(ii) Prove that $\Delta EDB \ ||| \ \Delta BCD$. **2**

(iii) Show that 8, p, q, 27 are the first four terms of a geometric series. **1**

(iv) Hence find the values of p and q. **2**

Marks

Question 9 (12 marks) Use a SEPARATE writing booklet.

(a) 3

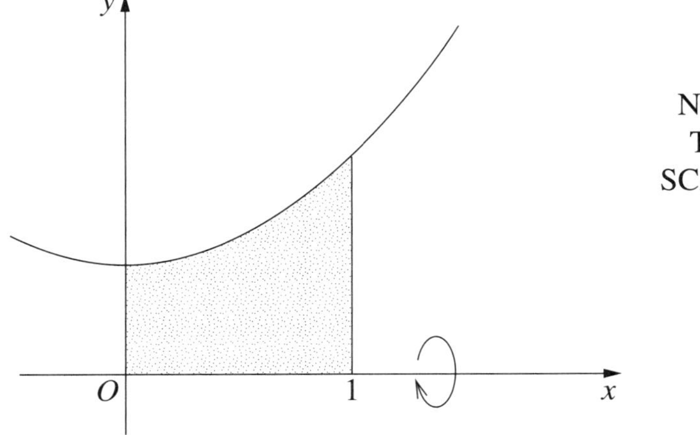

The shaded region in the diagram is bounded by the curve $y = x^2 + 1$, the x-axis, and the lines $x = 0$ and $x = 1$.

Find the volume of the solid of revolution formed when the shaded region is rotated about the x-axis.

(b) A pack of 52 cards consists of four suits with 13 cards in each suit.

(i) One card is drawn from the pack and kept on the table. A second card is drawn and placed beside it on the table. What is the probability that the second card is from a different suit to the first? 1

(ii) The two cards are replaced and the pack shuffled. Four cards are chosen from the pack and placed side by side on the table. What is the probability that these four cards are all from different suits? 2

Question 9 continues

Marks

Question 9 (continued)

(c) Mr and Mrs Caine each decide to invest some money each year to help pay for their son's university education. The parents choose different investment strategies.

(i) Mr Caine makes 18 yearly contributions of \$1000 into an investment fund. He makes his first contribution on the day his son is born, and his final contribution on his son's seventeenth birthday. His investment earns 6% compound interest per annum. **3**

Find the total value of Mr Caine's investment on his son's eighteenth birthday.

(ii) Mrs Caine makes her contributions into another fund. She contributes \$1000 on the day of her son's birth, and increases her annual contribution by 6% each year. Her investment also earns 6% compound interest per annum. **2**

Find the total value of Mrs Caine's investment on her son's third birthday (just before she makes her fourth contribution).

(iii) Mrs Caine also makes her final contribution on her son's seventeenth birthday. **1**

Find the total value of Mrs Caine's investment on her son's eighteenth birthday.

End of Question 9

Marks

Question 10 (12 marks) Use a SEPARATE writing booklet.

(a) An object is moving on the x-axis. The graph shows the velocity, $\frac{dx}{dt}$, of the object, as a function of time, t. The coordinates of the points shown on the graph are $A(2, 1)$, $B(4, 5)$, $C(5, 0)$ and $D(6, -5)$. The velocity is constant for $t \geq 6$.

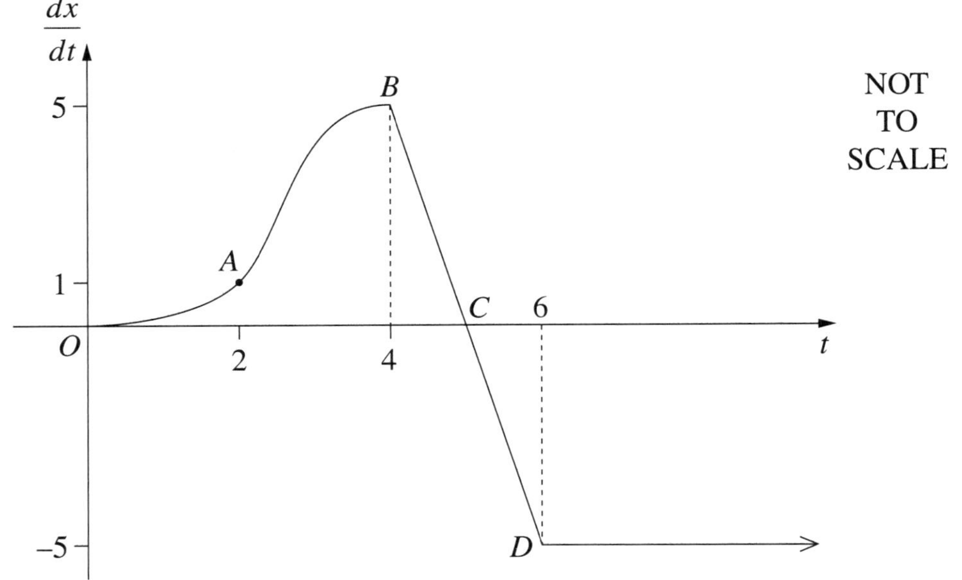

(i) Using Simpson's rule, estimate the distance travelled between $t = 0$ and $t = 4$. **2**

(ii) The object is initially at the origin. During which time(s) is the displacement of the object decreasing? **1**

(iii) Estimate the time at which the object returns to the origin. Justify your answer. **2**

(iv) Sketch the displacement, x, as a function of time. **2**

Question 10 continues

Marks

Question 10 (continued)

(b) The noise level, N, at a distance d metres from a single sound source of loudness L is given by the formula

$$N = \frac{L}{d^2}.$$

Two sound sources, of loudness L_1 and L_2 are placed m metres apart.

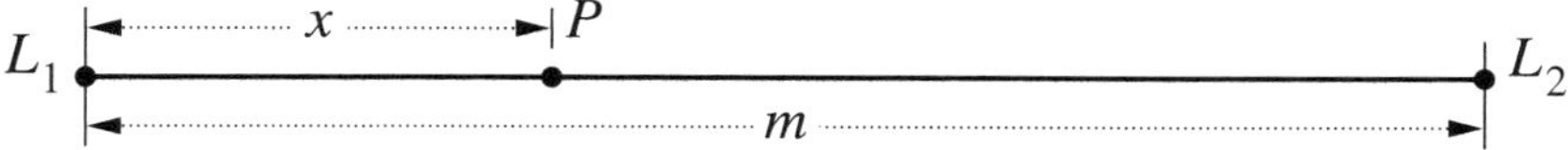

The point P lies on the line between the sound sources and is x metres from the sound source with loudness L_1.

(i) Write down a formula for the sum of the noise levels at P in terms of x. **1**

(ii) There is a point on the line between the sound sources where the sum of the noise levels is a minimum. **4**

Find an expression for x in terms of m, L_1 and L_2 if P is chosen to be this point.

End of paper

2007 Higher School Certificate Worked Answers

QUESTION 1

(a) $\sqrt{\pi^2 + 5} = 3.8561\ldots$ (by calc.)
$= 3.86$ to 2 decimal places.

(b) $2x - 5 > -3$
$2x > 2$
$\therefore x > 1.$

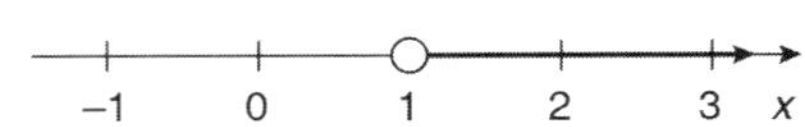

(c) $$\frac{1}{\sqrt{3}-1} = \frac{1}{\sqrt{3}-1} \times \frac{\sqrt{3}+1}{\sqrt{3}+1}$$
$$= \frac{\sqrt{3}+1}{(\sqrt{3})^2 - (1)^2}$$
$$= \frac{\sqrt{3}+1}{3-1}$$
$$= \frac{\sqrt{3}+1}{2}.$$

(d) $$\frac{3}{4} + \frac{3}{16} + \frac{3}{64} + \ldots$$

$$S_\infty = \frac{a}{1-r}, \text{ where } a = \frac{3}{4}, r = \frac{1}{4}$$
$$\therefore S_\infty = \frac{\frac{3}{4}}{1 - \frac{1}{4}}$$
$$= 1.$$

(e) $2x^2 + 5x - 12$
$= 2x^2 + 8x - 3x - 12$
$= 2x(x+4) - 3(x+4)$
$= (2x-3)(x+4).$

(f) $2x + y + 4 = 0$
$y = -2x - 4$
$\therefore m_1 = -2$

Now $m_1 \times m_2 = -1$
$\therefore$ $-2 \times m_2 = -1$
$\therefore m_2 = \frac{1}{2}$

$\therefore$ Equation of the line passing through $(-1, 3)$ and perpendicular to $2x + y + 4 = 0$ is given by
$y - y_1 = m(x - x_1)$

$$\therefore y - 3 = \frac{1}{2}\left(x - (-1)\right)$$
$$y - 3 = \frac{1}{2}x + \frac{1}{2}$$
$$y = \frac{1}{2}x + 3\frac{1}{2}$$
$$2y = x + 7$$
$$\therefore x - 2y + 7 = 0.$$

QUESTION 2

(a) (i) $$\frac{d}{dx}\left(\frac{2x}{e^x + 1}\right)$$
$$= \frac{(e^x + 1)\,.\,\frac{d}{dx}(2x) - 2x\,.\,\frac{d}{dx}(e^x + 1)}{(e^x + 1)^2}$$
$$= \frac{(e^x + 1)\,.\,2 - 2x\,.\,e^x}{(e^x + 1)^2}$$
$$= \frac{2e^x + 2 - 2xe^x}{(e^x + 1)^2}$$
$$= \frac{2(e^x + 1 - xe^x)}{(e^x + 1)^2}.$$

(ii) $$\frac{d}{dx}\left[(1 + \tan x)^{10}\right]$$
$$= 10(1 + \tan x)^9\,.\,\frac{d}{dx}(1 + \tan x)$$
$$= 10(1 + \tan x)^9\,.\,\sec^2 x$$
$$= 10\sec^2 x\,(1 + \tan x)^9.$$

(b) (i) $$\int (1 + \cos 3x)dx = x + \frac{1}{3}\sin 3x + C$$

(ii) $$\int_1^4 \frac{8}{x^2}\,dx$$
$$= 8\int_1^4 x^{-2}\,dx$$
$$= 8\left[\frac{x^{-1}}{-1}\right]_1^4$$
$$= 8\left[-\frac{1}{x}\right]_1^4$$
$$= 8\left[\left(-\frac{1}{4}\right) - (-1)\right]$$

$$= 8\left[-\frac{1}{4} + 1\right]$$

$$= 8 \times \frac{3}{4}$$

$$= 6.$$

(c) $y = x \sin x$

$$\frac{dy}{dx} = x \,.\, \frac{d}{dx}(\sin x) + \sin x \,.\, \frac{d}{dx}(x)$$

$$= x \,.\cos x + \sin x \,.\, 1$$

$$= x \cos x + \sin x$$

At $x = \pi$, $\frac{dy}{dx} = \pi \cos \pi + \sin \pi$

$$= -\pi + 0$$

$$= -\pi$$

$\therefore$ Equation of the tangent at $(\pi, 0)$ with gradient $= -\pi$ is

$y - 0 = -\pi(x - \pi)$

$\therefore\ y = -\pi x + \pi^2$.

QUESTION 3

(a)

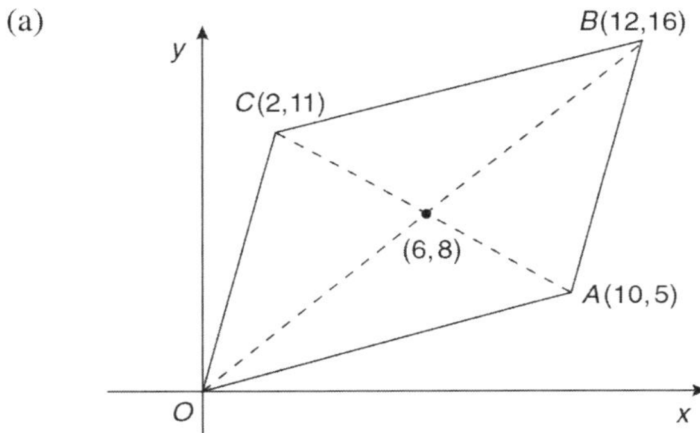

(i) A(10,5), C(2,11)

$$d_{AC} = \sqrt{(x_2 - x_1)^2 + (y_2 - y_1)^2}$$

$$= \sqrt{(2 - 10)^2 + (11 - 5)^2}$$

$$= \sqrt{64 + 36}$$

$$= \sqrt{100}$$

$\therefore\ d_{AC} = 10$ units.

(ii) A(10,5), C(2,11)

$$M_{AC} = \left(\frac{x_1 + x_2}{2}, \frac{y_1 + y_2}{2}\right)$$

$$M_{AC} = \left(\frac{10 + 2}{2}, \frac{5 + 11}{2}\right)$$

$$M_{AC} = (6, 8)$$

(iii) O(0,0), B(12,16)

$$m_{OB} = \frac{y_2 - y_1}{x_2 - x_1}$$

$$= \frac{16 - 0}{12 - 0}$$

$$\therefore\ m_{OB} = \frac{4}{3}$$

A(10,5), C(2,11)

$$m_{AC} = \frac{y_2 - y_1}{x_2 - x_1}$$

$$= \frac{11 - 5}{2 - 10}$$

$$\therefore\ m_{AC} = -\frac{3}{4}$$

If $m_1 \times m_2 = -1$

$\therefore\ m_{OB} \times m_{AC} = -1$

i.e. $\frac{4}{3} \times -\frac{3}{4} = -1$

$\therefore$ OB $\perp$ AC.

(iv) O(0,0), B(12,16)

$$M_{OB} = \left(\frac{x_1 + x_2}{2}, \frac{y_1 + y_2}{2}\right)$$

$$= \left(\frac{0 + 12}{2}, \frac{0 + 16}{2}\right)$$

$$\therefore\ M_{OB} = (6,8)$$

Since the diagonals are perpendicular to one another and they bisect each other (at a common midpoint), OABC is a rhombus.

(v) Area of OABC $= \frac{1}{2}xy$,

where $x = d_{AC}$
$= 10$ units.

and $y = d_{OB}$
$= \sqrt{(16^2) + (12^2)}$
$= 20$ units.

$\therefore$ Area of OABC $= \frac{1}{2} \times 10 \times 20$
$= 100$ units^2.

(b) (i) The distance Heather swims each day is an arithmetic sequence: 750, 850, 950, ... with $a = 750$, $d = 100$.

$$T_n = a + (n-1)d$$
$$= 750 + (n-1) \times 100$$
$$= 750 + 100n - 100$$
$$\therefore\ T_n = 650 + 100n.$$

(ii) $T_n = 650 + 100n$

$$\therefore\ T_{10} = 650 + 100 \times 10$$
$$= 1650$$

$\therefore$ Heather swims 1650 metres on the tenth day.

(iii) $S_n = \frac{n}{2}\left[2a + (n-1)d\right]$

$n = 10, a = 750, d = 100$

$$\therefore\ S_{10} = \frac{10}{2}\left[2(750) + (10-1) \times 100\right]$$
$$= 5(1500 + 900)$$
$$= 12\,000$$

or

$$S_n = \frac{n}{2}(a + l)$$

$n = 10, a = 750, l = 1650$

$$\therefore\ S_{10} = \frac{10}{2}(750 + 1650)$$
$$= 5 \times 2400$$
$$= 12\,000$$

$\therefore$ Heather swims 12 000 m or 12 km in the first ten days.

(iv) $S_n = \frac{n}{2}\left[2a + (n-1)d\right]$

$S_n = 34\,000, a = 750, d = 100$

$$\therefore\ 34\,000 = \frac{n}{2}\left[2(750) + (n-1) \times 100\right]$$
$$68\,000 = n\left[1500 + 100n - 100\right]$$
$$68\,000 = 1400n + 100n^2$$
$$n^2 + 14n - 680 = 0$$
$$\therefore\ n = \frac{-b \pm \sqrt{b^2 - 4ac}}{2a}$$
$$= \frac{-14 \pm \sqrt{14^2 - 4 \times (1) \times (-680)}}{2 \times 1}$$
$$= \frac{-14 \pm \sqrt{196 + 2720}}{2}$$
$$= \frac{-14 \pm \sqrt{2916}}{2}$$
$$= \frac{-14 \pm 54}{2}$$
$$\therefore\ n = -34 \text{ and } 20$$

Since $n = -34$ is not a solution,

$\therefore\ n = 20.$

$\therefore$ Heather will take 20 days to swim the length of the English Channel.

QUESTION 4

(a) $\sqrt{2}\sin x = 1 \quad$ for $0 \leqslant x \leqslant 2\pi$

$$\sin x = \frac{1}{\sqrt{2}}$$
$$\therefore\ x = \frac{\pi}{4} \text{ or } \pi - \frac{\pi}{4}$$
$$\therefore\ x = \frac{\pi}{4} \text{ or } \frac{3\pi}{4}.$$

(b)

Die 2 \ Die 1	1	2	3	4	5	6
1	2	3	4	5	6	7
2	3	4	5	6	7	8
3	4	5	6	7	8	9
4	5	6	7	8	9	10*
5	6	7	8	9	10*	11
6	7	8	9	10*	11	12

(i) $P(\text{score} = 10) = \frac{3}{36} = \frac{1}{12}.$

(ii) $P(\text{score} \neq 10) = 1 - P(\text{score} = 10)$

$$= 1 - \frac{1}{12}$$
$$= \frac{11}{12}.$$

(c)

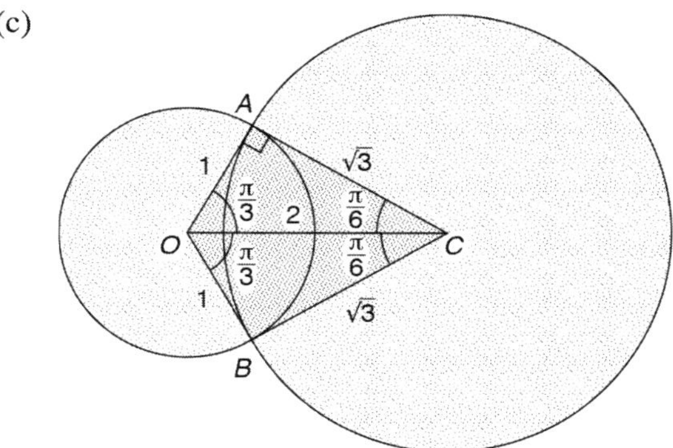

(i) In ∟ OAC,

$$OA = 1$$
$$AC = \sqrt{3}$$

By Pythagoras' theorem:

$OC^2 = OA^2 + AC^2$

$= 1^2 + (\sqrt{3})^2$

$= 4$

$\therefore OC = \sqrt{4}$

$= 2$

$\therefore \triangle OAC$ is right angled

$\therefore \angle OAC = \frac{\pi}{2}$.

(ii) In $\triangle OAC$,

$\tan \angle ACO = \frac{OA}{AC}$

$= \frac{1}{\sqrt{3}}$

$\therefore \angle ACO = \frac{\pi}{6}$

$\therefore \angle AOC = \pi - \frac{\pi}{2} - \frac{\pi}{6}$ (angle sum of $\triangle OAC$)

$= \frac{\pi}{3}$.

(iii) Area of $\triangle AOC = \frac{1}{2}(OA)(AC)$

$= \frac{1}{2} \times 1 \times \sqrt{3}$

$= \frac{\sqrt{3}}{2}$ m^2.

$\therefore$ Area of quadrilateral $AOBC$

$= \frac{\sqrt{3}}{2} \times 2$

$= \sqrt{3}$ m^2.

(iv) Now $\triangle OAC$ and $\triangle OBC$ are congruent (SSS),

$\therefore \angle AOC = \angle BOC = \frac{\pi}{3}$

and $\angle ACO = \angle BCO = \frac{\pi}{6}$

$\therefore \angle ACB = \frac{\pi}{6} + \frac{\pi}{6} = \frac{\pi}{3}$

Area of major sector ACB

$= \frac{1}{2}r^2\theta$ where $\theta = (2\pi - \angle ACB)$

$= \frac{1}{2}(\sqrt{3})^2\left(2\pi - \frac{\pi}{3}\right)$

$= \frac{1}{2} \times 3 \times \frac{5\pi}{3}$

$= \frac{5\pi}{2}$ m^2.

(v) Now $\angle AOB = \angle AOC + \angle BOC$

$= \frac{\pi}{3} + \frac{\pi}{3}$

$= \frac{2\pi}{3}$

Area of major sector AOB

$= \frac{1}{2}r^2\theta$ where $\theta = (2\pi - \angle AOB)$

$= \frac{1}{2}r^2\left(2\pi - \frac{2\pi}{3}\right)$

$= \frac{1}{2} \times 1 \times \frac{4\pi}{3}$

$= \frac{2\pi}{3}$ m^2.

Total area of the logo
= area of quadrilateral $AOBC$
+ area of major sector ACB
+ area of major sector AOB

$= \sqrt{3} + \frac{5\pi}{2} + \frac{2\pi}{3}$

$= \frac{6\sqrt{3} + 15\pi + 4\pi}{6}$

$= \left(\frac{6\sqrt{3} + 19\pi}{6}\right)$ m^2.

QUESTION 5

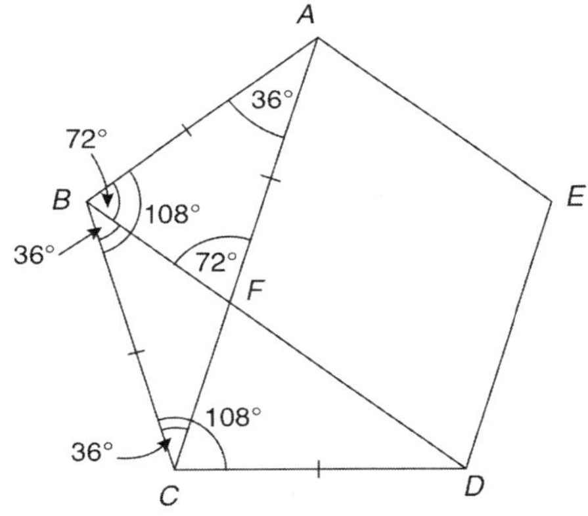

(a) (i) $S_{ABCDE} = (n - 2) \times 180°$

$= (5 - 2) \times 180°$

$= 540°$

$\therefore \angle ABC = 540° \div 5$

$= 108°$.

(ii) In $\triangle BAC$,

$AB = BC$ (equal sides of a regular pentagon)

$\therefore \triangle BAC$ is isosceles.

Now $\angle ABC = 108°$ (given)
and $\angle BAC = \angle BCA$
(equal angles of isosceles $\triangle$)
$\therefore \angle BAC + \angle BCA + \angle ABC = 180°$
(angle sum of $\triangle$)
$\therefore \angle BAC + \angle BCA + 108° = 180°$
$\angle BAC + \angle BCA = 72°$
$\therefore \angle BAC = 36°$.

(iii) In $\triangle BAC$ and $\triangle CBD$,
$BA = BC$
(equal sides of regular pentagon)
$\angle ABC = \angle BCD$
(interior angles of regular pentagon)
$BC = CD$
(equal sides of a regular pengaton)

$\therefore \triangle BAC \equiv \triangle CBD$ (SAS)

Since $\triangle BAC \equiv \triangle CBD$,
$\angle BAC = \angle CBD$
(corresponding angles in congruent triangles)
$= 36°$
and $\angle BAC = \angle BCA$
(equal angles in isosceles $\triangle$)
$= 36°$
$\therefore \angle BFA = \angle CBD + \angle BCA$
(exterior angle)
$= 72°$
and $\angle ABF = \angle ABC - \angle CBD$
$= 108° - 36°$
$= 72°$
$\therefore \angle BFA = \angle ABF$

$\therefore \triangle ABF$ is isosceles.

(b) (i) $v = \dfrac{2t}{16 + t^2}$

The initial velocity of the particles occurs when $t = 0$.

$$\therefore v = \frac{2 \times 0}{16 + 0^2} = 0$$

$\therefore$ The initial velocity of the particle is 0 m/s. The particle is stationary.

(ii)
$$a = \frac{dv}{dt}$$
$$= \frac{d}{dt}\left(\frac{2t}{16 + t^2}\right)$$
$$= \frac{(16 + t^2)\,.\,\frac{d}{dt}(2t) - (2t)\,.\,\frac{d}{dt}(16 + t^2)}{(16 + t^2)^2}$$
$$= \frac{(16 + t^2)(2) - (2t)(2t)}{(16 + t^2)^2}$$
$$= \frac{32 + 2t^2 - 4t^2}{(16 + t^2)^2}$$
$$\therefore a = \frac{32 - 2t^2}{(16 + t^2)^2}.$$

(iii) $a = \dfrac{32 - 2t^2}{(16 + t^2)^2}$

When $a = 0$, $0 = \dfrac{32 - 2t^2}{(16 + t^2)^2}$
$0 = 32 - 2t^2$
$0 = 2(16 - t^2)$
$16 - t^2 = 0$
$(4 + t)(4 - t) = 0$
$\therefore 4 + t = 0$ or $4 - t = 0$
$t = -4$ $\quad$ $t = 4$

Since $t = -4$ is not a possible solution, $t = 4$.

$\therefore$ The acceleration of the particle is zero after 4 seconds.

(iv) $v = \dfrac{2t}{16 + t^2}$
$$\therefore x = \int \frac{2t}{16 + t^2}\,dt$$
i.e. $x = \log_e\left(16 + t^2\right) + C$

When $t = 0, x = 0$
$\therefore 0 = \log_e\left(16 + (0)^2\right) + C$
$= \log_e 16 + C$
$\therefore C = -\log_e 16$

$\therefore x = \log_e\left(16 + t^2\right) - \log_e 16$

When $t = 4$,
$x = \log_e\left(16 + (4)^2\right) - \log_e 16$
$= \log_e 32 - \log_e 16$
$= \log_e\left(\dfrac{32}{16}\right)$
$= \log_e 2$
$= 0.6931\ldots$ (by calc.)
$= 0.693$ to 3 decimal places

$\therefore$ The position of the particle when $t = 4$ is $\log_e 2$ metres.

QUESTION 6

(a) $2e^{2x} - e^x = 0$

$\therefore 2(e^x)^2 - e^x = 0$

Let $m = e^x$,

$\therefore 2m^2 - m = 0$

$m(2m - 1) = 0$

$\therefore m = 0$ or $2m - 1 = 0$

no solution (as $e^x > 0$) $\quad m = \frac{1}{2}$

$\therefore e^x = \frac{1}{2}$

$\therefore \ln e^x = \ln\left(\frac{1}{2}\right)$

$\therefore x = \ln \frac{1}{2}$

$\therefore x = -\ln 2$

(b) (i) The curve crosses the x-axis when

$f(x) = 0$

$\therefore x^4 - 4x^3 = 0$

$x^3(x - 4) = 0$

$\therefore x = 0$ or 4.

Now $f(0) = 0$

and $f(4) = (4)^4 - 4(4)^3$

$= 0$

$\therefore$ Coordinates are $(0,0)$ and $(4,0)$.

(ii) $f(x) = x^4 - 4x^3$

$f'(x) = 4x^3 - 12x^2$

$= 4x^2(x - 3)$

Stationary points occur when

$f'(x) = 0$

$\therefore 4x^2(x - 3) = 0$

$\therefore x = 0$ or 3

Now $f(0) = 0$

and $f(3) = (3)^4 - 4(3)^3$

$= 81 - 108$

$= -27$

$\therefore$ Coordinates of the stationary points are $(0,0)$ and $(3,-27)$.

Now $f''(x) = 12x^2 - 24x$

$= 12x(x - 2)$

$\therefore f''(0) = 12(0)(0 - 2)$

$= 0$

Checking the change in concavity:

x	-0.1	0	0.1
$f''(x)$	>0	0	<0

$\therefore$ There is a change in concavity.

$\therefore$ A horizontal point of inflexion occurs at $(0,0)$.

Now $f''(3) = 12(3)(3 - 2)$

$= 36 > 0$

$\therefore$ A minimum turning point occurs at $(3,-27)$.

(iii) Possible points of inflexion occur when

$f''(x) = 0$

$12x^2 - 24x = 0$

$12x(x - 2) = 0$

$\therefore x = 0$ or 2

From (ii), a horizontal point of inflexion occurs at $(0,0)$.

Now at $x = 2$,
checking the change in concavity:

x	1.9	2	2.1
$f''(x)$	<0	0	>0

$\therefore$ There is a change in concavity and a point of inflexion at $x = 2$.

Now $f(2) = (2)^4 - 4(2)^3$

$= 16 - 32$

$= -16$

$\therefore$ A point of inflexion occurs at $(2,-16)$.

(iv)

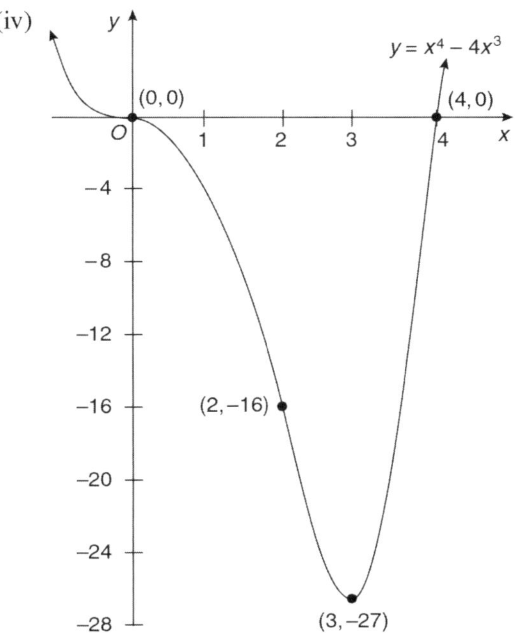

QUESTION 7

(a) (i) $y = x^2 + 4$

$\therefore x^2 = y - 4$

Now $(x - x_0)^2 = 4a(y - y_0)$,

where $x_0 = 0, a = \frac{1}{4}, y_0 = 4$

i.e. $(x - 0)^2 = 4 \times \frac{1}{4} \times (y - 4)$

Vertex: $V = (x_0, y_0)$
$= (0, 4)$

Focus: $S = (x_0, y_0 + a)$
$= \left(0, 4\frac{1}{4}\right)$

(ii) Since $y = x^2 + 4$
and $y = x + k$
$\therefore x^2 + 4 = x + k$
$\therefore x^2 - x + 4 - k = 0$

The x-coordinate of P satisfies
$x^2 - x + 4 - k = 0$.

(iii) $x^2 - x + 4 - k = 0$

$$\begin{aligned}\Delta &= b^2 - 4ac \\ &= (-1)^2 - 4(1)(4 - k) \\ &= 1 - 4(1)(4 - k) \\ &= 1 - 16 + 4k \\ \therefore \Delta &= 4k - 15\end{aligned}$$

For one point of intesection or equal roots,
$\Delta = 0$.
$\therefore 4k - 15 = 0$
$4k = 15$
$k = \frac{15}{4}$
$\therefore k = 3\frac{3}{4}$.

(iv) $x^2 - x + 4 - k = 0$

When $k = 3\frac{3}{4}$,

$x^2 - x + 4 - 3\frac{3}{4} = 0$

$x^2 - x + \frac{1}{4} = 0$

$4x^2 - 4x + 1 = 0$
$(2x - 1)^2 = 0$
$2x - 1 = 0$
$\therefore x = \frac{1}{2}$

Now $y = x + k$
$\therefore y = \frac{1}{2} + 3\frac{3}{4}$
$\therefore y = 4\frac{1}{4}$

$\therefore$ The coordinates of P are $\left(\frac{1}{2}, 4\frac{1}{4}\right)$.

(v) Directrix: $y = y_0 - a$
$= 4 - \frac{1}{4}$
$\therefore y = 3\frac{3}{4}$
$\therefore m_{directrix} = 0$

Since $S\left(0, 4\frac{1}{4}\right)$, $P\left(\frac{1}{2}, 4\frac{1}{4}\right)$.

$$\therefore m_{SP} = \frac{y_2 - y_1}{x_2 - x_1} = \frac{4\frac{1}{4} - 4\frac{1}{4}}{\frac{1}{2} - 0} = 0$$

$\therefore SP$ is parallel to the directrix of the parabola.

(b) (i) $\sqrt{3}\cos x = \sin x$

$\sqrt{3} = \frac{\sin x}{\cos x}$

$\tan x = \sqrt{3}$

$\therefore x = \frac{\pi}{3}, \pi + \frac{\pi}{3}$
$= \frac{\pi}{3}, \frac{4\pi}{3}$

$\therefore x$-coordinates of A and B are $\frac{\pi}{3}$ and $\frac{4\pi}{3}$ respectively.

(ii) $A = \int_{\frac{\pi}{3}}^{\frac{4\pi}{3}} (\sin x - \sqrt{3}\cos x)\,dx$

$$= \Big[-\cos x - \sqrt{3}\sin x \Big]_{\frac{\pi}{3}}^{\frac{4\pi}{3}}$$

$$= -\Big[\cos x + \sqrt{3}\sin x \Big]_{\frac{\pi}{3}}^{\frac{4\pi}{3}}$$

$$= -\left\{\left(\cos\frac{4\pi}{3} + \sqrt{3}\sin\frac{4\pi}{3}\right) - \left(\cos\frac{\pi}{3} + \sqrt{3}\sin\frac{\pi}{3}\right)\right\}$$

$$= -\left\{\left(-\cos\frac{\pi}{3} - \sqrt{3}\sin\frac{\pi}{3}\right) - \left(\cos\frac{\pi}{3} + \sqrt{3}\sin\frac{\pi}{3}\right)\right\}$$

$$= -\left(-\cos\frac{\pi}{3} - \sqrt{3}\sin\frac{\pi}{3} - \cos\frac{\pi}{3} - \sqrt{3}\sin\frac{\pi}{3}\right)$$

$$= -\left(-2\cos\frac{\pi}{3} - 2\sqrt{3}\sin\frac{\pi}{3}\right)$$

$$= 2\left(\frac{1}{2}\right) + 2\sqrt{3}\left(\frac{\sqrt{3}}{2}\right)$$

$= 1 + 3$

$= 4$ units^2.

QUESTION 8

(a) (i) $N = Ae^{kt}$

At $t = 1, N = 1600$

$\therefore\ 1600 = Ae^{k} \quad \ldots(1)$

At $t = 2, N = 2600$

$\therefore\ 2600 = Ae^{2k} \quad \ldots(2)$

Solving simultaneously:

(2) ÷ (1) $\dfrac{Ae^{2k}}{Ae^{k}} = \dfrac{2600}{1600}$

$$e^k = \left(\frac{13}{8}\right)$$

$$\ln(e^k) = \ln\left(\frac{13}{8}\right)$$

$$\therefore\ k = \ln\left(\frac{13}{8}\right)$$

$\doteqdot 0.4855\ldots$ (by calc.)

$= 0.49$ to 2 decimal places.

If $k = \ln\left(\frac{13}{8}\right)$, $Ae^k = 1600$

$$\therefore\ Ae^{\ln\left(\frac{13}{8}\right)} = 1600$$

$$\therefore\ A \times \left(\frac{13}{8}\right) = 1600$$

$$\therefore\ A = \frac{12\,800}{13}.$$

(ii) $N = \dfrac{12\,800}{13}e^{kt}$ $\left(\text{where } k = \ln\frac{13}{8}\right)$

If $N = 4000$,

$$4000 = \frac{12\,800}{13}e^{kt}$$

$$\frac{4000}{\frac{28\,000}{13}} = e^{kt}$$

$$\frac{65}{16} = e^{kt}$$

$$\ln\left(\frac{65}{16}\right) = \ln e^{kt}$$

$$\ln\left(\frac{65}{16}\right) = kt$$

$$t = \frac{\ln\left(\frac{65}{16}\right)}{k}$$

$$= \frac{\ln\left(\frac{65}{16}\right)}{\ln\left(\frac{13}{8}\right)}$$

$\doteqdot 2.8872\ldots$ (by calc.)
$\doteqdot$ 2 years 10.6 months.

$\therefore$ In November 2010, the number of mobile phones will exceed 4000 million.

(b)

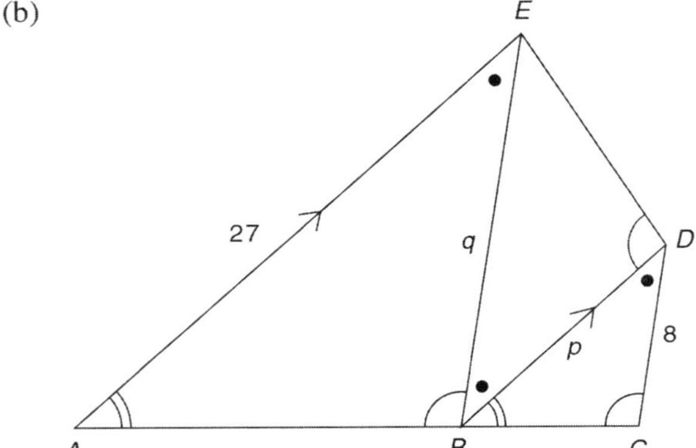

(i) In $\triangle ABE$ and $\triangle BCD$,

$\angle EAB = \angle DBC$
(corresponding angles, $AE \parallel BD$)
$\angle EBA = \angle DCB$ (given)

$\therefore \triangle ABE \,|||\, \triangle BCD$ (equiangular)

(ii) In $\triangle EDB$ and $\triangle BCD$,

$\angle EDB = \angle BCD$ (given)

Now $\angle ABE = \angle BCD$ (given)

and $\angle ABE + \angle EBD + \angle DBC = 180° \ldots (1)$
($\angle ABC$ is a straight angle, 180°)

and $\angle BDC + \angle BCD + \angle DBC = 180° \ldots (2)$
(angle sum of $\triangle BDC$ is 180°)

$\therefore \angle EBD = \angle BDC$ (equating (1) and (2))
$\therefore \triangle EDB \,|||\, \triangle BCD$ (equiangular)

(iii) Now $\frac{BD}{CD} = \frac{BE}{BD} = \frac{AE}{BE}$

(corresponding sides in the same ratio)

$\therefore \frac{p}{8} = \frac{q}{p} = \frac{27}{q}$

$\therefore 8, p, q, 27$ are the first four terms of a geometric series.

(iv) $U_n = ar^{n-1}$

$\therefore U_4 = ar^3$

$\therefore 27 = 8r^3$

$r^3 = \frac{27}{8}$

$r = \frac{3}{2}$

Now $U_2 = ar$

$\therefore p = (8)\left(\frac{3}{2}\right)$

$= 12$

and $U_3 = ar^2$

$\therefore q = 8\left(\frac{3}{2}\right)^2$

$= 18.$

QUESTION 9

(a) $y = x^2 + 1$

$\therefore y^2 = (x^2+1)^2$

$= x^4 + 2x^2 + 1$

$$V = \pi\int_a^b y^2\,dx$$

$$= \pi\int_0^1 (x^4 + 2x^2 + 1)\,dx$$

$$= \pi\left[\frac{x^5}{5} + \frac{2x^3}{3} + x\right]_0^1$$

$$= \pi\left\{\left(\frac{(1)^5}{5} + \frac{2(1)^3}{3} + 1\right) - \left(\frac{(0)^5}{5} + \frac{2(0)^3}{3} + 0\right)\right\}$$

$$= \pi\left\{\left(\frac{1}{5} + \frac{2}{3} + 1\right) - (0)\right\}$$

$$= \frac{28\pi}{15}\text{ units}^3.$$

(b) (i) $P = \frac{52}{52} \times \frac{39}{51}$

$= \frac{13}{17}$

$= 0.7647\ldots$ (by calc.)
$= 0.765$ to 3 decimal places

(ii) $P = \frac{52}{52} \times \frac{39}{51} \times \frac{26}{50} \times \frac{13}{49}$

$= \frac{13\,182}{124\,950}$

$= 0.1054\ldots$ (by calc.)
$= 0.105$ to 3 decimal places

(c) (i) $A_1 = 1000\left(1 + \dfrac{6}{100}\right)^{18}$

$A_2 = 1000\left(1 + \dfrac{6}{100}\right)^{17}$

$A_3 = 1000\left(1 + \dfrac{6}{100}\right)^{16}$

$\vdots$

$A_{18} = 1000\left(1 + \dfrac{6}{100}\right)^{1}$

The total value of the investment fund,

$$A = 1000\left(1 + \frac{6}{100}\right)^{18} + 1000\left(1 + \frac{6}{100}\right)^{17} + 1000\left(1 + \frac{6}{100}\right)^{16} \ldots + 1000\left(1 + \frac{6}{100}\right)^{1}$$

$$= 1000 \times \left[1.06^{18} + 1.06^{17} + 1.06^{16} + \ldots + 1.06^{1}\right]$$

$$= 1000 \times \left[1.06^{1} + 1.06^{2} + 1.06^{3} \ldots + 1.06^{18}\right]$$

Now $1.06^1 + 1.06^2 + 1.06^3 \ldots + 1.06^{18}$ is a geometric series with $a = 1.06,\ r = 1.06,\ n = 18$.

$$S_n = \frac{a(r^n - 1)}{r - 1}$$

$$\therefore S_{18} = \frac{1.06(1.06^{18} - 1)}{1.06 - 1}$$

$$= \frac{1.06(1.06^{18} - 1)}{0.06}$$

$= 32.7599 \ldots$ (by calc.)

$\therefore A = 1000 \times 32.7599 \ldots$
$= 32\,759.991 \ldots$ (by calc.)
$= \$32\,759.99$ to the nearest cent

$\therefore$ The total value of Mr Caine's investment is \$32 759.99.

(ii) $A_1 = 1000 \times (1.06)^3$

$A_2 = 1000 \times 1.06 \times 1.06^2$
$= 1000 \times (1.06)^3$

$A_3 = 1000 \times 1.06^2 \times 1.06$
$= 1000 \times (1.06)^3$

$\therefore A = 3 \times 1000 \times (1.06)^3$
$= \$3573.048 \ldots$ (by calc.)
$= \$3573.05$ to the nearest cent.

$\therefore$ The total value of Mrs Caine's investment on her son's third birthday is \$3573.05.

(iii) $A_{18} = 18 \times 1000 \times (1.06)^{18}$
$= \$51\,378.104 \ldots$ (by calc.)
$= \$51\,378.10$ to the nearest cent.

$\therefore$ The total value of Mrs Caine's investment is \$51 378.10.

QUESTION 10

(a) (i)

t	0	2	4
$\frac{dx}{dt}$	0	1	5
	y_0	y_1	y_z

$\therefore$ Distance travelled

$\doteqdot \frac{h}{3}(y_0 + 4y_1 + y_2)$

$= \frac{2}{3}[0 + 4(1) + 5]$

$= \frac{2}{3} \times 9$

$= 6$ units.

(ii) Displacement of the object is decreasing after 5 seconds as the velocity is negative.

(iii) At B, displacement is approximately 6 units. (from (i))

At C, displacement $= 6 + \left(\frac{1}{2} \times 1 \times 5\right)$

$= 8.5$ units.

At D, displacement $= 6$ units to the right.

$\therefore$ Time taken to return to the origin from $D(t = 6)$

$= \frac{d}{V}$

$= \frac{6}{5}$

$= 1.2$ seconds

$\therefore$ Time at which particle returns to the origin

$= 6 + 1.2$

$= 7.2$ seconds.

(iv)

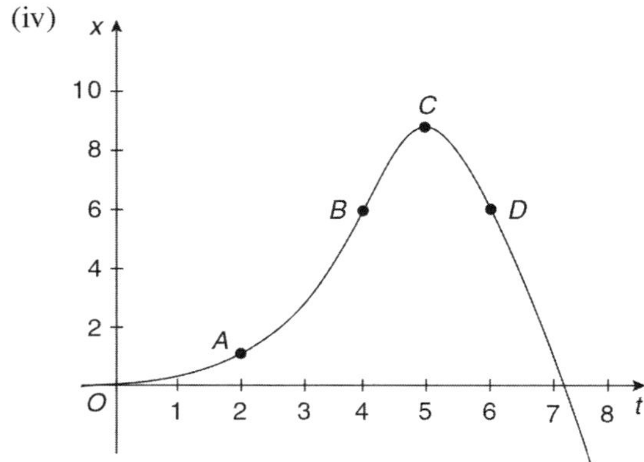

(b)

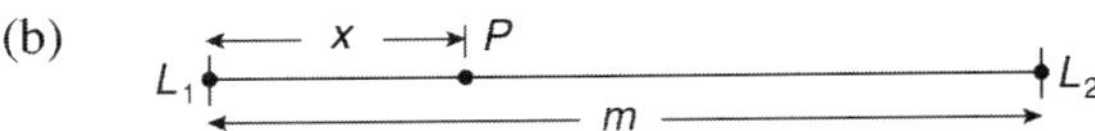

(i) Distance of P from L_2 is $(m - x)$.

Now $N = \frac{L}{d^2}$

$\therefore N = \frac{L_1}{x^2} + \frac{L_2}{(m-x)^2}$

(ii)

$$N = \frac{L_1}{x^2} + \frac{L_2}{(m-x)^2} \quad x \neq m,\ x \neq 0$$

$$= L_1x^{-2} + L_2(m-x)^{-2}$$

$$\frac{dN}{dx} = -2L_1x^{-3} - 2L_2(m-x)^{-3}.-1$$

$$= -2L_1x^{-3} + 2L_2(m-x)^{-3}$$

$$= \frac{-2L_1}{x^3} + \frac{2L_2}{(m-x)^3}$$

Stationary points occur when

$$\frac{dN}{dx} = 0$$

$$\therefore \frac{-2L_1}{x^3} + \frac{2L_2}{(m-x)^3} = 0$$

$$\frac{2L_2}{(m-x)^3} = \frac{2L_1}{x^3}$$

$$2L_2x^3 = 2L_1(m-x)^3$$

$$L_2x^3 = L_1(m-x)^3$$

$$\sqrt[3]{L_2}\,x = \sqrt[3]{L_1}\,(m-x)$$

$$\sqrt[3]{L_2}\,x + \sqrt[3]{L_1}\,x = \sqrt[3]{L_1}\,m$$

$$x\left(\sqrt[3]{L_2} + \sqrt[3]{L_1}\right) = \sqrt[3]{L_1}\,m$$

$$x = \frac{\sqrt[3]{L_1}m}{\sqrt[3]{L_1} + \sqrt[3]{L_2}}$$

Now $\frac{dN}{dx} = -2L_1x^{-3} + 2L_2(m-x)^{-3}$

$$\therefore \frac{d^2N}{dx^2} = 6L_1x^{-4} - 6L_2(m-x)^{4}.-1$$

$$= \frac{6L_1}{x^4} + \frac{6L_2}{(m-x)^4} > 0$$

$\therefore$ A minimum noise level occurs at P

when $x = \frac{\sqrt[3]{L_1}m}{\sqrt[3]{L_1} + \sqrt[3]{L_2}}$.

BOARD OF STUDIES
NEW SOUTH WALES

2008

HIGHER SCHOOL CERTIFICATE EXAMINATION

Mathematics

General Instructions

- Reading time – 5 minutes
- Working time – 3 hours
- Write using black or blue pen
- Board-approved calculators may be used
- A table of standard integrals is provided at the back of this paper
- All necessary working should be shown in every question

Total marks – 120

- Attempt Questions 1–10
- All questions are of equal value

Total marks – 120
Attempt Questions 1–10
All questions are of equal value

Answer each question in the appropriate writing booklet. Extra writing booklets are available.

Marks

Question 1 (12 marks) Use the Question 1 Writing Booklet.

(a) Evaluate $2\cos\frac{\pi}{5}$ correct to three significant figures. **2**

(b) Factorise $3x^2 + x - 2$. **2**

(c) Simplify $\frac{2}{n} - \frac{1}{n+1}$. **2**

(d) Solve $|4x - 3| = 7$. **2**

(e) Expand and simplify $(\sqrt{3} - 1)(2\sqrt{3} + 5)$. **2**

(f) Find the sum of the first 21 terms of the arithmetic series $3 + 7 + 11 + \cdots$. **2**

Marks

Question 2 (12 marks) Use the Question 2 Writing Booklet.

(a) Differentiate with respect to x:

(i) $(x^2 + 3)^9$ **2**

(ii) $x^2 \log_e x$ **2**

(iii) $\dfrac{\sin x}{x + 4}$. **2**

(b) Let M be the midpoint of $(-1, 4)$ and $(5, 8)$. **2**

Find the equation of the line through M with gradient $-\dfrac{1}{2}$.

(c) (i) Find $\displaystyle\int \frac{dx}{x + 5}$. **1**

(ii) Evaluate $\displaystyle\int_0^{\frac{\pi}{12}} \sec^2 3x \, dx$. **3**

Marks

Question 3 (12 marks) Use the Question 3 Writing Booklet.

(a)

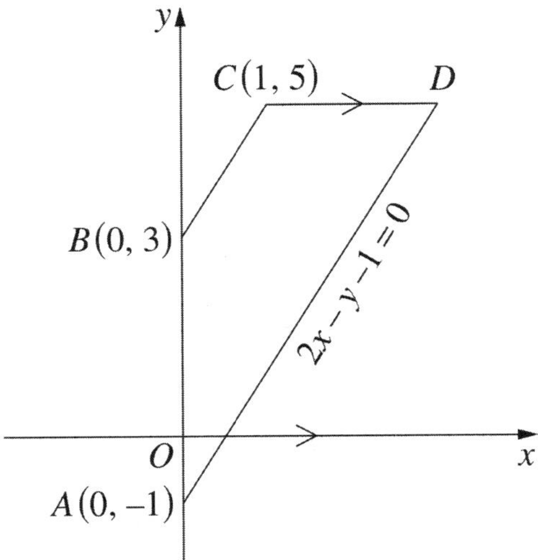

NOT TO SCALE

In the diagram, $ABCD$ is a quadrilateral. The equation of the line AD is $2x - y - 1 = 0$.

(i) Show that $ABCD$ is a trapezium by showing that BC is parallel to AD. **2**

(ii) The line CD is parallel to the x-axis. Find the coordinates of D. **1**

(iii) Find the length of BC. **1**

(iv) Show that the perpendicular distance from B to AD is $\dfrac{4}{\sqrt{5}}$. **2**

(v) Hence, or otherwise, find the area of the trapezium $ABCD$. **2**

(b) (i) Differentiate $\log_e(\cos x)$ with respect to x. **2**

(ii) Hence, or otherwise, evaluate $\displaystyle\int_0^{\frac{\pi}{4}} \tan x \, dx$. **2**

Marks

Question 4 (12 marks) Use the Question 4 Writing Booklet.

(a) **2**

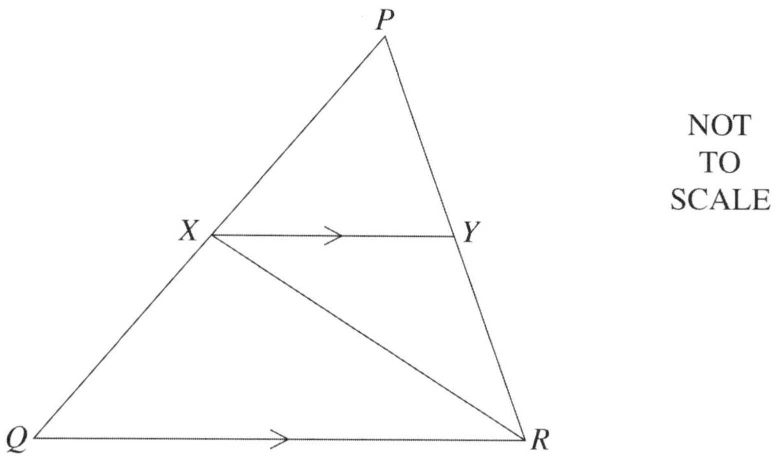

NOT TO SCALE

In the diagram, XR bisects $\angle PRQ$ and $XY \parallel QR$.

Copy or trace the diagram into your writing booklet.

Prove that ΔXYR is an isosceles triangle.

(b) The zoom function in a software package multiplies the dimensions of an image by 1.2. In an image, the height of a building is 50 mm. After the zoom function is applied once, the height of the building in the image is 60 mm. After a second application, its height is 72 mm.

(i) Calculate the height of the building in the image after the zoom function has been applied eight times. Give your answer to the nearest mm. **2**

(ii) The height of the building in the image is required to be more than 400 mm. Starting from the original image, what is the least number of times the zoom function must be applied? **2**

(c) Consider the parabola $x^2 = 8(y - 3)$.

(i) Write down the coordinates of the vertex. **1**

(ii) Find the coordinates of the focus. **1**

(iii) Sketch the parabola. **1**

(iv) Calculate the area bounded by the parabola and the line $y = 5$. **3**

Marks

Question 5 (12 marks) Use the Question 5 Writing Booklet.

(a) The gradient of a curve is given by $\frac{dy}{dx} = 1 - 6\sin 3x$. The curve passes through the point $(0, 7)$. **3**

What is the equation of the curve?

(b) Consider the geometric series

$$5 + 10x + 20x^2 + 40x^3 + \cdots.$$

(i) For what values of x does this series have a limiting sum? **2**

(ii) The limiting sum of this series is 100. **2**

Find the value of x.

(c) Light intensity is measured in lux. The light intensity at the surface of a lake is 6000 lux. The light intensity, I lux, a distance s metres below the surface of the lake is given by

$$I = Ae^{-ks}$$

where A and k are constants.

(i) Write down the value of A. **1**

(ii) The light intensity 6 metres below the surface of the lake is 1000 lux. **2**

Find the value of k.

(iii) At what rate, in lux per metre, is the light intensity decreasing 6 metres below the surface of the lake? **2**

Marks

Question 6 (12 marks) Use the Question 6 Writing Booklet.

(a) Solve $2\sin^2 \dfrac{x}{3} = 1$ for $-\pi \le x \le \pi$. **3**

(b) The graph shows the velocity of a particle, v metres per second, as a function of time, t seconds.

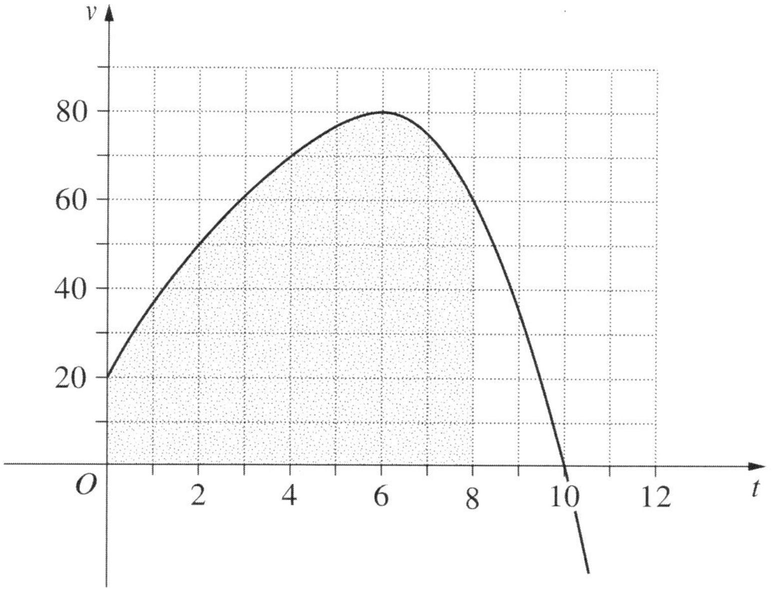

(i) What is the initial velocity of the particle? **1**

(ii) When is the velocity of the particle equal to zero? **1**

(iii) When is the acceleration of the particle equal to zero? **1**

(iv) By using Simpson's Rule with five function values, estimate the distance travelled by the particle between $t = 0$ and $t = 8$. **3**

Question 6 continues

Marks

Question 6 (continued)

(c) The graph of $y = \dfrac{5}{x-2}$ is shown below. **3**

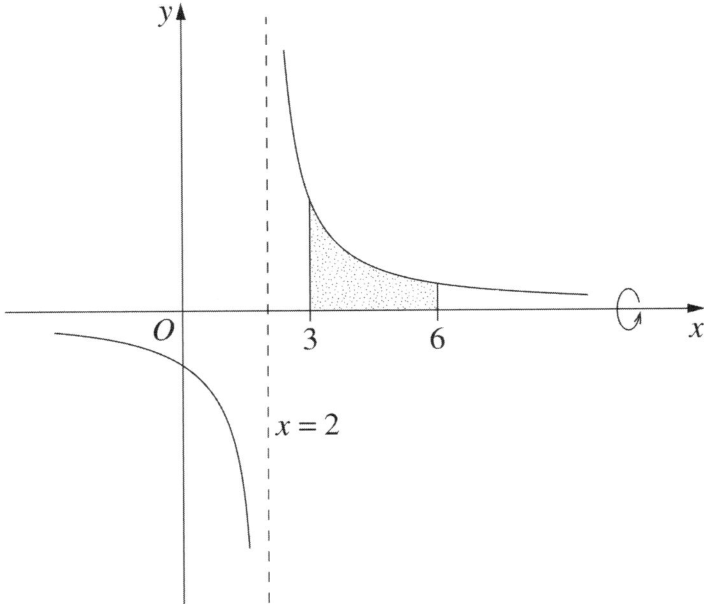

The shaded region in the diagram is bounded by the curve $y = \dfrac{5}{x-2}$, the x-axis and the lines $x = 3$ and $x = 6$.

Find the volume of the solid of revolution formed when the shaded region is rotated about the x-axis.

End of Question 6

Marks

Question 7 (12 marks) Use the Question 7 Writing Booklet.

(a) Solve $\log_e x - \dfrac{3}{\log_e x} = 2$. **3**

(b)

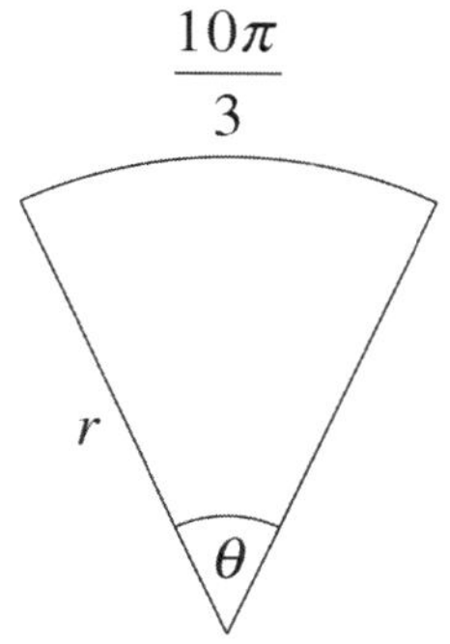

$\frac{10\pi}{3}$

r

θ

The diagram shows a sector with radius r and angle θ where $0 < \theta \leq 2\pi$.
The arc length is $\dfrac{10\pi}{3}$.

(i) Show that $r \geq \dfrac{5}{3}$. **2**

(ii) Calculate the area of the sector when $r = 4$. **2**

Question 7 continues

Marks

Question 7 (continued)

(c) Xena and Gabrielle compete in a series of games. The series finishes when one player has won two games. In any game, the probability that Xena wins is $\frac{2}{3}$ and the probability that Gabrielle wins is $\frac{1}{3}$.

Part of the tree diagram for this series of games is shown.

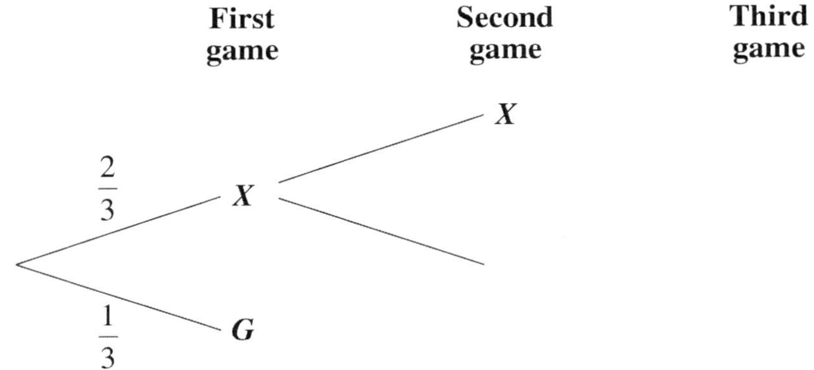

(i) Copy and complete the tree diagram showing the possible outcomes. **1**

(ii) What is the probability that Gabrielle wins the series? **2**

(iii) What is the probability that three games are played in the series? **2**

End of Question 7

Marks

Question 8 (12 marks) Use the Question 8 Writing Booklet.

(a) Let $f(x) = x^4 - 8x^2$.

(i) Find the coordinates of the points where the graph of $y = f(x)$ crosses the axes. **2**

(ii) Show that $f(x)$ is an even function. **1**

(iii) Find the coordinates of the stationary points of $f(x)$ and determine their nature. **4**

(iv) Sketch the graph of $y = f(x)$. **1**

(b)

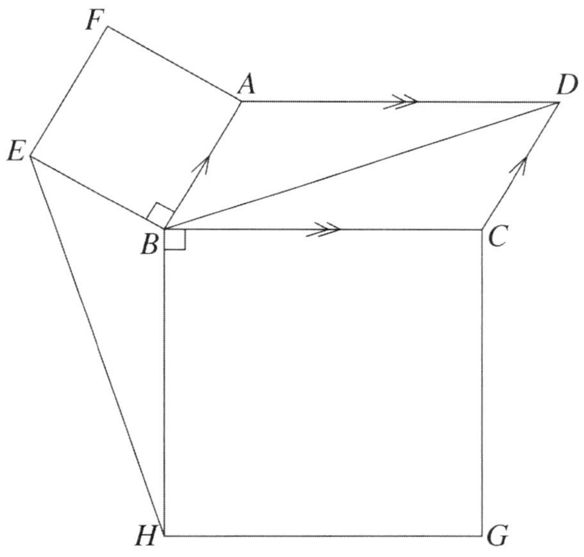

In the diagram, $ABCD$ is a parallelogram and $ABEF$ and $BCGH$ are both squares.

Copy or trace the diagram into your writing booklet.

(i) Prove that $CD = BE$. **1**

(ii) Prove that $BD = EH$. **3**

Marks

Question 9 (12 marks) Use the Question 9 Writing Booklet.

(a) It is estimated that 85% of students in Australia own a mobile phone.

(i) Two students are selected at random. What is the probability that neither of them owns a mobile phone? **2**

(ii) Based on a recent survey, 20% of the students who own a mobile phone have used their mobile phone during class time. A student is selected at random. What is the probability that the student owns a mobile phone and has used it during class time? **1**

(b) Peter retires with a lump sum of \$100 000. The money is invested in a fund which pays interest each month at a rate of 6% per annum, and Peter receives a fixed monthly payment of \$$M$ from the fund. Thus, the amount left in the fund after the first monthly payment is $\$(100\,500 - M)$.

(i) Find a formula for the amount, $\$A_n$, left in the fund after n monthly payments. **2**

(ii) Peter chooses the value of M so that there will be nothing left in the fund at the end of the 12th year (after 144 payments). Find the value of M. **3**

(c) A beam is supported at $(-b, 0)$ and $(b, 0)$ as shown in the diagram.

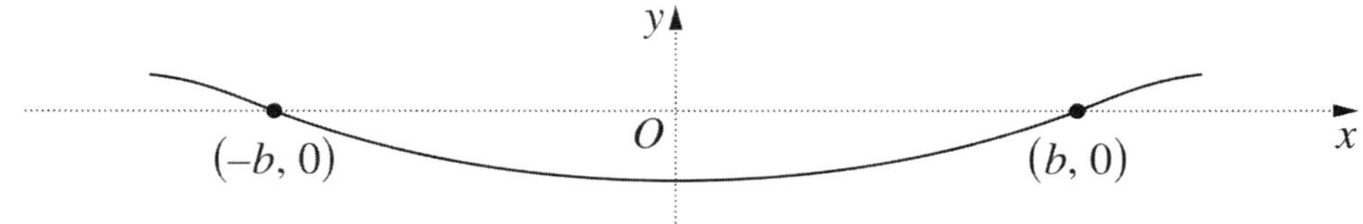

It is known that the shape formed by the beam has equation $y = f(x)$, where $f(x)$ satisfies

$$f''(x) = k\left(b^2 - x^2\right) \quad (k \text{ is a positive constant})$$

and $$f'(-b) = -f'(b).$$

(i) Show that **2**

$$f'(x) = k\left(b^2 x - \frac{x^3}{3}\right).$$

(ii) How far is the beam below the x-axis at $x = 0$? **2**

Marks

Question 10 (12 marks) Use the Question 10 Writing Booklet.

(a) **5**

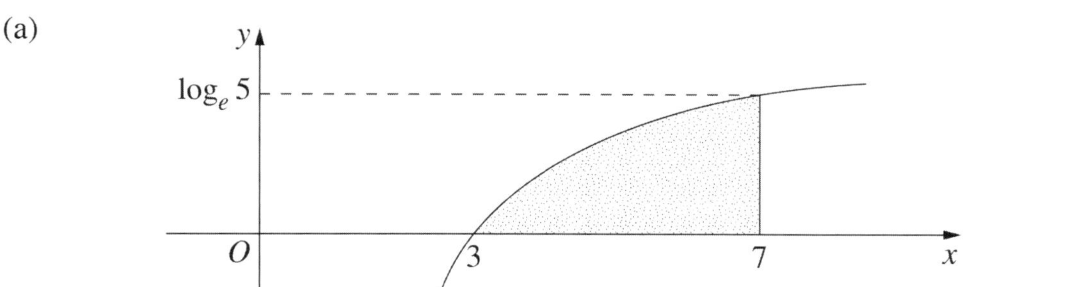

In the diagram, the shaded region is bounded by $y = \log_e(x - 2)$, the x-axis and the line $x = 7$.

Find the exact value of the area of the shaded region.

(b)

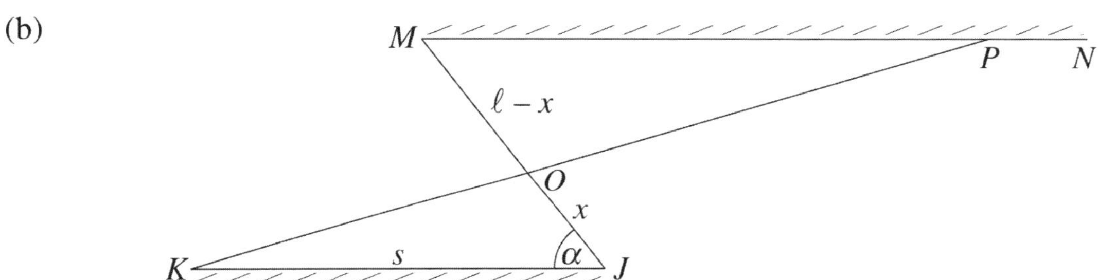

The diagram shows two parallel brick walls KJ and MN joined by a fence from J to M. The wall KJ is s metres long and $\angle KJM = \alpha$. The fence JM is ℓ metres long.

A new fence is to be built from K to a point P somewhere on MN. The new fence KP will cross the original fence JM at O.

Let $OJ = x$ metres, where $0 < x < \ell$.

(i) Show that the total area, A square metres, enclosed by ΔOKJ and ΔOMP is given by **3**

$$A = s\left(x - \ell + \frac{\ell^2}{2x}\right)\sin\alpha.$$

(ii) Find the value of x that makes A as small as possible. Justify the fact that this value of x gives the minimum value for A. **3**

(iii) Hence, find the length of MP when A is as small as possible. **1**

End of paper

2008 Higher School Certificate Worked Answers

QUESTION 1

(a) $2\cos\frac{\pi}{5} = 1.6180\ldots$ (by calc.)

$= 1.62$ to 3 significant figures.

(b) $3x^2 + x - 2 = 3x^2 + 3x - 2x - 2$
$= 3x(x+1) - 2(x+1)$
$= (3x-2)(x+1).$

(c) $\frac{2}{n} - \frac{1}{n+1} = \frac{2(n+1)-n}{n(n+1)}$
$= \frac{2n+2-n}{n(n+1)}$
$= \frac{n+2}{n(n+1)}.$

(d) $|4x-3| = 7$

$\therefore 4x - 3 = 7$ or $4x - 3 = -7$
$4x = 10$ $\qquad 4x = -4$
$x = \frac{10}{4}$ $\qquad \therefore x = -1$
$\therefore x = 2\frac{1}{2}$

Test $\left|4 \times 2\frac{1}{2} - 3\right| = 7$ $\qquad |4 \times -1 - 3| = 7$
$|10 - 3| = 7$ $\qquad |-4 - 3| = 7$

$\therefore x = 2\frac{1}{2}$ or $x = -1.$

(e) $(\sqrt{3} - 1)(2\sqrt{3} + 5)$
$= \sqrt{3} \times 2\sqrt{3} + \sqrt{3} \times 5 - 2\sqrt{3} - 5$
$= 6 + 5\sqrt{3} - 2\sqrt{3} - 5$
$= 1 + 3\sqrt{3}.$

(f) $3 + 7 + 11 + \ldots$

$S_n = \frac{n}{2}\left[2a + (n-1)d\right]$
$a = 3, d = 4, n = 21$

$\therefore S_{21} = \frac{21}{2}\left[2 \times 3 + (21-1) \times 4\right]$
$= 10\frac{1}{2} \times (6 + 20 \times 4)$
$= 903$ (by calc.)

QUESTION 2

(a) (i) $\frac{d}{dx}\left[(x^2+3)^9\right]$
$= 9(x^2+3)^8 \cdot \frac{d}{dx}(x^2+3)$
$= 9(x^2+3)^8 \cdot 2x$
$= 18x(x^2+3)^8.$

(ii) $\frac{d}{dx}(x^2 \log_e x)$
$= x^2 \cdot \frac{d}{dx}(\log_e x) + \log_e x \cdot \frac{d}{dx}(x^2)$
$= x^2 \cdot \frac{1}{x} + \log_e x \cdot 2x$
$= x + 2x\log_e x$
$= x(1 + 2\log_e x).$

(iii) $\frac{d}{dx}\left(\frac{\sin x}{x+4}\right)$

$$= \frac{(x+4) \cdot \frac{d}{dx}(\sin x) - \sin x \cdot \frac{d}{dx}(x+4)}{(x+4)^2}$$
$$= \frac{(x+4) \cdot \cos x - \sin x \cdot 1}{(x+4)^2}$$
$$= \frac{(x+4)\cos x - \sin x}{(x+4)^2}.$$

(b) $M = \left(\frac{x_1+x_2}{2}, \frac{y_1+y_2}{2}\right)$
$= \left(\frac{-1+5}{2}, \frac{4+8}{2}\right)$
$= (2, 6).$

Equation of the line with gradient $= -\frac{1}{2}$, through the point (2, 6) is given by

$$y - 6 = -\frac{1}{2}(x-2)$$
$$2y - 12 = -x + 2$$

i.e. $x + 2y - 14 = 0.$

(c) (i) $\int \frac{dx}{x+5} = \log_e(x+5) + C$

(ii) $\int_0^{\frac{\pi}{12}} \sec^2 3x \, dx$

$= \left[\frac{1}{3}\tan 3x\right]_0^{\frac{\pi}{12}}$ (using standard integrals)

$= \frac{1}{3}\left[\tan 3\left(\frac{\pi}{12}\right) - \tan 3(0)\right]$

$= \frac{1}{3}\left[\tan\frac{\pi}{4} - \tan 0\right]$

$= \frac{1}{3}(1-0)$

$= \frac{1}{3}.$

QUESTION 3

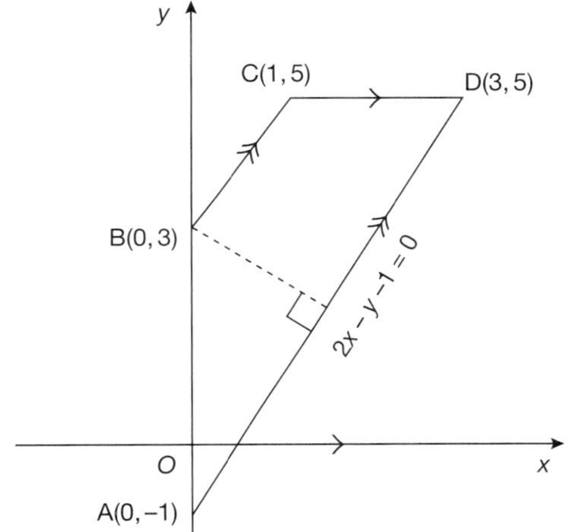

(a) (i) Gradient of $BC = \frac{y_2 - y_1}{x_2 - x_1}$

$= \frac{5-3}{1-0}$

$= 2$

Gradient of $AD = \frac{-A}{B}$

$= \frac{-2}{-1}$

$= 2$

Since $m_{BC} = m_{AD}$, $BC \parallel AD$
$\therefore ABCD$ is a trapezium with a pair of opposite sides parallel.

(ii) Since CD is parallel to the x-axis, at D, $y = 5$.
Now D lies on $2x - y - 1 = 0$

$\therefore 2x - 5 - 1 = 0$
$2x - 6 = 0$
$2x = 6$
$\therefore x = 3$

$\therefore D$ is the point $(3,5)$.

(iii) B(0, 3) C(1, 5)

$d_{BC} = \sqrt{(x_2 - x_1)^2 + (y_2 - y_1)^2}$

$= \sqrt{(1-0)^2 + (5-3)^2}$

$= \sqrt{1+4}$

$= \sqrt{5}.$

The length of BC = $\sqrt{5}$ units.

(iv) B(0, 3)
Equation of AD is $2x - y - 1 = 0$

$d = \left|\frac{ax_1 + by_1 + c}{\sqrt{a^2 + b^2}}\right|$

where $a = 2$, $b = -1$, $c = -1$, $x_1 = 0$, $y_1 = 3$

$\therefore d = \left|\frac{2 \times 0 + -1 \times 3 - 1}{\sqrt{2^2 + (-1)^2}}\right|$

$= \left|\frac{-4}{\sqrt{5}}\right|$

$= \frac{4}{\sqrt{5}}$ units.

(v) Area of trapezium ABCD

$= \frac{1}{2} \times h \times (a+b)$

Now $h = \frac{4}{\sqrt{5}}$

$a = d_{BC} = \sqrt{5}$

$b = d_{AD} = \sqrt{(3-0)^2 + (5+1)^2}$

$= \sqrt{9+36}$

$= \sqrt{45}$

$= 3\sqrt{5}$

$\therefore$ Area of trapezium ABCD

$= \frac{1}{2} \times \frac{4}{\sqrt{5}} \times (\sqrt{5} + 3\sqrt{5})$

$= \frac{1}{2} \times \frac{4}{\sqrt{5}} \times 4\sqrt{5}$

$= 8$ units^2.

(b) (i) $y = \log_e(\cos x)$

$$\frac{dy}{dx} = \frac{1}{\cos x} \times \frac{d}{dx}(\cos x)$$
$$= \frac{-\sin x}{\cos x}$$
$$= -\tan x.$$

(ii)
$$\int_0^{\frac{\pi}{4}} \tan x \, dx = -\Big[\log_e(\cos x)\Big]_0^{\frac{\pi}{4}}$$
$$= -\left[\log_e\left(\cos\frac{\pi}{4}\right) - \log_e(\cos 0)\right]$$
$$= -\left[\log_e\frac{1}{\sqrt{2}} - \log_e 1\right]$$
$$= -\left[\log_e\frac{1}{\sqrt{2}} - 0\right]$$
$$= -\log_e\frac{1}{\sqrt{2}}$$
$= 0.3465\ldots$ (by calc.)
$= 0.35$ to 2 decimal places.

QUESTION 4

(a)

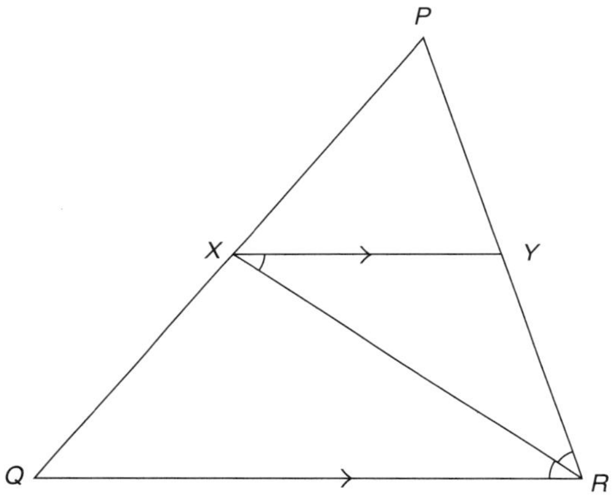

Now $\angle YRX = \angle XRQ$ (XR bisects $\angle PRQ$)
and $\angle YXR = \angle XRQ$ (alternate; $XY \parallel QR$)
$\therefore \angle YXR = \angle YRX$

$\therefore \triangle XYR$ is an isosceles triangle.

(b) (i) The heights of the building are a geometric series:
50, 60, 72 . . .

$T_n = ar^{n-1}$
$r = 1.2$, $a = 50$
Let $n = 9$ (zoom has been applied 8 times, *or* the 9th image)

$\therefore T_9 = 50(1.2)^{9-1}$
$= 50(1.2)^8$
$= 214.9908\ldots$ (by calc.)
$= 215$ mm to the nearest mm.

$\therefore$ The height of the building is 215 mm after 8 zoom function applications.

(ii) $T_n = ar^{n-1}$
$r = 1.2$, $a = 50$, $n =$ image no.

$$T_n > 400$$
$$\therefore ar^{n-1} > 400$$
$$50(1.2)^{n-1} > 400$$
$$(1.2)^{n-1} > \frac{400}{50}$$
$$(1.2)^{n-1} > 8$$
$$\ln(1.2)^{n-1} > \ln 8$$
$$(n-1)\ln 1.2 > \ln 8$$
$$n - 1 > \frac{\ln 8}{\ln 1.2}$$
$$n > \frac{\ln 8}{\ln 1.2} + 1$$
$> 12.4053\ldots$ (by calc.)
$\therefore n = 13$

The 13th image will be more than 400 mm in height. Therefore, the zoom function must be applied at least $13 - 1 = 12$ times.

(c) $x^2 = 8(y-3)$

(i) Coordinates of the vertex are (0, 3).

(ii) Now $4a = 8$
$\therefore a = 2$
i.e. The focal length is 2.

$\therefore$ Coordinates of the focus are (0, 5).

(iii)

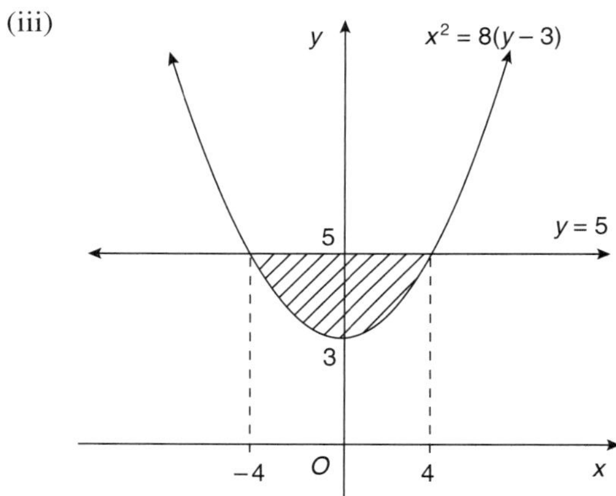

(iv) $x^2 = 8(y-3) \quad \therefore y = \dfrac{x^2}{8} + 3$

When $y = 5$,
$x^2 = 8(5-3)$
$= 16$
$\therefore x = \pm 4$

Required area
= Area of rectangle – Area under the curve

$$= l \times b - \int_{-4}^{4} \left(\frac{x^2}{8} + 3\right) dx$$

$$= 8 \times 5 - 2\int_{0}^{4} \left(\frac{x^2}{8} + 3\right) dx$$

$$= 40 - 2\left[\frac{x^3}{24} + 3x\right]_0^4$$

$$= 40 - 2\left[\left(\frac{4^3}{24} + 3 \times 4\right) - 0\right]$$

$$= 40 - 2\left[\frac{64}{24} + 12\right]$$

$$= 40 - 2 \times 14\tfrac{2}{3}$$

$$= 40 - 29\tfrac{1}{3}$$

$$= 10\tfrac{2}{3} \text{ units}^2.$$

QUESTION 5

(a) $\dfrac{dy}{dx} = 1 - 6\sin 3x$

$\therefore y = \int (1 - 6\sin 3x)\,dx$

$= x + 2\cos 3x + C$

Now the curve passes through (0, 7),

$\therefore 7 = 0 + 2\cos 3(0) + C$
$7 = 0 + 2\cos 0 + C$
$7 = 2 + C$
$\therefore C = 5$

$\therefore y = x + 2\cos 3x + 5.$

(b) $5 + 10x + 20x^2 + 40x^3 + \ldots$

(i) This is an infinite series with $a = 5$, $r = 2x$.
For a limiting sum: $-1 < r < 1$
$\therefore -1 < 2x < 1$
$\therefore -\frac{1}{2} < x < \frac{1}{2}$.

(ii) $S_\infty = \dfrac{a}{1 - r}$

If $S_\infty = 100$, $a = 5$ and $r = 2x$,

$\therefore 100 = \dfrac{5}{1 - 2x}$

$100 \times (1 - 2x) = 5$

$1 - 2x = \dfrac{5}{100}$

$-2x = -\dfrac{19}{20}$

$\therefore x = \dfrac{19}{40}$.

(c) (i) $I = Ae^{-ks}$

When $s = 0$, I = 6000

$\therefore 6000 = Ae^{-k \times 0}$
$6000 = Ae^0$
$6000 = A$

i.e. $A = 6000$.

(ii) $I = Ae^{-ks}$

When $s = 6$, $I = 1000$ and $A = 6000$

$\therefore 1000 = 6000 \times e^{-k \times 6}$

$\dfrac{1}{6} = e^{-6k}$

$\log_e \dfrac{1}{6} = -6k$

$k = -\dfrac{1}{6} \log_e \dfrac{1}{6}$

$= -\dfrac{1}{6} \log_e 6^{-1}$

$= \dfrac{1}{6} \log_e 6$

$= \dfrac{\log_e 6}{6}$

$= 0.2986\ldots$ (by calc.)

$\therefore k = 0.30$ to 2 decimal places.

(iii) $I = Ae^{-ks}$

$\therefore \dfrac{dI}{ds} = -kAe^{-ks}$

When $s = 6$, $A = 6000$
and $k = \dfrac{\log_e 6}{6}$ or $0.2986\ldots$,

$\therefore \dfrac{dI}{ds} = -\dfrac{\log_e 6}{6} \times 6000 \times e^{-\frac{\log_e 6}{6} \times 6}$

$= -\log_e 6 \times 1000 \times e^{-\log_e 6}$

$= -298.6265\ldots$ (by calc.)

$\doteqdot -299.$

$\therefore$ At 6 metres, the light intensity is decreasing at 299 lux per metre.

QUESTION 6

(a) $2\sin^2\frac{x}{3} = 1$ for $-\pi \leqslant x \leqslant \pi$

$\therefore \sin^2\frac{x}{3} = \frac{1}{2}$ for $-\frac{\pi}{3} \leqslant \frac{x}{3} \leqslant \frac{\pi}{3}$

$\therefore \sin\frac{x}{3} = \pm\frac{1}{\sqrt{2}}$

$\therefore \frac{x}{3} = \frac{\pi}{4}, \pi - \frac{\pi}{4}, -\frac{\pi}{4}, -\pi + \frac{\pi}{4}$

$= \frac{\pi}{4}, \frac{3\pi}{4}, -\frac{\pi}{4}, -\frac{3\pi}{4}$

$\therefore x = \frac{3\pi}{4}, \frac{9\pi}{4}, -\frac{3\pi}{4}, -\frac{9\pi}{4}$

$\therefore x = -\frac{3\pi}{4}, \frac{3\pi}{4}$ (by domain).

(b) (i) The initial velocity is when $t = 0$.
$\therefore$ Initial velocity is 20 m/s.

(ii) The velocity of the particle is equal to 0 when $t = 10$.

(iii) The acceleration of the particle is zero when $t = 6$.

(iv)

t	0	2	4	6	8
$v(t)$	20	50	70	80	60
	y_0	y_1	y_2	y_3	y_4

$\int_0^8 v(t)\,dt$

$\doteqdot \frac{h}{3}\left[y_0 + y_4 + 2y_2 + 4(y_1 + y_3)\right]$

$= \frac{2}{3}\left[20 + 60 + 2(70) + 4(50 + 80)\right]$

$= \frac{2}{3}\left[20 + 60 + 140 + 4 \times 130\right]$

$= \frac{2}{3} \times 740$

$= 493\frac{1}{3}$.

$\therefore$ Distance travelled is approximately $493\frac{1}{3}$ m.

(c) $y = \frac{5}{x-2}$

$\therefore y^2 = \left(\frac{5}{x-2}\right)^2$

$= \frac{25}{(x-2)^2}$

$v = \pi\int_a^b y^2\,dx$

$= \pi\int_3^6 \frac{25}{(x-2)^2}\,dx$

$= 25\pi\int_3^6 (x-2)^{-2}\,dx$

$= 25\pi\left[\frac{(x-2)^{-1}}{-1}\right]_3^6$

$= 25\pi\left[-\frac{1}{(x-2)}\right]_3^6$

$= 25\pi\left[\left(-\frac{1}{6-2}\right) - \left(-\frac{1}{3-2}\right)\right]$

$= 25\pi\left[-\frac{1}{4} + 1\right]$

$= 25\pi \times \frac{3}{4}$

$= \frac{75\pi}{4}$ units^3.

QUESTION 7

(a) $\log_e x - \frac{3}{\log_e x} = 2$

Let $m = \log_e x$

$\therefore m - \frac{3}{m} = 2$

$m^2 - 3 = 2m$

$m^2 - 2m - 3 = 0$

$(m-3)(m+1) = 0$

$\therefore m - 3 = 0$ *or* $m + 1 = 0$

$\therefore m = 3$ *or* $m = -1$

$\therefore \log_e x = 3$ *or* $\log_e x = -1$

$\therefore x = e^3$. $\quad x = e^{-1}$

$\therefore x = \frac{1}{e}$.

(b) (i) $l = r\theta$ where $0 < \theta \leqslant 2\pi$

$\therefore \frac{10\pi}{3} = r\theta$

$\therefore \theta = \frac{10\pi}{3r}$

Now $0 < \theta \leqslant 2\pi$

$\therefore 0 < \frac{10\pi}{3r} \leqslant 2\pi$

$\therefore 0 < 10\pi \leqslant 6\pi r$

$\therefore 0 < \dfrac{10\pi}{6\pi} \leqslant r$

$\therefore 0 < \dfrac{5}{3} \leqslant r$

$\therefore r \geqslant \dfrac{5}{3}$.

(ii) $A = \dfrac{1}{2}r^2\theta$

where $\theta = \dfrac{10\pi}{3r}$ and $r = 4$

$\therefore A = \dfrac{1}{2} \times 4^2 \times \dfrac{10\pi}{3 \times 4}$

$= \dfrac{1}{2} \times 16 \times \dfrac{10\pi}{12}$

$= \dfrac{80\pi}{12}$

$= \dfrac{20\pi}{3}$

$\therefore$ Area of sector is $\dfrac{20\pi}{3}$ units^2.

(c) (i)

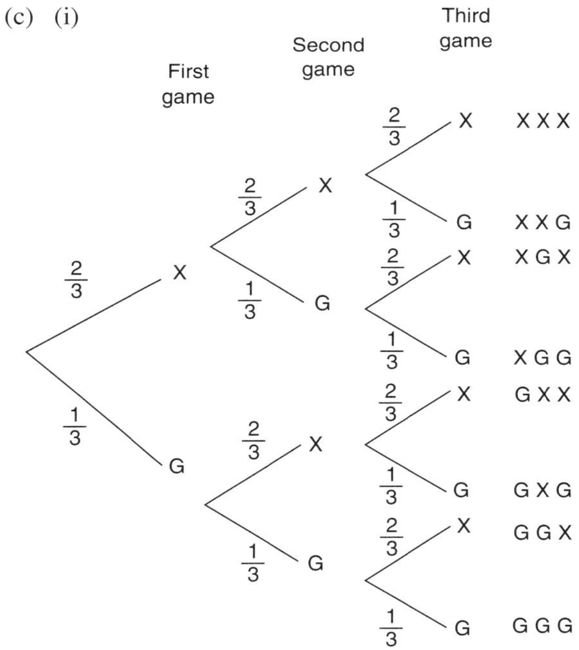

(ii) P(Gabrielle wins series)

$= P(XGG) + P(GXG) + P(GG)$

$= P\left(\dfrac{2}{3} \times \dfrac{1}{3} \times \dfrac{1}{3}\right) + P\left(\dfrac{1}{3} \times \dfrac{2}{3} \times \dfrac{1}{3}\right) + P\left(\dfrac{1}{3} \times \dfrac{1}{3}\right)$

$= \dfrac{2}{27} + \dfrac{2}{27} + \dfrac{1}{9}$

$= \dfrac{7}{27}$.

(iii) P(three games are played)

$= 1 - P(XX) - P(GG)$

$= 1 - P\left(\dfrac{2}{3} \times \dfrac{2}{3}\right) - P\left(\dfrac{1}{3} \times \dfrac{1}{3}\right)$

$= 1 - \dfrac{4}{9} - \dfrac{1}{9}$

$= \dfrac{4}{9}$.

QUESTION 8

(a) (i) The curve crosses the x-axis when

$f(x) = 0$

$\therefore x^4 - 8x^2 = 0$

$x^2(x^2 - 8) = 0$

$x^2(x - 2\sqrt{2})(x + 2\sqrt{2}) = 0$

$\therefore x = 0$ or $2\sqrt{2}$ or $-2\sqrt{2}$

$\therefore$ The coordinates are $(0, 0)$, $(2\sqrt{2}, 0)$ and $(-2\sqrt{2}, 0)$.

(ii) For an even function,

$f(x) = f(-x)$

Now $f(x) = x^4 - 8x^2$

and $f(-x) = (-x)^4 - 8(-x)^2$

$= x^4 - 8x^2$

$\therefore f(x) = f(-x)$

$\therefore f(x)$ is an even function.

(iii) $f(x) = x^4 - 8x^2$

$f'(x) = 4x^3 - 16x$

$= 4x(x^2 - 4)$

$= 4x(x - 2)(x + 2)$

Stationary points occur when

$f'(x) = 0$

$\therefore 4x(x - 2)(x + 2) = 0$

$\therefore x = 0$ or 2 or -2

Now $f(0) = 0$

and $f(2) = (2)^4 - 8(2)^2$

$= -16$

and $f(-2) = (-2)^4 - 8(-2)^2$

$= -16$

$\therefore$ Coordinates of the stationary points are $(0, 0)$ $(2, -16)$ and $(-2, -16)$.

Now $f''(x) = 12x^2 - 16$

$\therefore f''(0) = 12(0)^2 - 16$
$= -16 < 0$

$\therefore$ A maximum turning point occurs at (0, 0).

and $f''(2) = 12(2)^2 - 16$
$= 32 > 0$

$\therefore$ A minimum turning point occurs at (2, –16).

and $f''(-2) = 12(-2)^2 - 16$
$= 32 > 0$

$\therefore$ A minimum turning point occurs at (–2, –16).

(iv)

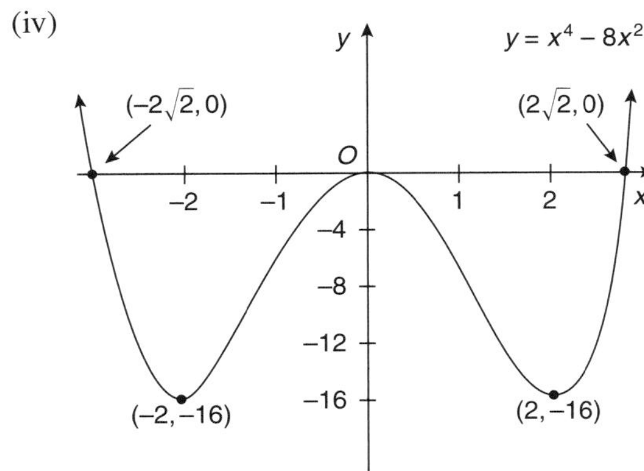

(b)

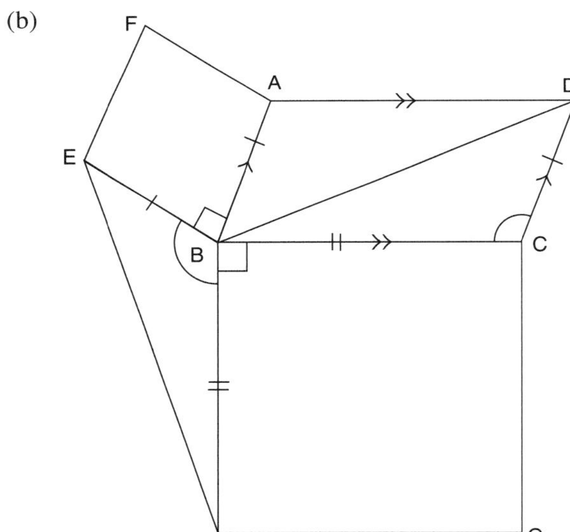

(i) Now $CD = AB$
(opp. sides of a parallelogram)
and $AB = BE$
(sides of a square are equal)
$\therefore CD = BE$.

(ii) Now $BC = BH$
(sides of a square are equal)
and $CD = BE$ (from (i))

Also $\angle BCD = 180° - \angle ABC$
(cointerior angles, $AB \parallel CD$)
and $\angle EBH + 90° + 90° + \angle ABC = 360°$
(angles about a point)
$\therefore \angle EBH = 180° - \angle ABC$
$= \angle BCD$
$\therefore \triangle EBH \equiv \triangle DCB$ (S.A.S.)
$\therefore EH = BD$
(corresponding sides in congruent triangles)

QUESTION 9

(a) (i)

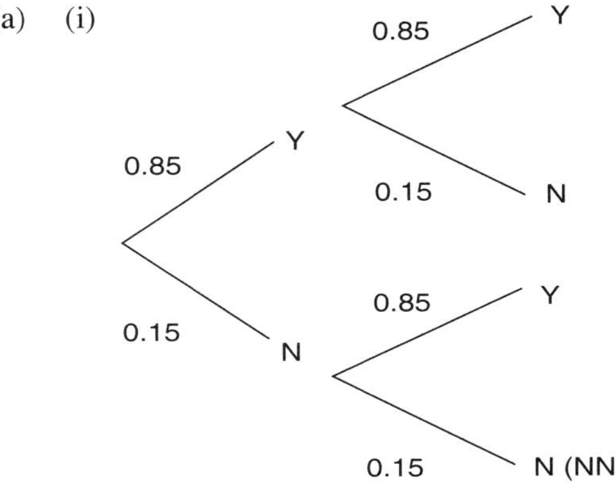

P(no phone) = P(NN)
$= 0.15 \times 0.15$
$= 0.0225$
$= 2.25\%$

(ii)

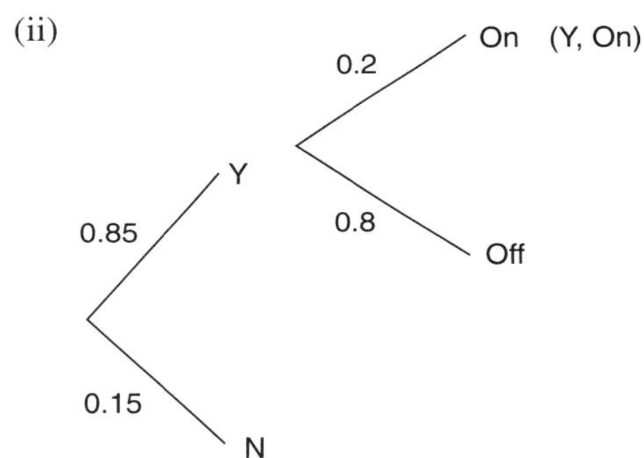

Required probability = P(Y, On)
$= 0.85 \times 0.20$
$= 0.17$
$= 17\%$

(b) (i) Let each monthly repayment be $\$M$

Interest = 6% per annum
= 0.06 per annum
= 0.005 per month.

$A_1 = 100\,000 \times 1.005 - M$

$$\begin{aligned} A_2 &= A_1 \times 1.005 - M \\ &= (100\,000 \times 1.005 - M) \times 1.005 - M \\ &= 100\,000 \times 1.005^2 - 1.005M - M \end{aligned}$$

$$\begin{aligned} A_3 &= A_2 \times 1.005 - M \\ &= (100\,000 \times 1.005^2 - 1.005M - M) \times 1.005 - M \\ &= 100\,000 \times 1.005^3 - 1.005^2M - 1.005M - M \\ &= 100\,000 \times 1.005^3 - M(1.005^2 + 1.005 + 1) \\ &= 100\,000 \times 1.005^3 - M(1 + 1.005 + 1.005^2) \end{aligned}$$

$\therefore A_n = 100\,000 \times 1.005^n - M(1 + 1.005 + 1.005^2 + \ldots + 1.005^{n-1})$

Now $(1 + 1.005 + 1.005^2 + \ldots + 1.005^{n-1})$ is a geometric series with $a = 1$, $r = 1.005$ and n terms.

$$\begin{aligned} \therefore A_n &= 100\,000 \times 1.005^n - M\left[\frac{a(r^n - 1)}{r - 1}\right] \\ &= 100\,000 \times 1.005^n - \text{M}\left[\frac{1(1.005^n - 1)}{1.005 - 1}\right] \\ &= 100\,000 \times 1.005^n - \frac{M(1.005^n - 1)}{0.005}. \end{aligned}$$

(ii) $$A_n = 100\,000 \times 1.005^n - \frac{M(1.005^n - 1)}{0.005}$$

After 144 payments, $A_n = 0$

$$\therefore 0 = 100\,000 \times 1.005^{144} - \frac{M(1.005^{144} - 1)}{0.005}$$

$$\begin{aligned} \frac{M(1.005^{144} - 1)}{0.005} &= 100\,000 \times 1.005^{144} \\ M(1.005^{144} - 1) &= 100\,000 \times 1.005^{144} \times 0.005 \\ M &= \frac{100\,000 \times 1.005^{144} \times 0.005}{1.005^{144} - 1} \\ &= \frac{1025.375408}{1.0507508} \quad \text{(by calc.)} \\ &= 975.8502136 \quad \text{(by calc.)} \\ &= 975.85 \end{aligned}$$

$\therefore$ The fixed monthly payment, $\$M$ is \$975.85.

(c) (i) $f''(x) = k(b^2 - x^2)$

$$\begin{aligned} \therefore f'(x) &= \int k(b^2 - x^2)\,dx \\ &= kb^2x - \frac{kx^3}{3} + C \end{aligned}$$

Now $$\begin{aligned} f'(b) &= kb^2(b) - k\frac{(b)^3}{3} + C \\ &= kb^3 - \frac{kb^3}{3} + C \end{aligned}$$

$$\therefore -f'(b) = -\left(kb^3 - \frac{kb^3}{3} + C \right)$$

$$= -kb^3 + \frac{kb^3}{3} - C$$

and $f'(-b) = kb^2(-b) - \frac{k(-b)^3}{3} + C$

$$= -kb^3 + \frac{kb^3}{3} + C$$

Now $\quad f'(-b) = -f'(b)$

$$\therefore -kb^3 + \frac{kb^3}{3} + C = -kb^3 + \frac{kb^3}{3} - C$$

$$\therefore C = -C$$
$$2C = 0$$
$$\therefore C = 0$$

$$\therefore f'(x) = kb^2x - \frac{kx^3}{3}$$

$$= k\left(b^2x - \frac{x^3}{3} \right).$$

(ii) $\quad f'(x) = k\left(b^2x - \frac{x^3}{3} \right)$

$$\therefore f(x) = \int k\left(b^2x - \frac{x^3}{3} \right) dx$$

$$= \frac{kb^2x^2}{2} - \frac{kx^4}{12} + C_2$$

Now $f(x)$ passes through $(b, 0)$,

$$\therefore f(b) = \frac{kb^2(b^2)}{2} - \frac{k(b^4)}{12} + C_2$$

$$= \frac{kb^4}{2} - \frac{kb^4}{12} + C_2$$

$$= \frac{6kb^4 - kb^4}{12} + C_2$$

$$= \frac{5kb^4}{12} + C_2$$

$$= 0$$

$$\therefore \frac{5kb^4}{12} + C_2 = 0$$

$$C_2 = -\frac{5kb^4}{12}$$

$$\therefore f(x) = \frac{kb^2x^2}{2} - \frac{kx^4}{12} - \frac{5kb^4}{12}$$

$$\therefore f(0) = \frac{kb^2(0)^2}{2} - \frac{k(0)^4}{12} - \frac{5kb^4}{12}$$

$$= -\frac{5kb^4}{12}$$

$\therefore$ The beam is $\frac{5kb^4}{12}$ units below the x-axis, at $x = 0$.

QUESTION 10

(a) $\quad y = \log_e(x - 2)$

(N.B. We cannot integrate $y = \log_e(x - 2)$)

$$\therefore (x - 2) = e^y$$
$$\therefore x = e^y + 2$$

Required area = Area of rectangle – Area under the curve bounded by the y-axis

$$x = \text{L} \times \text{B} - \int_0^{\log_e 5} e^y + 2\, dy$$

$$= 7 \times \log_e 5 - \left[e^y + 2y \right]_0^{\log_e 5}$$

$$= 7\log_e 5 - \left[\left(e^{\log e\, 5} + 2\log_e 5 \right) - \left(e^0 - 2(0) \right) \right]$$

$$= 7\log_e 5 - \left[5 + 2\log_e 5 - (1 - 0) \right]$$

$$= 7\log_e 5 - \left[4 + 2\log_e 5 \right]$$

$$= 7\log_e 5 - 4 - 2\log_e 5$$

$$= \left(5\log_e 5 - 4 \right) \text{ units}^2.$$

(b)

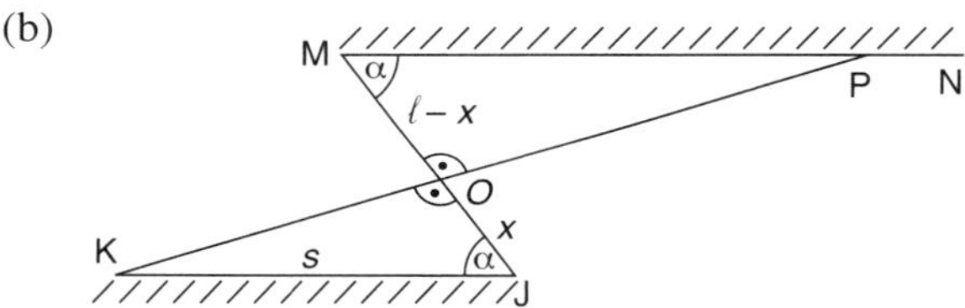

(i) In $\triangle OKJ$ and $\triangle OMP$,

$\angle KOJ = \angle POM$ (vertically opposite angles)

$\angle KJO = \angle PMO$ (alternate angles, $KJ \parallel MP$)

$\therefore \triangle OKJ \parallel\!| \triangle OMP$ (equiangular)

$$\therefore \frac{MP}{s} = \frac{\ell - x}{x}$$

$$\therefore MP = \frac{(\ell - x)s}{x}$$

Total area = Area $\triangle OKJ$ + Area $\triangle OMP$

$$A = \frac{1}{2}sx\sin\alpha + \frac{1}{2}MP(\ell - x)\sin\alpha$$

$$= \frac{1}{2}sx\sin\alpha + \frac{1}{2}\frac{(\ell - x)s}{x}(\ell - x)\sin\alpha$$

$$= \frac{1}{2}s\sin\alpha\left[x + \frac{(\ell - x)^2}{x} \right]$$

$$= \frac{1}{2} s \sin\alpha \left[x + \frac{\ell^2 - 2\ell x + x^2}{x} \right]$$

$$= \frac{1}{2} s \sin\alpha \left(x + \frac{\ell^2}{x} - 2\ell + x \right)$$

$$= \frac{1}{2} s \sin\alpha \left(2x - 2\ell + \frac{\ell^2}{x} \right)$$

$$= s \left(x - \ell + \frac{\ell^2}{2x} \right) \sin\alpha.$$

(ii) $A = s \left(x - \ell + \dfrac{\ell^2}{2x} \right) \sin\alpha$

$$\frac{dA}{dx} = s \left(1 - \frac{\ell^2}{2x^2} \right) \sin\alpha$$

Stationary points occur when

$$\frac{dA}{dx} = 0$$

$$\therefore\ s \left(1 - \frac{\ell^2}{2x^2} \right) \sin\alpha = 0$$

$$\therefore\ 1 - \frac{\ell^2}{2x^2} = 0$$

$$2x^2 = \ell^2$$

$$x^2 = \frac{\ell^2}{2}$$

$$\therefore\ x = \frac{\ell}{\sqrt{2}}, \quad x > 0$$

Now $\dfrac{dA}{dx} = s \left(1 - \dfrac{\ell^2}{2x^2} \right) \sin\alpha$

$$\therefore\ \frac{d^2A}{dx^2} = s \sin\alpha \left(\frac{\ell^2}{x^3} \right)$$

$$= s \sin\alpha \left[\frac{\ell^2}{\left(\frac{\ell}{\sqrt{2}} \right)^3} \right], \text{ when } x = \frac{\ell}{\sqrt{2}}$$

$$= \frac{2\sqrt{2}\, s \sin\alpha}{\ell} > 0 \quad \text{as } 0 < \alpha < 90°$$

$\therefore$ A minimum value of A occurs when $x = \dfrac{\ell}{\sqrt{2}}$.

(iii) $MP = \dfrac{(\ell - x)s}{x}$ (from (i))

When $x = \dfrac{\ell}{\sqrt{2}}$,

$$MP = \frac{\left(\ell - \frac{\ell}{\sqrt{2}} \right) \times s}{\frac{\ell}{\sqrt{2}}}$$

$$= \left(\ell - \frac{\ell}{\sqrt{2}} \right) \times s \times \frac{\sqrt{2}}{\ell}$$

$$= (\sqrt{2} - 1)s.$$

$\therefore$ The length of MP is $(\sqrt{2} - 1)s$ metres.

BOARD OF STUDIES
NEW SOUTH WALES

2009

HIGHER SCHOOL CERTIFICATE EXAMINATION

Mathematics

General Instructions

- Reading time – 5 minutes
- Working time – 3 hours
- Write using black or blue pen
- Board-approved calculators may be used
- A table of standard integrals is provided at the back of this paper
- All necessary working should be shown in every question

Total marks – 120

- Attempt Questions 1–10
- All questions are of equal value

Total marks – 120
Attempt Questions 1–10
All questions are of equal value

Answer each question in the appropriate writing booklet. Extra writing booklets are available.

Question 1 (12 marks) Use the Question 1 Writing Booklet.

(a) Sketch the graph of $y - 2x = 3$, showing the intercepts on both axes. **2**

(b) Solve $\dfrac{5x - 4}{x} = 2$. **2**

(c) Solve $|x + 1| = 5$. **2**

(d) Find the gradient of the tangent to the curve $y = x^4 - 3x$ at the point $(1, -2)$. **2**

(e) Find the exact value of θ such that $2\cos\theta = 1$, where $0 \le \theta \le \dfrac{\pi}{2}$. **2**

(f) Solve the equation $\ln x = 2$. Give your answer correct to four decimal places. **2**

Question 2 (12 marks) Use the Question 2 Writing Booklet.

(a) Differentiate with respect to x:

(i) $x \sin x$ **2**

(ii) $\left(e^x + 1\right)^2$. **2**

(b) (i) Find $\int 5\,dx$. **1**

(ii) Find $\int \frac{3}{(x-6)^2}\,dx$. **2**

(iii) Evaluate $\int_1^4 x^2 + \sqrt{x}\,dx$. **3**

(c) Evaluate $\sum_{k=1}^{4} (-1)^k k^2$. **2**

Question 3 (12 marks) Use the Question 3 Writing Booklet.

(a) An arithmetic series has 21 terms. The first term is 3 and the last term is 53. **2**

Find the sum of the series.

(b)

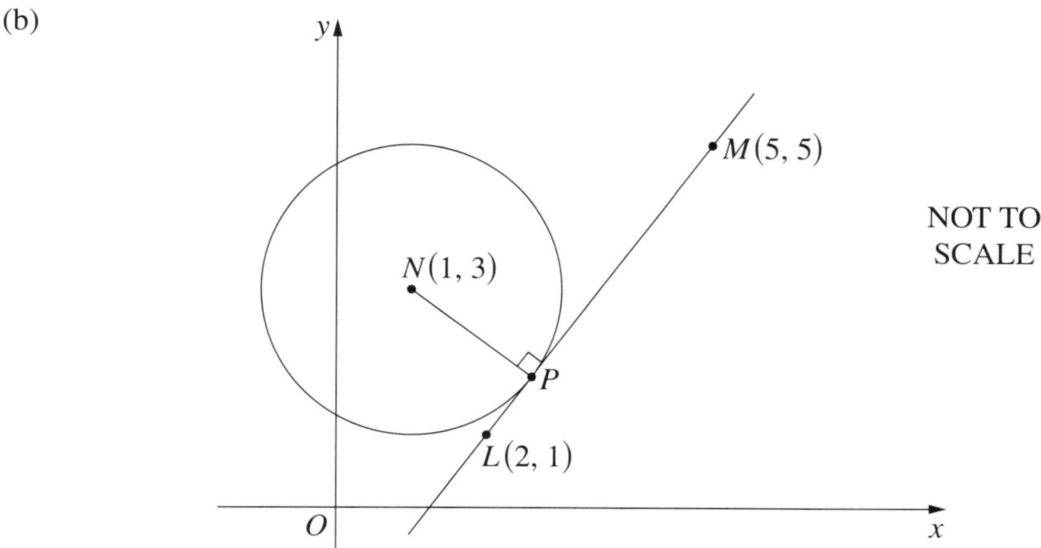

The circle in the diagram has centre *N*. The line *LM* is tangent to the circle at *P*.

(i) Find the equation of *LM* in the form $ax + by + c = 0$. **2**

(ii) Find the distance *NP*. **2**

(iii) Find the equation of the circle. **1**

(c) Shade the region in the plane defined by $y \geq 0$ and $y \leq 4 - x^2$. **2**

Question 3 continues

Question 3 (continued)

(d) The diagram shows a block of land and its dimensions, in metres. The block of land is bounded on one side by a river. Measurements are taken perpendicular to the line AB, from AB to the river, at equal intervals of 50 m. **3**

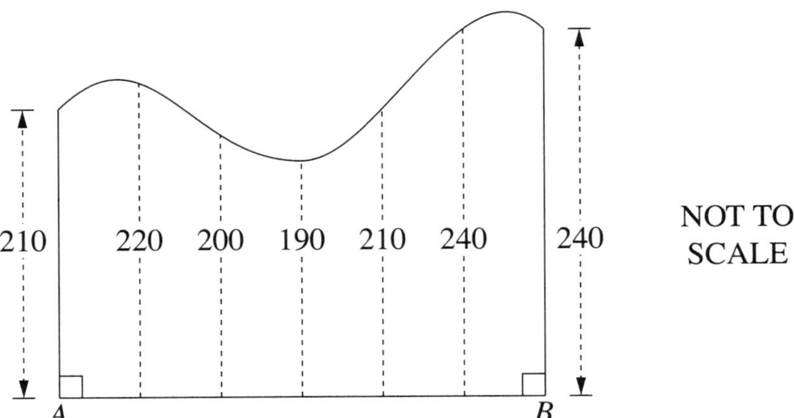

Use Simpson's rule with six subintervals to find an approximation to the area of the block of land.

End of Question 3

Question 4 (12 marks) Use the Question 4 Writing Booklet.

(a) A tree grows from ground level to a height of 1.2 metres in one year. In each subsequent year, it grows $\frac{9}{10}$ as much as it did in the previous year. **2**

Find the limiting height of the tree.

(b) Find the values of k for which the quadratic equation **3**

$$x^2 - (k+4)x + (k+7) = 0$$

has equal roots.

(c) In the diagram, $\triangle ABC$ is a right-angled triangle, with the right angle at C. The midpoint of AB is M, and $MP \perp AC$.

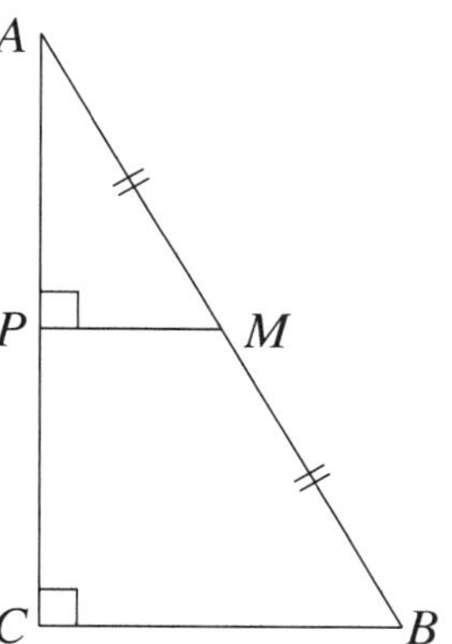

Copy or trace the diagram into your writing booklet.

(i) Prove that $\triangle AMP$ is similar to $\triangle ABC$. **2**

(ii) What is the ratio of AP to AC? **1**

(iii) Prove that $\triangle AMC$ is isosceles. **2**

(iv) Show that $\triangle ABC$ can be divided into two isosceles triangles. **1**

(v) **1**

Copy or trace this triangle into your writing booklet and show how to divide it into four isosceles triangles.

Question 5 (12 marks) Use the Question 5 Writing Booklet.

(a) In the diagram, the points A and C lie on the y-axis and the point B lies on the x-axis. The line AB has equation $y = \sqrt{3}x - 3$. The line BC is perpendicular to AB.

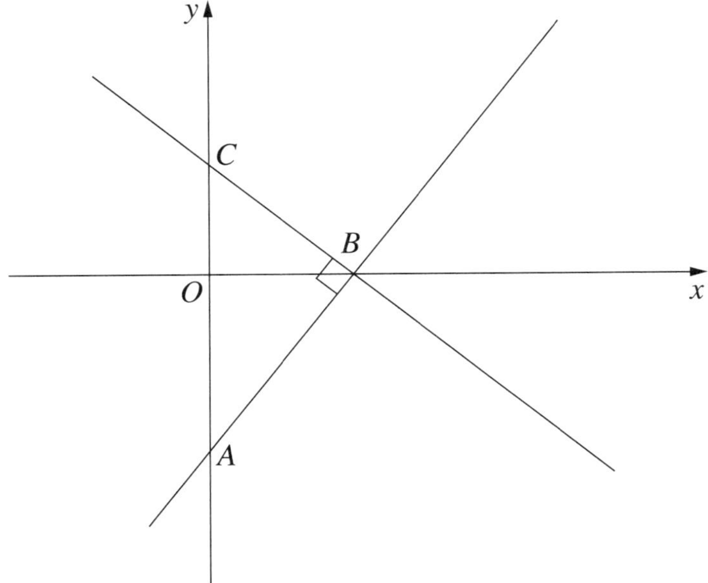

(i) Find the equation of the line BC. **2**

(ii) Find the area of the triangle ABC. **2**

(b) On each working day James parks his car in a parking station which has three levels. He parks his car on a randomly chosen level. He always forgets where he has parked, so when he leaves work he chooses a level at random and searches for his car. If his car is not on that level, he chooses a different level and continues in this way until he finds his car.

(i) What is the probability that his car is on the first level he searches? **1**

(ii) What is the probability that he must search all three levels before he finds his car? **1**

(iii) What is the probability that on every one of the five working days in a week, his car is not on the first level he searches? **1**

Question 5 continues

Question 5 (continued)

(c) The diagram shows a circle with centre O and radius 2 centimetres. The points A and B lie on the circumference of the circle and $\angle AOB = \theta$.

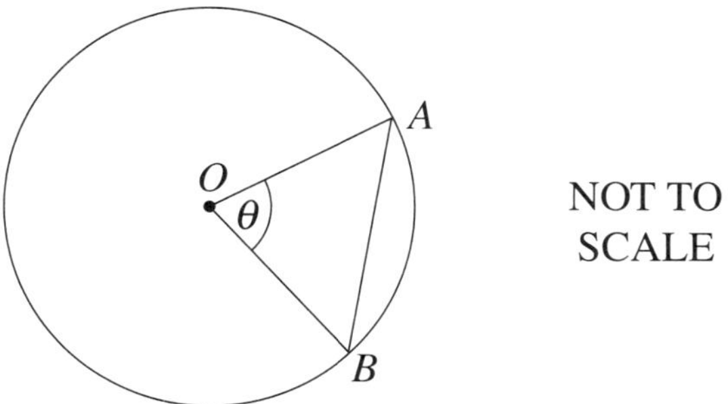

(i) There are two possible values of θ for which the area of $\triangle AOB$ is $\sqrt{3}$ square centimetres. One value is $\frac{\pi}{3}$. **2**

Find the other value.

(ii) Suppose that $\theta = \frac{\pi}{3}$.

(1) Find the area of the sector AOB. **1**

(2) Find the exact length of the perimeter of the minor segment bounded by the chord AB and the arc AB. **2**

End of Question 5

Question 6 (12 marks) Use the Question 6 Writing Booklet.

(a) The diagram shows the region bounded by the curve $y = \sec x$, the lines $x = \frac{\pi}{3}$ and $x = -\frac{\pi}{3}$, and the x-axis. **3**

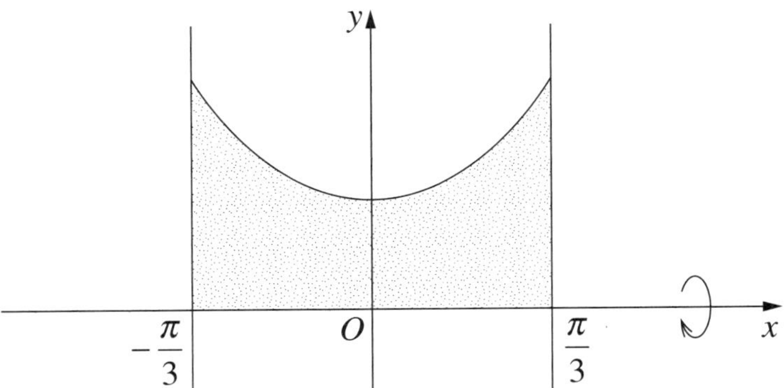

The region is rotated about the x-axis. Find the volume of the solid of revolution formed.

(b) Radium decays at a rate proportional to the amount of radium present. That is, if $Q(t)$ is the amount of radium present at time t, then $Q = Ae^{-kt}$, where k is a positive constant and A is the amount present at $t = 0$. It takes 1600 years for an amount of radium to reduce by half.

(i) Find the value of k. **2**

(ii) A factory site is contaminated with radium. The amount of radium on the site is currently three times the safe level. **2**

How many years will it be before the amount of radium reaches the safe level?

Question 6 continues

Question 6 (continued)

(c) The diagram illustrates the design for part of a roller-coaster track. The section RO is a straight line with slope 1.2, and the section PQ is a straight line with slope -1.8. The section OP is a parabola $y = ax^2 + bx$. The horizontal distance from the y-axis to P is 30 m.

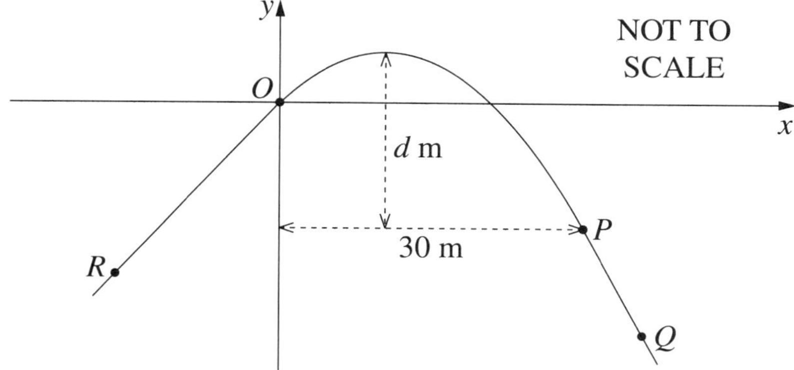

In order that the ride is smooth, the straight line sections must be tangent to the parabola at O and at P.

(i) Find the values of a and b so that the ride is smooth. **3**

(ii) Find the distance d, from the vertex of the parabola to the horizontal line through P, as shown on the diagram. **2**

End of Question 6

Question 7 (12 marks) Use the Question 7 Writing Booklet.

(a) The acceleration of a particle is given by

$$\ddot{x} = 8e^{-2t} + 3e^{-t},$$

where x is displacement in metres and t is time in seconds.

Initially its velocity is $-6\ \text{m s}^{-1}$ and its displacement is 5 m.

(i) Show that the displacement of the particle is given by **2**

$$x = 2e^{-2t} + 3e^{-t} + t.$$

(ii) Find the time when the particle comes to rest. **3**

(iii) Find the displacement when the particle comes to rest. **1**

(b) Between 5 am and 5 pm on 3 March 2009, the height, h, of the tide in a harbour was given by

$$h = 1 + 0.7\sin\frac{\pi}{6}t \quad \text{for } 0 \le t \le 12,$$

where h is in metres and t is in hours, with $t = 0$ at 5 am.

(i) What is the period of the function h? **1**

(ii) What was the value of h at low tide, and at what time did low tide occur? **2**

(iii) A ship is able to enter the harbour only if the height of the tide is at least 1.35 m. **3**

Find all times between 5 am and 5 pm on 3 March 2009 during which the ship was able to enter the harbour.

Question 8 (12 marks) Use the Question 8 Writing Booklet.

(a)

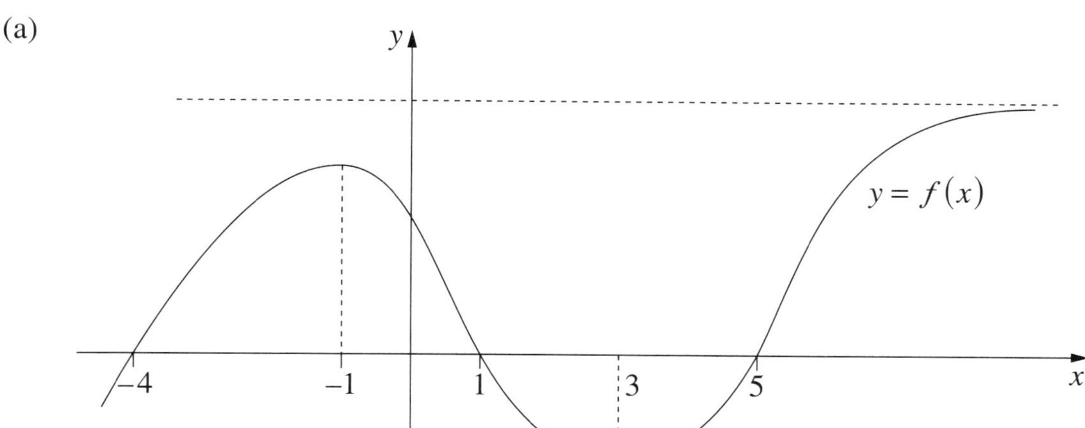

The diagram shows the graph of a function $y = f(x)$.

(i) For which values of x is the derivative, $f'(x)$, negative? **1**

(ii) What happens to $f'(x)$ for large values of x? **1**

(iii) Sketch the graph $y = f'(x)$. **2**

Question 8 continues

Question 8 (continued)

(b) One year ago Daniel borrowed \$350 000 to buy a house. The interest rate was 9% per annum, compounded monthly. He agreed to repay the loan in 25 years with equal monthly repayments of \$2937.

(i) Calculate how much Daniel owed after his first monthly repayment. **1**

(ii) Daniel has just made his 12th monthly repayment. He now owes \$346 095. The interest rate now decreases to 6% per annum, compounded monthly. **3**

The amount, $\$A_n$, owing on the loan after the nth monthly repayment is now calculated using the formula

$$A_n = 346\,095 \times 1.005^n - 1.005^{n-1}M - \cdots - 1.005M - M$$

where $\$M$ is the monthly repayment, and $n = 1, 2, \ldots, 288$. (Do NOT prove this formula.)

Calculate the monthly repayment if the loan is to be repaid over the remaining 24 years (288 months).

(iii) Daniel chooses to keep his monthly repayments at \$2937. Use the formula in part (ii) to calculate how long it will take him to repay the \$346 095. **3**

(iv) How much will Daniel save over the term of the loan by keeping his monthly repayments at \$2937, rather than reducing his repayments to the amount calculated in part (ii)? **1**

End of Question 8

Question 9 (12 marks) Use the Question 9 Writing Booklet.

(a) Each week Van and Marie take part in a raffle at their respective workplaces. The probability that Van wins a prize in his raffle is $\frac{1}{9}$. The probability that Marie wins a prize in her raffle is $\frac{1}{16}$. **2**

What is the probability that, during the next three weeks, at least one of them wins a prize?

(b) An oil rig, S, is 3 km offshore. A power station, P, is on the shore. A cable is to be laid from P to S. It costs \$1000 per kilometre to lay the cable along the shore and \$2600 per kilometre to lay the cable underwater from the shore to S.

The point R is the point on the shore closest to S, and the distance PR is 5 km.

The point Q is on the shore, at a distance of x km from R, as shown in the diagram.

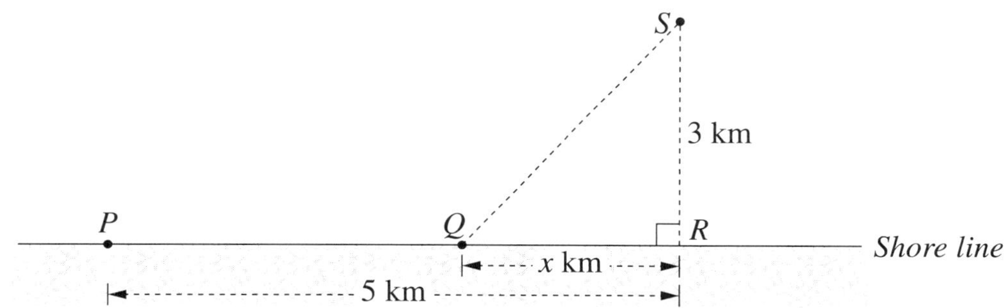

(i) Find the total cost of laying the cable in a straight line from P to R and then in a straight line from R to S. **1**

(ii) Find the cost of laying the cable in a straight line from P to S. **1**

(iii) Let $\$C$ be the total cost of laying the cable in a straight line from P to Q, and then in a straight line from Q to S. **2**

Show that $C = 1000\left(5 - x + 2.6\sqrt{x^2 + 9}\right)$.

(iv) Find the minimum cost of laying the cable. **4**

(v) New technology means that the cost of laying the cable underwater can be reduced to \$1100 per kilometre. **2**

Determine the path for laying the cable in order to minimise the cost in this case.

Question 10 (12 marks) Use the Question 10 Writing Booklet.

Let $f(x) = x - \dfrac{x^2}{2} + \dfrac{x^3}{3}$.

(a) Show that the graph of $y = f(x)$ has no turning points. **2**

(b) Find the point of inflexion of $y = f(x)$. **1**

(c) (i) Show that $1 - x + x^2 - \dfrac{1}{1+x} = \dfrac{x^3}{1+x}$ for $x \neq -1$. **1**

(ii) Let $g(x) = \ln(1+x)$. **2**

Use the result in part (c) (i) to show that $f'(x) \geq g'(x)$ for all $x \geq 0$.

(d) On the same set of axes, sketch the graphs of $y = f(x)$ and $y = g(x)$ for $x \geq 0$. **2**

(e) Show that $\dfrac{d}{dx}\left[(1+x)\ln(1+x) - (1+x)\right] = \ln(1+x)$. **2**

(f) Find the area enclosed by the graphs of $y = f(x)$ and $y = g(x)$, and the straight line $x = 1$. **2**

End of paper

2009 Higher School Certificate Worked Answers

QUESTION 1

(a) $y - 2x = 3$

$\therefore y = 2x + 3$

When $x = 0, y = 3$

$\therefore$ y-intercept of $y - 2x = 3$ is $(0, 3)$.

When $y = 0$, $0 = 2x + 3$

$2x = -3$

$x = -\frac{3}{2}$

$\therefore$ x-intercept of $y - 2x = 3$ is $\left(-\frac{3}{2}, 0\right)$.

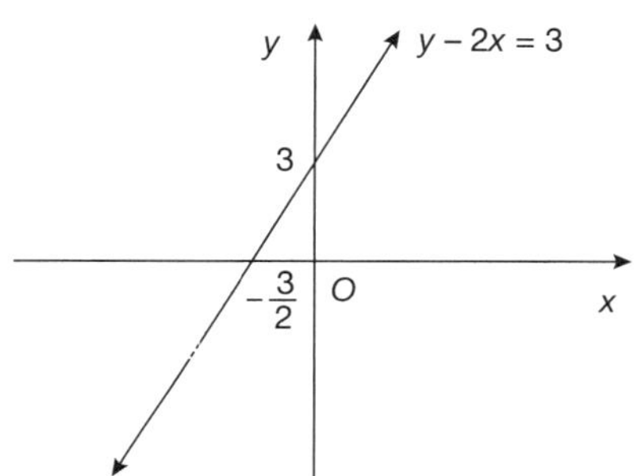

(b) $$\frac{5x-4}{x} = 2$$

$$\therefore x\left(\frac{5x-4}{x}\right) = 2x$$

$$5x - 4 = 2x$$

$$3x = 4$$

$$\therefore x = \frac{4}{3}.$$

(c) $|x + 1| = 5$

$\therefore x + 1 = \pm 5$

$\therefore x + 1 = 5$ or $x + 1 = -5$

$\therefore x = 4$ or $x = -6$

(d) $$y = x^4 - 3x$$

$$\frac{dy}{dx} = 4x^3 - 3$$

At $x = 1$, $\frac{dy}{dx} = 4 - 3$

$= 1$

$\therefore$ Gradient of tangent is 1.

(e) $2\cos\theta = 1 \quad 0 \leqslant \theta \leqslant \frac{\pi}{2}$

$$\cos\theta = \frac{1}{2}$$

$$\therefore \theta = \frac{\pi}{3}.$$

(f) $\ln x = 2$

$\therefore x = e^2$

$= 7.38905\ldots$ (by calc.)

$= 7.3891$ to 4 decimal places.

QUESTION 2

(a) (i) $\frac{d}{dx}(x \sin x)$

$$= x\,.\,\frac{d}{dx}(\sin x) + \sin x\,.\,\frac{d}{dx}(x)$$

$$= x\,.\cos x + \sin x\,.\,1$$

$$= x\cos x + \sin x.$$

(ii) $\frac{d}{dx}(e^x + 1)^2$

$$= 2(e^x + 1)\,.\,\frac{d}{dx}(e^x + 1)$$

$$= 2(e^x + 1)\,.\,e^x$$

$$= 2e^x(e^x + 1).$$

(b) (i) $\int 5\,dx = 5x + C$

(ii) $\int \frac{3}{(x-6)^2}\,dx$

$$= 3\int (x-6)^{-2}\,dx$$

$$= \frac{3(x-6)^{-1}}{-1} + C$$

$$= \frac{-3}{x-6} + C.$$

(iii) $\int_1^4 x^2 + \sqrt{x}\,dx$

$$= \int_1^4 x^2 + x^{\frac{1}{2}}\,dx$$

$$= \left[\frac{x^3}{3} + \frac{2}{3}x^{\frac{3}{2}}\right]_1^4$$

$$= \left(\frac{(4)^3}{3} + \frac{2}{3}(4)^{\frac{3}{2}} \right) - \left(\frac{(1)^3}{3} + \frac{2}{3}(1)^{\frac{3}{2}} \right)$$

$$= \left(\frac{64}{3} + 5\frac{1}{3} \right) - \left(\frac{1}{3} + \frac{2}{3} \right)$$

$$= 26\frac{2}{3} - 1$$

$$= 25\frac{2}{3} \quad \text{or} \quad 25{\cdot}\dot{6}.$$

(c) $\sum_{k=1}^{4} (-1)^k k^2$

$= (-1)^1(1)^2 + (-1)^2(2)^2 + (-1)^3(3)^2 + (-1)^4 4^2$
$= -1 + 4 - 9 + 16$
$= 10.$

QUESTION 3

(a) $S_n = \frac{n}{2}(a + L)$

$n = 21, \ a = 3, \ L = 53$

$$S_{21} = \frac{21}{2}(3 + 53)$$

$$= \frac{21}{2} \times 56$$

$$= 588.$$

(b)

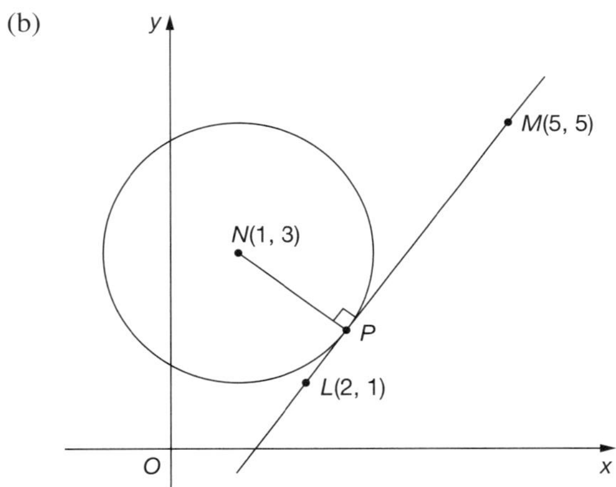

(i) Gradient of LM

$$= \frac{y_2 - y_1}{x_2 - x_1}$$

$$= \frac{5 - 1}{5 - 2}$$

$$= \frac{4}{3}.$$

$\therefore$ The equation of the line LM with gradient $= \frac{4}{3}$, passing through $L(2,1)$ is given by $y - 1 = \frac{4}{3}(x - 2)$

$$3y - 3 = 4x - 8$$

$\therefore \ 4x - 3y - 5 = 0.$

(ii) $N \equiv (1, 3)$
Line LM: $4x - 3y - 5 = 0$

$$d = \frac{|Ax_1 + By_1 + C|}{\sqrt{A^2 + B^2}}$$

$$= \frac{|4(1) - 3(3) - 5|}{\sqrt{4^2 + (-3)^2}}$$

$$= \frac{10}{5}$$

$$= 2.$$

(iii) Equation of a circle with centre $(1, 3)$ and radius $NP = 2$ units is given by

$$(x - 1)^2 + (y - 3)^2 = 2^2$$

$$x^2 - 2x + 1 + y^2 - 6y + 9 = 4$$

$$x^2 + y^2 - 2x - 6y + 6 = 0.$$

(c) Determine the region where each inequality holds:

$y \geqslant 0$ is the region above and including the line $y = 0$.

Test $(0, 0)$ in $y \leqslant 4 - x^2$
i.e. $0 \leqslant 4 - 0$

$\therefore$ Region is inside the parabola $y = 4 - x^2$.

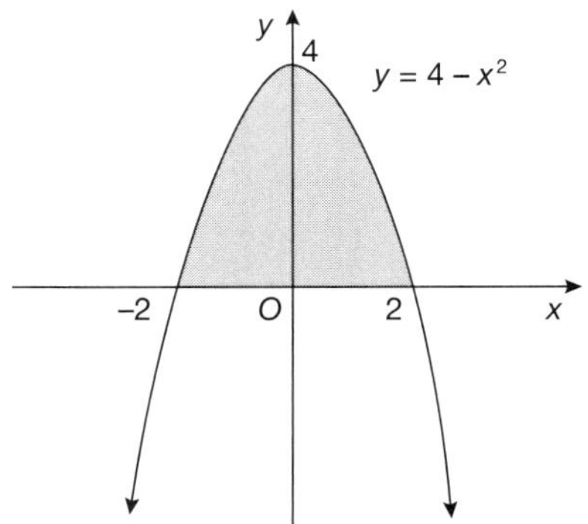

(d)

x	0	50	100	150	200	250	300
$f(x)$	210	220	200	190	210	240	240
	y_0	y_1	y_2	y_3	y_4	y_5	y_6

$$\begin{aligned}\text{Area of block of land} &\doteqdot \frac{h}{3}\big(y_0 + y_6 + 2(y_2 + y_4) + 4(y_1 + y_3 + y_5)\big)\\ &= \frac{50}{3}\big(210 + 240 + 2(200 + 210) + 4(220 + 190 + 240)\big)\\ &= \frac{50}{3}(210 + 240 + 400 + 420 + 880 + 760 + 960)\\ &= 64\,500 \quad \text{(by calc.)}\end{aligned}$$

$\therefore$ Area is approximately $64\,500$ m^2.

QUESTION 4

(a) $S_\infty = \dfrac{a}{1-r}$

$a = 1.2$, $r = \dfrac{9}{10}$

$$\therefore S_\infty = \frac{1.2}{1 - \frac{9}{10}} = 12$$

The limiting height of the tree is 12 m.

(b) $x^2 - (k+4)x + (k+7) = 0$

$$\begin{aligned}\Delta &= b^2 - 4ac\\ &= [-(k+4)]^2 - 4(1)(k+7)\\ &= k^2 + 8k + 16 - 4k - 28\\ &= k^2 + 4k - 12\end{aligned}$$

For equal roots, $\Delta = 0$.

$$\therefore k^2 + 4k - 12 = 0$$
$$(k+6)(k-2) = 0$$
$$\therefore k = -6 \text{ or } k = 2.$$

(c)

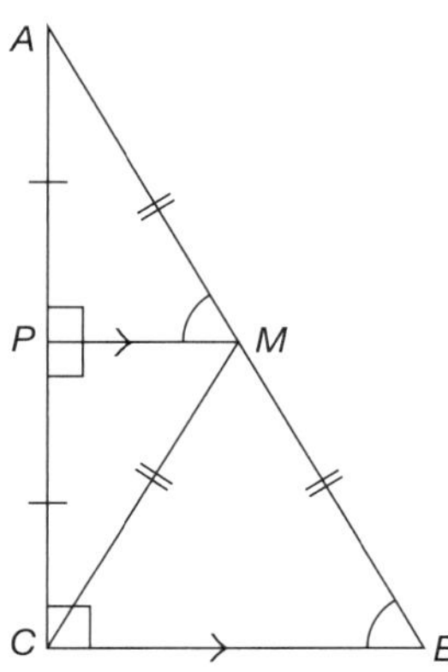

(i) In $\triangle AMP$ and $\triangle ABC$,

$\angle APM = \angle ACB = 90°$ (given)

$\angle A$ is common

$\angle AMP = \angle ABC$ (corresponding angles, $PM \parallel CB$)

$\therefore \triangle AMP \;|||\; \triangle ABC$ (AAA)

(ii) $\dfrac{AM}{AB} = \dfrac{1}{2}$ (given)

$\therefore \dfrac{AP}{AC} = \dfrac{1}{2}$

$\therefore AP : AC = 1 : 2$

(iii) In $\triangle AMP$ and $\triangle CMP$,

PM is common

$\angle CPM = 180° - \angle APM$ (angles in a straight line)
$= 180° - 90°$
$= 90°$

$\therefore \angle APM = \angle CPM$.

Since $\dfrac{AP}{AC} = \dfrac{1}{2}$

$2AP = AC$

Now $AC = AP + PC$

$\therefore 2AP = AP + PC$

$\therefore AP = PC$

$\therefore \triangle AMP \equiv \triangle CMP$ (SAS)

$\therefore AM = CM$ (corresponding sides in congruent triangles)

$\therefore \triangle AMC$ is isosceles.

(iv) Now $\triangle AMC$ is isosceles (proven in (iii))

Since $AM = CM$
and $AM = MB$ (given)
$\therefore CM = MB$

$\therefore \triangle MCB$ is isosceles.

Joining CM forms two isosceles triangles, $\triangle MCB$ and $\triangle AMC$.

(v)

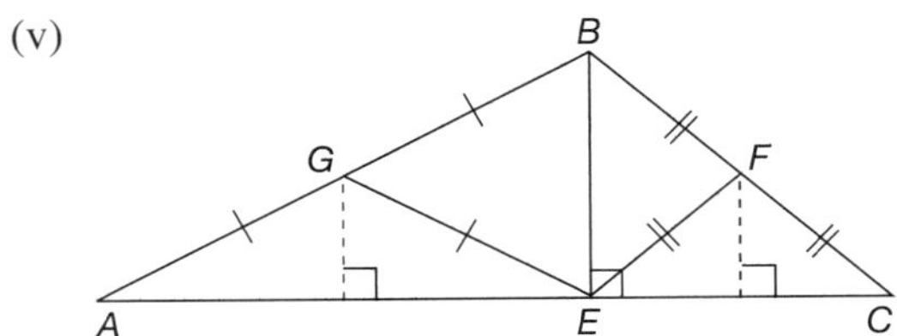

$\therefore\ \triangle EFC$ and $\triangle EFB$, and, $\triangle EGB$ and $\triangle EGA$ are isosceles triangles.

QUESTION 5

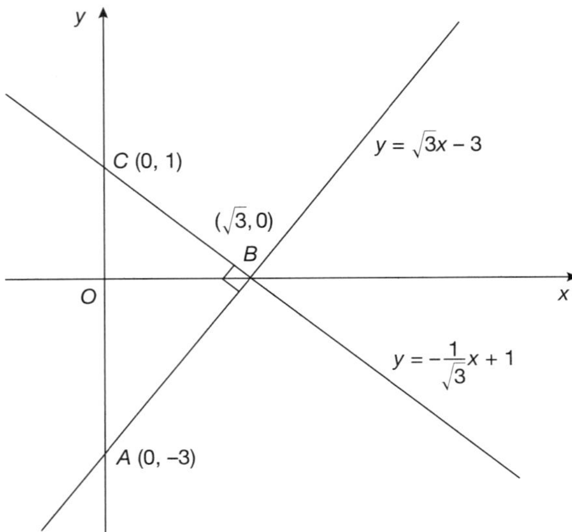

(a) (i) Equation of line AB is $y = \sqrt{3}\,x - 3$

$\therefore$ Gradient of $AB = \sqrt{3}$

$\therefore$ Gradient of $BC = -\dfrac{1}{\sqrt{3}}$ $\quad (AB \perp BC)$

Line AB cuts the x-axis at B where $y = 0$.

$$\therefore\ 0 = \sqrt{3}\,x - 3$$
$$x = \frac{3}{\sqrt{3}}$$
$$= \sqrt{3}$$

$\therefore$ Coordinates of B are $\left(\sqrt{3}\,, 0\right)$.

Equation of line BC, with gradient $= -\dfrac{1}{\sqrt{3}}$,

passing through $B\left(\sqrt{3}\,, 0\right)$ is given by

$$y - y_1 = m(x - x_1)$$
$$y - 0 = -\frac{1}{\sqrt{3}}\left(x - \sqrt{3}\right)$$
$$y = -\frac{1}{\sqrt{3}}\,x + 1$$

(ii) Line AB cuts the y-axis at A where $x = 0$,

$$\therefore\ y = \sqrt{3}\left(0\right) - 3$$
$$= -3$$

$\therefore$ Coordinates of A are $(0, -3)$.

Line BC cuts the y-axis at C where $x = 0$

$$\therefore\ y = -\frac{1}{\sqrt{3}}\left(0\right) + 1$$
$$= 1$$

$\therefore$ C has coordinates $(0, 1)$.

Area of triangle ABC

$$= \frac{1}{2} \times b \times h$$
$$= \frac{1}{2} \times AC \times OB$$
$$= \frac{1}{2} \times 4 \times \sqrt{3}$$
$$= 2\sqrt{3}\ \text{units}^2.$$

(b) (i) P(Correct level chosen)

$$= \frac{1}{3}.$$

(ii) P(All levels searched)

$$= \frac{2}{3} \times \frac{1}{2} \times 1$$
$$= \frac{1}{3}.$$

(iii) P(Not on first level)

$$= 1 - \frac{1}{3}$$
$$= \frac{2}{3}$$

$\therefore$ P(Not on first level each day)

$$= \left(\frac{2}{3}\right)^5$$
$$= \frac{32}{243}.$$

(c) (i) Area of $\triangle AOB = \dfrac{1}{2} \times AO \times OB \times \sin\theta$

$(0 < \theta < \pi)$

$$\therefore\ \sqrt{3} = \frac{1}{2} \times 2 \times 2 \times \sin\theta$$
$$\sqrt{3} = 2\sin\theta$$
$$\sin\theta = \frac{\sqrt{3}}{2}$$

$$\therefore\ \theta = \frac{\pi}{3},\ \theta - \frac{\pi}{3}$$
$$= \frac{\pi}{3},\ \frac{2\pi}{3}$$

$\therefore$ The other value is $\frac{2\pi}{3}$.

(ii) (1) $A = \frac{1}{2}r^2\theta$

where $\theta = \frac{\pi}{3}$ and $r = 2$.

$$\therefore\ A = \frac{1}{2} \times 2^2 \times \frac{\pi}{3}$$
$$= \frac{2\pi}{3}$$

$\therefore$ Area of sector $AOB = \frac{2\pi}{3}$ cm^2.

(2) $l = r\theta$

where $\theta = \frac{\pi}{3}$ and $r = 2$

$$\therefore\ l = 2 \times \frac{\pi}{3}$$
$$\therefore\ l = \frac{2\pi}{3}$$

By the cosine rule,

$$AB^2 = 2^2 + 2^2 - 2 \times 2^2 \times \cos\frac{\pi}{3}$$
$$= 4 + 4 - 8 \times \frac{1}{2}$$
$$= 4$$
$$\therefore\ AB = 2$$

$\therefore$ Perimeter $= \left(2 + \frac{2\pi}{3}\right)$ cm.

QUESTION 6

(a) $y = \sec x$

$\therefore\ y^2 = \sec^2 x$

Now $V = \pi\int_a^b y^2\,dx$

$$\therefore\ V = 2 \times \pi\int_0^{\frac{\pi}{3}} \sec^2 x\,dx$$
$$= 2\pi\Big[\tan x\Big]_0^{\frac{\pi}{3}}$$
$$= 2\pi\left(\tan\frac{\pi}{3} - \tan 0\right)$$
$$= 2\pi\left(\sqrt{3} - 0\right)$$
$$= 2\sqrt{3}\,\pi \text{ units}^3.$$

(b) (i) $Q = Ae^{-kt}$

When $t = 1600$, $Q = \frac{1}{2}A$

$$\therefore\ \frac{1}{2}A = Ae^{-1600k}$$
$$\frac{1}{2} = e^{-1600k}$$
$$\ln\frac{1}{2} = -1600k$$
$$\therefore\ k = \frac{\ln\frac{1}{2}}{-1600}$$
$$= 0.0004\ldots \quad \text{(by calc.)}$$

(ii) $Q = Ae^{-kt}$

Let $A = 3Q$,

$$\therefore\ Q = 3Qe^{-kt}$$
$$1 = 3e^{-kt}$$
$$\frac{1}{3} = e^{-kt}$$
$$\ln\frac{1}{3} = -kt$$
$$\therefore\ t = \ln\frac{1}{3} \div -k$$

where $k = -\frac{\ln\frac{1}{2}}{-1600} = 0.0004\ldots$

$\therefore\ t = 2535.94\ldots$ (by calc.)

It will take 2536 years for the amount of radium to reach a safe level.

(c) (i)
$$y = ax^2 + bx$$
$$\frac{dy}{dx} = 2ax + b$$

At O, $x = 0$ and $\frac{dy}{dx} = 1.2$

$$\therefore\ 2a(0) + b = 1.2$$
$$\therefore\ b = 1.2$$

At P, $x = 30$ and $\frac{dy}{dx} = -1.8$

$$\therefore\ 2ax + b = -1.8$$
$$\therefore\ 2a(30) + 1.2 = -1.8$$
$$60a = -3$$
$$\therefore\ a = -0.05.$$

(ii) $y = ax^2 + bx$

$\therefore y = -0.05x^2 + 1.2x$

$\frac{dy}{dx} = -0.1x + 1.2$

At the vertex,

$\frac{dy}{dx} = 0$

$-0.1x + 1.2 = 0$

$0.1x = 1.2$

$x = 12$

OR

At the vertex, $x = \frac{-b}{2a}$

$= \frac{-1.2}{2(-0.05)}$

$= \frac{-1.2}{-0.1}$

$= 12$

When $x = 12$, $y = -0.05(12)^2 + 1.2(12)$
$= 7.2$

$\therefore$ Coordinates of vertex are (12, 7.2)

At P, $x = 30$

$\therefore y = -0.05(30)^2 + 1.2(30)$
$= -9.$

$\therefore$ Coordinates of P are (30, –9)

$\therefore d = 7.2 + 9$
$= 16.2$

The distance, d, is 16.2 m.

QUESTION 7

(a) (i) $\ddot{x} = 8e^{-2t} + 3e^{-t}$

$\therefore \dot{x} = \int \left(8e^{-2t} + 3e^{-t}\right) dt$

$= -4e^{-2t} - 3e^{-t} + C$

When $t = 0$, $\dot{x} = -6$

$\therefore -6 = -4e^{-2(0)} - 3e^{-(0)} + C$

$-6 = -4 - 3 + C$

$\therefore C = 1$

$\therefore \dot{x} = -4e^{-2t} - 3e^{-t} + 1$

$\therefore x = \int \left(-4e^{-2t} - 3e^{-t} + 1\right) dt$

$= 2e^{-2t} + 3e^{-t} + t + K$

When $t = 0$, $x = 5$

$\therefore 5 = 2e^{-2(0)} + 3e^{0} + 0 + K$

$5 = 2 + 3 + K$

$5 = 5 + K$

$\therefore K = 0$

$\therefore x = 2e^{-2t} + 3e^{-t} + t$

(ii) The particle comes to rest when $\dot{x} = 0$

$\therefore -4e^{-2t} - 3e^{-t} + 1 = 0$

$-4(e^{-t})^2 - 3e^{-t} + 1 = 0$

Let $m = e^{-t}$,

$\therefore -4m^2 - 3m + 1 = 0$

$4m^2 + 3m - 1 = 0$

$(m + 1)(4m - 1) = 0$

$\therefore m + 1 = 0$ or $4m - 1 = 0$

$m = -1$ $\quad$ $m = \frac{1}{4}$

$\therefore e^{-t} = -1$ or $e^{-t} = \frac{1}{4}$

no solution

$\log_e e^{-t} = \log_e \frac{1}{4}$

$-t = \log_e \frac{1}{4}$

$-t = -\log_e 4$

$t = \log_e 4$

$\therefore$ Particle comes to rest when $t = \log_e 4$ seconds.

(iii) $x = 2e^{-2t} + 3e^{-t} + t$

When $t = \log_e 4$,

$x = 2e^{-2\log_e 4} + 3e^{-\log_e 4} + \log_e 4$

$= 2e^{\log_e 4^{-2}} + 3e^{\log_e 4^{-1}} + \log_e 4$

$= 2e^{\log_e\left(\frac{1}{16}\right)} + 3e^{\log_e\left(\frac{1}{4}\right)} + \log_e 4$

$= 2\left(\frac{1}{16}\right) + 3\left(\frac{1}{4}\right) + \log_e 4$

$= \frac{7}{8} + \log_e 4.$

(b) (i) Period $= \frac{2\pi}{n}$ where $n = \frac{\pi}{6}$

$= \frac{2\pi}{\frac{\pi}{6}}$

$= 12.$

(ii) The minimum value of h will occur when $\sin \frac{\pi}{6} t$ is a minimum.

The minimum value of $\sin \frac{\pi}{6} t$ is -1, as $-1 \leq \sin \frac{\pi}{6} t \leq 1$.

$h = 1 + 0.7 \sin \frac{\pi}{6} t$ for $0 \leq t \leq 12$.

$$\begin{aligned} \therefore h &= 1 + 0.7(-1) \\ &= 1 - 0.7 \\ &= 0.3 \end{aligned}$$

$\therefore$ At low tide, the value of h is 0.3.

$$\begin{aligned} \therefore 0.3 &= 1 + 0.7 \sin \frac{\pi}{6} t \\ -0.7 &= 0.7 \sin \frac{\pi}{6} t \\ \sin \frac{\pi}{6} t &= -1 \\ \frac{\pi}{6} t &= \frac{3\pi}{2} \\ t &= \frac{3\pi}{2} \times \frac{6}{\pi} \\ &= 9 \end{aligned}$$

$\therefore$ Low tide occurs after 9 hours, that is, 5 a.m. + 9 hours.
$\therefore$ Low tide occurs at 2 p.m.

(iii) The ship may enter the harbour when $h \geq 1.35$.

Solving for $h = 1.35$:

$$\begin{aligned} 1 + 0.7 \sin \frac{\pi}{6} t &= 1.35 \\ 0.7 \sin \frac{\pi}{6} t &= 0.35 \\ \sin \frac{\pi}{6} t &= \frac{1}{2} \\ \frac{\pi}{6} t &= \frac{\pi}{6}, \pi - \frac{\pi}{6} \\ &= \frac{\pi}{6}, \frac{5\pi}{6} \\ \therefore t &= 1, 5 \end{aligned}$$

By drawing the graphs of $y = \sin \frac{\pi}{6} t$ and $y = \frac{1}{2}$, it can be seen that the ship may enter the harbour between $t = 1$ and $t = 5$, i.e. $\sin \frac{\pi}{6} t \geq \frac{1}{2}$.

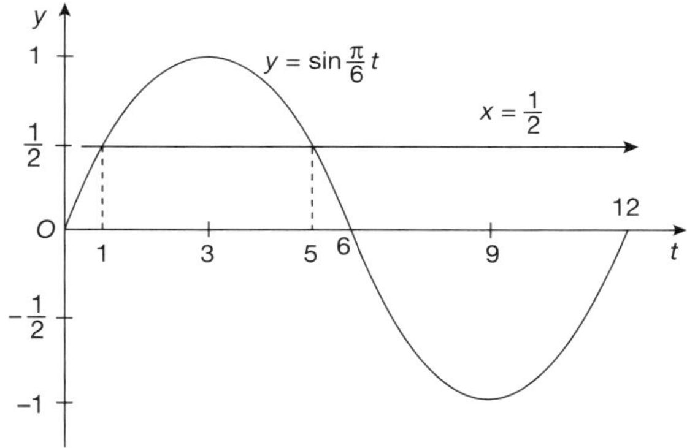

$\therefore$ The ship may enter the harbour between 6 a.m. and 11 a.m. as $t = 0$ at 5 a.m.

QUESTION 8

(a) (i) $-1 < x < 3$

(ii) As $x \to \infty, f'(x) \to 0$

(iii)

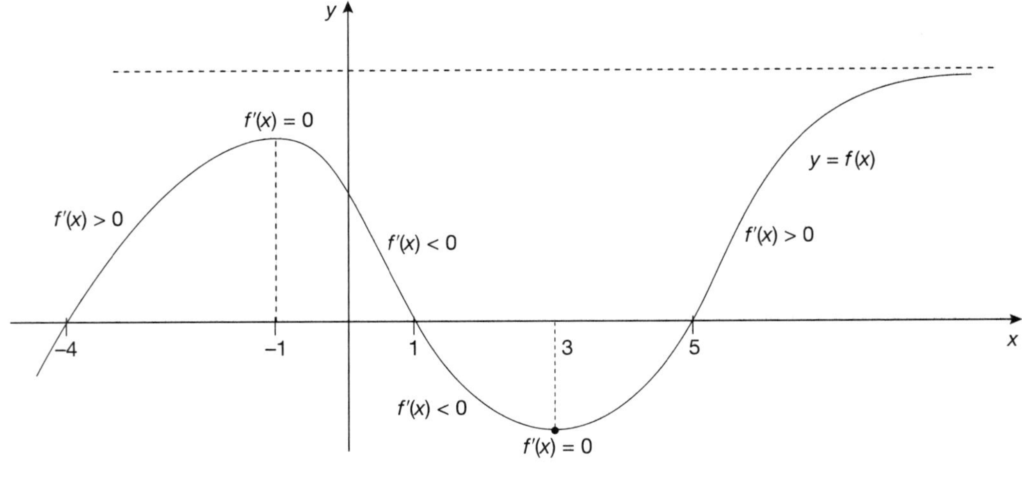

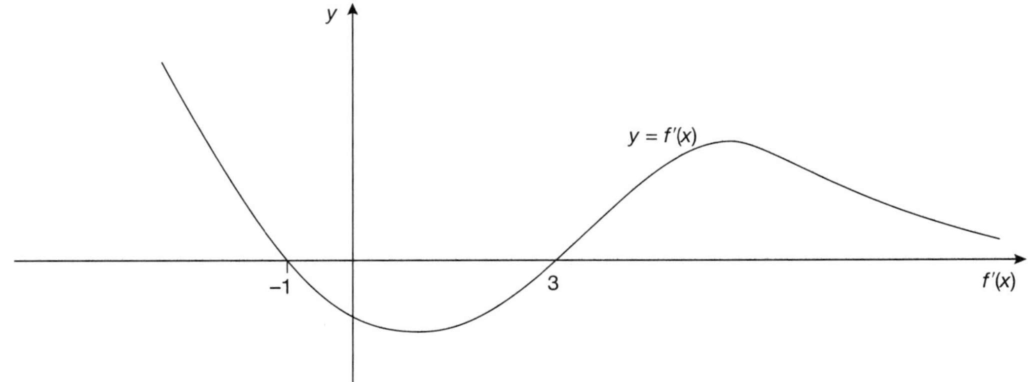

For $x < -1,\ f'(x) > 0$

At $x = -1,\ f'(x) = 0$

For $-1 < x < 3,\ f'(x) < 0$

At $x = 3,\ f'(x) = 0$

For $x > 3,\ f'(x) > 0$

As x approaches large values, $f'(x)$ approaches 0.

(b) (i) $A_n = P(1 + r)^n - M$

$P = \$350\,000, \quad n = 1, \quad M = \$2937,$
$r = 0.09$ p.a. $= 0.0075$ per month

$\therefore A_1 = 350\,000 \times (1 + 0.0075)^1 - 2937$
$= 349\,688$

$\therefore$ Daniel owed \$349 688 after his first monthly repayment.

(ii) $A_n = 346\,095 \times 1.005^n - 1.005^{n-1}M - \ldots - 1.005M - M$

After 288 months, $A_n = 0$.

$$\therefore\ 0 = 346\,095 \times 1.005^{288} - 1.005^{287}M - \ldots - 1.005M - M$$
$$0 = 346\,095 \times 1.005^{288} - M \times (1.005^{287} + \ldots + 1.005 + 1)$$
$$0 = 346\,095 \times 1.005^{288} - M \times (1 + 1.005 + \ldots + 1.005^{287})$$

Now $1 + 0.005 + \ldots + 0.005^{287}$ is a geometric series,
where $a = 1$, $r = 1.005$, $n = 288$

$$S_n = \frac{a(r^n - 1)}{r - 1}$$

$$\therefore\ S_{288} = \frac{1 \times (1.005^{288} - 1)}{1.005 - 1}$$

$$= \left[\frac{1.005^{288} - 1}{0.005}\right]$$

$$\therefore\ 0 = 346\,095 \times 1.005^{288} - M \times \left[\frac{1.005^{288} - 1}{0.005}\right]$$

$$M \times \left[\frac{1.005^{288} - 1}{0.005}\right] = 346\,095 \times 1.005^{288}$$

$$M = 346\,095 \times 1.005^{288} \div \left[\frac{1.005^{288} - 1}{0.005}\right]$$

$M = 1\,455\,529.832 \ldots \div 641.115 \ldots$ (by calc.)

$= 2270.310 \ldots$ (by calc.)

$\therefore$ The monthly repayment of the loan to be repaid over the remaining 24 years is \$2270.31 to the nearest cent.

(iii) $A_n = 346\,095 \times 1.005^n - 1.005^{n-1}M - \ldots - 1.005M - M$

$M = \$2937$, $r = 0.06$ p.a. $= 0.005$ per month, $A_n = 0$

$$\therefore\ 0 = 346\,095 \times 1.005^n - 1.005^{n-1}(2937) - \ldots - 1.005(2937) - 2937$$
$$= 346\,095 \times 1.005^n - 2937(1.005^{n-1} + \ldots + 1.005 + 1)$$
$$= 346\,095 \times 1.005^n - 2937(1 + 1.005 + \ldots + 1.005^{n-1})$$

Now $1 + 1.005 + \ldots + 1.005^{n-1}$ is a geometric series.
where $a = 1$, $r = 1.005$, $n = n$.

$$S_n = \frac{a(r^n - 1)}{r - 1}$$

$$\therefore\ S_n = \frac{1 \times (1.005^n - 1)}{1.005 - 1}$$

$$= \frac{1.005^n - 1}{0.005}$$

$$\therefore\ 0 = 346\,095 \times 1.005^n - 2937\left(\frac{1.005^n - 1}{0.005}\right)$$

$$= 346\,095 \times 1.005^n(0.005) - 2937(1.005^n - 1)$$
$$= 1730.475 \times 1.005^n - 2937(1.005^n) + 2937$$
$$= 1.005^n(1730.475 - 2937) + 2937$$
$$= -1206.525(1.005)^n + 2937$$

$$1206.525(1.005)^n = 2937$$
$$1.005^n = \frac{2937}{1206.525}$$
$$n \ln 1.005 = \ln \frac{2937}{1206.525}$$
$$n = \ln \frac{2937}{1206.525} \div \ln 1.005$$
$$= 178.373\ldots \quad \text{(by calc.)}$$

$\therefore$ Daniel will repay the loan in 179 months or 15 years.

(iv) Amount saved = \$2270.31 × 288 months – \$2937 × 179 months
= \$653 844.28 – \$525 723
= \$128 126.28 (by calc.)

Daniel will save \$128 126 over the term of the loan to the nearest dollar.

QUESTION 9

(a)

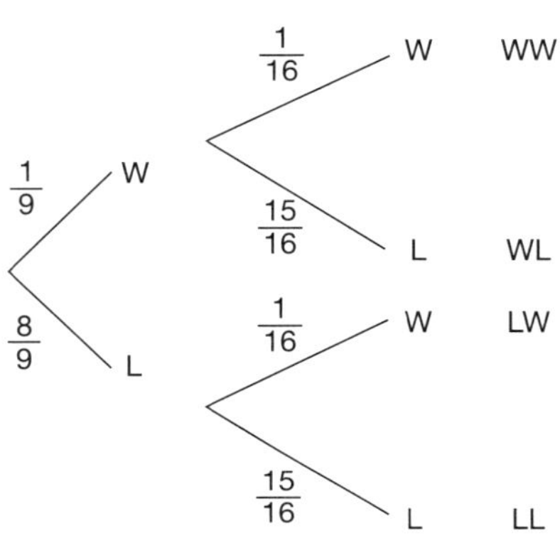

$$\text{P(No prize)} = \frac{8}{9} \times \frac{15}{16} = \frac{5}{6}$$

$$\therefore \text{P(No prize over 3 weeks)} = \left(\frac{5}{6}\right)^3 = \frac{125}{216}.$$

$$\therefore \text{P(At least one prize)} = 1 - \frac{125}{216} = \frac{91}{216}.$$

(b)

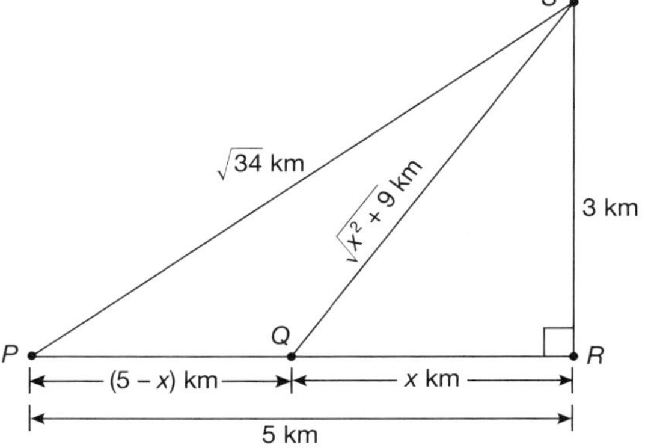

(i) Cost = \$1000 × 5 + \$2600 × 3
= \$5000 + \$7800
= \$12 800

(ii) By Pythagoras' Theorem in $\triangle PRS$,

$$PS^2 = PR^2 + RS^2$$
$$= 5^2 + 3^2$$
$$= 25 + 9$$
$$= 34$$
$$\therefore PS = \sqrt{34}$$

$\therefore$ Cost = \$2600 × $\sqrt{34}$
= \$15 160.474 . . . (by calc.)
= \$15 160 to the nearest dollar.

(iii) By Pythagoras' Theorem in $\triangle QRS$,

$$QS^2 = QR^2 + RS^2$$
$$= x^2 + 3^2$$
$$= x^2 + 9$$
$$\therefore QS = \sqrt{x^2 + 9}$$
and $PQ = 5 - x$

$$\therefore C = 1000 \times (5 - x) + 2600\sqrt{x^2 + 9}$$
$$= 1000\left(5 - x + 2.6\sqrt{x^2 + 9}\right)$$

(iv) $C = 1000\left(5 - x + 2.6\sqrt{x^2 + 9}\right)$

$$= 1000\left[5 - x + 2.6(x^2 + 9)^{\frac{1}{2}}\right]$$

$$\frac{dC}{dx} = 1000\left[-1 + 2.6\left(\frac{1}{2}\right)(x^2 + 9)^{-\frac{1}{2}} \times 2x\right]$$
$$= 1000\left[-1 + 2.6x(x^2 + 9)^{-\frac{1}{2}}\right]$$
$$= 1000\left(-1 + \frac{2.6x}{\sqrt{x^2 + 9}}\right)$$

Stationery points occur when

$$\frac{dC}{dx} = 0$$
$$1000\left(-1 + \frac{2.6x}{\sqrt{x^2 + 9}}\right) = 0$$
$$-1 + \frac{2.6x}{\sqrt{x^2 + 9}} = 0$$
$$\frac{2.6x}{\sqrt{x^2 + 9}} = 1$$
$$2.6x = \sqrt{x^2 + 9}$$
$$(2.6x)^2 = x^2 + 9$$
$$6.76x^2 = x^2 + 9$$
$$5.76x^2 = 9$$
$$x^2 = \frac{9}{5.76}$$
$$x = \frac{3}{2.4}$$
$$\therefore x = 1.25.$$

To determine nature of stationary point:

x	1.25^-	1.25	1.25^+
$\frac{dy}{dx}$	–	0	+

$\therefore$ A minimum value occurs when $x = 1.25$.

Now $C = 1000\left(5 - x + 2.6\sqrt{x^2 + 9}\right)$
When $x = 1.25$,

$$C = 1000\left(5 - 1.25 + 2.6\sqrt{1.25^2 + 9}\right)$$
$$= 12\,200 \quad \text{(by calc.)}$$

$\therefore$ The minimum cost of laying the cable is \$12 200.

(v) $C = 1000 \times (5 - x) + 1100\sqrt{x^2 + 9}$

$$= 1000\left(5 - x + 1.1\sqrt{x^2 + 9}\right)$$
$$= 1000\left[5 - x + 1.1(x^2 + 9)^{\frac{1}{2}}\right]$$

$$\frac{dC}{dx} = 1000\left[-1 + 1.1\left(\frac{1}{2}\right)(x^2 + 9)^{-\frac{1}{2}} \times 2x\right]$$
$$= 1000\left[-1 + 1.1x(x^2 + 9)^{-\frac{1}{2}}\right]$$
$$= 1000\left(-1 + \frac{1.1x}{\sqrt{x^2 + 9}}\right)$$

Stationary points occur when

$$\frac{dC}{dx} = 0$$
$$1000\left(-1 + \frac{1.1x}{\sqrt{x^2 + 9}}\right) = 0$$
$$\left(-1 + \frac{1.1x}{\sqrt{x^2 + 9}}\right) = 0$$
$$\frac{1.1x}{\sqrt{x^2 + 9}} = 1$$
$$1.1x = \sqrt{x^2 + 9}$$
$$(1.1x)^2 = x^2 + 9$$
$$1.21x^2 = x^2 + 9$$
$$0.21x^2 = 9$$
$$x^2 = \frac{9}{0.21}$$
$$\therefore x = \frac{3}{\sqrt{0.21}}$$
$$\doteqdot 6.5$$

Since $x < 5$, this solution for x is invalid. (See diagram.)

Consider cost of laying cable from P to R, then from R to S:

Cost = \$1000 × 5 + \$1100 × 3
= \$8300

Consider cost of laying cable in a straight line underwater directly from P to S:

Cost = \$1100 × PS
= \$1100 × $\sqrt{34}$

$= \$6414.047\ldots$ (by calc.)
$= \$6414$ to nearest dollar.

$\therefore$ Minimum cost of \$6414 can be achieved by laying cable underwater in a straight line directly from P to S.

QUESTION 10

(a) $f(x) = x - \dfrac{x^2}{2} + \dfrac{x^3}{3}$

$f'(x) = 1 - x + x^2$

Turning points may occur when

$f'(x) = 0.$

$\therefore\ 1 - x + x^2 = 0$

$x^2 - x + 1 = 0$

Now $\Delta = b^2 - 4ac$
$= (-1)^2 - 4(1)(1)$
$= -3$

Since $\Delta < 0$, $1 - x + x^2 = 0$ has no real solutions.

$\therefore\ y = f(x)$ has no turning points.

(b) $f''(x) = -1 + 2x$

Points of inflexion may occur when

$f''(x) = 0.$

$-1 + 2x = 0$

$2x = 1$

$x = \dfrac{1}{2}$

Test for change of concavity:

x	0	$\frac{1}{2}$	1
$f''(x)$	–	0	+

Since there is a change in concavity, there is a point of inflexion at $x = \dfrac{1}{2}$.

At $x = \dfrac{1}{2}$,

$$f(x) = \frac{1}{2} - \frac{\left(\frac{1}{2}\right)^2}{2} + \frac{\left(\frac{1}{2}\right)^3}{3}$$

$$= \frac{5}{12}$$

$\therefore$ A point of inflexion at $\left(\dfrac{1}{2}, \dfrac{5}{12}\right)$.

(c) (i) $1 - x + x^2 - \dfrac{1}{1+x} = \dfrac{x^3}{1+x}$ for $x \neq -1$

$$\text{LHS} = 1 - x + x^2 - \frac{1}{1+x}$$

$$= \frac{1(1+x) - x(1+x) + x^2(1+x) - 1}{1+x}$$

$$= \frac{1 + x - x - x^2 + x^2 + x^3 - 1}{1+x}$$

$$= \frac{x^3}{1+x}$$

$= \text{RHS}.$

(ii) $g(x) = \ln(1+x)$

$g'(x) = \dfrac{1}{1+x}$

Now $f'(x) = 1 - x + x^2$

$$\therefore\ f'(x) - g'(x) = 1 - x + x^2 - \frac{1}{1+x}$$

$$= \frac{x^3}{1+x} \text{ for } x \neq -1 \ \text{(from (i))}$$

$\therefore\ f'(x) - g'(x) \geqslant 0$

$\therefore\ f'(x) \geqslant g'(x)$ for $x \geqslant 0$.

(d)

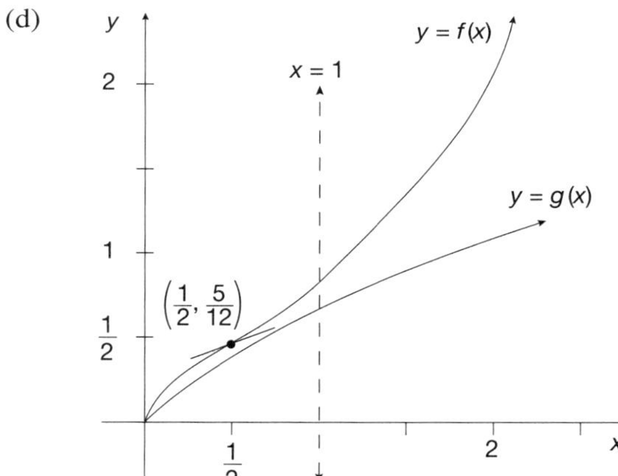

(e) $\dfrac{d}{dx}\left[(1+x)\ln(1+x) - (1+x)\right] = \ln(1+x)$

$$\text{LHS} = \frac{d}{dx}(1+x)\ln(1+x) - \frac{d}{dx}(1+x)$$

$$= (1+x)\cdot\frac{d}{dx}\ln(1+x) + \ln(1+x)\cdot\frac{d}{dx}(1+x) - \frac{d}{dx}(1+x)$$

$$= (1+x)\,.\,\frac{1}{1+x} + \ln(1+x)\,.\,1 - 1$$
$$= 1 + \ln(1+x) - 1$$
$$= \ln(1+x)$$
$$= \text{RHS}.$$

(f) $A = \displaystyle\int_a^b \big[f(x) - g(x)\big]dx$

$$= \int_0^1 \left(x - \frac{x^2}{2} + \frac{x^3}{3}\right)dx - \int_0^1 \ln(1+x)dx$$

$$= \left[\frac{x^2}{2} - \frac{x^3}{6} + \frac{x^4}{12}\right]_0^1 - \big[(1+x)\ln(1+x) - (1+x)\big]_0^1 \quad \big(\text{using (e)}\big)$$

$$= \left[\left(\frac{1}{2} - \frac{1}{6} + \frac{1}{12}\right) - (0 - 0 + 0)\right] - \big[(2\ln 2 - 2) - (\ln 1 - 1)\big]$$

$$= \frac{5}{12} - 2\ln 2 + 1$$

$$= 1\frac{5}{12} - 2\ln 2$$

$$\doteqdot 0.03 \quad \text{(by calc.)}$$

Area enclosed is approximately 0.03 units^2.

BOARD OF STUDIES

NEW SOUTH WALES

2010

HIGHER SCHOOL CERTIFICATE EXAMINATION

Mathematics

General Instructions

- Reading time – 5 minutes
- Working time – 3 hours
- Write using black or blue pen
- Board-approved calculators may be used
- A table of standard integrals is provided at the back of this paper
- All necessary working should be shown in every question

Total marks – 120

- Attempt Questions 1–10
- All questions are of equal value

Total marks – 120
Attempt Questions 1–10
All questions are of equal value

Answer each question in the appropriate writing booklet. Extra writing booklets are available.

Question 1 (12 marks) Use the Question 1 Writing Booklet.

(a) Solve $x^2 = 4x$. **2**

(b) Find integers a and b such that $\dfrac{1}{\sqrt{5} - 2} = a + b\sqrt{5}$. **2**

(c) Write down the equation of the circle with centre $(-1, 2)$ and radius 5. **1**

(d) Solve $|2x + 3| = 9$. **2**

(e) Differentiate $x^2 \tan x$ with respect to x. **2**

(f) Find the limiting sum of the geometric series $1 - \dfrac{1}{3} + \dfrac{1}{9} - \dfrac{1}{27} + \cdots$. **2**

(g) Let $f(x) = \sqrt{x - 8}$. What is the domain of $f(x)$? **1**

Question 2 (12 marks) Use the Question 2 Writing Booklet.

(a) Differentiate $\dfrac{\cos x}{x}$ with respect to x. **2**

(b) Solve the inequality $x^2 - x - 12 < 0$. **2**

(c) Find the gradient of the tangent to the curve $y = \ln(3x)$ at the point where $x = 2$. **2**

(d) (i) Find $\displaystyle\int \sqrt{5x+1}\, dx$. **2**

(ii) Find $\displaystyle\int \frac{x}{4+x^2}\, dx$. **2**

(e) Given that $\displaystyle\int_0^6 (x+k)\, dx = 30$, and k is a constant, find the value of k. **2**

Question 3 (12 marks) Use the Question 3 Writing Booklet.

(a) In the diagram A, B and C are the points $(-2, -4)$, $(12, 6)$ and $(6, 8)$ respectively. The point $N(2, 2)$ is the midpoint of AC. The point M is the midpoint of AB.

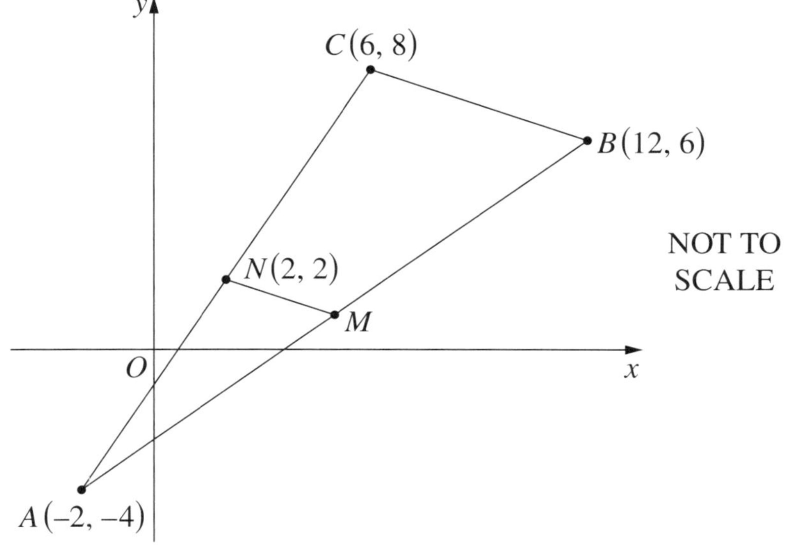

(i) Find the coordinates of M. **1**

(ii) Find the gradient of BC. **1**

(iii) Prove that $\triangle ABC$ is similar to $\triangle AMN$. **2**

(iv) Find the equation of MN. **2**

(v) Find the exact length of BC. **1**

(vi) Given that the area of $\triangle ABC$ is 44 square units, find the perpendicular distance from A to BC. **1**

Question 3 continues on following page

Question 3 (continued)

(b) (i) Sketch the curve $y = \ln x$. **1**

(ii) Use the trapezoidal rule with three function values to find an approximation to **2**

$$\int_1^3 \ln x \, dx.$$

(iii) State whether the approximation found in part (ii) is greater than or less than the exact value of $\int_1^3 \ln x \, dx$. Justify your answer. **1**

End of Question 3

Question 4 (12 marks) Use the Question 4 Writing Booklet.

(a) Susannah is training for a fun run by running every week for 26 weeks. She runs 1 km in the first week and each week after that she runs 750 m more than the previous week, until she reaches 10 km in a week. She then continues to run 10 km each week.

(i) How far does Susannah run in the 9th week? **1**

(ii) In which week does she first run 10 km? **1**

(iii) What is the total distance that Susannah runs in 26 weeks? **2**

(b) The curves $y = e^{2x}$ and $y = e^{-x}$ intersect at the point $(0, 1)$ as shown in the diagram. **3**

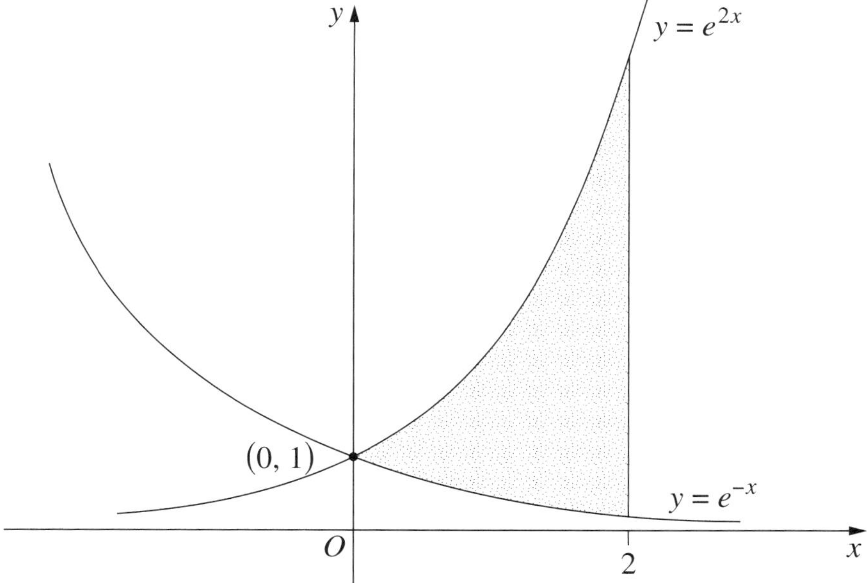

Find the exact area enclosed by the curves and the line $x = 2$.

Question 4 continues on following page

Question 4 (continued)

(c) There are twelve chocolates in a box. Four of the chocolates have mint centres, four have caramel centres and four have strawberry centres. Ali randomly selects two chocolates and eats them.

(i) What is the probability that the two chocolates have mint centres? **1**

(ii) What is the probability that the two chocolates have the same centre? **1**

(iii) What is the probability that the two chocolates have different centres? **1**

(d) Let $f(x) = 1 + e^x$. **2**

Show that $f(x) \times f(-x) = f(x) + f(-x)$.

End of Question 4

Question 5 (12 marks) Use the Question 5 Writing Booklet.

(a) A rainwater tank is to be designed in the shape of a cylinder with radius r metres and height h metres.

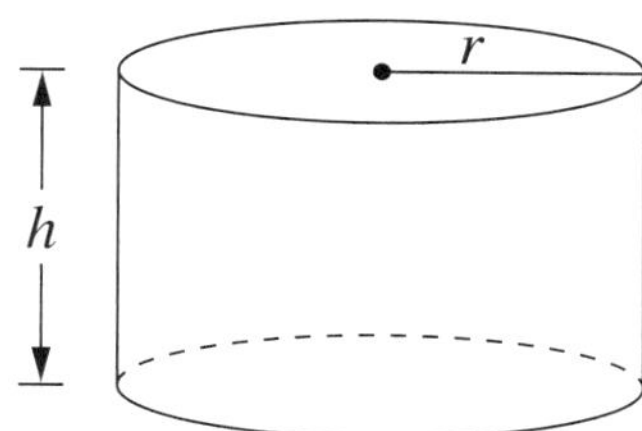

The volume of the tank is to be 10 cubic metres. Let A be the surface area of the tank, including its top and base, in square metres.

(i) Given that $A = 2\pi r^2 + 2\pi rh$, show that $A = 2\pi r^2 + \dfrac{20}{r}$. **2**

(ii) Show that A has a minimum value and find the value of r for which the minimum occurs. **3**

(b) (i) Prove that **1**

$$\sec^2 x + \sec x \tan x = \frac{1 + \sin x}{\cos^2 x}.$$

(ii) Hence prove that **1**

$$\sec^2 x + \sec x \tan x = \frac{1}{1 - \sin x}.$$

(iii) Hence use the table of standard integrals to find the exact value of **2**

$$\int_0^{\frac{\pi}{4}} \frac{1}{1 - \sin x}\, dx.$$

Question 5 continues on following page

Question 5 (continued)

(c) The diagram shows the curve $y = \frac{1}{x}$, for $x > 0$. **3**

The area under the curve between $x = a$ and $x = 1$ is A_1. The area under the curve between $x = 1$ and $x = b$ is A_2.

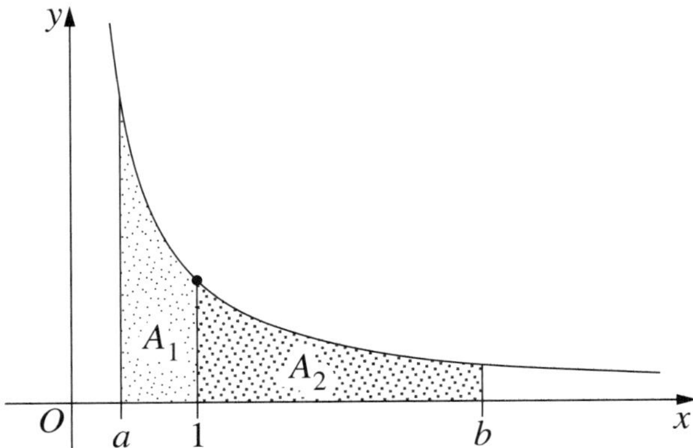

The areas A_1 and A_2 are each equal to 1 square unit.

Find the values of a and b.

End of Question 5

Question 6 (12 marks) Use the Question 6 Writing Booklet.

(a) Let $f(x) = (x + 2)(x^2 + 4)$.

(i) Show that the graph $y = f(x)$ has no stationary points. **2**

(ii) Find the values of x for which the graph $y = f(x)$ is concave down, and the values for which it is concave up. **2**

(iii) Sketch the graph $y = f(x)$, indicating the values of the x and y intercepts. **2**

(b) The diagram shows a circle with centre O and radius 5 cm.

The length of the arc PQ is 9 cm. Lines drawn perpendicular to OP and OQ at P and Q respectively meet at T.

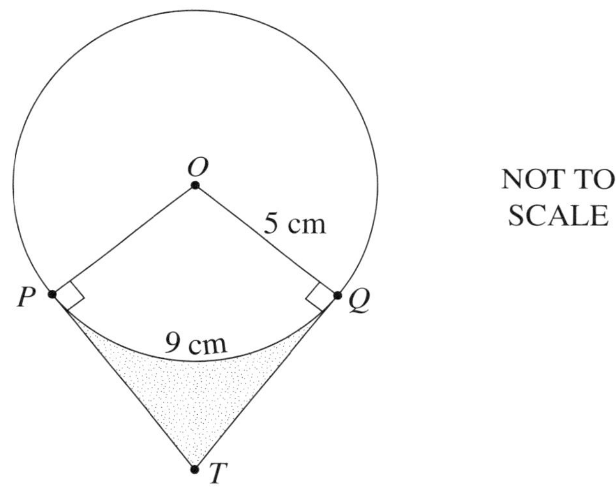

(i) Find $\angle POQ$ in radians. **1**

(ii) Prove that $\triangle OPT$ is congruent to $\triangle OQT$. **2**

(iii) Find the length of PT. **1**

(iv) Find the area of the shaded region. **2**

Question 7 (12 marks) Use the Question 7 Writing Booklet.

(a) The acceleration of a particle is given by

$$\ddot{x} = 4\cos 2t$$

where x is displacement in metres and t is time in seconds.

Initially the particle is at the origin with a velocity of 1 m s^{-1}.

(i) Show that the velocity of the particle is given by **2**

$$\dot{x} = 2\sin 2t + 1 .$$

(ii) Find the time when the particle first comes to rest. **2**

(iii) Find the displacement, x, of the particle in terms of t. **2**

(b) The parabola shown in the diagram is the graph $y = x^2$. The points $A(-1, 1)$ and $B(2, 4)$ are on the parabola.

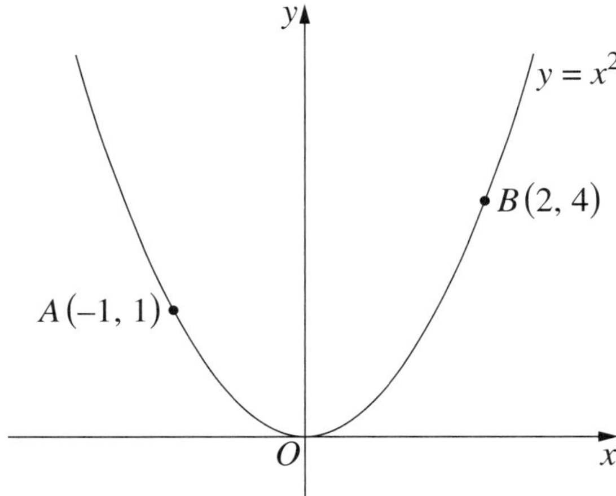

(i) Find the equation of the tangent to the parabola at A. **2**

(ii) Let M be the midpoint of AB. **2**
There is a point C on the parabola such that the tangent at C is parallel to AB.

Show that the line MC is vertical.

(iii) The tangent at A meets the line MC at T. **2**

Show that the line BT is a tangent to the parabola.

Question 8 (12 marks) Use the Question 8 Writing Booklet.

(a) Assume that the population, P, of cane toads in Australia has been growing at a rate proportional to P. That is, $\frac{dP}{dt} = kP$ where k is a positive constant. **4**

There were 102 cane toads brought to Australia from Hawaii in 1935.

Seventy-five years later, in 2010, it is estimated that there are 200 million cane toads in Australia.

If the population continues to grow at this rate, how many cane toads will there be in Australia in 2035?

(b) Two identical biased coins are tossed together, and the outcome is recorded. After a large number of trials it is observed that the probability that both coins land showing heads is 0.36. **2**

What is the probability that both coins land showing tails?

Question 8 continues on following page

Question 8 (continued)

(c) The graph shown is $y = A\sin bx$.

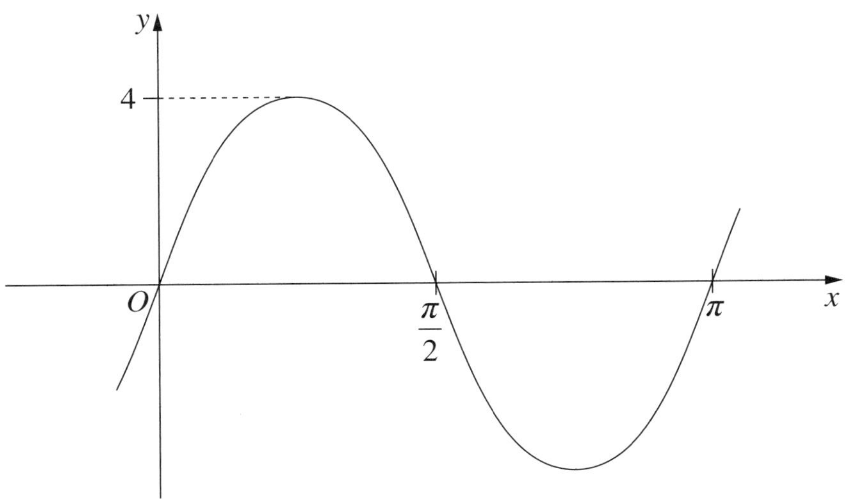

(i) Write down the value of A. **1**

(ii) Find the value of b. **1**

(iii) Copy or trace the graph into your writing booklet. **2**

On the same set of axes, draw the graph $y = 3\sin x + 1$, for $0 \le x \le \pi$.

(d) Let $f(x) = x^3 - 3x^2 + kx + 8$, where k is a constant. **2**

Find the values of k for which $f(x)$ is an increasing function.

End of Question 8

Question 9 (12 marks) Use the Question 9 Writing Booklet.

(a) (i) When Chris started a new job, \$500 was deposited into his superannuation fund at the beginning of each month. The money was invested at 0.5% per month, compounded monthly. **2**

Let \$$P$ be the value of the investment after 240 months, when Chris retires.

Show that $P = 232\,175.55$.

(ii) After retirement, Chris withdraws \$2000 from the account at the end of each month, without making any further deposits. The account continues to earn interest at 0.5% per month.

Let \$$A_n$ be the amount left in the account n months after Chris's retirement.

(1) Show that $A_n = (P - 400\,000) \times 1.005^n + 400\,000$. **3**

(2) For how many months after retirement will there be money left in the account? **2**

Question 9 continues on following page

Question 9 (continued)

(b) Let $y = f(x)$ be a function defined for $0 \leq x \leq 6$, with $f(0) = 0$.

The diagram shows the graph of the derivative of f, $y = f'(x)$.

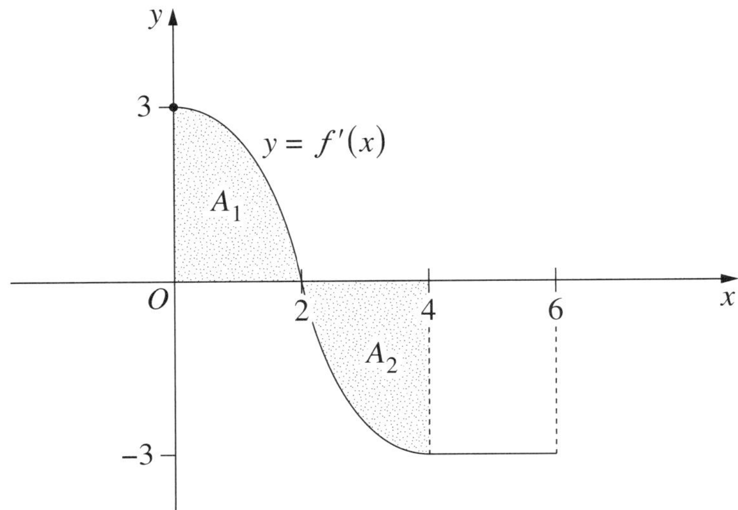

The shaded region A_1 has area 4 square units. The shaded region A_2 has area 4 square units.

(i) For which values of x is $f(x)$ increasing? **1**

(ii) What is the maximum value of $f(x)$? **1**

(iii) Find the value of $f(6)$. **1**

(iv) Draw a graph of $y = f(x)$ for $0 \leq x \leq 6$. **2**

End of Question 9

Question 10 (12 marks) Use the Question 10 Writing Booklet.

(a) In the diagram ABC is an isosceles triangle with $AC = BC = x$. The point D on the interval AB is chosen so that $AD = CD$. Let $AD = a$, $DB = y$ and $\angle ADC = \theta$.

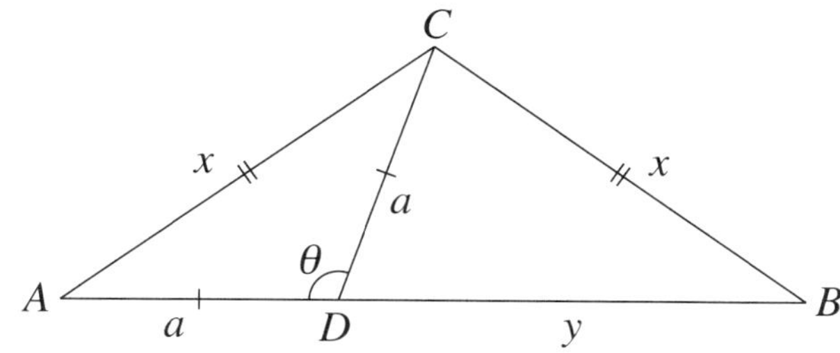

(i) Show that $\triangle ABC$ is similar to $\triangle ACD$. **2**

(ii) Show that $x^2 = a^2 + ay$. **1**

(iii) Show that $y = a(1 - 2\cos\theta)$. **2**

(iv) Deduce that $y \leq 3a$. **1**

Question 10 continues on following page

Question 10 (continued)

(b) The circle $x^2 + y^2 = r^2$ has radius r and centre O. The circle meets the positive x-axis at B. The point A is on the interval OB. A vertical line through A meets the circle at P. Let $\theta = \angle OPA$.

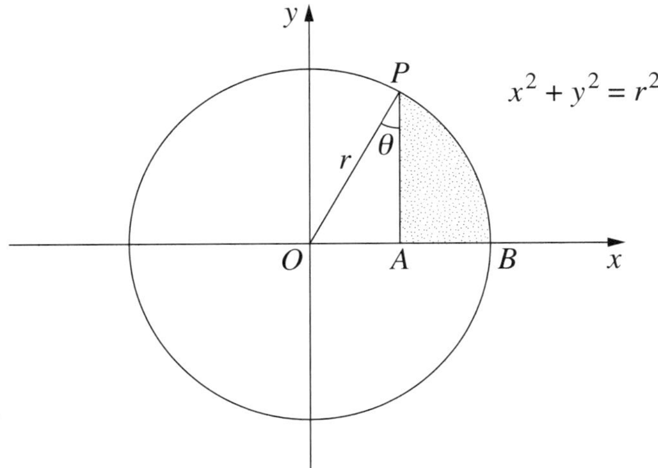

(i) The shaded region bounded by the arc PB and the intervals AB and AP is rotated about the x-axis. Show that the volume, V, formed is given by **3**

$$V = \frac{\pi r^3}{3}\left(2 - 3\sin\theta + \sin^3\theta\right).$$

(ii)

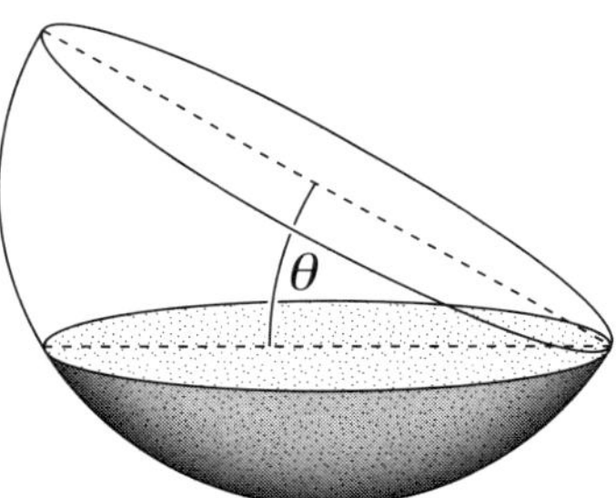

A container is in the shape of a hemisphere of radius r metres. The container is initially horizontal and full of water. The container is then tilted at an angle of θ to the horizontal so that some water spills out.

(1) Find θ so that the depth of water remaining is one half of the original depth. **1**

(2) What fraction of the original volume is left in the container? **2**

End of paper

2010 Higher School Certificate Worked Answers

QUESTION 1

(a) $x^2 = 4x$

$x^2 - 4x = 0$

$x(x - 4) = 0$

$\therefore x = 0$ or 4 *(2 marks)*

(b) $\dfrac{1}{\sqrt{5} - 2} = \dfrac{1}{\sqrt{5} - 2} \times \dfrac{\sqrt{5} + 2}{\sqrt{5} + 2}$

$= \dfrac{\sqrt{5} + 2}{5 - 4}$

$= \sqrt{5} + 2$

$= 2 + \sqrt{5}$, which is in the form $a + b\sqrt{5}$, where $a = 2$ and $b = 1$. *(2 marks)*

(c) Using the formula

$(x - h)^2 + (y - k)^2 = r^2$,

with centre (h, k) and radius r:

Now, $C(-1, 2)$ and $r = 5$:

$(x - (-1))^2 + (y - 2)^2 = 5^2$

$(x + 1)^2 + (y - 2)^2 = 25$ *(1 mark)*

(d) $|2x + 3| = 9$.

Two cases:

$2x + 3 = 9$	or	$-(2x + 3) = 9$
$2x = 9 - 3$	or	$-2x - 3 = 9$
$2x = 6$	or	$-2x = 9 + 3$
$x = 3$	or	$-2x = 12$
		$x = -6$

$\therefore x = 3$ or $x = -6$ *(2 marks)*

(e) Let $y = x^2 \tan x$.

Using the product rule,

Let $u = x^2, u' = 2x$

Let $v = \tan x, v' = \sec^2 x$

$\dfrac{dy}{dx} = u'.v + v'.u$

$= 2x.\tan x + \sec^2 x.x^2$

$= 2x \tan x + x^2 \sec^2 x$ *(2 marks)*

(f) $a = 1, r = -\dfrac{1}{3}$,

$S_\infty = \dfrac{a}{1 - r}$

$= \dfrac{1}{1 - \left(-\dfrac{1}{3}\right)}$

$= \dfrac{1}{1\frac{1}{3}}$

$= 1 \div 1\frac{1}{3}$

$= \dfrac{3}{4}$

$\therefore$ The limiting sum is $\dfrac{3}{4}$. *(2 marks)*

(g) $f(x) = \sqrt{x - 8}$.

Domain is $x - 8 \geq 0$

$x \geq 8$

$\therefore$ The domain is $x \geq 8$. *(1 mark)*

QUESTION 2

(a) Let $y = \dfrac{\cos x}{x}$.

Using the quotient rule,

Let $u = \cos x, u' = -\sin x$

Let $v = x, v' = 1$

$\dfrac{dy}{dx} = \dfrac{v.u' - u.v'}{v^2}$

$= \dfrac{x.(-\sin x) - \cos x.1}{x^2}$

$= \dfrac{-x\sin x - \cos x}{x^2}$ *(2 marks)*

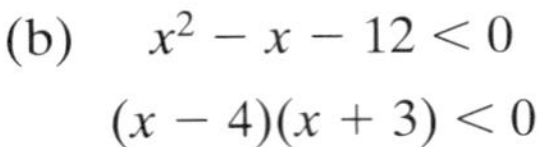

(b) $x^2 - x - 12 < 0$

$(x - 4)(x + 3) < 0$

Method 1:

Let $y = x^2 - x - 12$, so we need to consider values of x for which $y < 0$:

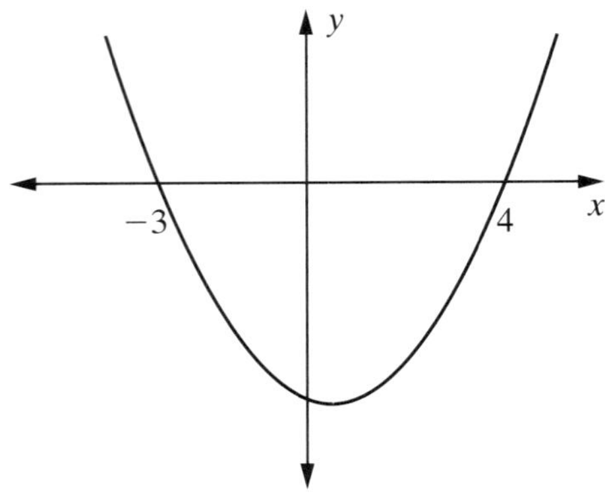

$\therefore\ -3 < x < 4$

Method 2:

Let $(x - 4)(x + 3) = 0$ gives $x = 4, -3$:

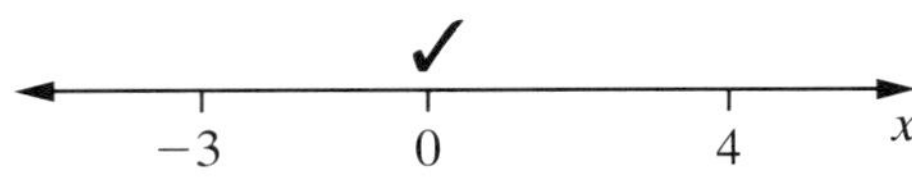

Choose 0: subs in $(x - 4)(x + 3) < 0$

$$\begin{aligned}\text{LHS} &= (0 - 4)(0 + 3)\\ &= -12 < 0, \quad \therefore \text{ true}\end{aligned}$$

$\therefore\ -3 < x < 4$ *(2 marks)*

(c) $y = \ln(3x)$

$$\frac{dy}{dx} = \frac{3}{3x} = \frac{1}{x}$$

At $x = 2$, $\dfrac{dy}{dx} = \dfrac{1}{2}$

$\therefore$ The gradient is $\dfrac{1}{2}$. *(2 marks)*

(d) (i)
$$\begin{aligned}&\int \sqrt{5x + 1}\, dx\\ &= \int (5x + 1)^{\frac{1}{2}}\, dx\\ &= \frac{(5x + 1)^{\frac{3}{2}}}{\frac{3}{2}.5} + c\\ &= \frac{2(5x + 1)^{\frac{3}{2}}}{15} + c\\ &= \frac{2\sqrt{(5x + 1)^3}}{15} + c\end{aligned}$$
(2 marks)

(ii)
$$\begin{aligned}&\int \frac{x}{4 + x^2}\, dx\\ &= \frac{1}{2}\int \frac{2x}{4 + x^2}\, dx\\ &= \frac{1}{2}\log_e(4 + x^2) + c\end{aligned}$$
(2 marks)

(e)
$$\begin{aligned}\int_0^6 (x + k)\, dx &= 30\\ \left[\frac{x^2}{2} + kx\right]_0^6 &= 30\\ \left[\frac{6^2}{2} + 6k\right] - \left[\frac{0^2}{2} + 0\right] &= 30\\ 18 + 6k &= 30\\ 6k &= 30 - 18\\ 6k &= 12\\ k &= 2\end{aligned}$$
(2 marks)

QUESTION 3

(a)

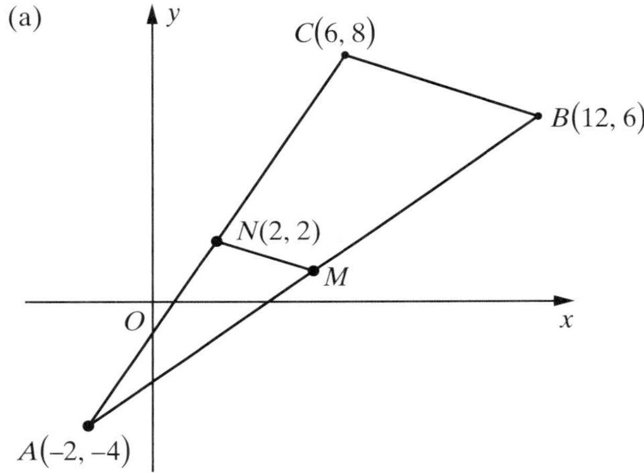

(i) Using $A(-2, -4)$ and $B(12, 6)$,

$$\begin{aligned}MP &= \left(\frac{x_1 + x_2}{2}, \frac{y_1 + y_2}{2}\right)\\ &= \left(\frac{-2 + 12}{2}, \frac{-4 + 6}{2}\right)\\ &= \left(\frac{10}{2}, \frac{2}{2}\right)\\ &= (5, 1) \quad \therefore M(5, 1)\end{aligned}$$
(1 mark)

(ii) Using $B(12, 6)$ and $C(6, 8)$,

$$m = \frac{y_2 - y_1}{x_2 - x_1} = \frac{8 - 6}{6 - 12} = \frac{2}{-6} = -\frac{1}{3}$$

$\therefore$ The gradient of BC is $-\frac{1}{3}$. *(1 mark)*

(iii) In $\triangle AMN$ and $\triangle ABC$:

$\angle A$ is common

As N is midpoint of AC (given)

and M is midpoint of AB (from (i))

$$\therefore \frac{AN}{AC} = \frac{AM}{AB}$$

$\therefore \triangle AMN \ ||| \ \triangle ABC$

(Two sides in proportion and included angle equal.) *(2 marks)*

(iv) Using $M(5, 1)$ and $N(2, 2)$,

$$m = \frac{y_2 - y_1}{x_2 - x_1} = \frac{2 - 1}{2 - 5} = -\frac{1}{3} \ **$$

Using $N(2, 2)$ and $m = -\frac{1}{3}$,

$$y - y_1 = m(x - x_1)$$
$$y - 2 = -\frac{1}{3}(x - 2)$$
$$3y - 6 = -x + 2$$

$\therefore x + 3y - 8 = 0$

[** Or, as $\triangle AMN \ ||| \ \triangle ABC$, then $MN \parallel BC$, so gradient $MN = -\frac{1}{3}$.]

(2 marks)

(v) Using $B(12, 6)$ and $C(6, 8)$,

$$d = \sqrt{(x_2 - x_1)^2 + (y_2 - y_1)^2} = \sqrt{(6 - 12)^2 + (8 - 6)^2} = \sqrt{36 + 4} = \sqrt{40} = 2\sqrt{10}$$

$\therefore$ The length of BC is $2\sqrt{10}$ units.

(1 mark)

(vi) Area $= \frac{1}{2} \times \text{base} \times \text{height}$

$$44 = \frac{1}{2} \times BC \times \text{perp. height}$$
$$44 = \frac{1}{2} \times 2\sqrt{10} \times \text{perp. height}$$
$$44 = \sqrt{10} \times \text{perp. height}$$

$$\therefore \text{perp. height} = \frac{44}{\sqrt{10}} = \frac{44}{\sqrt{10}} \times \frac{\sqrt{10}}{\sqrt{10}} = \frac{44\sqrt{10}}{10} = \frac{22\sqrt{10}}{5}$$

$\therefore$ The perpendicular distance from A to BC is $\frac{22\sqrt{10}}{5}$ units. *(1 mark)*

(b) (i)

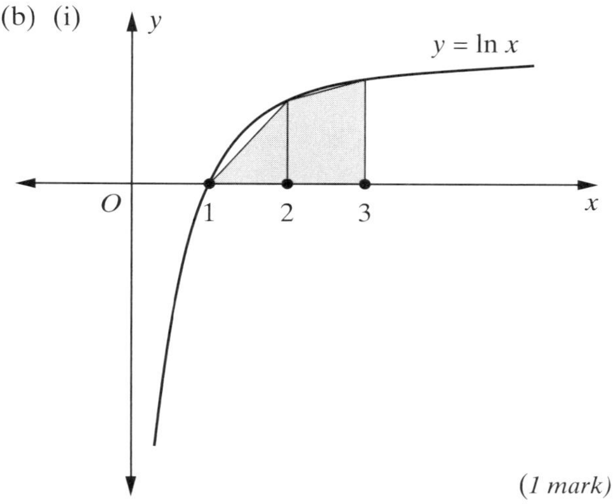

(1 mark)

(ii)

x	1	2	3
y	0	ln 2	ln 3

Trapezoidal rule:

$$\int_1^3 \ln x \, dx \approx \frac{1}{2}[0 + \ln 3 + 2 \ln 2] \approx \frac{1}{2}[\ln 3 + \ln 4] \approx \frac{1}{2}\ln 12$$

$\therefore$ The area is approximately $\frac{1}{2}\ln 12$ units^2. *(2 marks)*

(iii) The shaded region on the diagram in (i) is the area using the trapezoidal rule. It is less than the area under the curve. *(1 mark)*

QUESTION 4

(a) 1, 1.75, 2.5, …, 10, then 10, 10, …

(i) Using $a = 1, d = 0.75, n = 9$,

$$\begin{aligned} T_n &= a + (n-1)d \\ &= 1 + (9-1) \times 0.75 \\ &= 7 \end{aligned}$$

∴ She runs 7 km in the 9th week. *(1 mark)*

(ii) Using $a = 1, d = 0.75, T_n = 10$,

$$\begin{aligned} T_n &= a + (n-1)d \\ 10 &= 1 + (n-1) \times 0.75 \\ 9 &= 0.75(n-1) \\ n - 1 &= \frac{9}{0.75} \\ n - 1 &= 12 \\ n &= 13 \end{aligned}$$

∴ She first runs 10 km in the 13th week. *(1 mark)*

(iii) First find her total distance in the first 13 weeks:

Using $a = 1, d = 0.75, n = 13$,

$$\begin{aligned} S_n &= \frac{n}{2}[2a + (n-1)d] \\ &= \frac{13}{2}[2 + 12 \times 0.75] \\ &= 71.5 \end{aligned}$$

∴ She runs 71.5 km in the first 13 weeks.

For the following 13 weeks she runs 10 km each week.

As $71.5 + 130 = 201.5$, Susannah runs a total of 201.5 km in the 26 weeks.

(2 marks)

(b)
$$\begin{aligned} \text{Area} &= \int_0^2 e^{2x} - e^{-x}\,dx \\ &= \left[\frac{1}{2}e^{2x} + e^{-x}\right]_0^2 \\ &= \left[\frac{1}{2}e^4 + e^{-2}\right] - \left[\frac{1}{2}e^0 + e^0\right] \\ &= \left[\frac{1}{2}e^4 + e^{-2}\right] - 1\frac{1}{2} \\ &= \frac{1}{2}\left[e^4 + \frac{2}{e^2} - 3\right] \end{aligned}$$

∴ The area is $\frac{1}{2}\left[e^4 + \frac{2}{e^2} - 3\right]$ units^2.

(3 marks)

(c) 4M, 4C, 4S

(i)
$$\begin{aligned} P(MM) &= \frac{4}{12} \times \frac{3}{11} \\ &= \frac{1}{11} \end{aligned}$$
(1 mark)

(ii)
$$\begin{aligned} P(\text{MM or CC or SS}) &= \frac{1}{11} + \frac{1}{11} + \frac{1}{11} \\ &= \frac{3}{11} \end{aligned}$$
(1 mark)

(iii) Method 1:

$$\begin{aligned} P(\text{different}) &= 1 - P(\text{same}) \\ &= 1 - \frac{3}{11} \\ &= \frac{8}{11} \end{aligned}$$

Method 2:

P(MC or MS or CM or CS or SM or SC)

$$\begin{aligned} &= 6 \times \frac{4}{12} \times \frac{4}{11} \\ &= \frac{8}{11} \end{aligned}$$
(1 mark)

(d) $f(x) = 1 + e^x$

$$\begin{aligned} \text{LHS} &= f(x) \times f(-x) \\ &= (1 + e^x)(1 + e^{-x}) \\ &= 1 + e^{-x} + e^x + e^0 \\ &= 2 + e^{-x} + e^x \end{aligned}$$

$$\begin{aligned} \text{RHS} &= f(x) + f(-x) \\ &= 1 + e^x + 1 + e^{-x} \\ &= 2 + e^{-x} + e^x \end{aligned}$$

∴ LHS = RHS *(2 marks)*

QUESTION 5

(a) (i)
$$\begin{aligned} V &= \pi r^2 h \\ 10 &= \pi r^2 h \\ \therefore h &= \frac{10}{\pi r^2} \end{aligned}$$

Substitute h in A:

$$\begin{aligned} A &= 2\pi r^2 + 2\pi rh \\ &= 2\pi r^2 + 2\pi r \times \frac{10}{\pi r^2} \\ &= 2\pi r^2 + \frac{20}{r} \end{aligned}$$
(2 marks)

(ii)
$$A = 2\pi r^2 + \frac{20}{r}$$
$$= 2\pi r^2 + 20r^{-1}$$
$$\frac{dA}{dr} = 4\pi r - 20r^{-2}$$
$$\therefore 4\pi r - \frac{20}{r^2} = 0$$
$$4\pi r = \frac{20}{r^2}$$
$$4\pi r^3 = 20$$
$$r^3 = \frac{20}{4\pi}$$
$$r^3 = \frac{5}{\pi}$$
$$r = \sqrt[3]{\frac{5}{\pi}}$$

As $\frac{dA}{dr} = 4\pi r - 20r^{-2}$,

then $\frac{d^2A}{dr^2} = 4\pi + 40r^{-3}$

But as $r > 0$, then $\frac{d^2A}{dr^2} > 0$,

so A is minimum when $r = \sqrt[3]{\frac{5}{\pi}}$.

[Or, use first derivative test:

r	1	$\sqrt[3]{\frac{5}{\pi}}$	2
$\frac{dA}{dr}$	< 0	0	> 0

$\therefore$ A is minimum when

$r = \sqrt[3]{\frac{5}{\pi}}$]. *(3 marks)*

(b) (i) LHS $= \sec^2 x + \sec x \tan x$
$$= \frac{1}{\cos^2 x} + \frac{1}{\cos x}.\frac{\sin x}{\cos x}$$
$$= \frac{1}{\cos^2 x} + \frac{\sin x}{\cos^2 x}$$
$$= \frac{1 + \sin x}{\cos^2 x}$$
$$= \text{RHS}$$
(1 mark)

(ii) First, show $\frac{1 + \sin x}{\cos^2 x} = \frac{1}{1 - \sin x}$

LHS $= \frac{1 + \sin x}{\cos^2 x}$
$$= \frac{1 + \sin x}{1 - \sin^2 x}$$
$$= \frac{1 + \sin x}{(1 - \sin x)(1 + \sin x)}$$
$$= \frac{1}{1 - \sin x}$$
$$= \text{RHS}$$
$$\therefore \sec^2 x + \sec x \tan x = \frac{1}{1 - \sin x}$$
(1 mark)

(iii) $\int_0^{\frac{\pi}{4}} \frac{1}{1 - \sin x}\, dx$
$$= \int_0^{\frac{\pi}{4}} \sec^2 x + \sec x \tan x\, dx$$
$$= [\tan x + \sec x]_0^{\frac{\pi}{4}} \; ***$$
$$= \left[\tan\frac{\pi}{4} + \sec\frac{\pi}{4} - (\tan 0 + \sec 0)\right]$$
$$= 1 + \sqrt{2} - (0 + 1)$$
$$= \sqrt{2}$$

[*** from table of standard integrals:
$\int \sec ax \tan ax dx = \sec ax + c$,
where $a = 1$] *(2 marks)*

(c)
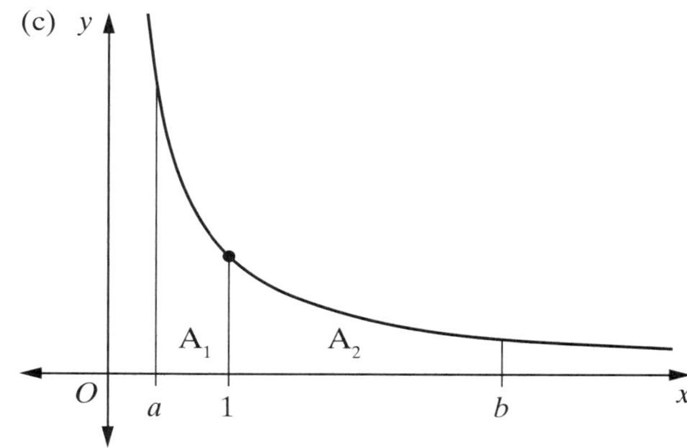

$$A_1 = \int_a^1 \frac{1}{x}\, dx = 1$$
$$\therefore [\log_e x]_a^1 = 1$$
$$\log_e 1 - \log_e a = 1$$
$$0 - \log_e a = 1$$
$$\log_e a = -1$$
$$e^{\log a} = e^{-1}$$
$$a = \frac{1}{e}$$

$$A_2 = \int_1^b \frac{1}{x}\,dx = 1$$

$$\therefore [\log_e x]_1^b = 1$$

$$\log_e b - \log_e 1 = 1$$

$$\log_e b - 0 = 1$$

$$\log_e b = 1$$

$$e^{\log b} = e^1$$

$$b = e$$

$$\therefore a = \frac{1}{e} \text{ and } b = e$$ *(3 marks)*

QUESTION 6

(a) $f(x) = (x + 2)(x^2 + 4)$

(i) Let $y = (x + 2)(x^2 + 4)$

$$= x^3 + 4x + 2x^2 + 8$$

$$= x^3 + 2x^2 + 4x + 8$$

$$\frac{dy}{dx} = 3x^2 + 4x + 4$$

For $3x^2 + 4x + 4$, check discriminant:

$$\Delta = b^2 - 4ac$$

$$= 4^2 - 4(3)(4)$$

$$= -32 < 0$$

$\therefore$ No real roots for $\frac{dy}{dx} = 0$

$\therefore$ No stationary points. *(2 marks)*

(ii) $\frac{dy}{dx} = 3x^2 + 4x + 4$

$$\frac{d^2y}{dx^2} = 6x + 4$$

Now, concave down when $\frac{d^2y}{dx^2} < 0$,

$$6x + 4 < 0$$

$$6x < -4$$

$$x < -\frac{2}{3}$$

$\therefore$ Concave down when $x < -\frac{2}{3}$.

Similarly, concave up when

$x > -\frac{2}{3}$. *(2 marks)*

(iii) $y = (x + 2)(x^2 + 4)$

x-intercept: let $y = 0$, then $x = -2$

y-intercept: let $x = 0$, then $y = 8$

Also, at $x = \frac{-2}{3}$,

$$\text{then } y = \left(\frac{-2}{3} + 2\right)\left(\left(\frac{-2}{3}\right)^2 + 4\right)$$

$$= 5.925925 \ldots$$

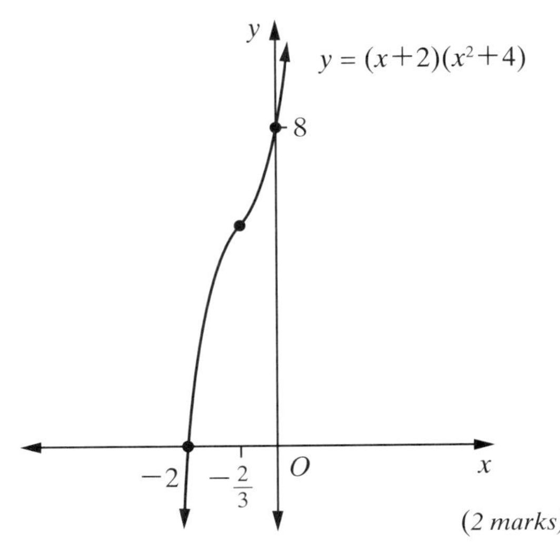

(2 marks)

(b) (i) Using $l = 9, r = 5$,

$$l = r\theta$$

$$9 = 5\theta$$

$$\theta = \frac{9}{5}$$

$\therefore \angle POQ$ is $\frac{9}{5}$ radians. *(1 mark)*

(ii)

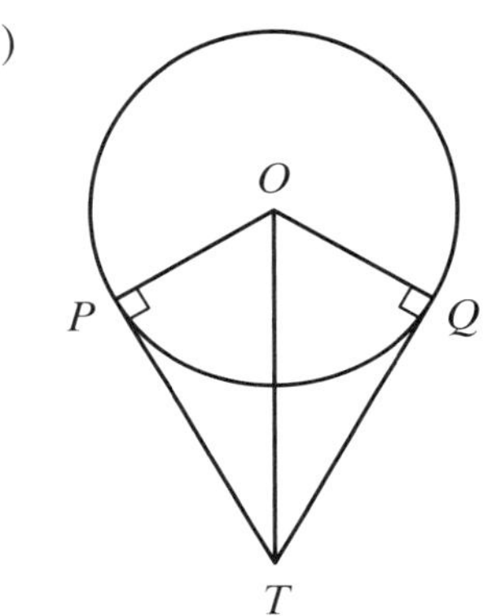

In $\triangle OPT$ and $\triangle OQT$,

OT is common

$\angle OPT = \angle OQT$ (given)

$OP = OQ$ (equal radii)

$\therefore \triangle OPT \equiv \triangle OQT$ (RHS test) *(2 marks)*

(iii) From (ii) $\angle POT = \angle QOT$ (matching $\angle$s of cong $\triangle$s)

Also, $\angle POQ = \frac{9}{5}$ radians (from (i)),

$$\text{then } \angle POT = \frac{1}{2} \times \frac{9}{5} \text{ radians}$$

$$= \frac{9}{10} \text{ radians}$$

$$= \frac{9}{10} \times \frac{180}{\pi} \text{ degrees}$$

$$= 51^\circ 34'$$

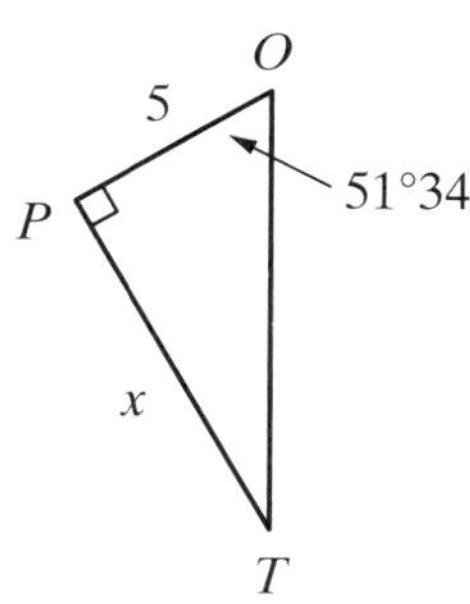

$$\frac{x}{5} = \tan 51°\,34'$$
$$x = 5 \times \tan 51°\,34'$$
$$= 6.300\ldots$$
$$= 6.3 \text{ (corr. to 1 dec. pl.)}$$

$\therefore$ length of PT is 6.3 cm *(1 mark)*

(iv) Area $OPTQ = 2 \times$ Area of $\triangle OPT$

$$= 2 \times \frac{1}{2} \times 5 \times 6.300..$$
$$= 31.503\ldots$$
$$= 31.5 \text{ (corr. to 1 dec. pl.)}$$

$\therefore$ The area of $OPTQ$ is 31.5 cm^2.

Shaded area = Area $OPTQ$ − Area sector

$$= 31.503\ldots - \frac{1}{2} \times 5^2 \times \frac{9}{5}$$
$$= 9.003\ldots$$
$$= 9 \text{ (nearest whole)}$$

$\therefore$ The shaded area is 9 cm^2. *(2 marks)*

QUESTION 7

(a) $\ddot{x} = 4\cos 2t$

(i) $\dot{x} = \int 4\cos 2t \, dt$

$\therefore \dot{x} = 2\sin 2t + c$

Subs $t = 0, v = \dot{x} = 1$:

$1 = 2\sin 0 + c$

$\therefore c = 1$

$\therefore \dot{x} = 2\sin 2t + 1$ *(2 marks)*

(ii) Subs $v = \dot{x} = 0$:

$0 = 2\sin 2t + 1$

$2\sin 2t = -1$

$\sin 2t = -\frac{1}{2}$

$2t = \frac{7\pi}{6}$

$t = \frac{7\pi}{12}$

$\therefore$ The particle comes to rest after $\frac{7\pi}{12}$ seconds. *(2 marks)*

(iii) $x = \int 2\sin 2t + 1 \, dt$

$\therefore x = -\cos 2t + t + c$

Subs $t = 0, x = 0$:

$0 = -\cos 0 + 0 + c$

$0 = -1 + c$

$c = 1$

$\therefore x = -\cos 2t + t + 1$

$\therefore$ The displacement is $x = 1 - \cos 2t + t$. *(2 marks)*

(b)

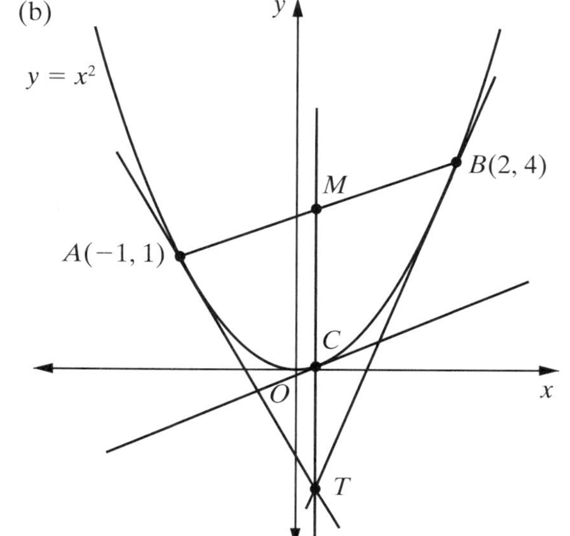

(i) $y = x^2$

$\frac{dy}{dx} = 2x$

At $x = -1$, $\frac{dy}{dx} = -2$

using $(-1, 1)$ and $m = -2$,

$y - y_1 = m(x - x_1)$

$y - 1 = -2(x - (-1))$

$y - 1 = -2(x + 1)$

$y - 1 = -2x - 2$

$2x + y + 1 = 0$

$\therefore$ The equation of the tangent is $2x + y + 1 = 0$. *(2 marks)*

(ii) Using $A(-1, 1)$ and $B(2, 4)$,

$$MP = \left(\frac{x_1 + x_2}{2}, \frac{y_1 + y_2}{2}\right)$$
$$= \left(\frac{-1 + 2}{2}, \frac{1 + 4}{2}\right)$$
$$= \left(\frac{1}{2}, \frac{5}{2}\right)$$
$$= (0.5, 2.5)$$

$\therefore$ The co-ordinates of M are $(0.5, 2.5)$.

Now, using $A(-1, 1)$ and $B(2, 4)$,

$$\begin{aligned} m &= \frac{y_2 - y_1}{x_2 - x_1} \\ &= \frac{4-1}{2+1} \\ &= \frac{3}{3} \\ &= 1 \end{aligned}$$

$\therefore$ The gradient of the tangent at C is 1.

Let $\frac{dy}{dx} = 2x = 1$

$\therefore x = \frac{1}{2}$

$\therefore$ As M and C both have an x-value of $\frac{1}{2}$, then MC is vertical.

(2 marks)

(iii) Tangent at A is $2x + y + 1 = 0$:

At T, subs $x = \frac{1}{2}$:

$$\begin{aligned} 2 \times \frac{1}{2} + y + 1 &= 0 \\ 1 + y + 1 &= 0 \\ y &= -2 \end{aligned}$$

$\therefore T(\frac{1}{2}, -2)$

Now, using $T(\frac{1}{2}, -2)$ and $B(2, 4)$,

$$\begin{aligned} m &= \frac{y_2 - y_1}{x_2 - x_1} \\ &= \frac{4+2}{2-\frac{1}{2}} \\ &= \frac{6}{1\frac{1}{2}} \\ &= 4 \end{aligned}$$

$\therefore$ The gradient of BT is 4.

But, the gradient of the tangent at B:

$$y = x^2$$

$$\frac{dy}{dx} = 2x$$

At $x = 2$, $\frac{dy}{dx} = 4$

As both have a gradient of 4, then BT is a tangent to the parabola. *(2 marks)*

QUESTION 8

(a) $\frac{dP}{dt} = kP$

Let $P = P_0 e^{kt}$,

where P_0 = initial population

When $t = 0$, $P_0 = 102$

$\therefore P = 102e^{kt}$,

From 1935 to 2010 is 75 years.

Let $P = 200\,000\,000$, $t = 75$: $P = 102e^{kt}$

$$\begin{aligned} 200\,000\,000 &= 102e^{75k} \\ e^{75k} &= \frac{200\,000\,000}{102} \\ 75k &= \log_e\left[\frac{200\,000\,000}{102}\right] \\ k &= \frac{\log_e\left[\frac{200\,000\,000}{102}\right]}{75} \\ &= 0.193184734\ldots \end{aligned}$$

From 2010 to 2035 is another 25 years:

$\therefore$ Let $t = 100$:

$$\begin{aligned} P &= 102e^{100k} \\ &= 2.503\ldots \times 10^{10} \end{aligned}$$

$\therefore$ The population will be approximately 2.5×10^{10} cane toads. *(4 marks)*

(b) P(both heads) = 0.36

$$\begin{aligned} \therefore \text{P(head)} &= \sqrt{0.36} \\ &= 0.6 \end{aligned}$$

$$\begin{aligned} \therefore \text{P(tail)} &= 1 - \text{P(head)} \\ &= 1 - 0.6 \\ &= 0.4 \end{aligned}$$

$$\begin{aligned} \therefore \text{P(both tails)} &= 0.4 \times 0.4 \\ &= 0.16 \end{aligned}$$

(2 marks)

(c) (i) From the graph,

A = amplitude = 4 *(1 mark)*

(ii) From the graph $\left(\frac{\pi}{4}, 4\right)$ lies on the curve.

Subs into $y = 4 \sin bx$:

$$\begin{aligned} 4 &= 4\sin\frac{\pi.b}{4} \\ \sin\frac{\pi b}{4} &= 1 \\ \frac{\pi b}{4} &= \frac{\pi}{2} \\ b &= \frac{\pi}{2} \div \frac{\pi}{4} \\ &= 2 \end{aligned}$$

(1 mark)

(iii)

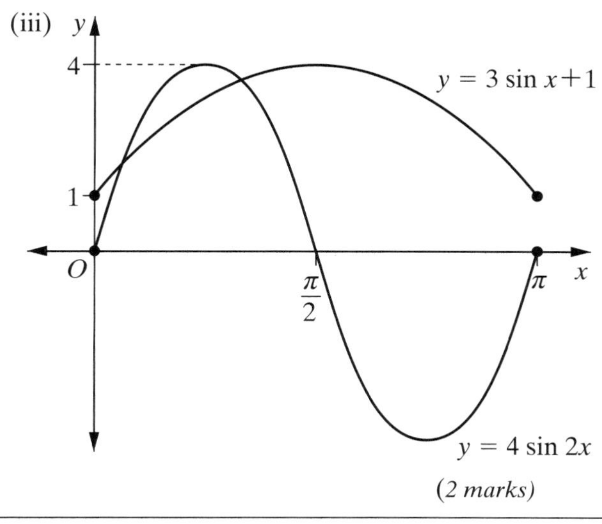

(2 marks)

(d) $f(x) = x^3 - 3x^2 + kx + 8$

$f'(x) = 3x^2 - 6x + k$

If $f(x)$ is increasing function, then $f'(x) > 0$:

Hence, find value of k where $3x^2 - 6x + k$ is positive definite.

$$\triangle = b^2 - 4ac < 0 \text{ and } a > 0$$
$$= (-6)^2 - 4(3)(k) < 0$$
$$= 36 - 12k < 0$$

$\therefore 12k > 36$

$k > 3$ [also, $a = 3 > 0$]

$\therefore$ When $k > 3$, $f(x)$ is an increasing function. *(2 marks)*

QUESTION 9

(a) (i) $P = 500 \times 1.005^{240} + 500 \times 1.005^{239} + 500 \times 1.005^{238} + \ldots + 500 \times 1.005$

$= 500[1.005 + 1.005^2 + \ldots + 1.005^{240}]$

Using $a = 1.005, r = 1.005, n = 240$,

$$S_n = \frac{a(r^n - 1)}{r - 1}$$

$$\therefore P = 500 \times \frac{1.005(1.005^{240} - 1)}{1.005 - 1}$$
$$= 232\,175.5498\ldots$$
$$= 232\,175.55 \text{ (to 2 decimal places)}$$

(2 marks)

(ii) (1) $A_1 = P \times 1.005 - 2000$

$A_2 = A_1 \times 1.005 - 2000$

$= (P \times 1.005 - 2000) \times 1.005 - 2000$

$= P \times 1.005^2 - 2000(1 + 1.005)$

Hence,

$A_n = P \times 1.005^n - 2000(1 + 1.005 + \ldots + 1.005^{n-1})$

Using $a = 1, r = 1.005, n = n$,

$$S_n = \frac{a(r^n - 1)}{r - 1}$$

$$A_n = P \times 1.005^n - 2000 \times \frac{1(1.005^n - 1)}{1.005 - 1}$$
$$= P \times 1.005^n - 400\,000 \times [1.005^n - 1]$$
$$= (P - 400\,000) \times 1.005^n + 400\,000$$

(3 marks)

(2) Let $A_n = 0$

$(P - 400000) \times 1.005^n + 400000 = 0$

$(P - 400000) \times 1.005^n = -400000$

$$1.005^n = \frac{-400000}{232175.55 - 400000}, \text{ using } P = 232175.55$$
$$= 2.383442937\ldots.$$
$$\log_e(1.005)^n = \log_e 2.383\ldots.$$
$$n = \frac{\log_e 2.383\ldots}{\log_e 1.005}$$
$$= 174.14312\ldots$$

$\therefore$ There will be money left for 175 months. *(2 marks)*

(b) (i) $f(x)$ is increasing when $f'(x) > 0$.
From the graph, $0 < x < 2$. *(1 mark)*

(ii) Maximum value of $f(x)$
when $f'(x) = 0$.
$\therefore$ when $x = 2$.

$$\text{Now, } \int_0^2 f'(x)\,dx = 4$$
$$[f(x)]_0^2 = 4$$
$$f(2) - f(0) = 4$$
But $f(0) = 0$, (given)
$$\therefore f(2) - 0 = 4$$
$$f(2) = 4$$
$\therefore$ The maximum value of $f(x)$ is 4. *(1 mark)*

(iii) Method 1:
As $f(0) = 0$ and $A_1 = A_2$, then $f(4) = 0$.
From $x = 4$ to $x = 6$, the gradient is -3.
As $m = -3$, then $f(6) = -6$.

Method 2:
$$\text{Now, } \int_2^4 f'(x)\,dx = -4$$
$$[f(x)]_2^4 = -4$$
$$f(4) - f(2) = -4$$
But $f(2) = 4$, (from ii)
$$\therefore f(4) - 4 = -4$$
$$f(4) = 0$$
From $x = 4$ to $x = 6$, the gradient is -3.
As $m = -3$, then $f(6) = -6$. *(1 mark)*

(iv)
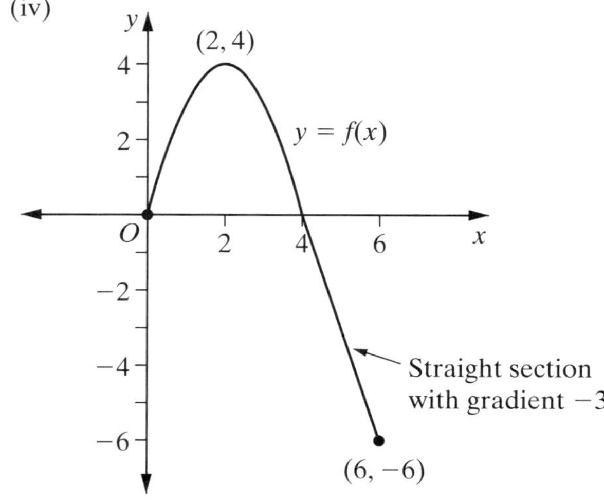

(2 marks)

QUESTION 10

(a)
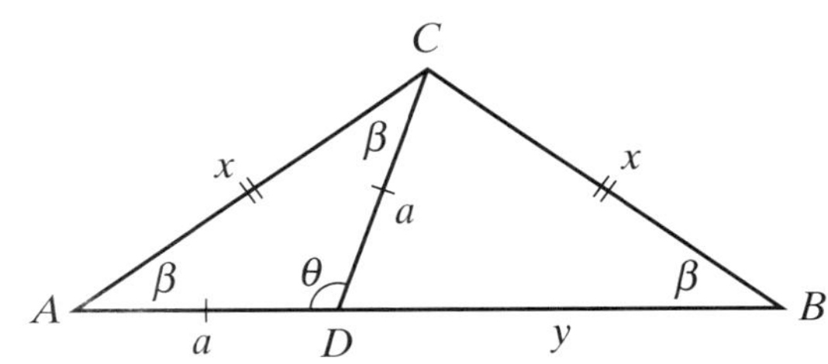

(i) In $\triangle ABC$,
Let $\angle CAB = \beta$
$\therefore \angle CBA = \beta$ (base $\angle$s of isosceles $\triangle$s)
$\therefore \angle CAB = \angle CBA = \beta$

Also, in $\triangle ACD$:
$\angle ACD = \beta$ (base $\angle$s of isosceles $\triangle$s)
$\therefore \angle CAD = \angle ACD = \beta$
$\therefore \triangle ABC$ and $\triangle ACD$ are similar (2 $\angle$s equal). *(2 marks)*

(ii) $\dfrac{AB}{AC} = \dfrac{AC}{AD}$ (matching sides of similar $\triangle$s in proportion)
$$\frac{a+y}{x} = \frac{x}{a}$$
$$x^2 = a(a+y) \;\ldots\ldots\ldots\ldots (1)$$
$$\therefore x^2 = a^2 + ay$$ *(1 mark)*

(iii) In $\triangle ACD$, use cosine rule:
$$x^2 = a^2 + a^2 - 2 \times a \times a \times \cos\theta$$
$$\therefore x^2 = 2a^2 - 2a^2\cos\theta$$
$$x^2 = 2a^2(1 - \cos\theta) \;\ldots\ldots\ldots (2)$$
Let (1) = (2)
$$a(a+y) = 2a^2(1 - \cos\theta)$$
$$a^2 + ay = 2a^2 - 2a^2\cos\theta$$
$$ay = a^2 - 2a^2\cos\theta$$
$$y = a - 2a\cos\theta$$
$$\therefore y = a(1 - 2\cos\theta)$$ *(2 marks)*

(iv) Method 1:
$$\text{As } \cos\theta \geq -1$$
$$2\cos\theta \geq -2$$
$$-2\cos\theta \leq 2$$
$$1 - 2\cos\theta \leq 3$$
$$a(1 - 2\cos\theta) \leq 3a$$
But, $y = a(1 - 2\cos\theta)$ from (iii)
$$\therefore y \leq 3\text{a}$$

Method 2:

From $\triangle ACD$, $x \le a + a$

$\therefore x \le 2a$

$\therefore 2x \le 4a$ .... (3)

From $\triangle ABC$, $a + y \le x + x$

$a + y \le 2x$ .... (4)

From (3) and (4):

$\therefore a + y \le 4a$

$y \le 4a - a$

$y \le 3a$ *(1 mark)*

(b) (i)

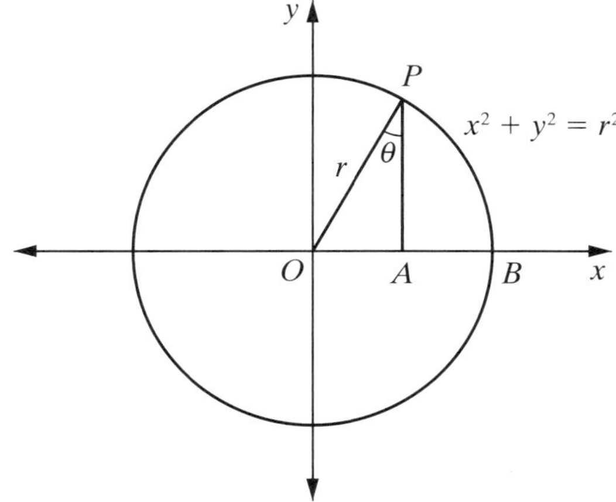

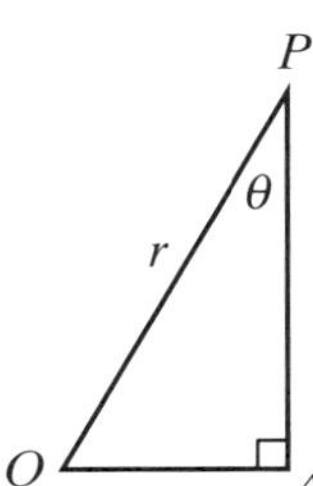

Using $\triangle PAO$, $\dfrac{OA}{r} = \sin\theta$

$OA = r\sin\theta$

Also, as $x^2 + y^2 = r^2$

$\therefore y^2 = r^2 - x^2$

Now $V = \pi \displaystyle\int_{r\sin\theta}^{r} r^2 - x^2 dx$

$$= \pi\left[r^2x - \frac{x^3}{3}\right]_{r\sin\theta}^{r}$$

$$= \pi\left[r^3 - \frac{r^3}{3} - \left(r^3\sin\theta - \frac{r^3\sin^3\theta}{3}\right)\right]$$

$$= \pi\left[\frac{2r^3}{3} - r^3\sin\theta - \frac{r^3\sin^3\theta}{3}\right]$$

$$= \frac{\pi r^3}{3}[2 - 3\sin\theta + \sin^3\theta]$$

(3 marks)

(ii) (1)

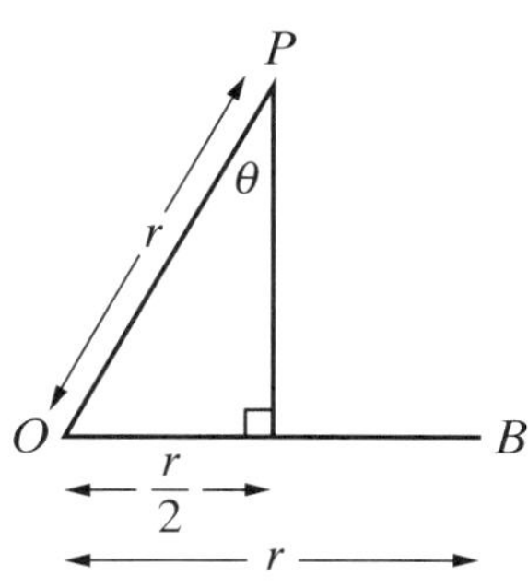

As $OB = r$, then $OA = \dfrac{r}{2}$ (half-depth)

$$\therefore \sin\theta = \frac{\frac{r}{2}}{r}$$

$$= \frac{r}{2} \div r$$

$$= \frac{1}{2}$$

$\therefore \theta = 30°$ *(1 mark)*

(2) Volume of hemisphere $= \dfrac{1}{2} \times \dfrac{4}{3}\pi r^3$

$$= \frac{2}{3}\pi r^3$$

$\therefore$ Volume of original hemisphere is $\dfrac{2}{3}\pi r^3$ units³.

When water is one half of the original depth:

subs $\theta = 30°$ in V:

$$V = \frac{\pi r^3}{3}[2 - 3\sin 30° + \sin^3 30°]$$

$$= \frac{\pi r^3}{3}\left[2 - \frac{3}{2} + \frac{1}{8}\right]$$

$$= \frac{5\pi r^3}{24}$$

$\therefore$ New volume is $\dfrac{5\pi r^3}{24}$ units³.

$$\therefore \text{Fraction} = \frac{5\pi r^3}{24} \div \frac{2}{3}\pi r^3$$

$$= \frac{5\pi r^3}{24} \times \frac{3}{2\pi r^3}$$

$$= \frac{5}{16}$$

(2 marks)

BOARD OF STUDIES
NEW SOUTH WALES

2011
HIGHER SCHOOL CERTIFICATE EXAMINATION

Mathematics

General Instructions

- Reading time – 5 minutes
- Working time – 3 hours
- Write using black or blue pen
  Black pen is preferred
- Board-approved calculators may be used
- A table of standard integrals is provided at the back of this paper
- All necessary working should be shown in every question

Total marks – 120

- Attempt Questions 1–10
- All questions are of equal value

Total marks – 120
Attempt Questions 1–10
All questions are of equal value

Answer each question in the appropriate writing booklet. Extra writing booklets are available.

Question 1 (12 marks) Use the Question 1 Writing Booklet.

(a) Evaluate $\sqrt[3]{\dfrac{651}{4\pi}}$ correct to four significant figures. **2**

(b) Simplify $\dfrac{n^2 - 25}{n - 5}$. **1**

(c) Solve $2^{2x+1} = 32$. **2**

(d) Differentiate $\ln(5x + 2)$ with respect to x. **2**

(e) Solve $2 - 3x \leq 8$. **2**

(f) Rationalise the denominator of $\dfrac{4}{\sqrt{5} - \sqrt{3}}$. **2**

Give your answer in the simplest form.

(g) A batch of 800 items is examined. The probability that an item from this batch is defective is 0.02. **1**

How many items from this batch are defective?

Question 2 (12 marks) Use the Question 2 Writing Booklet.

(a) The quadratic equation $x^2 - 6x + 2 = 0$ has roots α and β.

(i) Find $\alpha + \beta$. **1**

(ii) Find $\alpha\beta$. **1**

(iii) Find $\dfrac{1}{\alpha} + \dfrac{1}{\beta}$. **1**

(b) Find the exact values of x such that $2\sin x = -\sqrt{3}$, where $0 \le x \le 2\pi$. **2**

(c) Find the equation of the tangent to the curve $y = (2x+1)^4$ at the point where $x = -1$. **3**

(d) Find the derivative of $y = x^2 e^x$ with respect to x. **2**

(e) Find $\displaystyle\int \frac{1}{3x^2}\, dx$. **2**

Question 3 (12 marks) Use the Question 3 Writing Booklet.

(a) A skyscraper of 110 floors is to be built. The first floor to be built will cost \$3 million. The cost of building each subsequent floor will be \$0.5 million more than the floor immediately below.

(i) What will be the cost of building the 25th floor? **2**

(ii) What will be the cost of building all 110 floors of the skyscraper? **2**

(b) A parabola has focus $(3, 2)$ and directrix $y = -4$. Find the coordinates of the vertex. **2**

(c) The diagram shows a line ℓ_1, with equation $3x + 4y - 12 = 0$, which intersects the y-axis at B.

A second line ℓ_2, with equation $4x - 3y = 0$, passes through the origin O and intersects ℓ_1 at E.

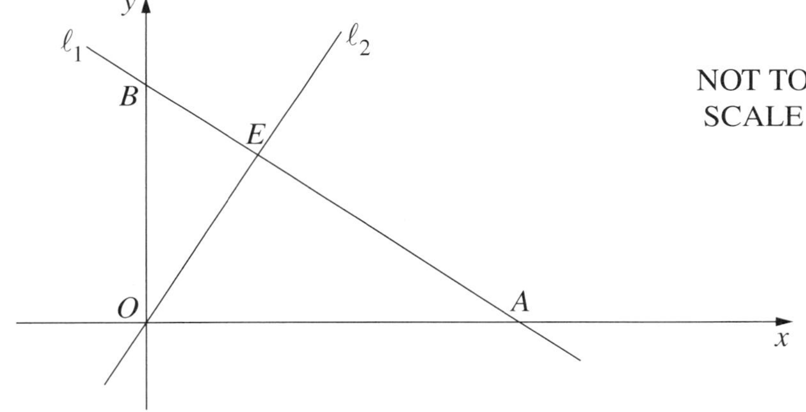

(i) Show that the coordinates of B are $(0, 3)$. **1**

(ii) Show that ℓ_1 is perpendicular to ℓ_2. **2**

(iii) Show that the perpendicular distance from O to ℓ_1 is $\dfrac{12}{5}$. **1**

(iv) Using Pythagoras' theorem, or otherwise, find the length of the interval BE. **1**

(v) Hence, or otherwise, find the area of $\triangle BOE$. **1**

Question 4 (12 marks) Use the Question 4 Writing Booklet.

(a) Differentiate $\dfrac{x}{\sin x}$ with respect to x. **2**

(b) Evaluate $\displaystyle\int_e^{e^3} \frac{5}{x}\,dx$. **2**

(c) The gradient of a curve is given by $\dfrac{dy}{dx} = 6x - 2$. The curve passes through the point $(-1, 4)$. **2**

What is the equation of the curve?

(d) (i) Differentiate $y = \sqrt{9 - x^2}$ with respect to x. **2**

(ii) Hence, or otherwise, find $\displaystyle\int \frac{6x}{\sqrt{9 - x^2}}\,dx$. **2**

(e) The diagram shows the graphs $y = |x| - 2$ and $y = 4 - x^2$. **2**

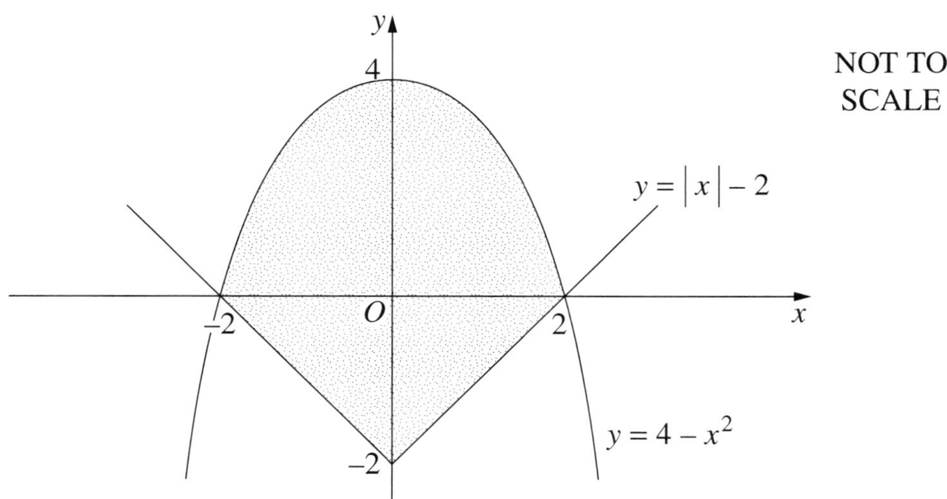

Write down inequalities that together describe the shaded region.

Question 5 (12 marks) Use the Question 5 Writing Booklet.

(a) The number of members of a new social networking site doubles every day. On Day 1 there were 27 members and on Day 2 there were 54 members.

(i) How many members were there on Day 12? **1**

(ii) On which day was the number of members first greater than 10 million? **2**

(iii) The site earns 0.5 cents per member per day. How much money did the site earn in the first 12 days? Give your answer to the nearest dollar. **2**

(b) Kim has three red shirts and two yellow shirts. On each of the three days, Monday, Tuesday and Wednesday, she selects one shirt at random to wear. Kim wears each shirt that she selects only once.

(i) What is the probability that Kim wears a red shirt on Monday? **1**

(ii) What is the probability that Kim wears a shirt of the same colour on all three days? **1**

(iii) What is the probability that Kim does not wear a shirt of the same colour on consecutive days? **2**

(c) The table gives the speed v of a jogger at time t in minutes over a 20-minute period. The speed v is measured in metres per minute, in intervals of 5 minutes. **3**

t	0	5	10	15	20
v	173	81	127	195	168

The distance covered by the jogger over the 20-minute period is given by

$$\int_0^{20} v\,dt.$$

Use Simpson's rule and the speed at each of the five time values to find the approximate distance the jogger covers in the 20-minute period.

Question 6 (12 marks) Use the Question 6 Writing Booklet.

(a) The diagram shows a regular pentagon $ABCDE$. Sides ED and BC are produced to meet at P.

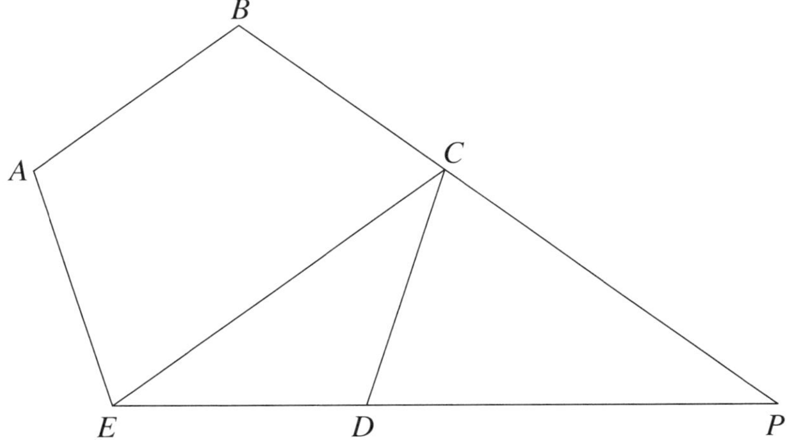

Copy or trace the diagram into your writing booklet.

(i) Find the size of $\angle CDE$. **1**

(ii) Hence, show that $\triangle EPC$ is isosceles. **2**

(b) A point $P(x, y)$ moves so that the sum of the squares of its distance from each of the points $A(-1, 0)$ and $B(3, 0)$ is equal to 40. **3**

Show that the locus of $P(x, y)$ is a circle, and state its radius and centre.

Question 6 continues on following page

Question 6 (continued)

(c) The diagram shows the graph $y = 2\cos x$.

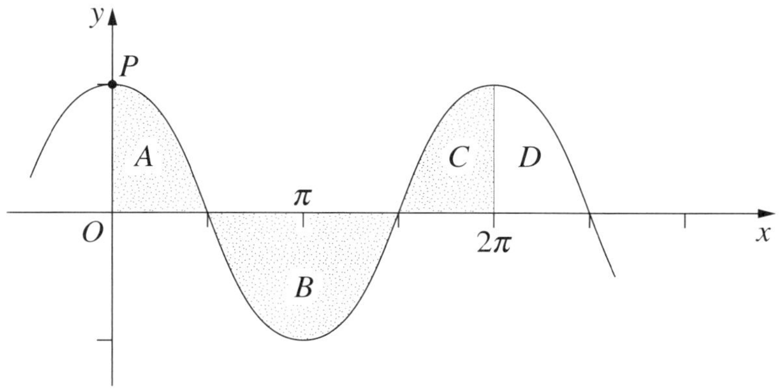

(i) State the coordinates of P. **1**

(ii) Evaluate the integral $\int_0^{\frac{\pi}{2}} 2\cos x \, dx$. **2**

(iii) Indicate which area in the diagram, A, B, C or D, is represented by the integral $\int_{\frac{3\pi}{2}}^{2\pi} 2\cos x \, dx$. **1**

(iv) Using parts (ii) and (iii), or otherwise, find the area of the region bounded by the curve $y = 2\cos x$ and the x-axis, between $x = 0$ and $x = 2\pi$. **1**

(v) Using the parts above, write down the value of $\int_{\frac{\pi}{2}}^{2\pi} 2\cos x \, dx$. **1**

End of Question 6

Question 7 (12 marks) Use the Question 7 Writing Booklet.

(a) Let $f(x) = x^3 - 3x + 2$.

(i) Find the coordinates of the stationary points of $y = f(x)$, and determine their nature. **3**

(ii) Hence, sketch the graph $y = f(x)$ showing all stationary points and the y-intercept. **2**

(b) The velocity of a particle moving along the x-axis is given by

$$\dot{x} = 8 - 8e^{-2t},$$

where t is the time in seconds and x is the displacement in metres.

(i) Show that the particle is initially at rest. **1**

(ii) Show that the acceleration of the particle is always positive. **1**

(iii) Explain why the particle is moving in the positive direction for all $t > 0$. **2**

(iv) As $t \to \infty$, the velocity of the particle approaches a constant. **1**

Find the value of this constant.

(v) Sketch the graph of the particle's velocity as a function of time. **2**

Question 8 (12 marks) Use the Question 8 Writing Booklet.

(a) In the diagram, the shop at S is 20 kilometres across the bay from the post office at P. The distance from the shop to the lighthouse at L is 22 kilometres and $\angle SPL$ is $60°$.

Let the distance PL be x kilometres.

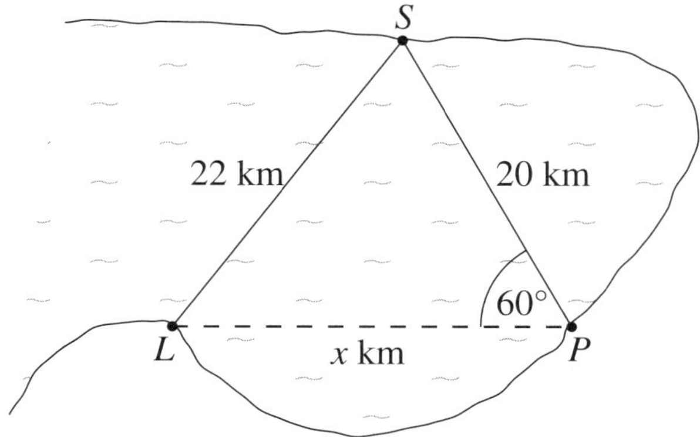

(i) Use the cosine rule to show that $x^2 - 20x - 84 = 0$. **1**

(ii) Hence, find the distance from the post office to the lighthouse. Give your answer correct to the nearest kilometre. **2**

(b) The diagram shows the region enclosed by the parabola $y = x^2$, the y-axis and the line $y = h$, where $h > 0$. This region is rotated about the y-axis to form a solid called a paraboloid. The point C is the intersection of $y = x^2$ and $y = h$. The point H has coordinates $(0, h)$.

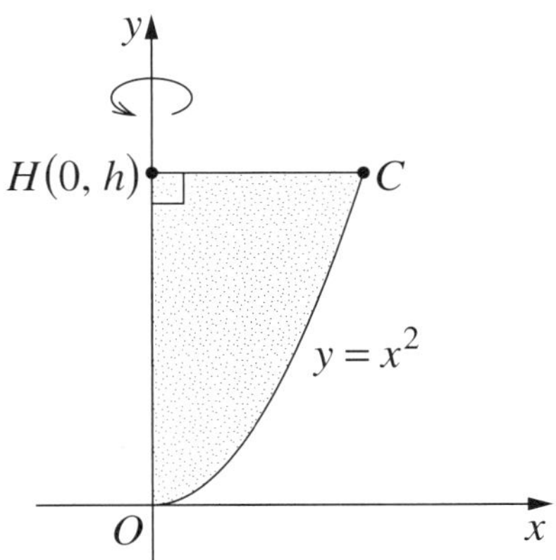

(i) Find the exact volume of the paraboloid in terms of h. **2**

(ii) A cylinder has radius HC and height h. **1**

What is the ratio of the volume of the paraboloid to the volume of the cylinder?

Question 8 (continued)

(c) When Jules started working she began paying \$100 at the beginning of each month into a superannuation fund.

The contributions are compounded monthly at an interest rate of 6% per annum.

She intends to retire after having worked for 35 years.

(i) Let $\$P$ be the final value of Jules's superannuation when she retires after 35 years (420 months). **2**

Show that $\$P = \$143\,183$ to the nearest dollar.

(ii) Fifteen years after she started working Jules read a magazine article about retirement, and realised that she would need \$800 000 in her fund when she retires. At the time of reading the magazine article she had \$29 227 in her fund. For the remaining 20 years she intends to work, she decides to pay a total of $\$M$ into her fund at the beginning of each month. The contributions continue to attract the same interest rate of 6% per annum, compounded monthly.

At the end of n months after starting the new contributions, the amount in the fund is $\$A_n$.

(1) Show that $A_2 = 29\,227 \times 1.005^2 + M(1.005 + 1.005^2)$. **1**

(2) Find the value of M so that Jules will have \$800 000 in her fund after the remaining 20 years (240 months). **3**

End of Question 8

Question 9 (12 marks) Use the Question 9 Writing Booklet.

(a) The diagram shows $\triangle ADE$, where B is the midpoint of AD and C is the midpoint of AE. The intervals BE and CD meet at F.

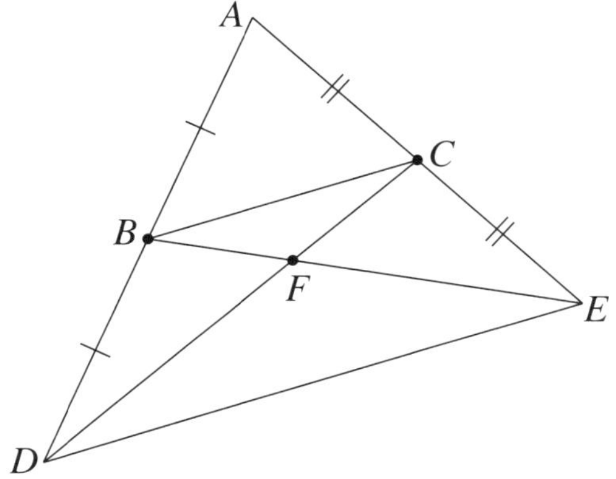

(i) Explain why $\triangle ABC$ is similar to $\triangle ADE$. **1**

(ii) Hence, or otherwise, prove that the ratio $BF : FE = 1 : 2$. **2**

(b) A tap releases liquid A into a tank at the rate of $\left(2 + \dfrac{t^2}{t+1}\right)$ litres per minute, where t is time in minutes. A second tap releases liquid B into the same tank at the rate of $\left(1 + \dfrac{1}{t+1}\right)$ litres per minute. The taps are opened at the same time and release the liquids into an empty tank.

(i) Show that the rate of flow of liquid A is greater than the rate of flow of liquid B by t litres per minute. **1**

(ii) The taps are closed after 4 minutes. By how many litres is the volume of liquid A greater than the volume of liquid B in the tank when the taps are closed? **2**

Question 9 continues on following page

Question 9 (continued)

(c) The graph $y = f(x)$ in the diagram has a stationary point when $x = 1$, a point of inflexion when $x = 3$, and a horizontal asymptote $y = -2$. **3**

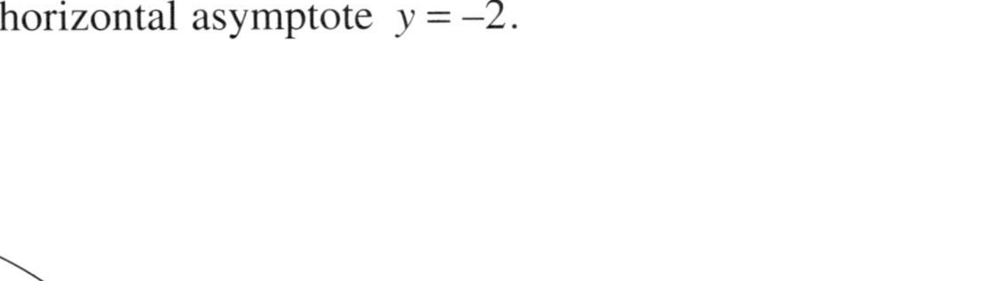
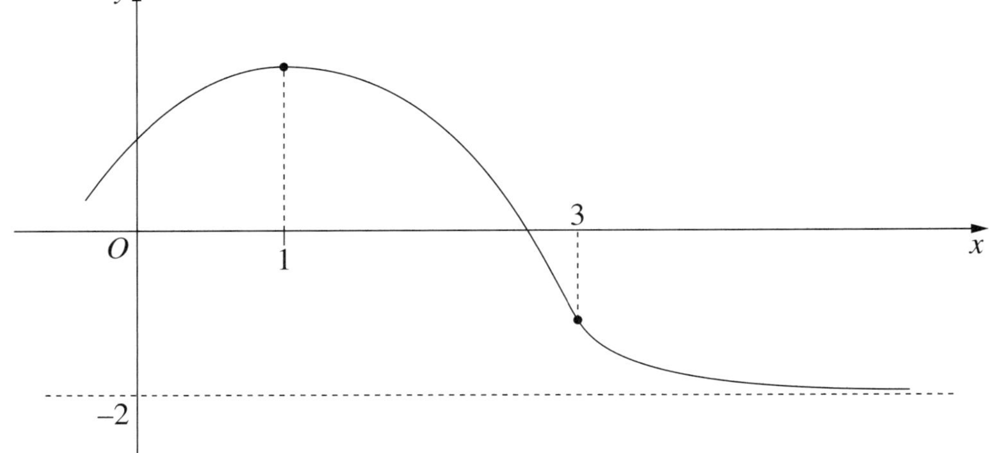

Sketch the graph $y = f'(x)$, clearly indicating its features at $x = 1$ and at $x = 3$, and the shape of the graph as $x \to \infty$.

(d) (i) Rationalise the denominator in the expression **1**

$$\frac{1}{\sqrt{n} + \sqrt{n+1}},$$

where n is an integer and $n \geq 1$.

(ii) Using your result from part (i), or otherwise, find the value of the sum **2**

$$\frac{1}{\sqrt{1} + \sqrt{2}} + \frac{1}{\sqrt{2} + \sqrt{3}} + \frac{1}{\sqrt{3} + \sqrt{4}} + \cdots + \frac{1}{\sqrt{99} + \sqrt{100}}.$$

End of Question 9

Question 10 (12 marks) Use the Question 10 Writing Booklet.

(a) The intensity I, measured in watt/m^2, of a sound is given by

$$I = 10^{-12} \times e^{0.1L},$$

where L is the loudness of the sound in decibels.

(i) If the loudness of a sound at a concert is 110 decibels, find the intensity of the sound. Give your answer in scientific notation. **1**

(ii) Ear damage occurs if the intensity of a sound is greater than 8.1×10^{-9} watt/m^2. **2**

What is the maximum loudness of a sound so that no ear damage occurs?

(iii) By how much will the loudness of a sound have increased if its intensity has doubled? **2**

Question 10 continues on following page

Question 10 (continued)

(b) A farmer is fencing a paddock using P metres of fencing. The paddock is to be in the shape of a sector of a circle with radius r and sector angle θ in radians, as shown in the diagram.

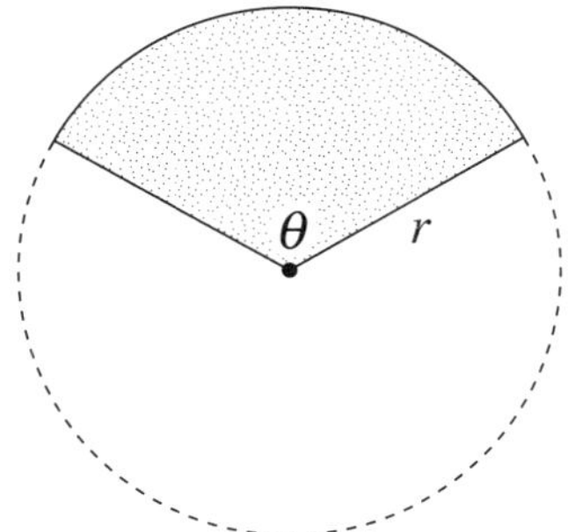

(i) Show that the length of fencing required to fence the perimeter of the paddock is $P = r(\theta + 2)$. **1**

(ii) Show that the area of the sector is $A = \frac{1}{2}Pr - r^2$. **1**

(iii) Find the radius of the sector, in terms of P, that will maximise the area of the paddock. **2**

(iv) Find the angle θ that gives the maximum area of the paddock. **1**

(v) Explain why it is only possible to construct a paddock in the shape of a sector if $\frac{P}{2(\pi + 1)} < r < \frac{P}{2}$. **2**

End of paper

2011 Higher School Certificate

Worked Answers

QUESTION 1

(a) $\sqrt[3]{\dfrac{651}{4\pi}} = \sqrt[3]{(651 \div (4 \times \pi))}$

$= 3.727\,838\,09 \ldots$

$= 3.728$ (to 4 sig. figs) *(2 marks)*

(b) $\dfrac{n^2 - 25}{n - 5} = \dfrac{(n-5)(n+5)}{n-5}$

$= n + 5$ *(1 mark)*

(c) $2^{2x+1} = 32$

$2^{2x+1} = 2^5$

$\therefore\ 2x + 1 = 5$

$2x = 4$

$x = 2$ *(2 marks)*

(d) $\dfrac{d}{dx}[\ln(5x+2)] = \dfrac{5}{5x+2}$ *(2 marks)*

(e) $2 - 3x \leq 8$

$-3x \leq 8 - 2$

$\dfrac{-3x}{-3} \geq \dfrac{6}{-3}$

$x \geq -2$ *(2 marks)*

(f) $\dfrac{4}{\sqrt{5} - \sqrt{3}} = \dfrac{4}{\sqrt{5} - \sqrt{3}} \times \dfrac{\sqrt{5} + \sqrt{3}}{\sqrt{5} + \sqrt{3}}$

$= \dfrac{4(\sqrt{5} + \sqrt{3})}{5 - 3}$

$= \dfrac{4(\sqrt{5} + \sqrt{3})}{2}$

$= 2(\sqrt{5} + \sqrt{3})$ *(2 marks)*

(g) Defective $= 0.02 \times 800$

$= 16$

$\therefore$ there are 16 items defective. *(1 mark)*

QUESTION 2

(a) $x^2 - 6x + 2 = 0$

(i) $\alpha + \beta = -\dfrac{b}{a}$

$= -\dfrac{(-6)}{1}$

$= 6$ *(1 mark)*

(ii) $\alpha\beta = \dfrac{c}{a}$

$= \dfrac{2}{1}$

$= 2$ *(1 mark)*

(iii) $\dfrac{1}{\alpha} + \dfrac{1}{\beta} = \dfrac{\beta + \alpha}{\alpha\beta}$

$= \dfrac{6}{2}$

$= 3$ *(1 mark)*

(b) $2 \sin x = -\sqrt{3}$

$\sin x = -\dfrac{\sqrt{3}}{2}$

$x = \dfrac{4\pi}{3}, \dfrac{5\pi}{3}$ *(2 marks)*

(c) $y = (2x+1)^4$

Using the function of a function rule:

$\dfrac{dy}{dx} = 4(2x+1)^3.\dfrac{d}{dx}(2x+1)$

$= 4(2x+1)^3.\,2$

$= 8(2x+1)^3$

At $x = -1$, gradient $m = 8(2(-1)+1)^3$

$= 8 \times -1$

$= -8$

At $x = -1$, $y = (2(-1)+1)^4$

$= 1 \quad \therefore\ (-1, 1)$

Using $y - y_1 = m(x - x_1)$

$y - 1 = -8(x - (-1))$

$y - 1 = -8(x + 1)$

$y - 1 = -8x - 8$

$8x + y + 7 = 0$ *(3 marks)*

(d) $y = x^2 e^x$

Using the product rule,

Let $u = x^2$, $\quad u' = 2x$

Let $v = e^x$, $\quad v' = e^x$

$\dfrac{dy}{dx} = u'.v + v'.u$

$= 2x.e^x + e^x.x^2$

$= x\,e^x(2 + x)$ *(2 marks)*

(e) $\int \frac{1}{3x^2}\,dx = \int \frac{1}{3}x^{-2}\,dx$

$= \frac{1}{-3}x^{-1} + c$

$= -\frac{1}{3x} + c$ *(2 marks)*

QUESTION 3

(a) Using 3 to represent \$3 million, … the series is 3, 3.5, 4, …

Arithmetic series with $a = 3, d = 0.5$

(i) $n = 25, \quad T_n = a + (n-1)d$

$T_{25} = 3 + (25 - 1) \times 0.5$

$= 3 + 24 \times 0.5$

$= 15$

$\therefore$ the cost is \$15 million. *(2 marks)*

(ii) $n = 110, S_n = \frac{n}{2}[2a + (n-1)d]$

$S_{110} = \frac{110}{2}[2(3) + (110 - 1) \times 0.5]$

$= 55[6 + 109 \times 0.5]$

$= 3327.5$

$\therefore$ the cost is \$3327.5 million. *(2 marks)*

(b)

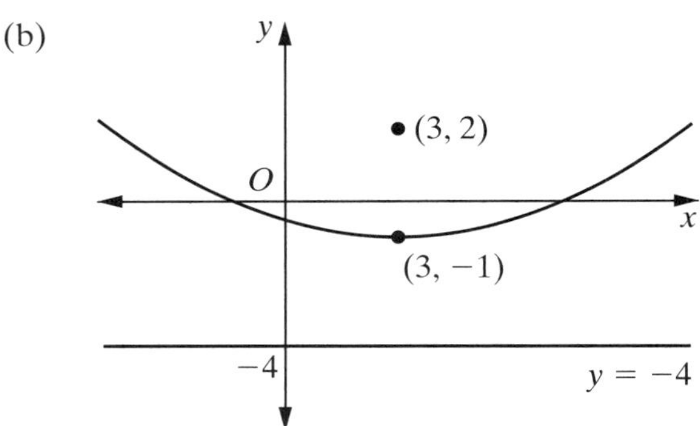

Focal length $a = 3$.

$\therefore$ vertex is $(3, -1)$. *(2 marks)*

(c)

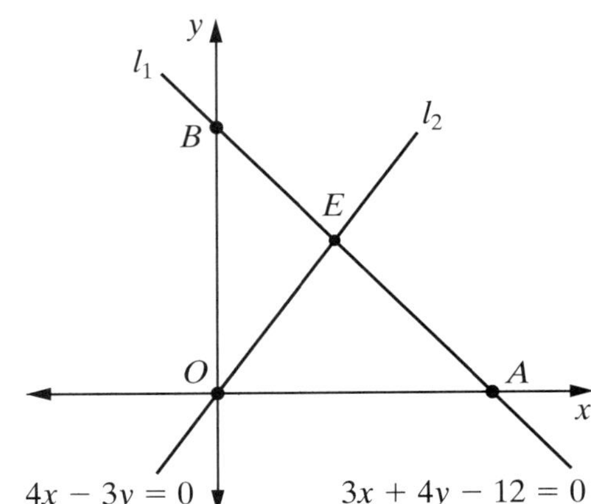

(i) B is on y-axis which has x-value of 0.

Now, substitute $(0, 3)$ in $3x + 4y - 12$:

$3(0) + 4(3) - 12 = 0.$

$\therefore\ B(0, 3)$ *(1 mark)*

(ii) gradient $l_1 : \frac{-3}{4}$

gradient $l_2 : \frac{4}{3}$

As $\frac{-3}{4} \times \frac{4}{3} = -1,$

$\therefore$ perpendicular *(2 marks)*

(iii) As $l_1 \perp l_2,$
use $(0, 0), 3x + 4y - 12 = 0,$

$d = \left|\frac{ax_1 + by_1 + c}{\sqrt{a^2 + b^2}}\right|$

$= \left|\frac{0 + 0 + (-12)}{\sqrt{3^2 + 4^2}}\right|$

$= \frac{12}{5}$ *(1 mark)*

(iv) As $OB = 3, OE = \frac{12}{5} = 2.4$, then

$BE^2 = OB^2 - OE^2$

$= 3^2 - 2.4^2$

$= 3.24$

$BE = \sqrt{3.24}$

$= 1.8$ *(1 mark)*

(v) Area $= \frac{1}{2} \times OE \times BE$

$= \frac{1}{2} \times 2.4 \times 1.8$

$= 2.16$

$\therefore$ area is 2.16 units^2 *(1 mark)*

QUESTION 4

(a) Let $y = \frac{x}{\sin x}$

Using the quotient rule,

Let $u = x$, $u' = 1$

Let $v = \sin x$, $v' = \cos x$

$$\frac{dy}{dx} = \frac{v.u' - u.v'}{v^2}$$
$$= \frac{\sin x.1 - x.\cos x}{\sin^2 x}$$
$$= \frac{\sin x - x\cos x}{\sin^2 x}$$

(2 marks)

(b) $$\int_e^{e^3} \frac{5}{x}\,dx = 5\int_e^{e^3} \frac{1}{x}\,dx$$
$$= 5[\log_e e^3 - \log_e e]$$
$$= 5[3 - 1]$$
$$= 10$$

(2 marks)

(c) $$\frac{dy}{dx} = 6x - 2$$
$$y = 3x^2 - 2x + c$$

Subs in $(-1, 4)$:

$$4 = 3 \times (-1)^2 - 2 \times (-1) + c$$
$$4 = 3 + 2 + c$$
$$c = 4 - 5$$
$$= -1$$

$\therefore$ the equation is $y = 3x^2 - 2x - 1$

(2 marks)

(d) (i) $$y = \sqrt{9 - x^2}$$
$$= (9 - x^2)^{\frac{1}{2}}$$

Using the function of a function rule:

$$\frac{dy}{dx} = \frac{1}{2}(9 - x^2)^{-\frac{1}{2}}.(-2x)$$
$$= \frac{-x}{\sqrt{9 - x^2}}$$

(2 marks)

(ii) From (i),

$$\int \frac{-x}{\sqrt{9 - x^2}}\,dx = \sqrt{9 - x^2} + c$$
$$\therefore \int \frac{6x}{\sqrt{9 - x^2}}\,dx = -6\sqrt{9 - x^2} + c$$

(2 marks)

(e) $y \le 4 - x^2$, $y \ge |x| - 2$ *(2 marks)*

QUESTION 5

(a) A geometric series: 27, 54, 108, …

$$a = 27, r = 2$$

(i) $n = 12$, $T_n = ar^{n-1}$

$$T_{12} = 27 \times 2^{11}$$
$$= 55\,296$$

$\therefore$ 55 296 members *(1 mark)*

(ii) Let $T_n = 10\,000\,000$

$$10\,000\,000 = 27 \times 2^{n-1}$$
$$2^{n-1} = \frac{10000000}{27}$$
$$\log_e 2^{n-1} = \log_e \frac{10000000}{27}$$
$$(n - 1)\log_e 2 = \log_e \frac{10000000}{27}$$
$$n - 1 = \frac{\log_e \frac{10000000}{27}}{\log_e 2}$$
$$n - 1 = 18.498\ldots$$
$$n = 19.498\ldots$$

$\therefore$ on the 20th day the number exceeds 10 million. *(2 marks)*

(iii) $n = 12$, $S_n = \frac{a(r^n - 1)}{r - 1}$

$$S_{12} = \frac{27(2^{12} - 1)}{2 - 1}$$
$$= 110\,565$$

$$\text{Total money} = 110\,565 \times \$0.005$$
$$= \$552.825$$
$$= \$553 \text{ (nearest \$)}$$

(2 marks)

(b) (i) $P(\text{red}) = \frac{3}{5}$ *(1 mark)*

(ii) $P(\text{same colour}) = P(\text{red, red, red})$

$$= \frac{3}{5} \times \frac{2}{4} \times \frac{1}{3}$$
$$= \frac{1}{10}$$

(1 mark)

(iii) P(not wear shirt of same colour over consecutive days)

$$= P(\text{RYR or YRY})$$
$$= \frac{3}{5} \times \frac{2}{4} \times \frac{2}{3} + \frac{2}{5} \times \frac{3}{4} \times \frac{1}{3}$$
$$= \frac{3}{10}$$

(2 marks)

(c) $\text{Dist} = \int_0^{20} v\,dt$

$\approx \frac{5}{3}[173 + 168 + 2 \times 127 + 4 \times (81 + 195)]$

$= 2831.666\ldots$

$\therefore$ jogger covers approximately 2832 metres. *(3 marks)*

QUESTION 6

(a)

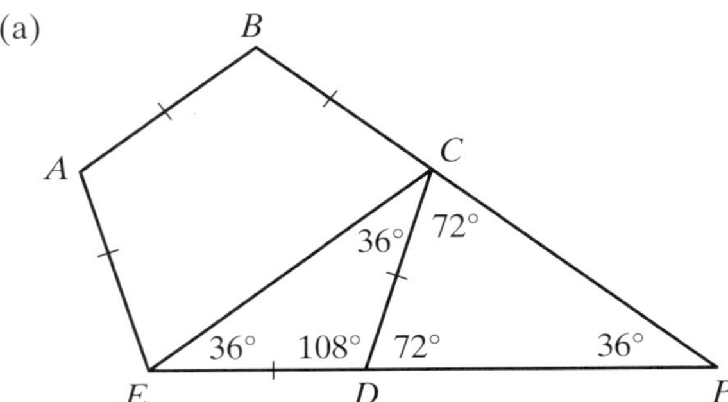

(i) Angle sum of pentagon is 540°

$\therefore \angle CDE = \frac{540^o}{5}$

$= 108^o$ *(1 mark)*

(ii) $DE = DC$ (sides of regular pentagon)

$\angle DEC = 36^o$ (base $\angle$s isos $\triangle$)

$\angle CDP = \angle DCP = 72^o$ (ext $\angle$s of regular pentagon)

$\angle CPD = 36^o$ ($\angle$ sum of $\triangle$)

$\therefore \triangle EPC$ is isosceles (base $\angle$s equal). *(2 marks)*

(b) Using $d = \sqrt{(x_2 - x_1)^2 + (y_2 - y_1)^2}$,

$$[\sqrt{(x - (-1))^2 + (y - 0)^2}]^2 + [\sqrt{(x - 3)^2 + (y - 0)^2}]^2 = 40$$
$$(x + 1)^2 + y^2 + (x - 3)^2 + y^2 = 40$$
$$x^2 + 2x + 1 + y^2 + x^2 - 6x + 9 + y^2 = 40$$
$$2x^2 - 4x + 2y^2 + 10 = 40$$
$$2x^2 - 4x + 2y^2 = 30$$
$$x^2 - 2x + y^2 = 15$$
$$x^2 - 2x + 1 + y^2 = 15 + 1$$
$$(x - 1)^2 + y^2 = 16$$

which is a circle with centre at $(1, 0)$ and radius 4 *(3 marks)*

(c) (i) Subs $x = 0$ in $y = 2\cos x$

$y = 2\cos 0$

$= 2 \times 1$

$= 2 \quad \therefore P(0, 2)$ *(1 mark)*

(ii) $\int_0^{\frac{\pi}{2}} 2\cos x\, dx = \left[2\sin x\right]_0^{\frac{\pi}{2}}$

$= 2(\sin \frac{\pi}{2} - \sin 0)$

$= 2(1 - 0)$

$= 2$ *(2 marks)*

(iii) C *(1 mark)*

(iv) Area A = Area C

Area B is twice Area A

$\therefore$ Total area $= 4 \times$ Area A

$= 4 \times 2$

$= 8$

The area is 8 units^2 *(1 mark)*

(v) Area B is 4 units^2 below x-axis,
Area C is 2 units^2.

$\therefore$ value of integral $= -4 + 2$

$= -2$ *(1 mark)*

QUESTION 7

(a) (i) $f(x) = x^3 - 3x + 2$

$f'(x) = 3x^2 - 3 = 0$

$3x^2 = 3$

$x = \pm 1$

$f(1) = (1)^3 - 3(1) + 2 = 0$

$f(-1) = (-1)^3 - 3(-1) + 2 = 4$

Stat points at $(1, 0)$ and $(-1, 4)$

$f''(x) = 6x$

$f''(1) = 6(1) = 6 > 0 \quad \therefore$ min $(1, 0)$

$f''(-1) = 6(-1) = -6 < 0$

$\therefore$ max $(-1, 4)$

(3 marks)

(ii) y-intercept of 2

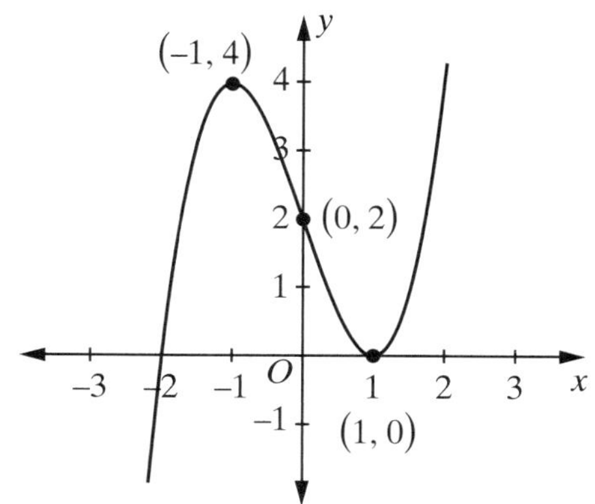

(2 marks)

(b) (i) $\dot{x} = 8 - 8e^{-2t}$

Subs $t = 0$,

$\dot{x} = 8 - 8e^{-2(0)}$

$= 8 - 8e^0$

$= 8 - 8$

$= 0$

$\therefore$ particle initially at rest. *(1 mark)*

(ii) $\ddot{x} = 16e^{-2t}$

As $e^{-2t} > 0$, for all values of t,
then $\ddot{x}$ is always positive. *(1 mark)*

(iii) Since particle is initially at rest (from (i)) and always a positive acceleration applied (from (ii)), then the velocity must be positive for all $t > 0$. This means that it is moving in the positive direction. *(2 marks)*

(iv) $\dot{x} = 8 - 8e^{-2t}$

As $t \to \infty$, $e^{-2t} \to 0$,

$\therefore$ $\dot{x}$ approaches 8.

The constant is 8. *(1 mark)*

(v)

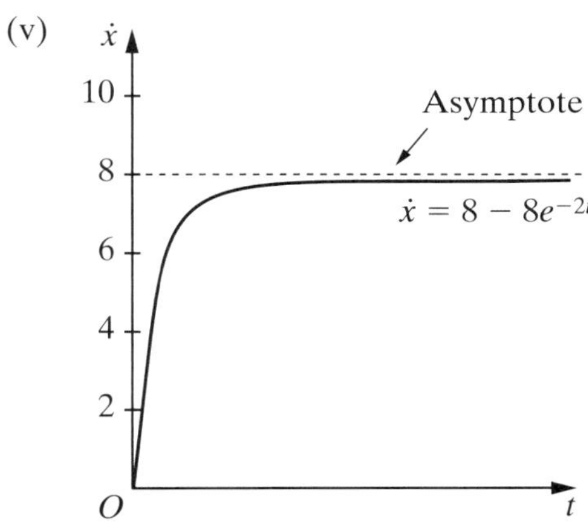

(2 marks)

QUESTION 8

(a) (i) $$22^2 = x^2 + 20^2 - 2 \times x \times 20 \times \cos 60^{\circ}$$
$$484 = x^2 + 400 - 20x$$
$$x^2 - 20x - 84 = 0$$ *(1 mark)*

(ii) $$x = \frac{20 \pm \sqrt{400 - 4 \times 1 \times (-84)}}{2 \times 1}$$
$$= \frac{20 \pm \sqrt{736}}{2}$$
$$= 23.564\,65 \ldots, -3.564\,65 \ldots$$

$\therefore$ the distance is 24 km. *(2 marks)*

(b) (i) $$V = \pi \int x^2 \, dy$$
$$= \pi \int_0^h y \, dy$$
$$= \pi \left[\frac{y^2}{2} \right]_0^h$$
$$= \pi \left[\frac{h^2}{2} - 0 \right]$$
$$= \frac{\pi h^2}{2}$$

$\therefore$ the volume is $\frac{\pi h^2}{2}$ units^3. *(2 marks)*

(ii) If $y = h$, then $x^2 = h$
$$x = \sqrt{h}$$
$\therefore$ $C(\sqrt{h}, h)$

V of cylinder $= \pi \times (\sqrt{h})^2 \times h$
$$= \pi h^2$$

$$\text{Ratio} = \frac{\pi h^2}{2} : \pi h^2$$
$$= 1 : 2$$ *(1 mark)*

(c) (i) 6% pa = 0.5% per month
$$P = 100 \times 1.005^{420} + 100 \times 1.005^{419} + \ldots + 100 \times 1.005$$
$$= 100[1.005 + 1.005^2 + \ldots + 1.005^{420}]$$

Using $a = 1.005, r = 1.005, n = 420$,
$$P = 100 \times \frac{1.005(1.005^{420} - 1)}{1.005 - 1}$$
$$= 143\,183.385$$

$\therefore$ \$143 183 to nearest dollar. *(2 marks)*

(ii) (1) $A_1 = (29\,227 + M) \times 1.005$

$= 29\,227 \times 1.005 + 1.005M$

$A_2 = [(29\,227 \times 1.005 + 1.005M) + M] \times 1.005$

$= 29\,227 \times 1.005^2 + 1.005^2M + 1.005M$

$= 29\,227 \times 1.005^2 + M(1.005 + 1.005^2)$ *(1 mark)*

(2) $A_{240} = 29\,227 \times 1.005^{240} + M(1.005 + 1.005^2 + \ldots + 1.005^{240})$

Using $a = 1.005, r = 1.005, n = 240,$

$$800\,000 = 29\,227 \times 1.005^{240} + M \times \frac{1.005(1.005^{240} - 1)}{1.005 - 1}$$

$$800\,000 = 96\,747.34621 + 464.3510996M$$

$$464.3510996M = 703\,252.6538$$

$$M = 1514.48 \text{ (2 dec. places)}$$

$\therefore$ she needs to contribute \$1514.48 each month. *(3 marks)*

QUESTION 9

(a)

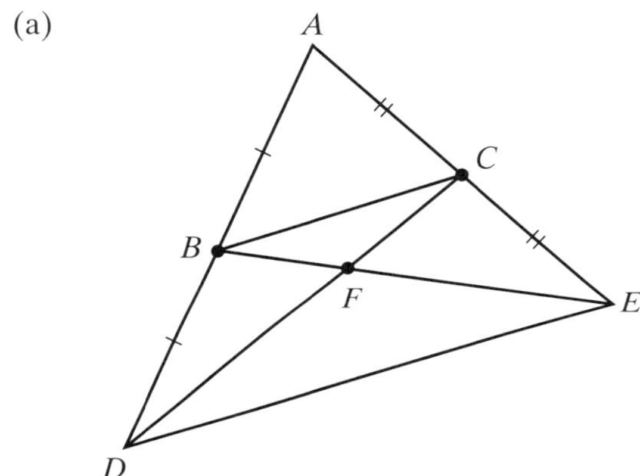

(i) $\angle A$ is common

$\frac{AB}{AD} = \frac{AC}{AE}$ (given)

$\therefore$ $\triangle ABC$ and $\triangle ADE$ are similar because 2 sides in proportion and included angle equal. *(1 mark)*

(ii) $\angle ABC = \angle ADE$ (matching $\angle$s of similar $\triangle$s equal)

$\therefore$ $BC \parallel DE$ (corr $\angle$s equal)

$\therefore$ $\angle CBF = \angle DEF$ (alt $\angle$s equal, $BC \parallel DE$)

$\angle BCF = \angle EDF$ (alt $\angle$s equal, $BC \parallel DE$)

$\therefore$ $\triangle BCF$ and $\triangle EDF$ are similar (2 $\angle$s equal)

But $\frac{BC}{DE} = \frac{1}{2}$ (from similar $\triangle$s in (i))

$\therefore$ $\frac{BF}{FE} = \frac{1}{2}$ or, $BF : FE = 1:2$ (matching sides of sim $\triangle$s in proportion) *(2 marks)*

(b) (i) Difference $= 2 + \frac{t^2}{t+1} - \left(1 + \frac{1}{t+1}\right)$

$$= 1 + \frac{t^2}{t+1} - \frac{1}{t+1}$$
$$= 1 + \frac{t^2 - 1}{t+1}$$
$$= 1 + \frac{(t-1)(t+1)}{t+1}$$
$$= 1 + t - 1$$
$$= t$$

$\therefore$ *A* is greater by *t* litres per minute. *(1 mark)*

(ii) Volume difference

$$= \int_0^4 \left(2 + \frac{t^2}{t+1}\right) - \left(1 + \frac{1}{t+1}\right) dt$$
$$= \int_0^4 t\, dt \text{ (from (i))}$$
$$= \left[\frac{t^2}{2}\right]_0^4$$
$$= 8 - 0$$
$$= 8$$

$\therefore$ *A* has 8 L more than *B*. *(2 marks)*

(c)

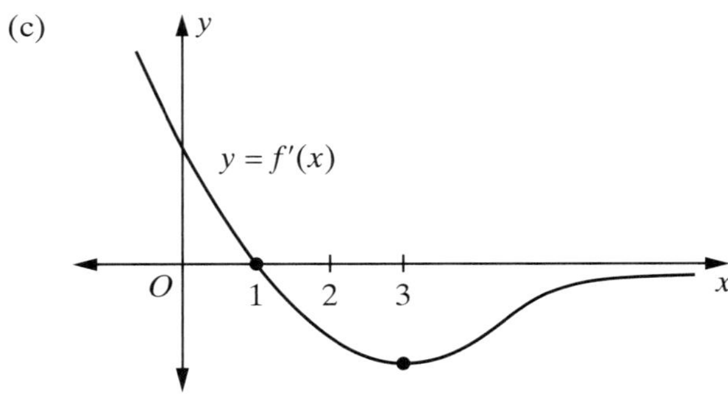

(3 marks)

(d) (i) $\frac{1}{\sqrt{n} + \sqrt{n+1}} \times \frac{\sqrt{n} - \sqrt{n+1}}{\sqrt{n} - \sqrt{n+1}}$

$$= \frac{\sqrt{n} - \sqrt{n+1}}{n - (n+1)}$$
$$= \frac{\sqrt{n} - \sqrt{n+1}}{-1}$$
$$= \sqrt{n+1} - \sqrt{n}$$ *(1 mark)*

(ii) From (i),

$$\frac{1}{\sqrt{n} + \sqrt{n+1}} = \sqrt{n+1} - \sqrt{n}$$

$\therefore$ As $n = 1$, $\frac{1}{\sqrt{1} + \sqrt{2}} = \sqrt{2} - \sqrt{1}$

$n = 2$, $\frac{1}{\sqrt{2} + \sqrt{3}} = \sqrt{3} - \sqrt{2}$

$n = 3$, $\frac{1}{\sqrt{3} + \sqrt{4}} = \sqrt{4} - \sqrt{3}$

$\therefore$ series is $(\sqrt{2} - \sqrt{1}) + (\sqrt{3} - \sqrt{2}) + (\sqrt{4} - \sqrt{3}) + \ldots + (\sqrt{100} - \sqrt{99})$

$$= -\sqrt{1} + \sqrt{100}$$
$$= -1 + 10$$
$$= 9$$ *(2 marks)*

QUESTION 10

(a) (i) $I = 10^{-12} \times e^{0.1L}$

Subs $L = 110$,

$$I = 10^{-12} \times e^{0.1(110)}$$
$$= 0.000\,000\,059\ldots$$
$$= 6 \times 10^{-8} \text{ (1 sig. fig.)}$$ *(1 mark)*

(ii) Let $I = 8.1 \times 10^{-9}$

$$8.1 \times 10^{-9} = 10^{-12} \times e^{0.1L}$$
$$e^{0.1L} = (8.1 \times 10^{-9}) \div 10^{-12}$$
$$= 8100$$
$$0.1L \log_e e = \log_e 8100$$
$$L = \frac{\log_e 8100}{0.1}$$
$$= 89.996\,19\ldots$$
$$= 90 \text{ (nearest whole)}$$

$\therefore$ max loudness is 90 decibels. *(2 marks)*

(iii) *Method 1:* From (i),

$I = 6 \times 10^{-8}$, when $L = 110$.

We can double the value of *I*:

$$\text{Let } I = 12 \times 10^{-8}$$
$$= 1.2 \times 10^{-7}$$
$$1.2 \times 10^{-7} = 10^{-12} \times e^{0.1L}$$
$$e^{0.1L} = (1.2 \times 10^{-7}) \div 10^{-12}$$
$$= 120\,000$$
$$0.1L \log_e e = \log_e 120\,000$$
$$= 116.952\,47\ldots$$
$$= 117 \text{ (nearest whole)}$$

$\therefore$ As $117 - 110 = 7$, the loudness has increased by 7 decibels. *(2 marks)*

Method 2:

Using exponential growth and decay:

$$I = I_0 e^{0.1L}$$

Let $I = 2I_0$

$$\therefore\ 2I_0 = I_0 e^{0.1L}$$
$$e^{0.1L} = 2$$
$$0.1L = \log_e 2$$
$$L = 10\log_e 2$$
$$= 6.931\,47\ldots$$
$$= 7 \text{ (nearest whole)}$$

$\therefore$ increased by 7 decibels.

(b) (i) Using arc length $= r\theta$

$$P = r + r + r\theta$$
$$= 2r + r\theta$$
$$= r(\theta + 2)$$

(1 mark)

(ii)

$$P = r(\theta + 2) \quad \text{from (i)}$$
$$P = r\theta + 2r$$
$$r\theta = P - 2r$$
$$\theta = \frac{P - 2r}{r}$$

Using area of sector $= \frac{1}{2}r^2\theta$

$$A = \frac{1}{2}r^2 \times \frac{P - 2r}{r}$$
$$= \frac{1}{2}Pr - r^2$$
$$\therefore\ A = \frac{1}{2}Pr - r^2$$

(1 mark)

(iii)

$$A = \frac{1}{2}Pr - r^2$$
$$\frac{dA}{dr} = \frac{1}{2}P - 2r = 0$$
$$2r = \frac{P}{2}$$
$$r = \frac{P}{4}$$
$$\frac{d^2A}{dr^2} = -2 < 0, \ \therefore \text{ maximum}$$

$\therefore$ max when $r = \frac{P}{4}$ *(2 marks)*

(iv) Subs $r = \frac{P}{4}$ in $P = r(\theta + 2)$

$$P = \frac{P}{4}(\theta + 2)$$
$$4 = \theta + 2$$
$$\theta = 2$$

$\therefore$ angle is 2 radians. *(1 mark)*

(v) Using $0 < \theta < 2\pi$

From (ii),

$$0 < \frac{P - 2r}{r} < 2\pi$$
$$0 < \frac{P}{r} - 2 < 2\pi$$
$$2 < \frac{P}{r} < 2\pi + 2$$
$$\frac{1}{2} > \frac{r}{P} > \frac{1}{2\pi + 2}$$
$$\frac{P}{2} > r > \frac{P}{2(\pi + 1)}$$

i.e. $\frac{P}{2(\pi + 1)} < r < \frac{P}{2}$ *(2 marks)*

Alternatively:

From part (i), $P = r(\theta + 2)$

$$\therefore P = r\theta + 2r > 2r$$
$$r < \frac{P}{2}$$

Also from part (i), since $\theta < 2\pi$,

$$\therefore P = r(\theta + 2) < r(2\pi + 2)$$
$$r > \frac{P}{2(\pi + 1)}$$
$$\therefore \frac{P}{2(\pi + 1)} < r < \frac{P}{2}$$

BOARD OF STUDIES
NEW SOUTH WALES

2012

HIGHER SCHOOL CERTIFICATE EXAMINATION

Mathematics

General Instructions

- Reading time – 5 minutes
- Working time – 3 hours
- Write using black or blue pen
  Black pen is preferred
- Board-approved calculators may be used
- A table of standard integrals is provided at the back of this paper
- In Questions 11–16, show relevant mathematical reasoning and/or calculations

Total marks – 100

Section I

10 marks

- Attempt Questions 1–10
- Allow about 15 minutes for this section

Section II

90 marks

- Attempt Questions 11–16
- Allow about 2 hours and 45 minutes for this section

Section I

10 marks
Attempt Questions 1–10
Allow about 15 minutes for this section

Use the multiple-choice answer sheet for Questions 1–10.

---

1 What is 4.097 84 correct to three significant figures?

(A) 4.09

(B) 4.10

(C) 4.097

(D) 4.098

2 Which of the following is equal to $\dfrac{1}{2\sqrt{5}-\sqrt{3}}$?

(A) $\dfrac{2\sqrt{5}-\sqrt{3}}{7}$

(B) $\dfrac{2\sqrt{5}+\sqrt{3}}{7}$

(C) $\dfrac{2\sqrt{5}-\sqrt{3}}{17}$

(D) $\dfrac{2\sqrt{5}+\sqrt{3}}{17}$

3 The quadratic equation $x^2 + 3x - 1 = 0$ has roots α and β.

What is the value of $\alpha\beta + (\alpha + \beta)$?

(A) 4

(B) 2

(C) −4

(D) −2

4 The diagram shows the graph $y = f(x)$.

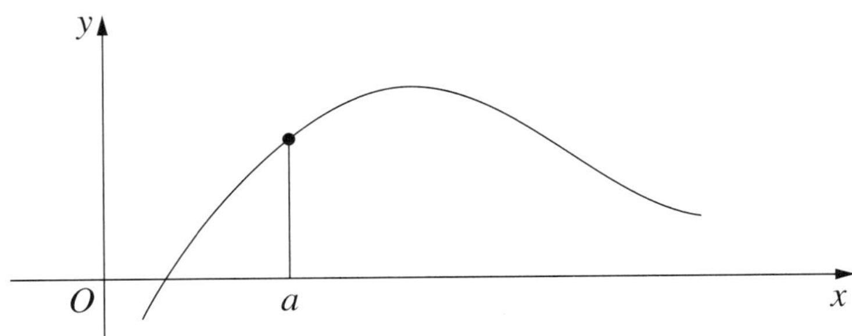

Which of the following statements is true?

(A) $f'(a) > 0$ and $f''(a) < 0$

(B) $f'(a) > 0$ and $f''(a) > 0$

(C) $f'(a) < 0$ and $f''(a) < 0$

(D) $f'(a) < 0$ and $f''(a) > 0$

5 What is the perpendicular distance of the point $(2, -1)$ from the line $y = 3x + 1$?

(A) $\dfrac{6}{\sqrt{10}}$

(B) $\dfrac{6}{\sqrt{5}}$

(C) $\dfrac{8}{\sqrt{10}}$

(D) $\dfrac{8}{\sqrt{5}}$

6 What are the solutions of $\sqrt{3}\tan x = -1$ for $0 \le x \le 2\pi$?

(A) $\frac{2\pi}{3}$ and $\frac{4\pi}{3}$

(B) $\frac{2\pi}{3}$ and $\frac{5\pi}{3}$

(C) $\frac{5\pi}{6}$ and $\frac{7\pi}{6}$

(D) $\frac{5\pi}{6}$ and $\frac{11\pi}{6}$

7 Let $a = e^x$.

Which expression is equal to $\log_e\left(a^2\right)$?

(A) e^{2x}

(B) e^{x^2}

(C) $2x$

(D) x^2

8 The diagram shows the region enclosed by $y = x - 2$ and $y^2 = 4 - x$.

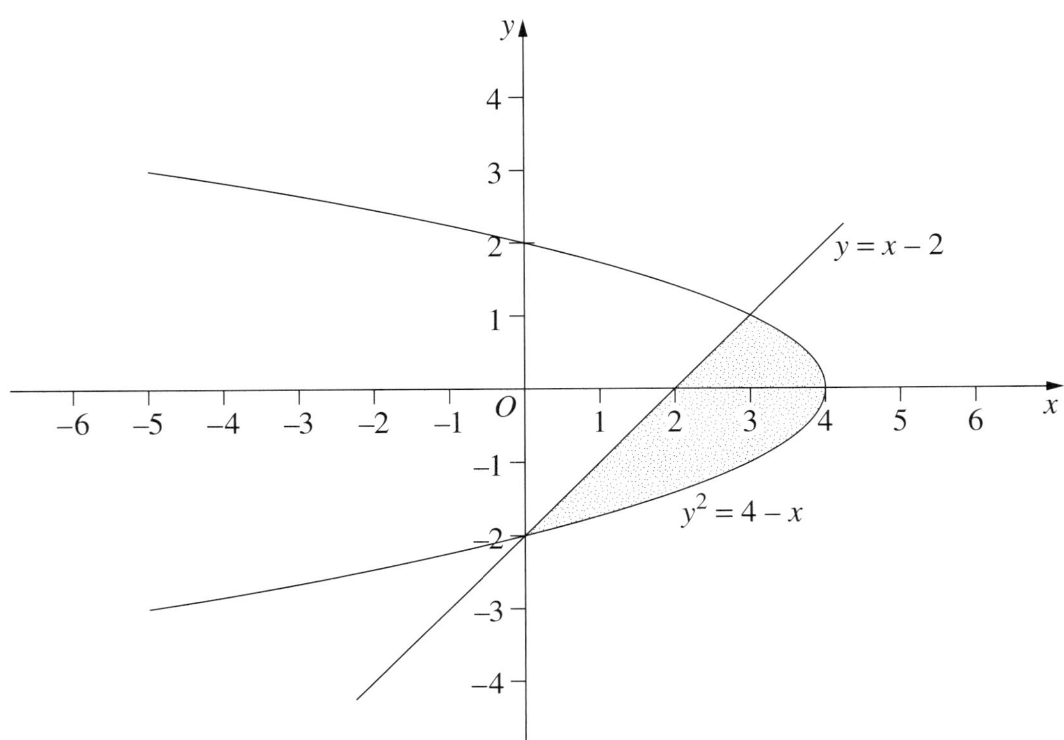

Which of the following pairs of inequalities describes the shaded region in the diagram?

(A) $y^2 \le 4 - x$ and $y \le x - 2$

(B) $y^2 \le 4 - x$ and $y \ge x - 2$

(C) $y^2 \ge 4 - x$ and $y \le x - 2$

(D) $y^2 \ge 4 - x$ and $y \ge x - 2$

9 What is the value of $\displaystyle\int_1^4 \frac{1}{3x}\,dx$?

(A) $\frac{1}{3}\ln 3$

(B) $\frac{1}{3}\ln 4$

(C) $\ln 9$

(D) $\ln 12$

10 The graph of $y = f(x)$ has been drawn to scale for $0 \leq x \leq 8$.

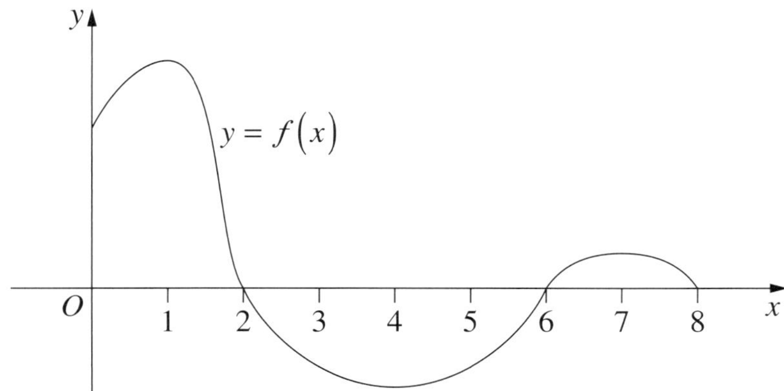

Which of the following integrals has the greatest value?

(A) $\int_0^1 f(x)\,dx$

(B) $\int_0^2 f(x)\,dx$

(C) $\int_0^7 f(x)\,dx$

(D) $\int_0^8 f(x)\,dx$

Section II

90 marks
Attempt Questions 11–16
Allow about 2 hours and 45 minutes for this section

Answer each question in the appropriate writing booklet. Extra writing booklets are available.

In Questions 11–16, your responses should include relevant mathematical reasoning and/or calculations.

Question 11 (15 marks) Use the Question 11 Writing Booklet.

(a) Factorise $2x^2 - 7x + 3$. **2**

(b) Solve $|3x - 1| < 2$. **2**

(c) Find the equation of the tangent to the curve $y = x^2$ at the point where $x = 3$. **2**

(d) Differentiate $\left(3 + e^{2x}\right)^5$. **2**

(e) Find the coordinates of the focus of the parabola $x^2 = 16(y - 2)$. **2**

(f) The area of a sector of a circle of radius 6 cm is 50 cm^2. **2**

Find the length of the arc of the sector.

(g) Find $\int_0^{\frac{\pi}{2}} \sec^2 \frac{x}{2}\, dx$. **3**

Question 12 (15 marks) Use the Question 12 Writing Booklet.

(a) Differentiate with respect to x.

(i) $(x-1)\log_e x$ **2**

(ii) $\dfrac{\cos x}{x^2}$ **2**

(b) Find $\displaystyle\int \frac{4x}{x^2+6}\,dx$. **2**

(c) Jay is making a pattern using triangular tiles. The pattern has 3 tiles in the first row, 5 tiles in the second row, and each successive row has 2 more tiles than the previous row.

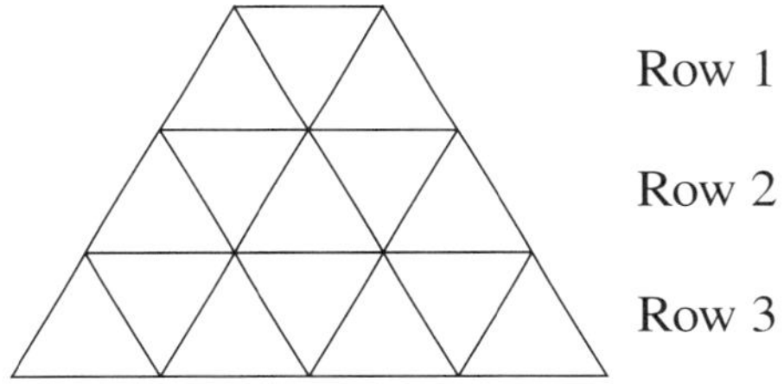

(i) How many tiles would Jay use in row 20? **2**

(ii) How many tiles would Jay use altogether to make the first 20 rows? **1**

(iii) Jay has only 200 tiles. **2**

How many complete rows of the pattern can Jay make?

Question 12 continues on following page

Question 12 (continued)

(d) At a certain location a river is 12 metres wide. At this location the depth of the river, in metres, has been measured at 3 metre intervals. The cross-section is shown below.

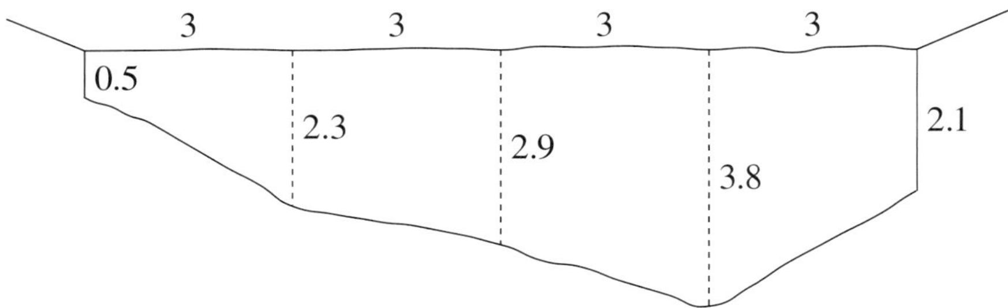

(i) Use Simpson's rule with the five depth measurements to calculate the approximate area of the cross-section. **3**

(ii) The river flows at 0.4 metres per second. **1**

Calculate the approximate volume of water flowing through the cross-section in 10 seconds.

End of Question 12

Question 13 (15 marks) Use the Question 13 Writing Booklet.

(a) The diagram shows a triangle ABC. The line $2x + y = 8$ meets the x and y axes at the points A and B respectively. The point C has coordinates $(7, 4)$.

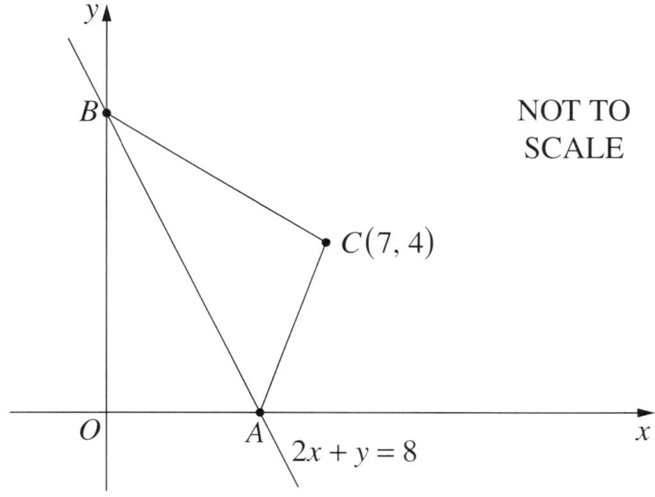

(i) Calculate the distance AB. **2**

(ii) It is known that $AC = 5$ and $BC = \sqrt{65}$. (Do NOT prove this.) **2**

Calculate the size of $\angle ABC$ to the nearest degree.

(iii) The point N lies on AB such that CN is perpendicular to AB. **3**

Find the coordinates of N.

Question 13 continues on following page

Question 13 (continued)

(b) The diagram shows the parabolas $y = 5x - x^2$ and $y = x^2 - 3x$. The parabolas intersect at the origin O and the point A. The region between the two parabolas is shaded.

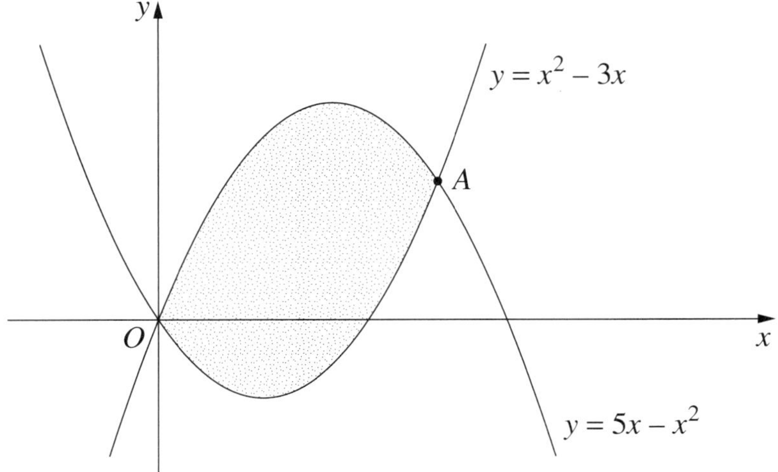

(i) Find the x-coordinate of the point A. **1**

(ii) Find the area of the shaded region. **3**

(c) Two buckets each contain red marbles and white marbles. Bucket A contains 3 red and 2 white marbles. Bucket B contains 3 red and 4 white marbles.

Chris randomly chooses one marble from each bucket.

(i) What is the probability that both marbles are red? **1**

(ii) What is the probability that at least one of the marbles is white? **1**

(iii) What is the probability that both marbles are the same colour? **2**

End of Question 13

Question 14 (15 marks) Use the Question 14 Writing Booklet.

(a) A function is given by $f(x) = 3x^4 + 4x^3 - 12x^2$.

(i) Find the coordinates of the stationary points of $f(x)$ and determine their nature. **3**

(ii) Hence, sketch the graph $y = f(x)$ showing the stationary points. **2**

(iii) For what values of x is the function increasing? **1**

(iv) For what values of k will $3x^4 + 4x^3 - 12x^2 + k = 0$ have no solution? **1**

(b) The diagram shows the region bounded by $y = \dfrac{3}{(x+2)^2}$, the x-axis, the y-axis, and the line $x = 1$. **3**

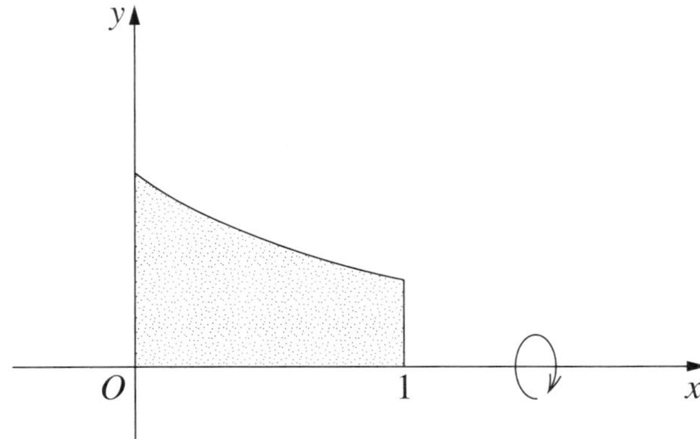

The region is rotated about the x-axis to form a solid.

Find the volume of the solid.

Question 14 continues on following page

Question 14 (continued)

(c) Professor Smith has a colony of bacteria. Initially there are 1000 bacteria. The number of bacteria, $N(t)$, after t minutes is given by

$$N(t) = 1000e^{kt}.$$

(i) After 20 minutes there are 2000 bacteria. **1**

Show that $k = 0.0347$ correct to four decimal places.

(ii) How many bacteria are there when $t = 120$? **1**

(iii) What is the rate of change of the number of bacteria per minute, when $t = 120$? **1**

(iv) How long does it take for the number of bacteria to increase from 1000 to 100 000? **2**

End of Question 14

Question 15 (15 marks) Use the Question 15 Writing Booklet.

(a) Rectangles of the same height are cut from a strip and arranged in a row. The first rectangle has width 10 cm. The width of each subsequent rectangle is 96% of the width of the previous rectangle.

10 cm

NOT TO SCALE

(i) Find the length of the strip required to make the first ten rectangles. **2**

(ii) Explain why a strip of length 3 m is sufficient to make any number of rectangles. **1**

(b) The velocity of a particle is given by

$$\dot{x} = 1 - 2\cos t,$$

where x is the displacement in metres and t is the time in seconds. Initially the particle is 3 m to the right of the origin.

(i) Find the initial velocity of the particle. **1**

(ii) Find the maximum velocity of the particle. **1**

(iii) Find the displacement, x, of the particle in terms of t. **2**

(iv) Find the position of the particle when it is at rest for the first time. **2**

Question 15 continues on following page

Question 15 (continued)

(c) Ari takes out a loan of \$360 000. The loan is to be repaid in equal monthly repayments, $\$M$, at the end of each month, over 25 years (300 months). Reducible interest is charged at 6% per annum, calculated monthly.

Let $\$A_n$ be the amount owing after the nth repayment.

(i) Write down an expression for the amount owing after two months, $\$A_2$. **1**

(ii) Show that the monthly repayment is approximately \$2319.50. **2**

(iii) After how many months will the amount owing, $\$A_n$, become less than \$180 000? **3**

End of Question 15

Question 16 (15 marks) Use the Question 16 Writing Booklet.

(a) The diagram shows a triangle ABC with sides $BC = a$ and $AC = b$. The points D, E and F lie on the sides AC, AB and BC, respectively, so that $CDEF$ is a rhombus with sides of length x.

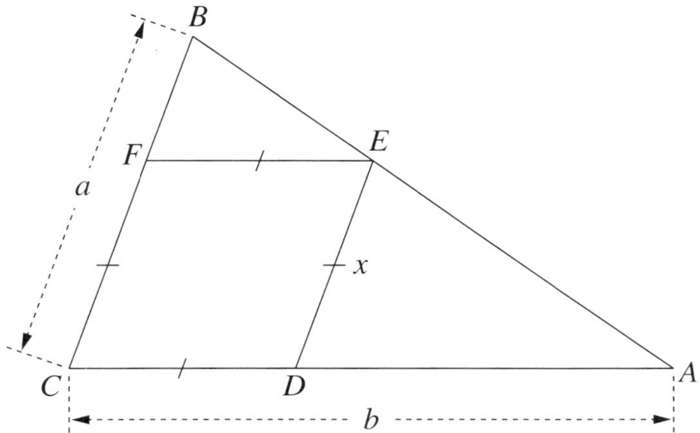

(i) Prove that $\triangle EBF$ is similar to $\triangle AED$. **2**

(ii) Find an expression for x in terms of a and b. **2**

Question 16 continues on following page

Question 16 (continued)

(b) The diagram shows a point T on the unit circle $x^2 + y^2 = 1$ at angle θ from the positive x-axis, where $0 < \theta < \dfrac{\pi}{2}$.

The tangent to the circle at T is perpendicular to OT, and intersects the x-axis at P, and the line $y = 1$ at Q. The line $y = 1$ intersects the y-axis at B.

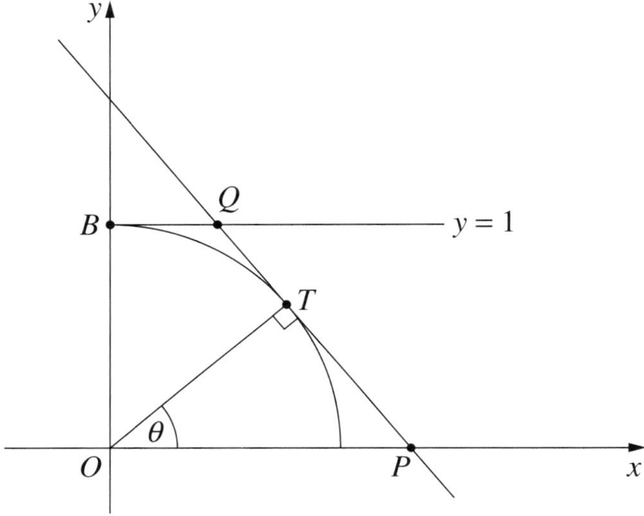

(i) Show that the equation of the line PT is **2**

$$x\cos\theta + y\sin\theta = 1.$$

(ii) Find the length of BQ in terms of θ. **1**

(iii) Show that the area, A, of the trapezium $OPQB$ is given by **2**

$$A = \frac{2 - \sin\theta}{2\cos\theta}.$$

(iv) Find the angle θ that gives the minimum area of the trapezium. **3**

Question 16 continues on following page

Question 16 (continued)

(c) The circle $x^2 + (y - c)^2 = r^2$, where $c > 0$ and $r > 0$, lies inside the parabola $y = x^2$. The circle touches the parabola at exactly two points located symmetrically on opposite sides of the y-axis, as shown in the diagram.

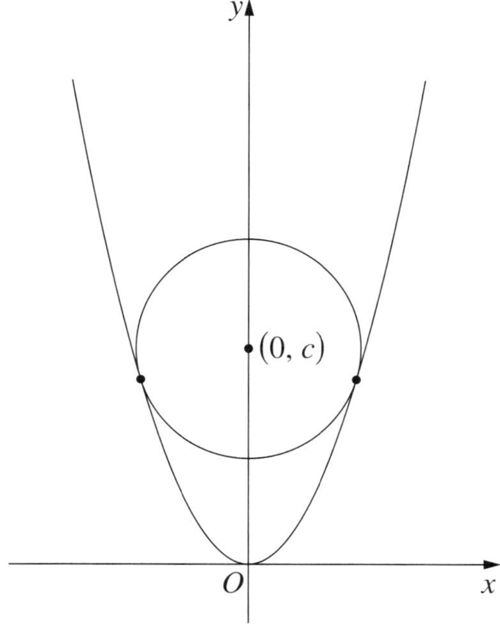

(i) Show that $4c = 1 + 4r^2$. **2**

(ii) Deduce that $c > \frac{1}{2}$. **1**

End of paper

2012 Higher School Certificate

Worked Answers

Section I

QUESTION 1

4.097 84 = 4.10 (to 3 sig. figs.)

Answer B *(1 mark)*

QUESTION 2

$$\frac{1}{2\sqrt{5}-\sqrt{3}} = \frac{1}{2\sqrt{5}-\sqrt{3}} \times \frac{2\sqrt{5}+\sqrt{3}}{2\sqrt{5}+\sqrt{3}}$$

$$= \frac{2\sqrt{5}+\sqrt{3}}{20-3}$$

$$= \frac{2\sqrt{5}+\sqrt{3}}{17}$$

Answer D *(1 mark)*

QUESTION 3

$x^2 + 3x - 1 = 0$

$$\alpha + \beta = \frac{-b}{a}$$

$$= -3$$

$$\alpha\beta = \frac{c}{a}$$

$$= -1$$

$$\alpha\beta + (\alpha + \beta) = -1 + (-3)$$
$$= -4$$

Answer C *(1 mark)*

QUESTION 4

At $x = a$, the curve is *increasing* and *concave down.*

$\therefore\ f'(a) > 0$ and $f''(a) < 0$

Answer A *(1 mark)*

QUESTION 5

Use $(2, -1), 3x - y + 1 = 0$,

$$d = \left|\frac{ax_1 + by_1 + c}{\sqrt{a^2 + b^2}}\right|$$

$$= \left|\frac{3(2) - 1(-1) + 1}{\sqrt{3^2 + (-1)^2}}\right|$$

$$= \left|\frac{8}{\sqrt{10}}\right|$$

$$= \frac{8}{\sqrt{10}}$$

Answer C *(1 mark)*

QUESTION 6

$$\sqrt{3}\tan x = -1$$

$$\tan x = \frac{-1}{\sqrt{3}}$$

$$x = \frac{5\pi}{6}, \frac{11\pi}{6}$$

Answer D *(1 mark)*

QUESTION 7

$$a = e^x$$
$$a^2 = (e^x)^2$$
$$= e^{2x}$$

Taking logs of both sides:

$$\log_e(a^2) = \log_e(e^{2x})$$
$$= 2x\log_e e$$
$$= 2x$$

Answer C *(1 mark)*

QUESTION 8

Choose a point inside the shaded region: e.g. (3, 0).

For $y^2 = 4 - x, 0 = 4 - 3$?

$0 = 1$?

But $0 \leq 1$

$\therefore y^2 \leq 4 - x$

For $y = x - 2, 0 = 3 - 2$?

$0 = 1$?

But $0 \leq 1$

$\therefore y \leq x - 2$

$\therefore y^2 \leq 4 - x$ and $y \leq x - 2$

Answer A *(1 mark)*

QUESTION 9

$$\int_1^4 \frac{1}{3x}\,dx = \frac{1}{3}\int_1^4 \frac{1}{x}\,dx$$
$$= \frac{1}{3}[\ln x]_1^4$$
$$= \frac{1}{3}[\ln 4 - \ln 1]$$
$$= \frac{1}{3}\ln 4$$

Answer B *(1 mark)*

QUESTION 10

As $\int_2^6 f(x)\,dx < 0$ and

$\int_6^8 f(x)\,dx < \left|\int_2^6 f(x)\right|dx$, then the greatest

value is $\int_0^2 f(x)\,dx$.

Answer B *(1 mark)*

Section II

QUESTION 11

(a) $2x^2 - 7x + 3 = (2x - 1)(x - 3)$ *(2 marks)*

(b) $|3x - 1| < 2$

Two cases:

$3x - 1 < 2$ $\qquad -(3x - 1) < 2$

$3x < 2 + 1$ $\qquad 3x - 1 > -2$

$3x < 3$ $\qquad 3x > -2 + 1$

$x < 1$ $\qquad 3x > -1$

$\qquad x > -\frac{1}{3}$

$\therefore -\frac{1}{3} < x < 1$ *(2 marks)*

(c) $y = x^2$

$\frac{dy}{dx} = 2x$

At $x = 3$, gradient $m = 2 \times 3$
$= 6$

At $x = 3$, $y = (3)^2$
$= 9 \quad \therefore (3, 9)$

Using $y - y_1 = m(x - x_1)$

$y - 9 = 6(x - 3)$

$y - 9 = 6x - 18$

$\therefore 6x - y - 9 = 0$ *(2 marks)*

(d) $y = (3 + e^{2x})^5$

Using the function of a function rule:

$$\frac{dy}{dx} = 5(3 + e^{2x})^4 . \frac{d}{dx}(3 + e^{2x})$$
$$= 5(3 + e^{2x})^4 . 2e^{2x}$$
$$= 10e^{2x}(3 + e^{2x})^4$$

(2 marks)

(e) $x^2 = 16(y - 2)$

which is of the form $(x - h)^2 = 4a(y - k)$

$\therefore$ vertex $(0, 2)$, and $a = 4$,

As parabola is concave up then focus $(0, 6)$.

(2 marks)

(f) Area $= \frac{1}{2}r^2\theta$

$50 = \frac{1}{2} \times 6^2 \times \theta$

$\therefore 50 = 18\theta$

$\theta = \frac{25}{9}$

Now, $\ell = r\theta$

$= 6 \times \frac{25}{9}$

$= \frac{50}{3}$

$\therefore$ length is $\frac{50}{3}$ cm *(2 marks)*

(g) $\int_0^{\frac{\pi}{2}} \sec^2\frac{x}{2}\,dx$

$= \left[2\tan\frac{x}{2}\right]_0^{\frac{\pi}{2}}$

$= 2\left[\tan\frac{\pi}{4} - \tan 0\right]$

$= 2[1 - 0]$

$= 2$ *(3 marks)*

QUESTION 12

(a) (i) $(x - 1)\log_e x$

Using the product rule,

Let $u = x - 1$, $u' = 1$

Let $v = \log_e x$, $v' = \frac{1}{x}$

$\frac{dy}{dx} = u'.v + v'.u$

$= 1.\log_e x + \frac{1}{x}.(x - 1)$

$= \log_e x + \frac{x - 1}{x}$

$\left[\text{or } \frac{x\log_e x + x - 1}{x}\right]$ *(2 marks)*

(ii) $\frac{\cos x}{x^2}$

Using the quotient rule,

Let $u = \cos x$, $u' = -\sin x$

Let $v = x^2$, $v' = 2x$

$\frac{dy}{dx} = \frac{v.u' - u.v'}{v^2}$

$= \frac{x^2.(-\sin x) - \cos x.2x}{(x^2)^2}$

$= \frac{-x^2\sin x - 2x\cos x}{x^4}$

$= \frac{-x\sin x - 2\cos x}{x^3}$ *(2 marks)*

(b) $\int \frac{4x}{x^2 + 6}\,dx = 2\int \frac{2x}{x^2 + 6}\,dx$

$= 2\log_e(x^2 + 6) + c$

(2 marks)

(c) Arithmetic series: $3 + 5 + 7 + \ldots$

$a = 3, d = 2$

(i) Use $n = 20$,

$T_n = a + (n - 1)d$

$T_{20} = 3 + (20 - 1) \times 2$

$= 3 + 38$

$= 41$ *(2 marks)*

(ii) $S_n = \frac{n}{2}[a + \ell]$

$= \frac{20}{2}[3 + 41]$

$= 440$ *(1 mark)*

(iii) $S_n = \frac{n}{2}[2a + (n - 1)d] = 200$

$\frac{n}{2}[2 \times 3 + (n - 1) \times 2] = 200$

$n[6 + 2n - 2] = 400$

$n[2n + 4] = 400$

$2n^2 + 4n - 400 = 0$

$n^2 + 2n - 200 = 0$

$\therefore n = \frac{-2 \pm \sqrt{2^2 - 4 \times 1 \times (-200)}}{2 \times 1}$

$= \frac{-2 \pm \sqrt{804}}{2}$

$= 13.18, -15.18$ (corr. 2 dec. pl.)

$\therefore$ only 13 complete rows could be made *(2 marks)*

(d) (i) Area $\approx \frac{h}{3}[y_0 + y_4 + 4(y_1 + y_3) + 2(y_2)]$

$= \frac{3}{3}[0.5 + 2.1 + 4(2.3 + 3.8) + 2(2.9)]$

$= 32.8$

$\therefore$ area is approximately 32.8 m^2

(3 marks)

(ii) As $10 \times 0.4 = 4$, then

Volume $= 32.8 \times 4$

$= 131.2$

$\therefore$ volume is approximately 131.2 m^3

(1 mark)

QUESTION 13

(a) (i) $A(x, 0)$: subs $y = 0$ in $2x + y = 8$

$$2x = 8$$
$$x = 4$$

$B(0, y)$: subs $x = 0$ in $2x + y = 8$

$$y = 8$$

$\therefore A(4, 0), B(0, 8)$

$AB^2 = 4^2 + 8^2$
$= 16 + 64$
$= 80$

$AB = \sqrt{80}$

$\therefore$ distance is $4\sqrt{5}$ units

(2 marks)

(ii)

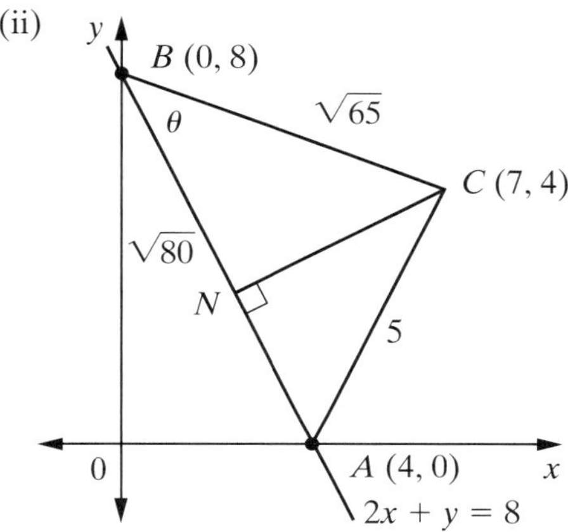

Let $\angle ABC = \theta$

Using $\cos\theta = \dfrac{(\sqrt{65})^2 + (\sqrt{80})^2 - 5^2}{2 \times \sqrt{65} \times \sqrt{80}}$

$= \dfrac{120}{144.222051\ldots}$

$= \dfrac{120}{144.222051\ldots}$

$= 0.83205\ldots$

$\theta = 33.690067751\ldots$

$\therefore \angle ABC = 34^\circ$ (nearest degree)

(2 marks)

(iii) Gradient of $2x + y = 8$ is -2

$\therefore$ gradient of perpendicular $= \dfrac{1}{2}$

equation of CN:

$y - y_1 = m(x - x_1)$

$y - 4 = \dfrac{1}{2}(x - 7)$

$2y - 8 = x - 7$

$x - 2y = -1$

Solve simultaneously:

$2x + y = 8$ **1**

$x - 2y = -1$ **2**

$2 \times$ **2**: $2x - 4y = -2$ **3**

1 $-$ **3**: $5y = 10$

$y = 2$

Subs in **1**:

$2x + 2 = 8$

$2x = 8 - 2$

$2x = 6$

$x = 3$

$\therefore N(3, 2)$ *(3 marks)*

(b) (i) Solve simultaneously:

$x^2 - 3x = 5x - x^2$

$2x^2 - 8x = 0$

$2x(x - 4) = 0$

$x = 0, 4$

$\therefore A$ has x-coordinate of 4 *(1 mark)*

(ii) Area $= \displaystyle\int_0^4 5x - x^2 - (x^2 - 3x)\,dx$

$= \displaystyle\int_0^4 8x - 2x^2\,dx$

$= \left[4x^2 - \dfrac{2x^3}{3}\right]_0^4$

$= 4 \times 4^2 - \dfrac{2(4)^3}{3} - 0$

$= 64 - \dfrac{128}{3}$

$= \dfrac{64}{3}$

$\therefore$ area is $21\frac{1}{3}$ units^2 *(3 marks)*

(c) (i) $P(RR) = \dfrac{3}{5} \times \dfrac{3}{7}$

$= \dfrac{9}{35}$ *(1 mark)*

(ii) P(at least one white) $= 1 - P(RR)$

$= 1 - \dfrac{9}{35}$

$= \dfrac{26}{35}$

(1 mark)

(iii) P(both same colour)
$= P(RR) + P(WW)$
$= \frac{3}{5} \times \frac{3}{7} + \frac{2}{5} \times \frac{4}{7}$
$= \frac{17}{35}$ *(2 marks)*

QUESTION 14

(a) $f(x) = 3x^4 + 4x^3 - 12x^2$

(i) $f'(x) = 12x^3 + 12x^2 - 24x = 0$
$12x(x^2 + x - 2) = 0$
$12x(x+2)(x-1) = 0$
$\therefore$ Stat. points at $x = 0, -2, 1$
$f(0) = 3(0)^4 + 4(0)^3 - 12(0)^2$
$= 0 \quad \therefore (0, 0)$
$f(-2) = 3(-2)^4 + 4(-2)^3 - 12(-2)^2$
$= -32 \quad \therefore (-2, -32)$
$f(1) = 3(1)^4 + 4(1)^3 - 12(1)^2$
$= -5 \quad \therefore (1, -5)$
$\therefore$ Stat. points at $(0, 0), (-2, -32), (1, -5)$
$f''(x) = 36x^2 + 24x - 24$
For $(0, 0)$:
$f''(0) = 36(0)^2 + 24(0) - 24$
$< 0 \quad \therefore$ Max $(0, 0)$
For $(-2, -32)$:
$f''(-2) = 36(-2)^2 + 24(-2) - 24$
$> 0 \quad \therefore$ Min $(-2, -32)$
For $(1, -5)$:
$f''(1) = 36(1)^2 + 24(1) - 24$
$> 0 \quad \therefore$ Min $(1, -5)$
$\therefore$ Max $(0, 0)$, Min $(-2, -32)$, Min $(1, -5)$
(3 marks)

(ii)

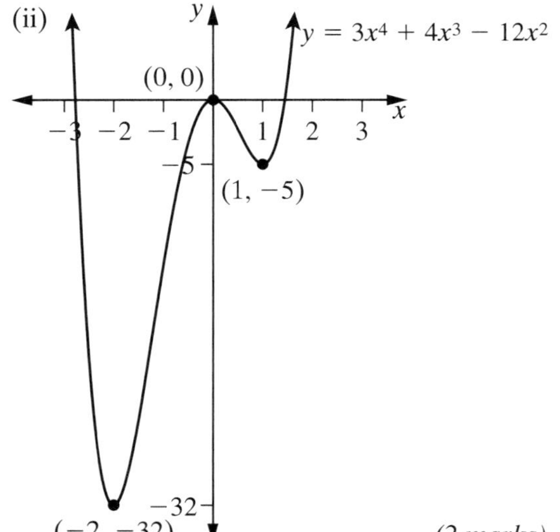

(2 marks)

(iii) From (ii), the function is increasing for $-2 < x < 0$ and $x > 1$
(1 mark)

(iv) $3x^4 + 4x^3 - 12x^2 + k = 0$
$3x^4 + 4x^3 - 12x^2 = -k$
Solution is point(s) of intersection of $y = 3x^4 + 4x^3 - 12x^2$ and $y = -k$.
Using part (ii), no solutions if $-k < -32$
i.e. $k > 32$
(1 mark)

(b) $V = \pi \int_0^1 y^2 \, dx$
$= \pi \int_0^1 \left[\frac{3}{(x+2)^2}\right]^2 dx$
$= \pi \int_0^1 \frac{9}{(x+2)^4} \, dx$
$= 9\pi \int_0^1 (x+2)^{-4} \, dx$
$= 9\pi \left[\frac{(x+2)^{-3}}{-3}\right]_0^1$
$= -3\pi \left[\frac{1}{(x+2)^3}\right]_0^1$
$= -3\pi \left[\frac{1}{27} - \frac{1}{8}\right]$
$= -3\pi \left[\frac{-19}{216}\right]$
$= \frac{19\pi}{72}$
$\therefore$ volume is $\frac{19\pi}{72}$ units^3 *(3 marks)*

(c) (i) $N = 1000e^{kt}, t = 20, N = 2000$
$2000 = 1000e^{20k}$
$e^{20k} = 2$
$\log_e e^{20k} = \log_e 2$
$20k = \log_e 2$
$k = \frac{\log_e 2}{20}$
$= 0.034\,657\,359 \ldots$
$= 0.0347$ (to 4 dec. pl.)
(1 mark)

(ii) $N = 1000e^{kt}, t = 120,$
$= 1000e^{k(120)}$
$= 64000$

$\therefore$ 64 000 bacteria
[or, using $k = 0.0347, N = 64328$]

(1 mark)

(iii) $N = 1000e^{kt}$

$$\frac{dN}{dt} = k.1000e^{kt}$$
$= k.64000$
$= 64000k$
$= 2218.070978\ldots$
$= 2218$ (nearest whole)

$\therefore$ 2218 bacteria per minute
[or, using $k = 0.0347$, 2232 per minute]

(1 mark)

(iv) $N = 1000e^{kt}, N = 100000$
$100000 = 1000e^{kt}$
$e^{kt} = 100$
$\log_e e^{kt} = \log_e 100$
$kt = \log_e 100$
$$t = \frac{\log_e 100}{k}$$
$= 132.8771238\ldots$
$= 133$ (nearest whole)

$\therefore$ 2h 13 min *(2 marks)*

QUESTION 15

(a) $10, 9.6, \ldots$

(i) $a = 10, r = 0.96, n = 10$

$$S_n = \frac{a(1-r^n)}{1-r}$$
$$= \frac{10(1-0.96^{10})}{1-0.96}$$
$= 83.791841\ldots$
$= 83.79$ (2 dec. pl.)

$\therefore$ length of the strip is 83.79 cm

(2 marks)

(ii) $$\text{lim sum} = \frac{a}{1-r}$$
$$= \frac{10}{1-0.96}$$
$= 250$

As limiting sum is 250, then the strip has a maximum length of 250 cm, or 2.5 m, which means that 3 m is sufficient.

(1 mark)

(b) (i) $\dot{x} = 1 - 2\cos t$

Subs $t = 0$,
$\dot{x}(0) = 1 - 2\cos(0)$
$= 1 - 2$
$= -1$

$\therefore$ initial velocity is -1 ms^{-1} *(1 mark)*

(ii) As $-1 \leq \cos t \leq 1$, then
$-2 \leq 2\cos t \leq 2$
$\therefore 2 \geq -2\cos t \geq -2$
$\therefore 3 \geq 1 - 2\cos t \geq -1$

$\therefore$ maximum velocity is 3 ms^{-1}

OR: Maximum velocity when $\ddot{x} = 0$
$\ddot{x} = 2\sin t = 0$
$t = 0, \pi, 2\pi, \ldots$
$\dot{x}(0) = -1$ from (i)
$\dot{x}(\pi) = 1 - 2\cos\pi$
$= 1 - 2(-1)$
$= 3$

$\therefore$ maximum velocity is 3 ms^{-1} *(1 mark)*

(iii) $\dot{x} = 1 - 2\cos t$
$x = t - 2\sin t + c$
When $t = 0, x = 3$,
$3 = 0 - 2\sin 0 + c$
$\therefore c = 3$
$\therefore x = t - 2\sin t + 3$ *(2 marks)*

(iv) $\dot{x} = 1 - 2\cos t = 0$

$$\cos t = \frac{1}{2}$$
$$t = \frac{\pi}{3}$$
$$x\left(\frac{\pi}{3}\right) = \frac{\pi}{3} - 2\sin\frac{\pi}{3} + 3$$
$$= \frac{\pi}{3} - 2\left(\frac{\sqrt{3}}{2}\right) + 3$$
$$= \frac{\pi}{3} - \sqrt{3} + 3$$ *(2 marks)*

(c) (i) 6% pa = 0.5% per month

As 0.5% = 0.005,
$A_1 = 360000 \times 1.005 - M$
$A_2 = A_1 \times 1.005 - M$
$= (360000 \times 1.005 - M) \times 1.005 - M$
$= 360000 \times 1.005^2 - M(1 + 1.005)$

(1 mark)

(ii) $A_{300} = 360\,000 \times 1.005^{300}$
$- M(1 + 1.005 + 1.005^2 + \ldots + 1.005^{299})$

Now, $A_{300} = 0$

Using geometric series with
$a = 1, r = 1.005, n = 300$

$$M = 360\,000 \times 1.005^{300} \div \frac{1(1.005^{300} - 1)}{1.005 - 1}$$

$= 2319.485045$

$\therefore$ approximately $2319.50 *(2 marks)*

(iii) Let $A_n = 180\,000$

$360\,000 \times 1.005^n$
$- 2319.50\,(1 + 1.005 + \ldots + 1.005^{n-1})$
$= 180\,000$

$$360\,000 \times 1.005^n - 2319.50 \times \frac{1(1.005^n - 1)}{0.005} = 180\,000$$

1800×1.005^n
$- 2319.5 \times (1.005^n - 1) = 900$

1800×1.005^n
$- 2319.5 \times 1.005^n + 2319.5 = 900$

$519.5 \times 1.005^n = 1419.5$

$519.5 \times 1.005^n = 1419.5$

$$1.005^n = \frac{1419.5}{519.5}$$

$$\log_e 1.005^n = \log_e\left[\frac{1419.5}{519.5}\right]$$

$$n \log_e 1.005 = \log_e\left[\frac{1419.5}{519.5}\right]$$

$$n = \frac{\log_e\left[\frac{1419.5}{519.5}\right]}{\log_e 1.005}$$

$= 201.5408119\ldots$

$\therefore$ after 202 months the amount will be less than $180 000. *(3 marks)*

QUESTION 16

(a)

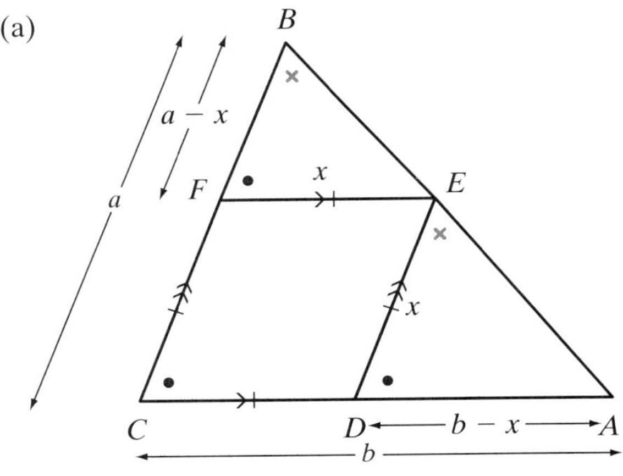

(i) $\angle BFE = \angle FCD$ (corr $\angle$s, $FE \| CD$)
$\angle FCD = \angle EDA$ (corr $\angle$s, $FC \| ED$)
$\therefore \angle BFE = \angle EDA$
$\angle FBE = \angle DEA$ (corr $\angle$s, $BC \| ED$)
$\therefore \triangle EBF$ is similar to $\triangle AED$ (2 $\angle$s equal)
(2 marks)

(ii) $\dfrac{FE}{DA} = \dfrac{BF}{ED}$ (matching sides of sim. $\triangle$s in proportion)

$$\frac{x}{b - x} = \frac{a - x}{x}$$

$x^2 = (a - x)(b - x)$

$x^2 = ab - ax - bx + x^2$

$ax + bx = ab$

$x(a + b) = ab$

$$x = \frac{ab}{a + b}$$ *(2 marks)*

(b)

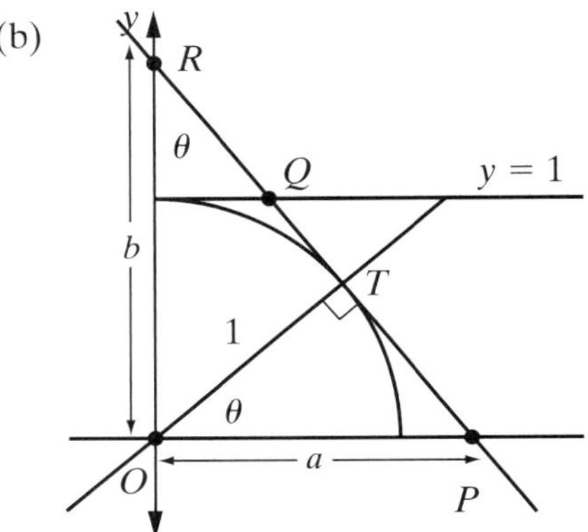

(i) Using $\frac{x}{a}+\frac{y}{b}=1$:

In $\triangle OPT$, is $\frac{OT}{OP}=\cos\theta$

$$\frac{1}{a}=\cos\theta$$

$$a=\frac{1}{\cos\theta}$$

In $\triangle ORT$, is $\frac{OT}{OR}=\sin\theta$

$$\frac{1}{b}=\sin\theta$$

$$b=\frac{1}{\sin\theta}$$

$\therefore$ equation is $\frac{x}{\frac{1}{\cos\theta}}+\frac{y}{\frac{1}{\sin\theta}}=1$

$x\cos\theta+y\sin\theta=1$ **1**

(2 marks)

(ii) Subs $y=1$ in **1**:

$x\cos\theta+\sin\theta=1$

$$x\cos\theta=1-\sin\theta$$

$$x=\frac{1-\sin\theta}{\cos\theta}$$

$\therefore Q(\frac{1-\sin\theta}{\cos\theta},1)$

As $B(0,1)$, then $BQ=\frac{1-\sin\theta}{\cos\theta}$

(1 mark)

(iii) Area $=\frac{1}{2}h(a+b)$

$$A=\frac{1}{2}\times1\times(\frac{1-\sin\theta}{\cos\theta}+\frac{1}{\cos\theta})$$

$$=\frac{1}{2}\times(\frac{2-\sin\theta}{\cos\theta})$$

$$=\frac{2-\sin\theta}{2\cos\theta}$$

$\therefore A=\frac{2-\sin\theta}{2\cos\theta}$ units^2

(2 marks)

(iv) $A=\frac{2-\sin\theta}{2\cos\theta}$

$$\frac{dA}{d\theta}=\frac{2\cos\theta(-\cos\theta)-(2-\sin\theta).(-2\sin\theta)}{4\cos^2\theta}$$

$$=\frac{-2\cos^2\theta+4\sin\theta-2\sin^2\theta}{4\cos^2\theta}$$

$$=\frac{4\sin\theta-2(\sin^2\theta+\cos^2\theta)}{4\cos^2\theta}$$

$$=\frac{4\sin\theta-2}{4\cos^2\theta}$$

$$=\frac{2(2\sin\theta-1)}{4\cos^2\theta}$$

$$=\frac{(2\sin\theta-1)}{2\cos^2\theta}=0$$

$2\sin\theta-1=0$

$$\sin\theta=\frac{1}{2}$$

$$\theta=\frac{\pi}{6}\quad(\text{as } 0<\theta<\frac{\pi}{2})$$

Test for min:

θ	$\frac{\pi}{12}$	$\frac{\pi}{6}$	$\frac{\pi}{3}$
$\frac{dA}{d\theta}$	<0	0	>0

$\therefore\theta=\frac{\pi}{6}$ for minimum *(3 marks)*

(c) (i) $y=x^2$ **1**

$x^2+(y-c)^2=r^2$ **2**

$\therefore$ Subs **1** in **2**:

$y+(y-c)^2=r^2$

$y+y^2-2cy+c^2=r^2$

$y^2+(1-2c)y+c^2-r^2=0$

$\Delta=0$ for equal roots:

$\therefore\ \Delta=(1-2c)^2-4\times1\times(c^2-r^2)$

$=1-4c+4c^2-4c^2+4r^2=0$

$4r^2-4c+1=0$

$4c=1+4r^2$ *(2 marks)*

(ii) From diagram, $c>r$

$\therefore 4c>4r$

$16c^2>16r^2$

$16c^2>4(4r^2)$

$16c^2>4(4c-1)$ (using part i)

$16c^2>16c-4$

$16c^2-16c+4>0$

$4c^2-4c+1>0$

$(2c-1)^2>0$

$2c-1>0$ (since $c>0$)

$2c>1$

$c>\frac{1}{2}$ *(1 mark)*

2013

HIGHER SCHOOL CERTIFICATE EXAMINATION

Mathematics

General Instructions

- Reading time – 5 minutes
- Working time – 3 hours
- Write using black or blue pen
  Black pen is preferred
- Board-approved calculators may be used
- A table of standard integrals is provided at the back of this paper
- In Questions 11–16, show relevant mathematical reasoning and/or calculations

Total marks – 100

Section I

10 marks

- Attempt Questions 1–10
- Allow about 15 minutes for this section

Section II

90 marks

- Attempt Questions 11–16
- Allow about 2 hours and 45 minutes for this section

Section I

10 marks
Attempt Questions 1–10
Allow about 15 minutes for this section

Use the multiple-choice answer sheet for Questions 1–10.

---

1 What are the solutions of $2x^2 - 5x - 1 = 0$?

(A) $x = \dfrac{-5 \pm \sqrt{17}}{4}$

(B) $x = \dfrac{5 \pm \sqrt{17}}{4}$

(C) $x = \dfrac{-5 \pm \sqrt{33}}{4}$

(D) $x = \dfrac{5 \pm \sqrt{33}}{4}$

2 The diagram shows the line ℓ.

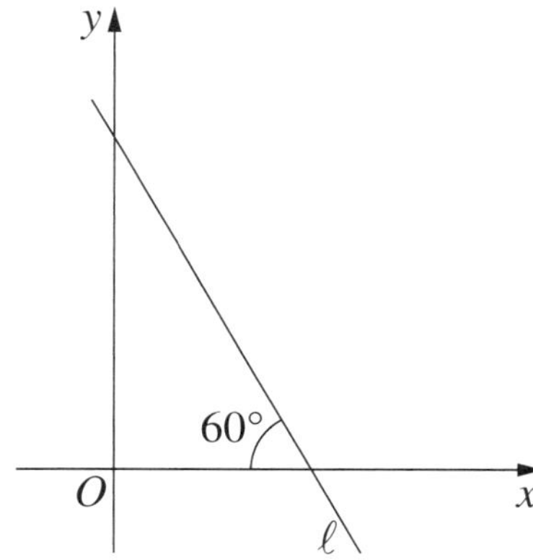

What is the slope of the line ℓ?

(A) $\sqrt{3}$

(B) $-\sqrt{3}$

(C) $\dfrac{1}{\sqrt{3}}$

(D) $-\dfrac{1}{\sqrt{3}}$

3 Which inequality defines the domain of the function $f(x) = \dfrac{1}{\sqrt{x+3}}$?

(A) $x > -3$

(B) $x \geq -3$

(C) $x < -3$

(D) $x \leq -3$

4 What is the derivative of $\dfrac{x}{\cos x}$?

(A) $\dfrac{\cos x + x \sin x}{\cos^2 x}$

(B) $\dfrac{\cos x - x \sin x}{\cos^2 x}$

(C) $\dfrac{x \sin x - \cos x}{\cos^2 x}$

(D) $\dfrac{-x \sin x - \cos x}{\cos^2 x}$

5 A bag contains 4 red marbles and 6 blue marbles. Three marbles are selected at random without replacement.

What is the probability that at least one of the marbles selected is red?

(A) $\dfrac{1}{6}$

(B) $\dfrac{1}{2}$

(C) $\dfrac{5}{6}$

(D) $\dfrac{29}{30}$

6 Which diagram shows the graph $y = \sin\left(2x + \frac{\pi}{3}\right)$?

(A)

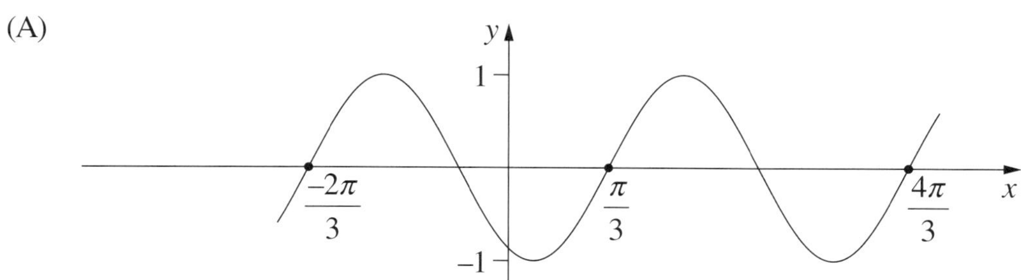

(B)

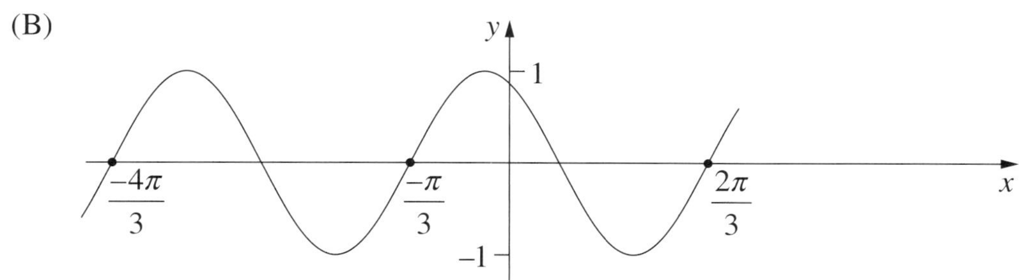

(C)

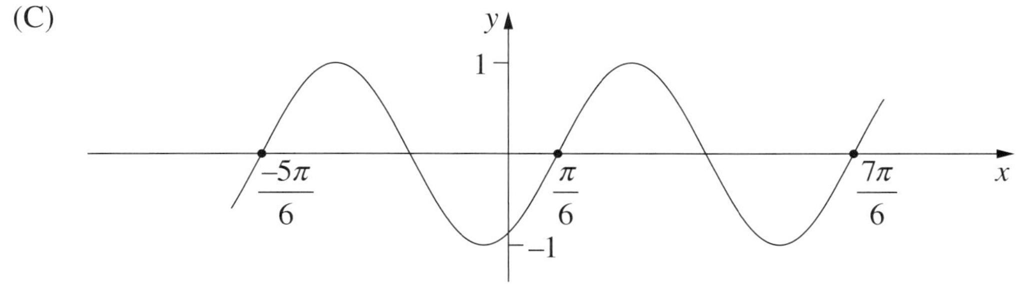

(D)

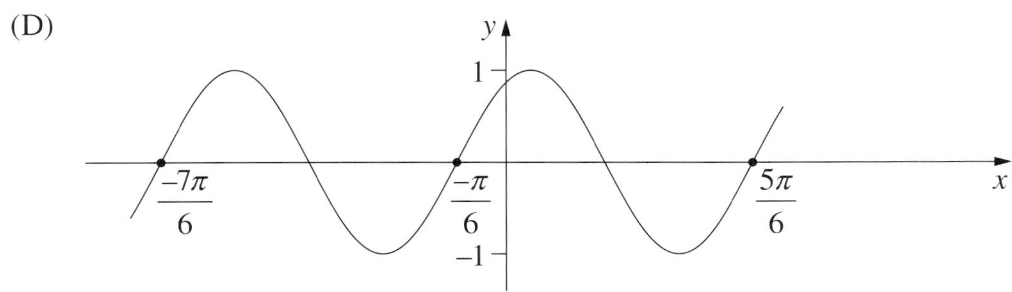

7 A parabola has focus $(5, 0)$ and directrix $x = 1$.

What is the equation of the parabola?

(A) $y^2 = 16(x-5)$

(B) $y^2 = 8(x-3)$

(C) $y^2 = -16(x-5)$

(D) $y^2 = -8(x-3)$

8 The diagram shows points A, B, C and D on the graph $y = f(x)$.

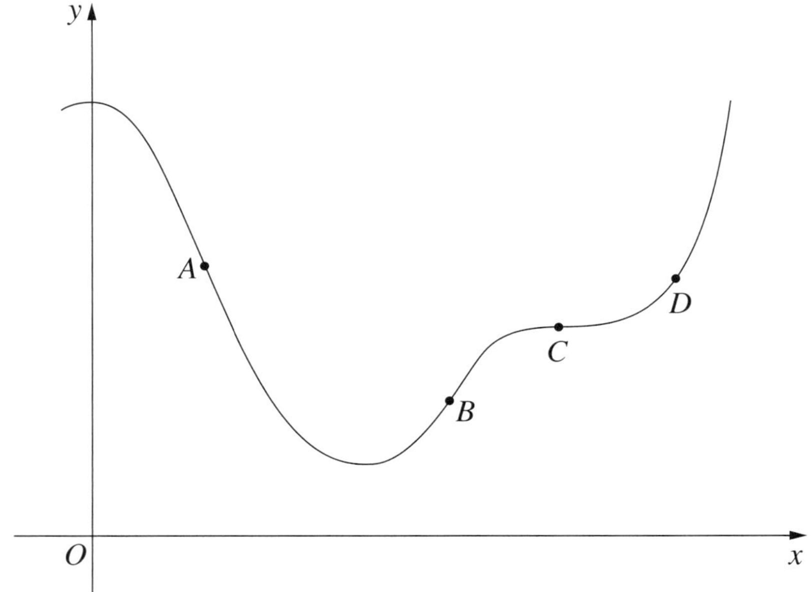

At which point is $f'(x) > 0$ and $f''(x) = 0$?

(A) A

(B) B

(C) C

(D) D

9 What is the solution of $5^x = 4$?

(A) $x = \dfrac{\log_e 4}{5}$

(B) $x = \dfrac{4}{\log_e 5}$

(C) $x = \dfrac{\log_e 4}{\log_e 5}$

(D) $x = \log_e \left(\dfrac{4}{5} \right)$

10 A particle is moving along the x-axis. The displacement of the particle at time t seconds is x metres.

At a certain time, $\dot{x} = -3\,\text{m s}^{-1}$ and $\ddot{x} = 2\,\text{m s}^{-2}$.

Which statement describes the motion of the particle at that time?

(A) The particle is moving to the right with increasing speed.

(B) The particle is moving to the left with increasing speed.

(C) The particle is moving to the right with decreasing speed.

(D) The particle is moving to the left with decreasing speed.

Section II

90 marks
Attempt Questions 11–16
Allow about 2 hours and 45 minutes for this section

Answer each question in the appropriate writing booklet. Extra writing booklets are available.

In Questions 11–16, your responses should include relevant mathematical reasoning and/or calculations.

Question 11 (15 marks) Use the Question 11 Writing Booklet.

(a) Evaluate $\ln 3$ correct to three significant figures. **1**

(b) Evaluate $\lim_{x \to 2} \dfrac{x^3 - 8}{x^2 - 4}$. **2**

(c) Differentiate $(\sin x - 1)^8$. **2**

(d) Differentiate $x^2 e^x$. **2**

(e) Find $\int e^{4x+1}\,dx$. **2**

(f) Evaluate $\int_0^1 \dfrac{x^2}{x^3 + 1}\,dx$. **3**

(g) Sketch the region defined by $(x - 2)^2 + (y - 3)^2 \geq 4$. **3**

Question 12 (15 marks) Use the Question 12 Writing Booklet.

(a) The cubic $y = ax^3 + bx^2 + cx + d$ has a point of inflexion at $x = p$. **2**

Show that $p = -\dfrac{b}{3a}$.

(b) The points $A(-2, -1)$, $B(-2, 24)$, $C(22, 42)$ and $D(22, 17)$ form a parallelogram as shown. The point $E(18, 39)$ lies on BC. The point F is the midpoint of AD.

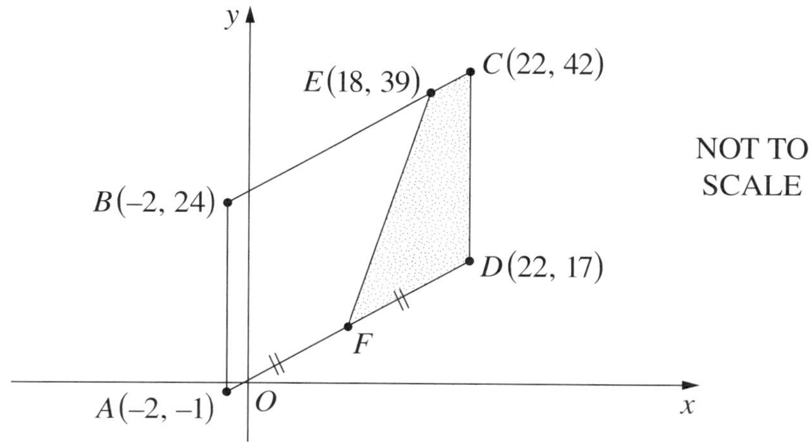

(i) Show that the equation of the line through A and D is $3x - 4y + 2 = 0$. **2**

(ii) Show that the perpendicular distance from B to the line through A and D is 20 units. **1**

(iii) Find the length of EC. **1**

(iv) Find the area of the trapezium $EFDC$. **2**

(c) Kim and Alex start jobs at the beginning of the same year. Kim's annual salary in the first year is \$30 000, and increases by 5% at the beginning of each subsequent year. Alex's annual salary in the first year is \$33 000, and increases by \$1500 at the beginning of each subsequent year.

(i) Show that in the 10th year Kim's annual salary is higher than Alex's annual salary. **2**

(ii) In the first 10 years how much, in total, does Kim earn? **2**

(iii) Every year, Alex saves $\frac{1}{3}$ of her annual salary. How many years does it take her to save \$87 500? **3**

Question 13 (15 marks) Use the Question 13 Writing Booklet.

(a) The population of a herd of wild horses is given by

$$P(t) = 400 + 50\cos\left(\frac{\pi}{6}t\right),$$

where t is time in months.

(i) Find all times during the first 12 months when the population equals 375 horses. **2**

(ii) Sketch the graph of $P(t)$ for $0 \le t \le 12$. **2**

(b) The diagram shows the graphs of the functions $f(x) = 4x^3 - 4x^2 + 3x$ and $g(x) = 2x$. The graphs meet at O and at T.

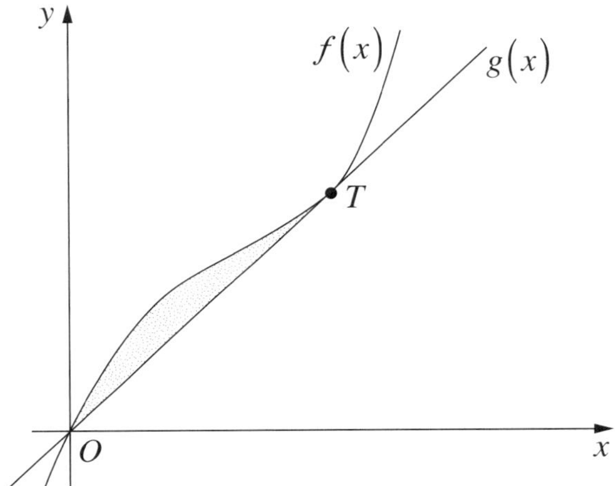

(i) Find the x-coordinate of T. **1**

(ii) Find the area of the shaded region between the graphs of the functions $f(x)$ and $g(x)$. **3**

Question 13 continues on the following page

Question 13 (continued)

(c) The region ABC is a sector of a circle with radius 30 cm, centred at C. The angle of the sector is θ. The arc DE lies on a circle also centred at C, as shown in the diagram. **2**

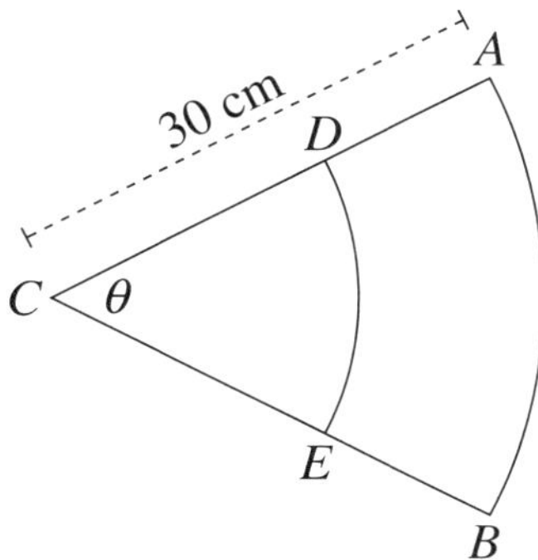

The arc DE divides the sector ABC into two regions of equal area.

Find the exact length of the interval CD.

(d) A family borrows \$500 000 to buy a house. The loan is to be repaid in equal monthly instalments. The interest, which is charged at 6% per annum, is reducible and calculated monthly. The amount owing after n months, $\$A_n$, is given by

$$A_n = Pr^n - M\left(1 + r + r^2 + \cdots + r^{n-1}\right), \qquad \text{(Do NOT prove this)}$$

where $\$P$ is the amount borrowed, $r = 1.005$ and $\$M$ is the monthly repayment.

(i) The loan is to be repaid over 30 years. Show that the monthly repayment is \$2998 to the nearest dollar. **2**

(ii) Show that the balance owing after 20 years is \$270 000 to the nearest thousand dollars. **1**

(iii) After 20 years the family borrows an extra amount, so that the family then owes a total of \$370 000. The monthly repayment remains \$2998, and the interest rate remains the same. **2**

How long will it take to repay the \$370 000?

End of Question 13

Question 14 (15 marks) Use the Question 14 Writing Booklet.

(a) The velocity of a particle moving along the x-axis is given by $\dot{x} = 10 - 2t$, where x is the displacement from the origin in metres and t is the time in seconds. Initially the particle is 5 metres to the right of the origin.

(i) Show that the acceleration of the particle is constant. **1**

(ii) Find the time when the particle is at rest. **1**

(iii) Show that the position of the particle after 7 seconds is 26 metres to the right of the origin. **2**

(iv) Find the distance travelled by the particle during the first 7 seconds. **2**

(b) Two straight roads meet at R at an angle of 60°. At time $t = 0$ car A leaves R on one road, and car B is 100 km from R on the other road. Car A travels away from R at a speed of 80 km/h, and car B travels towards R at a speed of 50 km/h.

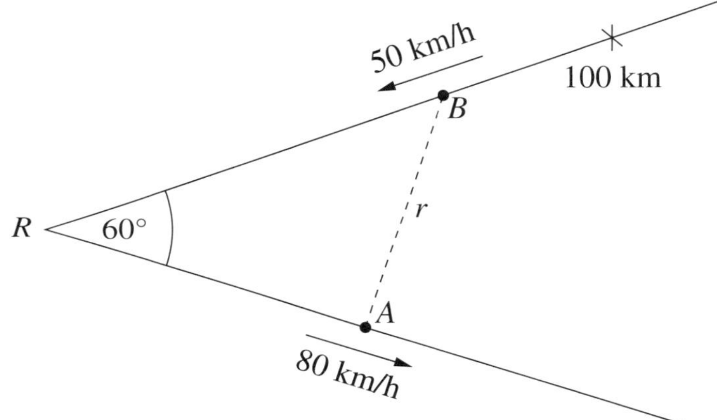

The distance between the cars at time t hours is r km.

(i) Show that $r^2 = 12\,900t^2 - 18\,000t + 10\,000$. **2**

(ii) Find the minimum distance between the cars. **3**

Question 14 continues on the following page

Question 14 (continued)

(c) The right-angled triangle ABC has hypotenuse $AB = 13$. The point D is on AC such that $DC = 4$, $\angle DBC = \frac{\pi}{6}$ and $\angle ABD = x$. **3**

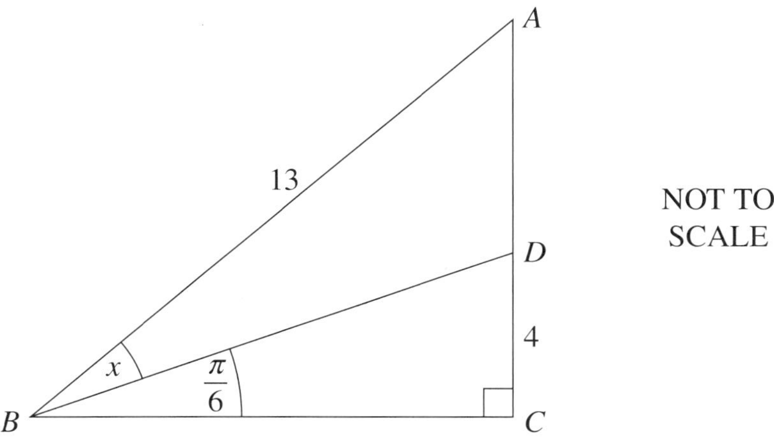

NOT TO SCALE

Using the sine rule, or otherwise, find the exact value of $\sin x$.

(d) The diagram shows the graph $y = f(x)$. **1**

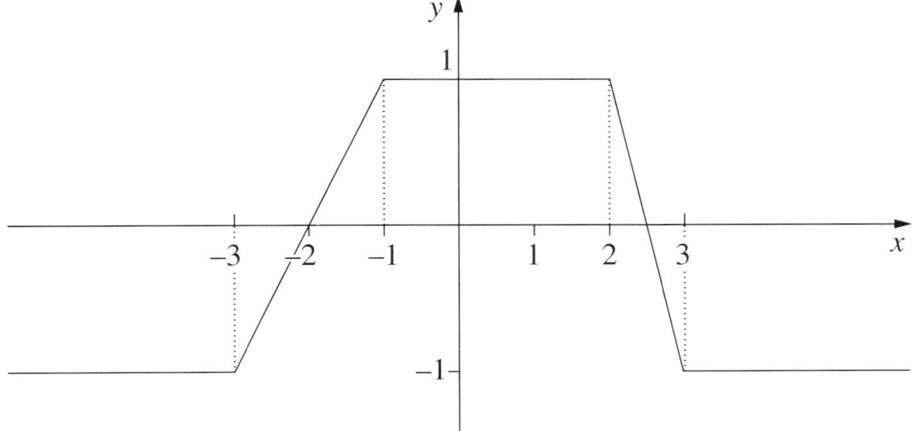

What is the value of a, where $a > 0$, so that $\int_{-a}^{a} f(x)\,dx = 0$?

End of Question 14

Question 15 (15 marks) Use the Question 15 Writing Booklet.

(a) The diagram shows the front of a tent supported by three vertical poles. The poles are 1.2 m apart. The height of each outer pole is 1.5 m, and the height of the middle pole is 1.8 m. The roof hangs between the poles.

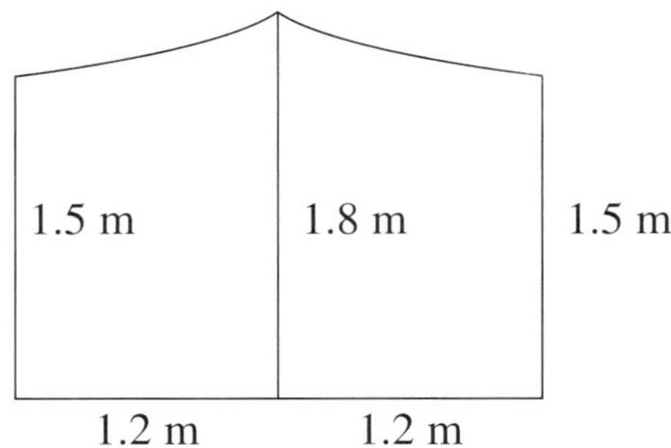

The front of the tent has area A m^2.

(i) Use the trapezoidal rule to estimate A. **1**

(ii) Use Simpson's rule to estimate A. **1**

(iii) Explain why the trapezoidal rule gives the better estimate of A. **1**

(b) The region bounded by the x-axis, the y-axis and the parabola $y = (x-2)^2$ is rotated about the y-axis to form a solid.  **4**

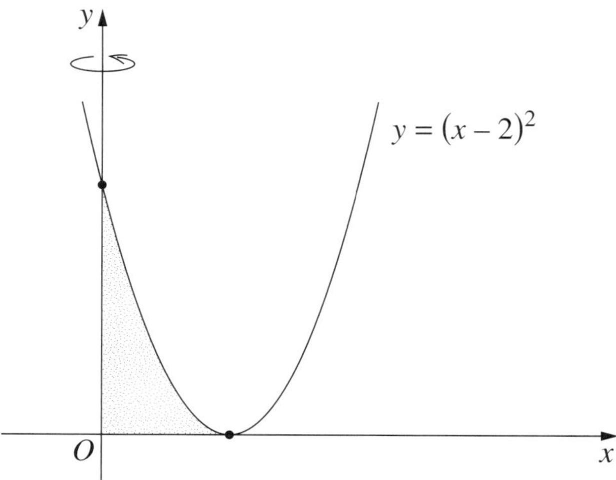

Find the volume of the solid.

Question 15 continues on the following page

Question 15 (continued)

(c) (i) Sketch the graph $y = |2x - 3|$. **1**

(ii) Using the graph from part (i), or otherwise, find all values of m for which the equation $|2x - 3| = mx + 1$ has exactly one solution. **2**

(d) Pat and Chandra are playing a game. They take turns throwing two dice. The game is won by the first player to throw a double six. Pat starts the game.

(i) Find the probability that Pat wins the game on the first throw. **1**

(ii) What is the probability that Pat wins the game on the first or on the second throw? **2**

(iii) Find the probability that Pat eventually wins the game. **2**

End of Question 15

Question 16 (15 marks) Use the Question 16 Writing Booklet.

(a) The derivative of a function $f(x)$ is $f'(x) = 4x - 3$. The line $y = 5x - 7$ is tangent to the graph of $f(x)$. **3**

Find the function $f(x)$.

(b) Trout and carp are types of fish. A lake contains a number of trout. At a certain time 10 carp are introduced into the lake and start eating the trout. As a consequence, the number of trout, N, decreases according to

$$N = 375 - e^{0.04t},$$

where t is the time in months after the carp are introduced.

The population of carp, P, increases according to

$$\frac{dP}{dt} = 0.02P.$$

(i) How many trout were in the lake when the carp were introduced? **1**

(ii) When will the population of trout be zero? **1**

(iii) Sketch the number of trout as a function of time. **1**

(iv) When is the rate of increase of carp equal to the rate of decrease of trout? **3**

(v) When is the number of carp equal to the number of trout? **2**

(c) The diagram shows triangles ABC and ABD with AD parallel to BC. The sides AC and BD intersect at Y. The point X lies on AB such that XY is parallel to AD and BC.

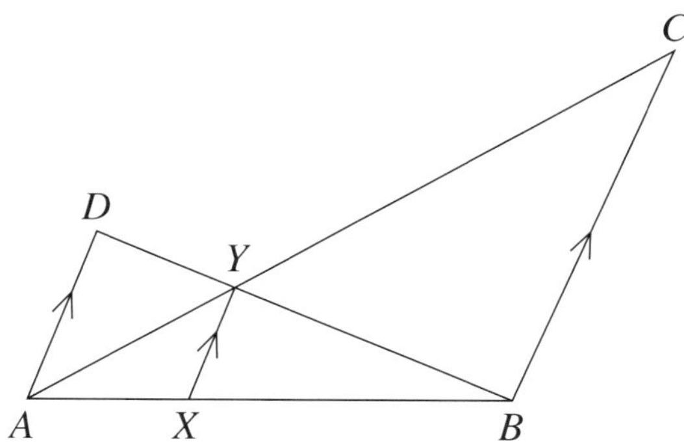

(i) Prove that $\triangle ABC$ is similar to $\triangle AXY$. **2**

(ii) Hence, or otherwise, prove that $\dfrac{1}{XY} = \dfrac{1}{AD} + \dfrac{1}{BC}$. **2**

End of paper

2013 Higher School Certificate

Worked Answers

Section I

QUESTION 1

$2x^2 - 5x - 1 = 0$

$$x = \frac{5 \pm \sqrt{(-5)^2 - 4(2)(-1)}}{2(2)}$$

$$= \frac{5 \pm \sqrt{33}}{4}$$

Answer D

(1 mark)

QUESTION 2

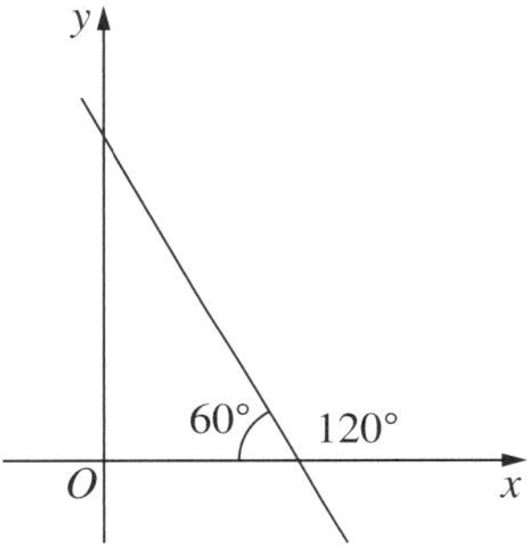

The line forms an angle of 120° with the positive direction of the x-axis:

$\therefore m = \tan\theta$

$= \tan 120°$

$= -\tan 60°$

$= -\sqrt{3}$

Answer B

(1 mark)

QUESTION 3

$$f(x) = \frac{1}{\sqrt{x+3}}$$

Consider $x + 3 > 0$

$x > -3$

Answer A

(1 mark)

QUESTION 4

Using the quotient rule,

Let $u = x$, $\quad u' = 1$

Let $v = \cos x$, $\quad v' = -\sin x$

$$\frac{d}{dx}\left[\frac{x}{\cos x}\right] = \frac{v.u' - u.v'}{v^2}$$

$$= \frac{\cos x.1 - x.(-\sin x)}{(\cos x)^2}$$

$$= \frac{\cos x + x\sin x}{\cos^2 x}$$

Answer A

(1 mark)

QUESTION 5

$$\text{P(at least one is red)} = 1 - \text{P(none red)}$$

$$= 1 - \frac{6}{10} \times \frac{5}{9} \times \frac{4}{8}$$

$$= \frac{5}{6}$$

Answer C

(1 mark)

QUESTION 6

Solve $\sin\left(2x + \frac{\pi}{3}\right) = 0$

$$2x + \frac{\pi}{3} = 0, \pi, 2\pi \ldots -\pi, -2\pi \ldots$$

$$2x = -\frac{\pi}{3}, \frac{2\pi}{3}, \frac{5\pi}{3} \ldots \frac{-4\pi}{3}, \frac{-7\pi}{3} \ldots$$

$$x = -\frac{\pi}{6}, \frac{\pi}{3}, \frac{5\pi}{6}, \frac{-2\pi}{3}, \frac{-7\pi}{6} \ldots$$

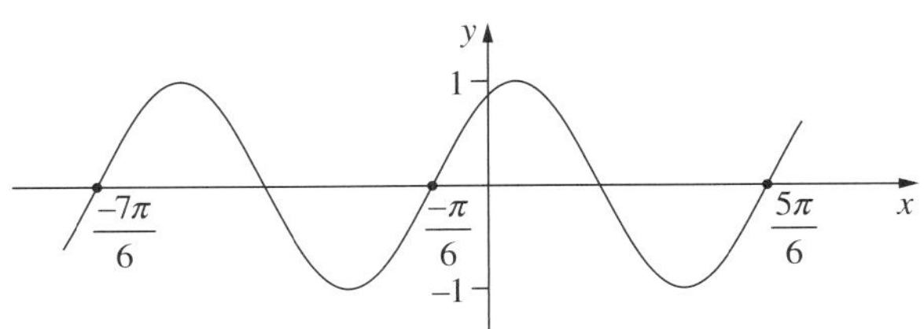

Answer D

(1 mark)

QUESTION 7

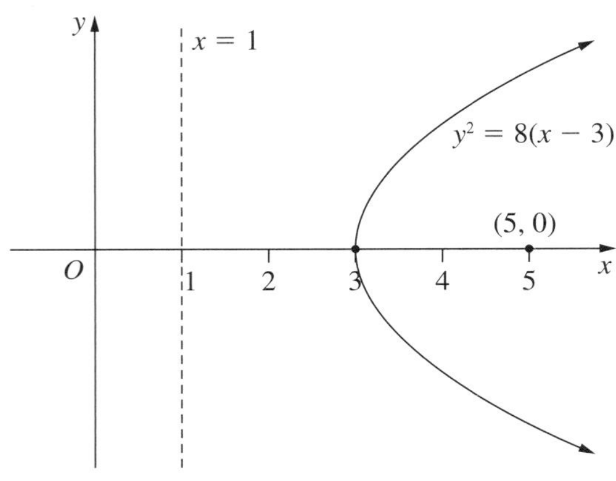

$V(3, 0), a = 2;$

$$(y - k)^2 = 4a(x - h)$$
$$(y - 0)^2 = 8(x - 3)$$
$$y^2 = 8(x - 3)$$

Answer B

(1 mark)

QUESTION 8

At B: increasing ($f'(x) > 0$), and point of inflexion ($f''(x) = 0$)

[Note: At A, decreasing ($f'(x) < 0$)
At C: stationary ($f'(x) = 0$)
At D: concave up ($f''(x) > 0$)]

Answer B

(1 mark)

QUESTION 9

$$5^x = 4$$
$$\log_e 5^x = \log_e 4$$
$$x \log_e 5 = \log_e 4$$
$$x = \frac{\log_e 4}{\log_e 5}$$

Answer C

(1 mark)

QUESTION 10

$\dot{x} = -3$ means the particle is moving to the left; as $\ddot{x} = +2$, the particle is slowing down.

$\therefore$ particle is moving to the left with decreasing speed.

Answer D

(1 mark)

Section II

QUESTION 11

(a) $\ln 3 = 1.0986 \ldots$
$= 1.10$ (3 sig. figs)

(1 mark)

(b)
$$\lim_{x \to 2} \frac{x^3 - 8}{x^2 - 4} = \lim_{x \to 2} \frac{(x - 2)(x^2 + 2x + 4)}{(x - 2)(x + 2)}$$
$$= \lim_{x \to 2} \frac{x^2 + 2x + 4}{x + 2}$$
$$= \frac{2^2 + 2(2) + 4}{2 + 2}$$
$$= 3$$

(2 marks)

(c) $(\sin x - 1)^8$

Using the function of a function rule:

$$\frac{d}{dx}[(\sin x - 1)^8] = 8(\sin x - 1)^7 . \frac{d}{dx}(\sin x - 1)$$
$$= 8(\sin x - 1)^7 . \cos x$$
$$= 8 \cos x \, (\sin x - 1)^7$$

(2 marks)

(d) $x^2 e^x$

Using the product rule,

Let $u = x^2$, $u' = 2x$

Let $v = e^x$, $v' = e^x$

$$\begin{aligned}\frac{dy}{dx} &= u'.v + v'.u \\ &= 2x.e^x + e^x.x^2 \\ &= 2xe^x + x^2e^x\end{aligned}$$

(2 marks)

(e) $\int e^{4x+1}\,dx = \frac{1}{4}e^{4x+1} + c$ *(2 marks)*

(f)
$$\begin{aligned}\int_0^1 \frac{x^2}{x^3+1}\,dx &= \frac{1}{3}\int_0^1 \frac{3x^2}{x^3+1}\,dx \\ &= \frac{1}{3}[\log_e(x^3+1)]_0^1 \\ &= \frac{1}{3}[\log_e(1+1) - \log_e(0+1)] \\ &= \frac{1}{3}[\log_e 2 - \log_e 1] \\ &= \frac{1}{3}\log_e 2\end{aligned}$$

(3 marks)

(g) $(x-2)^2 + (y-3)^2 \geq 4$

Circle, centre $(2, 3)$, radius $= 2$

Check $(0, 0)$:

$\therefore (0-2)^2 + (0-3)^2 \geq 4$?

Yes!

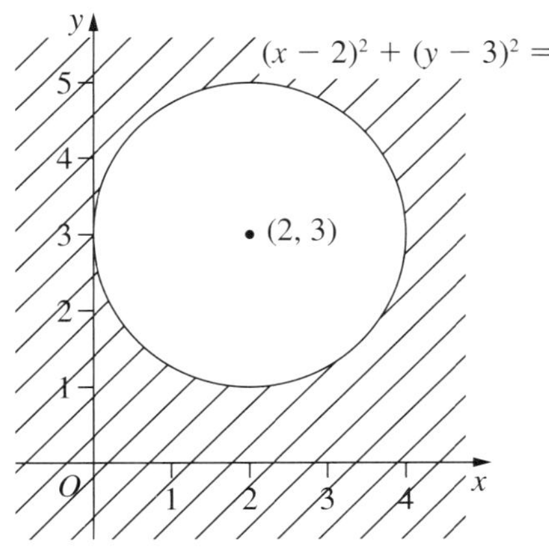

(3 marks)

QUESTION 12

(a)
$$\begin{aligned}y &= ax^3 + bx^2 + cx + d \\ y' &= 3ax^2 + 2bx + c \\ y'' &= 6ax + 2b \\ y''(p) &= 6ap + 2b = 0 \\ 6ap &= -2b \\ p &= \frac{-2b}{6a} \\ &= \frac{-b}{3a}\end{aligned}$$

(2 marks)

(b) (i)
$$\begin{aligned}\text{grad } AD &= \frac{17+1}{22+2} \\ &= \frac{18}{24} \\ &= \frac{3}{4} \\ y + 1 &= \frac{3}{4}(x+2) \\ 4y + 4 &= 3x + 6 \\ 3x - 4y + 2 &= 0\end{aligned}$$

(2 marks)

(ii) Use $B(-2, 24)$ and $3x - 4y + 2 = 0$

$$\begin{aligned}d &= \left|\frac{ax_1 + by_1 + c}{\sqrt{a^2+b^2}}\right| \\ &= \left|\frac{3(-2) - 4(24) + 2}{\sqrt{3^2 + (-4)^2}}\right| \\ &= \left|\frac{-100}{\sqrt{25}}\right| \\ &= |-20| \\ &= 20 \quad \therefore 20 \text{ units}\end{aligned}$$

(1 mark)

(iii)
$$\begin{aligned}d &= \sqrt{(22-18)^2 + (42-39)^2} \\ &= \sqrt{16+9} \\ &= \sqrt{25} \\ &= 5 \quad \therefore 5 \text{ units}\end{aligned}$$

(1 mark)

(iv) Midpoint of AD

$$\begin{aligned}&= \left(\frac{22-2}{2}, \frac{17-1}{2}\right) \\ &= (10, 8)\end{aligned}$$

Now, distance (10, 8) to (22, 17):

$$d = \sqrt{(22-10)^2 + (17-8)^2}$$
$$= \sqrt{144 + 81}$$
$$= \sqrt{225}$$
$$= 15 \quad \therefore 15 \text{ units}$$

$\therefore$ Area of trapezium:

$$A = \frac{1}{2} \times 20(5 + 15)$$
$$= 10(20)$$
$$= 200 \quad \therefore 200 \text{ units}^2$$

(2 marks)

(c) (i) Kim: 30 000 + 31 500 + …

$a = 30\,000, r = 1.05, n = 10$

$$T_n = ar^{n-1}$$
$$T_{10} = 30\,000 \times (1.05)^{10-1}$$
$$= 30\,000 \times (1.05)^9$$
$$= 46\,539.846\,48 \ldots$$
$$= 46\,540 \text{ (nearest whole)}$$

Alex: 33 000 + 34 500 + …

$a = 33\,000, d = 1500, n = 10$

$$T_n = a + (n-1)d$$
$$T_{10} = 33\,000 + (10-1) \times 1500$$
$$= 46\,500$$

$\therefore$ Kim's salary is \$40 more than Alex. *(2 marks)*

(ii)
$$S_n = \frac{a(r^n - 1)}{r - 1}$$
$$S_{10} = \frac{30\,000(1.05^{10} - 1)}{1.05 - 1}$$
$$= 377\,336.7761 \ldots$$
$$= 377\,336.78 \text{ (2 dec. pl.)}$$

$\therefore$ Kim earns \$377 336.78 *(2 marks)*

(iii) Total salary = \$87 500 × 3
= \$262 500

$$S_n = \frac{n}{2}[2a + (n-1)d]$$
$$= \frac{n}{2}[2(33\,000) + (n-1)1500]$$
$$= 262\,500$$
$$n[33\,000 + (n-1)750] = 262\,500$$
$$33\,000n + 750n^2 - 750n - 262\,500 = 0$$
$$750n^2 + 32\,250n - 262\,500 = 0$$
$$n^2 + 43n - 350 = 0$$
$$(n-7)(n+50) = 0$$
$$= 7, -50 \quad (\text{but}, n > 0)$$

$\therefore$ takes 7 years *(3 marks)*

QUESTION 13

(a) $P(t) = 400 + 50\cos\left(\frac{\pi}{6}t\right)$

(i)
$$400 + 50\cos\left(\frac{\pi}{6}t\right) = 375$$
$$50\cos\left(\frac{\pi}{6}t\right) = 375 - 400$$
$$50\cos\left(\frac{\pi}{6}t\right) = -25$$
$$\cos\left(\frac{\pi}{6}t\right) = -\frac{1}{2}$$
$$\frac{\pi}{6}t = \frac{2\pi}{3}, \frac{4\pi}{3}, \frac{8\pi}{3} \ldots$$
$$t = 4, 8, 16 \ldots$$

After 4 months and 8 months. *(2 marks)*

(ii)
$$P(0) = 400 + 50\cos\left(\frac{\pi}{6} \times 0\right)$$
$$= 400 + 50\cos 0$$
$$= 450$$

$$P(12) = 400 + 50\cos\left(\frac{\pi}{6} \times 12\right)$$
$$= 400 + 50\cos 2\pi$$
$$= 450$$

Also, $P(6) = 400 + 50\cos\left(\frac{\pi}{6} \times 6\right)$
$$= 400 + 50\cos\pi$$
$$= 350$$

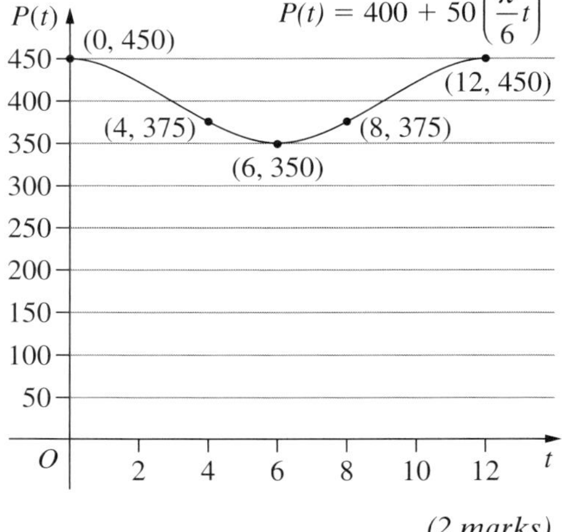

(2 marks)

(b) (i)

$$4x^3 - 4x^2 + 3x = 2x$$
$$4x^3 - 4x^2 + x = 0$$
$$x(4x^2 - 4x + 1) = 0$$
$$x(2x - 1)^2 = 0$$
$$x = 0, \frac{1}{2}$$

$\therefore T$ has x-coordinate of $\frac{1}{2}$ *(1 mark)*

(ii)

$$\text{Area} = \int_0^{\frac{1}{2}} 4x^3 - 4x^2 + 3x - 2x\, dx$$
$$= \int_0^{\frac{1}{2}} 4x^3 - 4x^2 + x\, dx$$
$$= \left[x^4 - \frac{4x^3}{3} + \frac{x^2}{2}\right]_0^{\frac{1}{2}}$$
$$= \left(\frac{1}{2}\right)^4 - \frac{4\left(\frac{1}{2}\right)^3}{3} + \frac{\left(\frac{1}{2}\right)^2}{2} - 0$$
$$= \frac{1}{16} - \frac{1}{6} + \frac{1}{8}$$
$$= \frac{1}{48}$$

$\therefore$ area is $\frac{1}{48}$ units^2 *(3 marks)*

(c) Area of sector ACB

$$= \frac{1}{2}r^2\theta$$
$$= \frac{1}{2} \times 30^2 \times \theta$$
$$= 450\theta$$

Area of sector DCE

$$= \frac{1}{2} \times 450\theta$$
$$= 225\theta$$

$$\therefore \frac{1}{2} \times r^2 \times \theta = 225\theta$$
$$\therefore r^2 = 450$$
$$r = \sqrt{450}$$
$$= 15\sqrt{2}$$

$\therefore$ length of CD is $15\sqrt{2}$ cm

(2 marks)

(d) (i)

$$A_n = Pr^n - M(1 + r + r^2 + \ldots + r^{n-1})$$
$$n = 30 \times 12 = 360$$
$$A_{360} = 500\,000 \times 1.005^{360} - M(1 + 1.005 + \ldots + 1.005^{359})$$

Now, $1 + 1.005 + \ldots + 1.005^{359}$ is Geom. series, with $a = 1, r = 1.005, n = 360$.

Using $S_n = \dfrac{a(r^n - 1)}{r - 1}$,

$$A_{360} = 500\,000 \times 1.005^{360} - M \times \frac{1(1.005^{360} - 1)}{1.005 - 1}$$

Let $A_{360} = 0$:

$$M = 500\,000 \times 1.005^{360} \div \frac{1(1.005^{360} - 1)}{1.005 - 1}$$
$$= 2997.752\,626\ldots$$
$$= 2998 \text{ (nearest whole)}$$

$\therefore \$2998$

(2 marks)

(ii) Let $n = 240, M = 2998$:

$$A_{240} = 500\,000 \times 1.005^{240} - 2998 \times \frac{1(1.005^{240} - 1)}{1.005 - 1}$$
$$= 269\,903.6342\ldots$$
$$= 270\,000 \text{ (nearest thousand)}$$

$\therefore \$270\,000$

(1 mark)

(iii) Let $P = 370\,000$, $M = 2998$:

$$A_n = 370\,000 \times 1.005^n - 2998 \times \frac{1(1.005^n - 1)}{1.005 - 1}$$

$$A_n = 370\,000 \times 1.005^n - 599\,600(1.005^n - 1)$$
$$= 0$$

$$370\,000 \times 1.005^n - 599\,600 \times 1.005^n + 599\,600 = 0$$

$$229\,600 \times 1.005^n = 599\,600$$

$$1.005^n = \frac{599\,600}{229\,600}$$

$$n \log_e 1.005 = \log_e \frac{5996}{2296}$$

$$n = \log_e \frac{5996}{2296} \div \log_e 1.005$$

$$= 192.464\,3834 \ldots$$

$\therefore$ will take 193 repayments or 16 years 1 month

(2 marks)

QUESTION 14

(a) $\dot{x} = 10 - 2t$

(i) $\ddot{x} = -2$, which is constant. *(1 mark)*

(ii) $\dot{x} = 10 - 2t = 0$

$$2t = 10$$
$$t = 5$$

$\therefore$ at rest after 5 seconds *(1 mark)*

(iii) $\dot{x} = 10 - 2t$

$x = 10t - t^2 + c$

When $t = 0, x = 5$:

$$5 = 10(0) - (0)^2 + c$$
$$c = 5$$
$$\therefore x = 10t - t^2 + 5$$

When $t = 7$:

$$x = 10(7) - (7)^2 + 5$$
$$= 26$$

$\therefore$ after 7 seconds, 26 metres to the right.

(2 marks)

(iv) $x = 10t - t^2 + 5$

Subs. $t = 0$, $x = 5$

Subs. $t = 5$, $x = 10(5) - (5)^2 + 5 = 30$

Subs. $t = 7$, $x = 26$

From $x = 5$ to $x = 30$ is 25 metres, then back to $x = 26$ is 4 metres.

$\therefore$ particle travels 29 metres. *(2 marks)*

(b) (i) As $D = ST$, then:

Car A travels $80t$ km; car B travels $50t$ km.

$\therefore AR = 80t$

and $BR = 100 - 50t$

Using cosine rule:

$$r^2 = (100 - 50t)^2 + (80t)^2 - 2(100 - 50t)(80t) \times \cos 60°$$
$$= 10\,000 - 10\,000t + 2500t^2 + 6400t^2 - 8000t + 4000t^2$$
$$\therefore r^2 = 12\,900t^2 - 18\,000t + 10\,000$$

(2 marks)

(ii) Min. when $\dfrac{d(r^2)}{dt} = 0$

$$\frac{d(r^2)}{dt} = 25\,800t - 18\,000 = 0$$
$$t = \frac{18\,000}{25\,800}$$
$$= \frac{30}{43}$$

Consider nature:

$$\frac{d^2(r^2)}{dt^2} = 25\,800 > 0 \quad \therefore \text{Minimum}$$

Subs. in r^2:

$$r^2 = 12\,900\left(\frac{30}{43}\right)^2 - 18\,000\left(\frac{30}{43}\right) + 10\,000$$
$$r = 60.999\,428\,13\ldots$$
$$= 61 \text{ (nearest whole)}$$

$\therefore$ minimum distance is 61 km *(3 marks)*

(c)

$$\frac{4}{BD} = \sin\frac{\pi}{6}$$
$$\frac{4}{BD} = \frac{1}{2}$$
$$\therefore BD = 8$$

By Pythagoras,

$$BC^2 = 8^2 - 4^2$$
$$= 64 - 16$$
$$= 48$$
$$BC = \sqrt{48}$$

By Pythagoras,

$$AC^2 = 13^2 - (\sqrt{48})^2$$
$$= 121$$
$$AC = 11$$
$$\therefore AD = 7$$

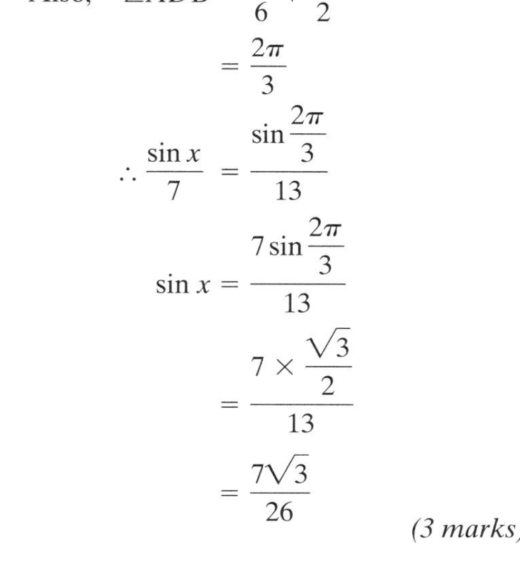

Also, $\angle ADB = \dfrac{\pi}{6} + \dfrac{\pi}{2}$

$$= \frac{2\pi}{3}$$
$$\therefore \frac{\sin x}{7} = \frac{\sin\frac{2\pi}{3}}{13}$$
$$\sin x = \frac{7\sin\frac{2\pi}{3}}{13}$$
$$= \frac{7 \times \frac{\sqrt{3}}{2}}{13}$$
$$= \frac{7\sqrt{3}}{26}$$

(3 marks)

(d)

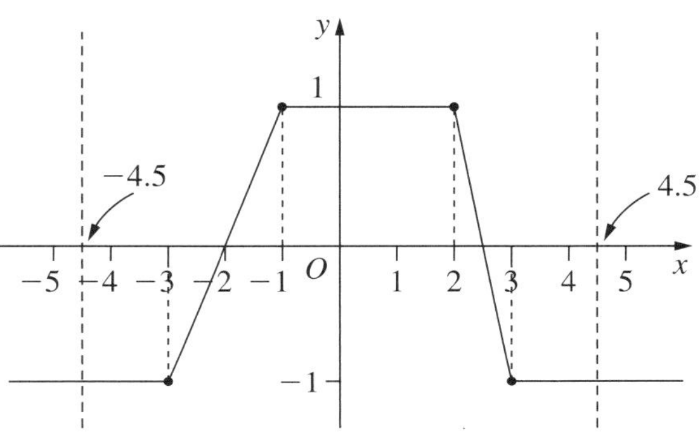

Consider the area between the x-axis and the graph $y = f(x)$. You need to find a value of a, such that the area above the x-axis is identical to the area below the x-axis.

When $a = 4.5$, $\displaystyle\int_{-4.5}^{4.5} f(x)\,dx = 0.$

$\therefore a = 4.5$

OR:

Firstly, $\displaystyle\int_{-3}^{-1} f(x)\,dx = 0$ and $\displaystyle\int_{2}^{3} f(x)\,dx = 0.$

Now $\displaystyle\int_{-1}^{2} f(x)\,dx = 3$, so

$$\int_{-4.5}^{-3} f(x)\,dx + \int_{3}^{4.5} f(x)\,dx = 3.$$

$\therefore a = 4.5$ *(1 mark)*

QUESTION 15

(a) (i) $\frac{1.2}{2}[1.5 + 1.5 + 2(1.8)] = 3.96$

(1 mark)

(ii) $\frac{1.2}{3}[1.5 + 1.5 + 4(1.8)] = 4.08$

(1 mark)

(iii) Simpson's Rule is based on the parabola passing through the three points at the top of the tent. As the Trapezoidal Rule uses straight lines, it would give a better estimate.

(1 mark)

(b) $y = (x - 2)^2$

$(x - 2)^2 = y$

$x - 2 = \pm\sqrt{y}$

$x = 2 \pm \sqrt{y}$

When $y = 4, x = 0$, so $x = 2 - \sqrt{y}$

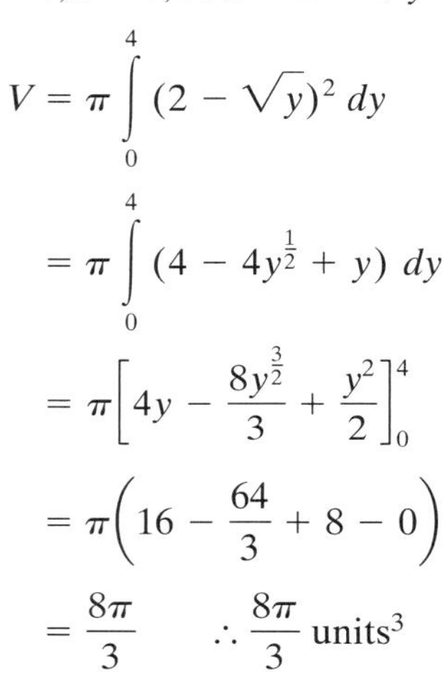

$$V = \pi \int_0^4 (2 - \sqrt{y})^2 \, dy$$

$$= \pi \int_0^4 (4 - 4y^{\frac{1}{2}} + y) \, dy$$

$$= \pi \left[4y - \frac{8y^{\frac{3}{2}}}{3} + \frac{y^2}{2} \right]_0^4$$

$$= \pi \left(16 - \frac{64}{3} + 8 - 0 \right)$$

$$= \frac{8\pi}{3} \qquad \therefore \frac{8\pi}{3} \text{ units}^3$$

(4 marks)

(c) (i)

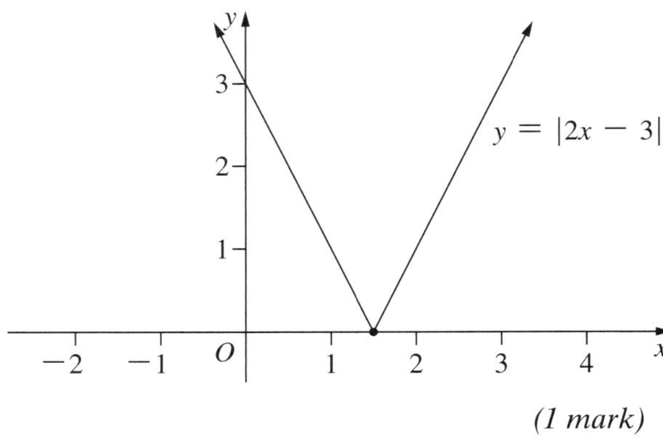

(1 mark)

(ii) Gradient of $y = 2x - 3$ is 2, and gradient of $y = -(2x - 3)$ is -2.

If only one point of intersection, then $m < -2, m = -\frac{2}{3}$ or $m \geq 2$.

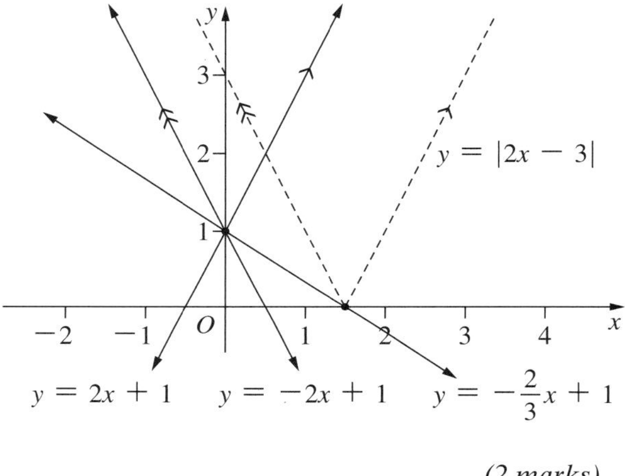

(2 marks)

(d) (i) $P(\text{Pat: double 6}) = \frac{1}{36}$

(1 mark)

(ii) Pat wins on first or second

$= P(\text{Pat: double 6}) + \text{P(Pat: no double 6, Chandra: no double 6, Pat: double 6)}$

$= \frac{1}{36} + \frac{35}{36} \times \frac{35}{36} \times \frac{1}{36}$

$= \frac{2521}{46\,656}$

(2 marks)

(iii) $P(\text{Pat eventually wins}) = \frac{1}{36} + \left(\frac{35}{36}\right)^2 \times \frac{1}{36} + \left(\frac{35}{36}\right)^4 \times \frac{1}{36} + \ldots$

$= \frac{1}{36}[1 + \left(\frac{35}{36}\right)^2 + \left(\frac{35}{36}\right)^4 + \ldots]$

Geometric series, $a = 1, r = \left(\frac{35}{36}\right)^2$,

$$\text{limiting sum} = \frac{1}{1-r}$$
$$= \frac{1}{36}\left[\frac{1}{1-\left(\frac{35}{36}\right)^2}\right]$$
$$= \frac{36}{71}$$

(2 marks)

QUESTION 16

(a)
$$f'(x) = 4x - 3$$
$$f(x) = 2x^2 - 3x + c$$
$$2x^2 - 3x + c = 5x - 7$$
$$2x^2 - 8x + c + 7 = 0$$

If tangent, then $\Delta = 0$:

$$\Delta = (-8)^2 - 4(2)(c+7) = 0$$
$$64 - 8(c+7) = 0$$
$$64 - 8c - 56 = 0$$
$$8 - 8c = 0$$
$$8c = 8$$
$$c = 1$$
$$\therefore f(x) = 2x^2 - 3x + 1$$

OR:

Gradient of tangent = 5

$$\therefore 4x - 3 = 5$$
$$4x = 8$$
$$x = 2$$

Subs. in $y = 5x - 7$:

$$y = 5(2) - 7$$
$$= 3$$

$\therefore$ tangent is at the point $(2, 3)$

As $f'(x) = 4x - 3$

$$f(x) = 2x^2 - 3x + c$$

Subs. $(2, 3)$:

$$3 = 2(2)^2 - 3(2) + c$$
$$3 = 8 - 6 + c$$
$$c = 1$$
$$\therefore f(x) = 2x^2 - 3x + 1$$

(3 marks)

(b) (i)
$$N = 375 - e^{0.04t}$$

Let $t = 0$:

$$N = 375 - e^{0.04(0)}$$
$$= 375 - 1$$
$$= 374$$

$\therefore$ 374 trout

(1 mark)

(ii) Let $N = 0$:

$$0 = 375 - e^{0.04t}$$
$$e^{0.04t} = 375$$
$$\log_e e^{0.04t} = \log_e 375$$
$$0.04t = \log_e 375$$
$$t = \frac{\log_e 375}{0.04}$$
$$= 148.173\,1506\ldots$$
$$= 148 \text{ (nearest whole)}$$

$\therefore$ after about 148 months

(1 mark)

(iii)

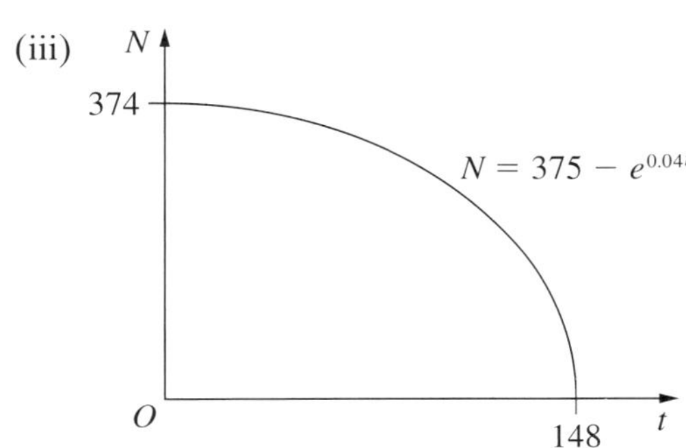

(1 mark)

(iv)
$$N = 375 - e^{0.04t}$$
$$\frac{dN}{dt} = -0.04e^{0.04t}$$

If $\frac{dP}{dt} = 0.02P$, then

$$P = Ae^{0.02t}$$

Now, when $t = 0, P = 10$:

$$\therefore 10 = Ae^{0.02(0)} \qquad \therefore A = 10$$
$$\therefore P = 10e^{0.02t}$$

Now, $\frac{dP}{dt} = 0.2e^{0.02t}$

If rates equal, then:

$$0.04e^{0.04t} = 0.2e^{0.02t}$$
$$e^{0.04t} = 5e^{0.02t}$$
$$e^{0.04t} \div e^{0.02t} = 5$$
$$e^{0.02t} = 5$$
$$\log_e e^{0.02t} = \log_e 5$$
$$0.02t = \log_e 5$$
$$t = \frac{\log_e 5}{0.02}$$
$$= 80.471\,895\,62\ldots$$
$$= 80 \text{ (nearest whole)}$$

$\therefore$ after about 80 months

(3 marks)

(v)

$$N = P$$
$$375 - e^{0.04t} = 10e^{0.02t}$$
$$e^{0.04t} + 10e^{0.02t} - 375 = 0$$

Let $m = e^{0.02t}$

$$m^2 + 10m - 375 = 0$$
$$(m + 25)(m - 15) = 0$$
$$m = -25, 15$$
$$e^{0.02t} = 15 \quad (\text{as } e^{0.02t} > 0)$$
$$\log_e e^{0.02t} = \log_e 15$$
$$0.02t = \log_e 15$$
$$t = \frac{\log_e 15}{0.02}$$
$$= 135.402\,5101\ldots$$
$$= 135 \text{ (nearest whole)}$$

$\therefore$ after about 135 months

(2 marks)

(c) (i)

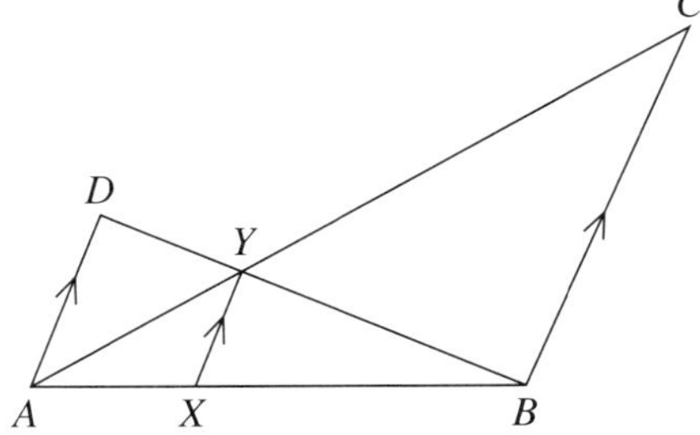

In $\triangle$s ABC, AXY:

$\angle A$ is common

$\angle AXY = \angle ABC$ (corr. $\angle$s, XY || BC)

$\therefore \triangle ABC$ similar to $\triangle AXY$ (equiangular)

(2 marks)

(ii) As $\triangle ABC$ similar to $\triangle AXY$,

$$\frac{XY}{BC} = \frac{AX}{AB}$$

$$\therefore AX = \frac{XY.AB}{BC} \quad \ldots\ldots\ldots\ldots\ldots\ldots\ldots \mathbf{1}$$

(matching sides of sim. $\triangle$s in prop^n)

Similarly, $\triangle BXY$ is similar to $\triangle BAD$ (equiangular)

$$\therefore \frac{XY}{AD} = \frac{XB}{AB}$$

$$\therefore XB = \frac{XY.AB}{AD} \quad \ldots\ldots\ldots\ldots\ldots\ldots\ldots \mathbf{2}$$

As $AB = AX + XB$, then using **1** and **2**:

$$AB = \frac{XY.AB}{BC} + \frac{XY.AB}{AD}$$

Dividing by $XY.AB$:

$$\frac{1}{XY} = \frac{1}{BC} + \frac{1}{AD}$$

(2 marks)

2014 **HIGHER SCHOOL CERTIFICATE EXAMINATION**

Mathematics

General Instructions

- Reading time – 5 minutes
- Working time – 3 hours
- Write using black or blue pen
  Black pen is preferred
- Board-approved calculators may be used
- A table of standard integrals is provided at the back of this paper
- In Questions 11–16, show relevant mathematical reasoning and/or calculations

Total marks – 100

Section I

10 marks

- Attempt Questions 1–10
- Allow about 15 minutes for this section

Section II

90 marks

- Attempt Questions 11–16
- Allow about 2 hours and 45 minutes for this section

Section I

10 marks
Attempt Questions 1–10
Allow about 15 minutes for this section

Use the multiple-choice answer sheet for Questions 1–10.

---

1 What is the value of $\dfrac{\pi^2}{6}$, correct to 3 significant figures?

(A) 1.64

(B) 1.65

(C) 1.644

(D) 1.645

2 Which graph best represents $y = (x - 1)^2$?

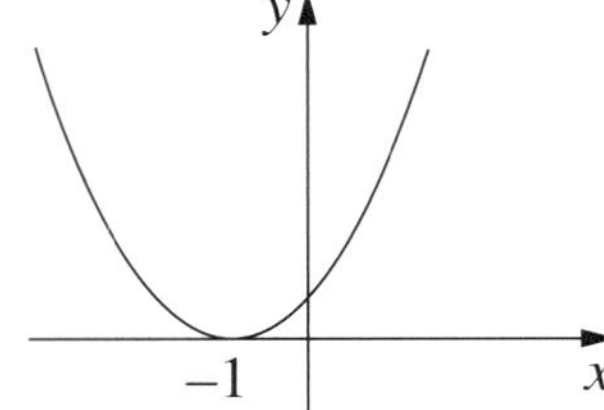

(B)

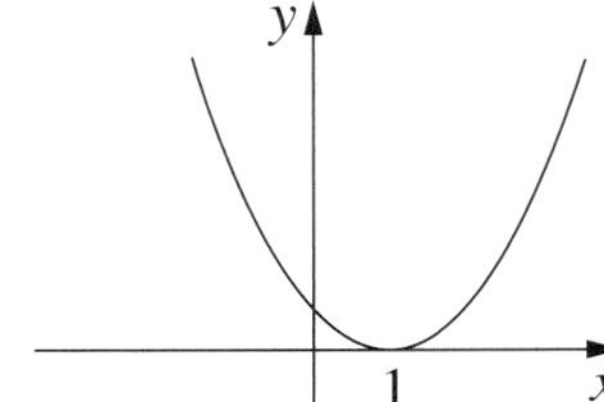

(C)

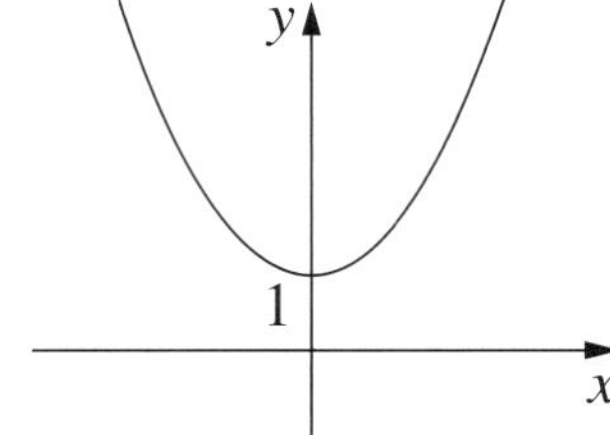

(D)

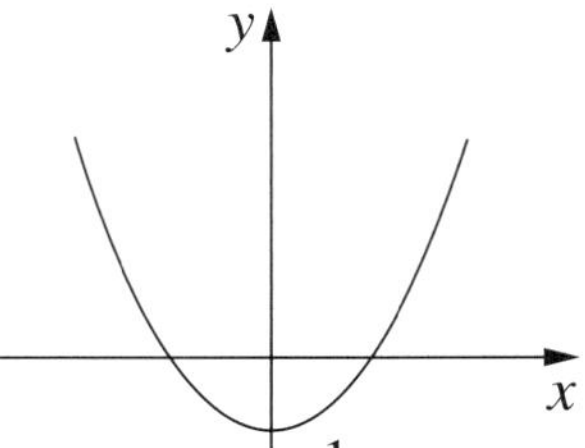

3 What is the solution to the equation $\log_2(x - 1) = 8$?

(A) 4

(B) 17

(C) 65

(D) 257

4 Which expression is equal to $\int e^{2x}\,dx$?

(A) $e^{2x} + c$

(B) $2e^{2x} + c$

(C) $\dfrac{e^{2x}}{2} + c$

(D) $\dfrac{e^{2x+1}}{2x+1} + c$

5 Which equation represents the line perpendicular to $2x - 3y = 8$, passing through the point $(2, 0)$?

(A) $3x + 2y = 4$

(B) $3x + 2y = 6$

(C) $3x - 2y = -4$

(D) $3x - 2y = 6$

6 Which expression is a factorisation of $8x^3 + 27$?

(A) $(2x - 3)(4x^2 + 12x - 9)$

(B) $(2x + 3)(4x^2 - 12x + 9)$

(C) $(2x - 3)(4x^2 + 6x - 9)$

(D) $(2x + 3)(4x^2 - 6x + 9)$

7 How many solutions of the equation $(\sin x - 1)(\tan x + 2) = 0$ lie between 0 and 2π?

(A) 1

(B) 2

(C) 3

(D) 4

8 Which expression is a term of the geometric series $3x - 6x^2 + 12x^3 - \cdots$?

(A) $3072x^{10}$

(B) $-3072x^{10}$

(C) $3072x^{11}$

(D) $-3072x^{11}$

9 The graph shows the displacement x of a particle moving along a straight line as a function of time t.

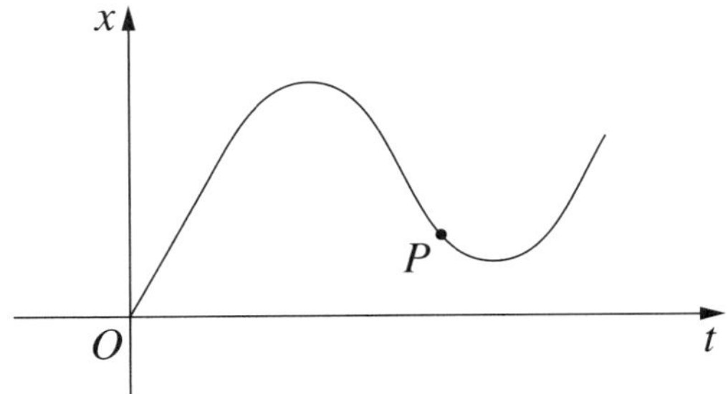

Which statement describes the motion of the particle at the point P?

(A) The velocity is negative and the acceleration is positive.

(B) The velocity is negative and the acceleration is negative.

(C) The velocity is positive and the acceleration is positive.

(D) The velocity is positive and the acceleration is negative.

10 Three runners compete in a race. The probabilities that the three runners finish the race in under 10 seconds are $\frac{1}{4}$, $\frac{1}{6}$ and $\frac{2}{5}$ respectively.

What is the probability that at least one of the three runners will finish the race in under 10 seconds?

(A) $\frac{1}{60}$

(B) $\frac{37}{60}$

(C) $\frac{3}{8}$

(D) $\frac{5}{8}$

Section II

90 marks
Attempt Questions 11–16
Allow about 2 hours and 45 minutes for this section

Answer each question in the appropriate writing booklet. Extra writing booklets are available.

In Questions 11–16, your responses should include relevant mathematical reasoning and/or calculations.

Question 11 (15 marks) Use the Question 11 Writing Booklet.

(a) Rationalise the denominator of $\dfrac{1}{\sqrt{5}-2}$. **2**

(b) Factorise $3x^2 + x - 2$. **2**

(c) Differentiate $\dfrac{x^3}{x+1}$. **2**

(d) Find $\displaystyle\int \frac{1}{(x+3)^2}\,dx$. **2**

(e) Evaluate $\displaystyle\int_0^{\frac{\pi}{2}} \sin\frac{x}{2}\,dx$. **3**

Question 11 continues on the following page

Question 11 (continued)

(f) The gradient function of a curve $y = f(x)$ is given by $f'(x) = 4x - 5$. **2**
The curve passes through the point $(2, 3)$.

Find the equation of the curve.

(g) The angle of a sector in a circle of radius 8 cm is $\frac{\pi}{7}$ radians, as shown in the diagram. **2**

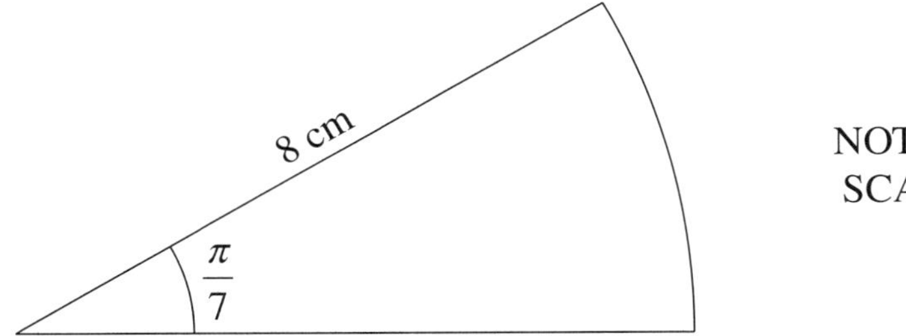

Find the exact value of the perimeter of the sector.

End of Question 11

Question 12 (15 marks) Use the Question 12 Writing Booklet.

(a) Evaluate the arithmetic series $2 + 5 + 8 + 11 + \cdots + 1094$. **2**

(b) The points $A(0, 4)$, $B(3, 0)$ and $C(6, 1)$ form a triangle, as shown in the diagram.

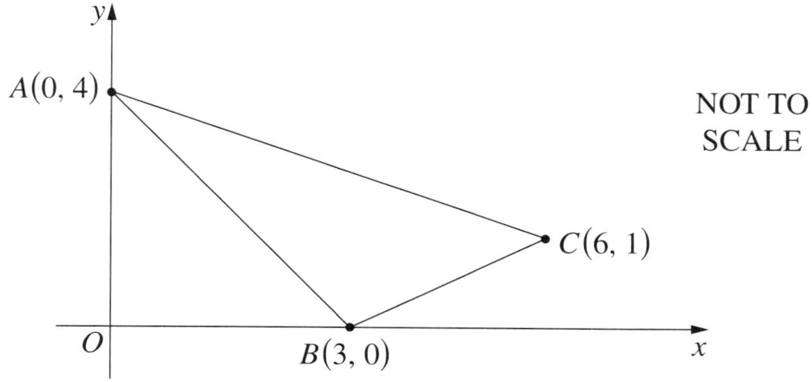

(i) Show that the equation of AC is $x + 2y - 8 = 0$. **2**

(ii) Find the perpendicular distance from B to AC. **2**

(iii) Hence, or otherwise, find the area of $\triangle ABC$. **2**

(c) A packet of lollies contains 5 red lollies and 14 green lollies. Two lollies are selected at random without replacement.

(i) Draw a tree diagram to show the possible outcomes. Include the probability on each branch. **2**

(ii) What is the probability that the two lollies are of different colours? **1**

(d) The parabola $y = -2x^2 + 8x$ and the line $y = 2x$ intersect at the origin and at the point A.

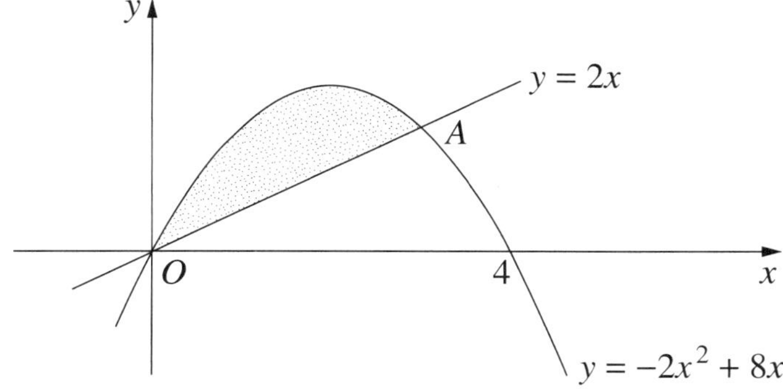

(i) Find the x-coordinate of the point A. **1**

(ii) Calculate the area enclosed by the parabola and the line. **3**

Question 13 (15 marks) Use the Question 13 Writing Booklet.

(a) (i) Differentiate $3 + \sin 2x$. **1**

(ii) Hence, or otherwise, find $\int \frac{\cos 2x}{3 + \sin 2x}\, dx$. **2**

(b) A quantity of radioactive material decays according to the equation

$$\frac{dM}{dt} = -kM,$$

where M is the mass of the material in kg, t is the time in years and k is a constant.

(i) Show that $M = Ae^{-kt}$ is a solution to the equation, where A is a constant. **1**

(ii) The time for half of the material to decay is 300 years. If the initial amount of material is 20 kg, find the amount remaining after 1000 years. **3**

(c) The displacement of a particle moving along the x-axis is given by

$$x = t - \frac{1}{1 + t},$$

where x is the displacement from the origin in metres, t is the time in seconds, and $t \geq 0$.

(i) Show that the acceleration of the particle is always negative. **2**

(ii) What value does the velocity approach as t increases indefinitely? **1**

Question 13 continues on the following page

Question 13 (continued)

(d) Chris leaves island A in a boat and sails 142 km on a bearing of 078° to island B. Chris then sails on a bearing of 191° for 220 km to island C, as shown in the diagram.

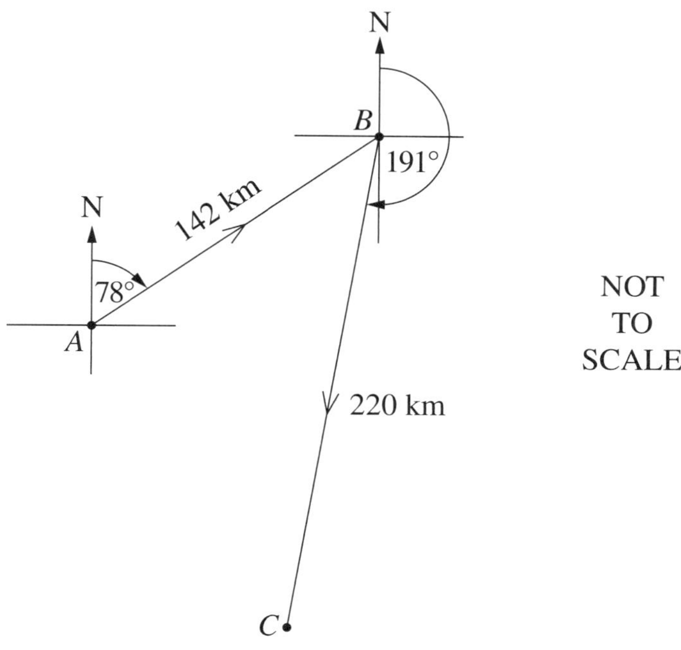

(i) Show that the distance from island C to island A is approximately 210 km. **2**

(ii) Chris wants to sail from island C directly to island A. On what bearing should Chris sail? Give your answer correct to the nearest degree. **3**

End of Question 13

Question 14 (15 marks) Use the Question 14 Writing Booklet.

(a) Find the coordinates of the stationary point on the graph $y = e^x - ex$, and determine its nature. **3**

(b) The roots of the quadratic equation $2x^2 + 8x + k = 0$ are α and β.

(i) Find the value of $\alpha + \beta$. **1**

(ii) Given that $\alpha^2\beta + \alpha\beta^2 = 6$, find the value of k. **2**

(c) The region bounded by the curve $y = 1 + \sqrt{x}$ and the x-axis between $x = 0$ and $x = 4$ is rotated about the x-axis to form a solid. **3**

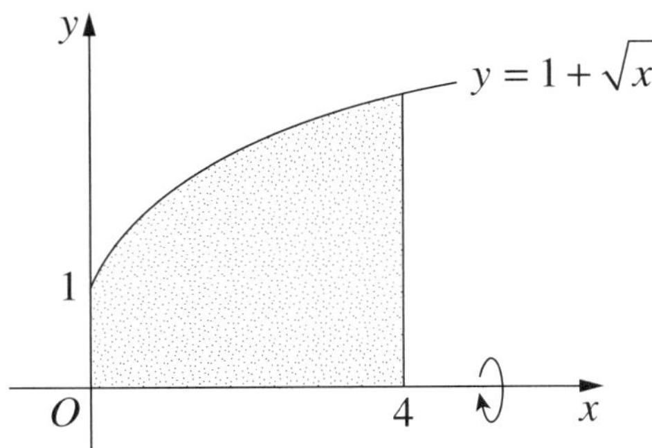

Find the volume of the solid.

(d) At the beginning of every 8-hour period, a patient is given 10 mL of a particular drug.

During each of these 8-hour periods, the patient's body partially breaks down the drug. Only $\frac{1}{3}$ of the total amount of the drug present in the patient's body at the beginning of each 8-hour period remains at the end of that period.

(i) How much of the drug is in the patient's body immediately after the second dose is given? **1**

(ii) Show that the total amount of the drug in the patient's body never exceeds 15 mL. **2**

Question 14 continues on the following page

Question 14 (continued)

(e) The diagram shows the graph of a function $f(x)$. **3**

The graph has a horizontal point of inflexion at A, a point of inflexion at B and a maximum turning point at C.

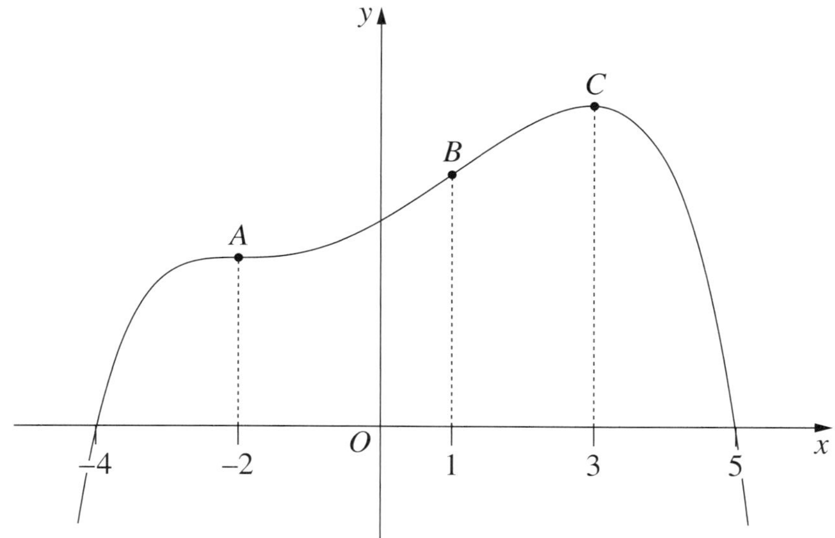

Sketch the graph of the derivative $f'(x)$.

End of Question 14

Question 15 (15 marks) Use the Question 15 Writing Booklet.

(a) Find all solutions of $2\sin^2 x + \cos x - 2 = 0$, where $0 \leq x \leq 2\pi$. **3**

(b) In $\triangle DEF$, a point S is chosen on the side DE. The length of DS is x, and the length of ES is y. The line through S parallel to DF meets EF at Q. The line through S parallel to EF meets DF at R.

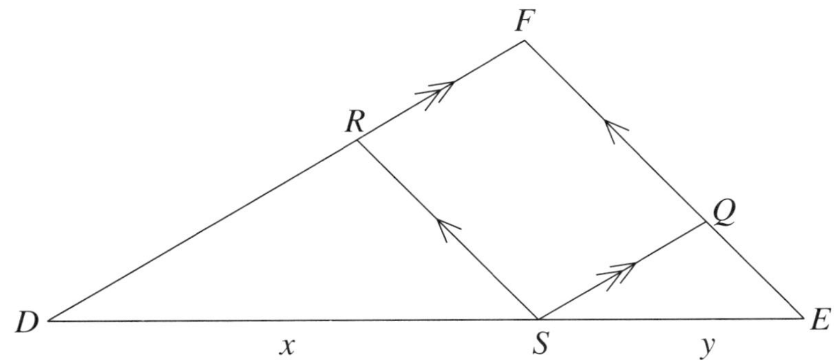

The area of $\triangle DEF$ is A. The areas of $\triangle DSR$ and $\triangle SEQ$ are A_1 and A_2 respectively.

(i) Show that $\triangle DEF$ is similar to $\triangle DSR$. **2**

(ii) Explain why $\dfrac{DR}{DF} = \dfrac{x}{x+y}$. **1**

(iii) Show that $\sqrt{\dfrac{A_1}{A}} = \dfrac{x}{x+y}$. **2**

(iv) Using the result from part (iii) and a similar expression for $\sqrt{\dfrac{A_2}{A}}$, deduce that $\sqrt{A} = \sqrt{A_1} + \sqrt{A_2}$. **2**

Question 15 continues on the following page

Question 15 (continued)

(c) The line $y = mx$ is a tangent to the curve $y = e^{2x}$ at a point P.

(i) Sketch the line and the curve on one diagram. **1**

(ii) Find the coordinates of P. **3**

(iii) Find the value of m. **1**

End of Question 15

Question 16 (15 marks) Use the Question 16 Writing Booklet.

(a) Use Simpson's Rule with five function values to show that **3**

$$\int_{-\frac{\pi}{3}}^{\frac{\pi}{3}} \sec x \, dx \approx \frac{\pi}{9}\left(3 + \frac{8}{\sqrt{3}}\right).$$

(b) At the start of a month, Jo opens a bank account and makes a deposit of \$500. At the start of each subsequent month, Jo makes a deposit which is 1% more than the previous deposit.

At the end of every month, the bank pays interest of 0.3% (per month) on the balance of the account.

(i) Explain why the balance of the account at the end of the second month is **2**

$$\$500(1.003)^2 + \$500(1.01)(1.003).$$

(ii) Find the balance of the account at the end of the 60th month, correct to the nearest dollar. **3**

Question 16 continues on the following page

Question 16 (continued)

(c) The diagram shows a window consisting of two sections. The top section is a semicircle of diameter x m. The bottom section is a rectangle of width x m and height y m.

The entire frame of the window, including the piece that separates the two sections, is made using 10 m of thin metal.

The semicircular section is made of coloured glass and the rectangular section is made of clear glass.

Under test conditions the amount of light coming through one square metre of the coloured glass is 1 unit and the amount of light coming through one square metre of the clear glass is 3 units.

The total amount of light coming through the window under test conditions is L units.

(i) Show that $y = 5 - x\left(1 + \frac{\pi}{4}\right)$. **2**

(ii) Show that $L = 15x - x^2\left(3 + \frac{5\pi}{8}\right)$. **2**

(iii) Find the values of x and y that maximise the amount of light coming through the window under test conditions. **3**

End of paper

2014 Higher School Certificate

Worked Answers

Section I

QUESTION 1

$$\frac{\pi^2}{6} = 1.644\,934\,067\ldots$$
$$= 1.64 \text{ (3 sig. figs)}$$

Answer A

(1 mark)

QUESTION 2

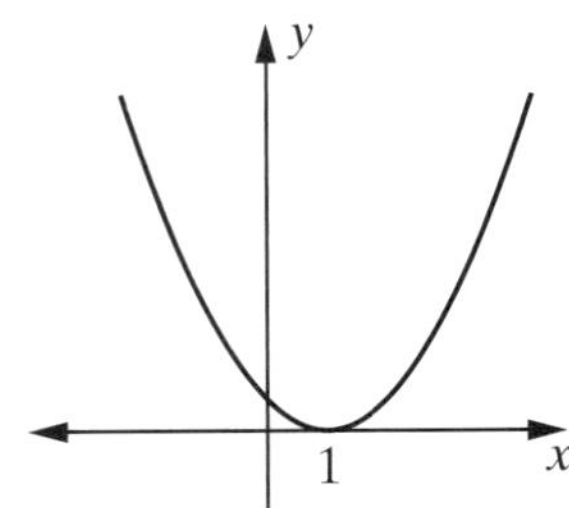

Answer B

(1 mark)

QUESTION 3

$$\log_2(x - 1) = 8$$
$$x - 1 = 2^8$$
$$x - 1 = 256$$
$$x = 257$$

Answer D

(1 mark)

QUESTION 4

$$\int e^{2x}\,dx = \frac{1}{2}e^{2x} + c$$
$$= \frac{e^{2x}}{2} + c$$

Answer C

(1 mark)

QUESTION 5

$$2x - 3y = 8$$
$$3y = 2x - 8$$
$$y = \frac{2x - 8}{3}$$

$\therefore$ gradient of perpendicular is $\frac{-3}{2}$.

$$y - 0 = \frac{-3}{2}(x - 2)$$
$$2y = -3x + 6$$
$$3x + 2y = 6$$

Answer B

(1 mark)

QUESTION 6

$$8x^3 + 27 = (2x + 3)(4x^2 - 6x + 9)$$

Answer D

(1 mark)

QUESTION 7

$$(\sin x - 1)(\tan x + 2) = 0$$
$$\sin x - 1 = 0$$
$$\sin x = 1$$
$$x = \frac{\pi}{2}$$
$$\tan x + 2 = 0$$

$\tan x = -2$ has two solutions.

But $x = \frac{\pi}{2}$ is not a solution of $\tan x + 2 = 0$, as $\tan\frac{\pi}{2}$ is undefined.

$\therefore$ there are only 2 solutions.

Answer B

(1 mark)

QUESTION 8

Geometric series, where $a = 3x, r = -2x$,

Consider $n = 11$:

$$T_n = ar^{n-1}$$
$$T_{11} = 3x \times (-2x)^{10}$$
$$= 3x \times 1024x^{10}$$
$$= 3072x^{11}$$

Answer C

(1 mark)

QUESTION 9

At P: decreasing ($f'(x) < 0$), and concave up ($f''(x) > 0$)

$\therefore$velocity is negative and acceleration is positive.

Answer A

(1 mark)

QUESTION 10

P(at least one is under 10 s)

$$= 1 - \text{P(no runner under 10 s)}$$
$$= 1 - \left[\frac{3}{4} \times \frac{5}{6} \times \frac{3}{5}\right]$$
$$= \frac{5}{8}$$

Answer D

(1 mark)

Section II

QUESTION 11

(a) $$\frac{1}{\sqrt{5} - 2} = \frac{1}{\sqrt{5} - 2} \times \frac{\sqrt{5} + 2}{\sqrt{5} + 2}$$
$$= \frac{\sqrt{5} + 2}{1}$$
$$= \sqrt{5} + 2$$

(2 marks)

(b) $3x^2 + x - 2 = (3x - 2)(x + 1)$

(2 marks)

(c) $\dfrac{x^3}{x + 1}$

Using the quotient rule,

Let $u = x^3,\ u' = 3x^2$

Let $v = x + 1,\ v' = 1$

$$\frac{dy}{dx} = \frac{vu' - uv'}{v^2}$$
$$= \frac{(x + 1).3x^2 - x^3.1}{(x + 1)^2}$$
$$= \frac{3x^3 + 3x^2 - x^3}{(x + 1)^2}$$
$$= \frac{2x^3 + 3x^2}{(x + 1)^2}$$

(2 marks)

(d) $$\int \frac{1}{(x + 3)^2}\,dx = \int (x + 3)^{-2}\,dx$$
$$= \frac{(x + 3)^{-1}}{-1} + c$$
$$= \frac{-1}{x + 3} + c$$

(2 marks)

(e) $$\int_0^{\frac{\pi}{2}} \sin\frac{x}{2}\,dx = \left[-2\cos\frac{x}{2}\right]_0^{\frac{\pi}{2}}$$
$$= -2\left(\cos\frac{\pi}{4} - \cos 0\right)$$
$$= -2\left(\frac{1}{\sqrt{2}} - 1\right)$$
$$= 2 - \frac{2}{\sqrt{2}}$$
$$= 2 - \frac{2}{\sqrt{2}} \times \frac{\sqrt{2}}{\sqrt{2}}$$
$$= 2 - \sqrt{2}$$

(3 marks)

(f) As $f'(x) = 4x - 5$

$$f(x) = 2x^2 - 5x + c$$

Subs. (2, 3):

$$3 = 2(2)^2 - 5(2) + c$$

$$3 = 8 - 10 + c$$

$$c = 5$$

$\therefore\ f(x) = 2x^2 - 5x + 5$

(2 marks)

(g) Arc length $= r\theta$

$$= 8 \times \frac{\pi}{7}$$

$$= \frac{8\pi}{7}$$

$$\text{Perimeter} = 8 + 8 + \frac{8\pi}{7}$$

$$= 16 + \frac{8\pi}{7}$$

$\therefore \left(16 + \frac{8\pi}{7}\right)$ cm.

(2 marks)

QUESTION 12

(a) Arithmetic series:

$a = 2, d = 3, T_n = 1094$

To find n: $T_n = a + (n - 1)d$

$$1094 = 2 + (n - 1)3$$

$$1094 = 3n - 1$$

$$3n = 1095$$

$$n = 365$$

To find sum: $S_n = \frac{n}{2}(a + l)$

$$= \frac{365}{2}(2 + 1094)$$

$$= 200\,020$$

(2 marks)

(b) (i) grad. $AC = \frac{1 - 4}{6 - 0}$

$$= \frac{-3}{6}$$

$$= -\frac{1}{2}$$

$$y - 4 = -\frac{1}{2}(x - 0)$$

$$2y - 8 = -x$$

$\therefore x + 2y - 8 = 0$

(2 marks)

(ii) Use $B(3, 0)$ and $x + 2y - 8 = 0$

$$d = \left|\frac{ax_1 + by_1 + c}{\sqrt{a^2 + b^2}}\right|$$

$$= \left|\frac{1(3) + 2(0) - 8}{\sqrt{1^2 + 2^2}}\right|$$

$$= \left|\frac{-5}{\sqrt{5}}\right|$$

$$= \frac{5}{\sqrt{5}}$$

$$= \sqrt{5}$$

$\therefore\ \sqrt{5}$ units

(2 marks)

(iii) Length of AC:

$$d = \sqrt{(6 - 0)^2 + (1 - 4)^2}$$

$$= \sqrt{36 + 9}$$

$$= \sqrt{45}$$

$$= 3\sqrt{5}$$

$\therefore 3\sqrt{5}$ units

$$\text{Area} = \frac{1}{2} \times 3\sqrt{5} \times \sqrt{5}$$

$$= 7.5$$

$\therefore 7.5$ units^2

(2 marks)

(c) (i)

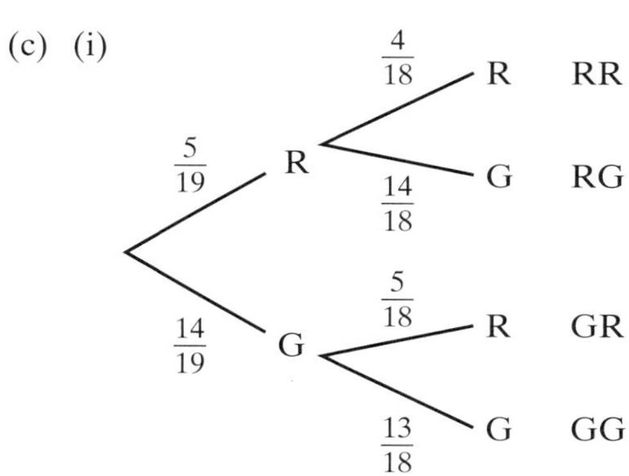

(2 marks)

(ii) P(different colours)

$$= P(RG) + P(GR)$$
$$= \frac{5}{19} \times \frac{14}{18} + \frac{14}{19} \times \frac{5}{18}$$
$$= \frac{70}{171}$$

(1 mark)

(d) (i)

$$2x = -2x^2 + 8x$$
$$2x^2 + 2x - 8x = 0$$
$$2x^2 - 6x = 0$$
$$2x(x - 3) = 0$$
$$x = 0, 3$$

$\therefore$ At $A, x = 3$.

(1 mark)

(ii)

$$\text{Area} = \int_0^3 -2x^2 + 8x - 2x \, dx$$
$$= \int_0^3 -2x^2 + 6x \, dx$$
$$= \left[\frac{-2x^3}{3} + 3x^2\right]_0^3$$
$$= \frac{-2(3)^3}{3} + 3(3)^2 - 0$$
$$= -18 + 27$$
$$= 9$$

$\therefore$ area is 9 units^2.

(3 marks)

QUESTION 13

(a) (i) $\frac{d}{dx}[3 + \sin 2x] = 2\cos 2x$

(1 mark)

(ii)

$$\int \frac{\cos 2x}{3 + \sin 2x} dx = \frac{1}{2}\int \frac{2\cos 2x}{3 + \sin 2x} dx$$
$$= \frac{1}{2}\log_e(3 + \sin 2x) + c$$

(using result in part (i))

(2 marks)

(b) (i)

$$M = Ae^{-kt}$$
$$\frac{dM}{dt} = -k.Ae^{-kt}$$
$$= -kM$$

$\therefore$ $M = Ae^{-kt}$ is a solution.

(1 mark)

(ii) Let $A = 1$, then $M = 0.5$ and $t = 300$:

$$0.5 = 1e^{-k(300)}$$
$$e^{-300k} = 0.5$$
$$-300k = \log_e 0.5$$
$$k = -\frac{\log_e 0.5}{300}$$

Let $A = 20, t = 1000$:

$$M = 20e^{-k(1000)}$$
$$= 1.984\,251\,315\ldots$$
$$= 1.98 \text{ (2 dec. pl.)}$$

$\therefore$ 1.98 kg remain.

(3 marks)

(c) $x = t - \dfrac{1}{1 + t}$

(i) $x = t - (1 + t)^{-1}$

$\dot{x} = 1 + 1(1 + t)^{-2}$

$= 1 + \dfrac{1}{(1 + t)^2}$

$\ddot{x} = -2(1 + t)^{-3}$

$= \dfrac{-2}{(1 + t)^3}$

As $t > 0$, then $(1 + t)^3 > 0$,

then $\ddot{x} < 0$ for all t.

(2 marks)

(ii) Use $\dot{x} = 1 + \dfrac{1}{(1 + t)^2}$.

As $t \to \infty$, $(1 + t)^2 \to \infty$,

and $\dfrac{1}{(1 + t)^2} \to 0$.

Hence, $1 + \dfrac{1}{(1 + t)^2} \to 1$.

$\therefore$ velocity approaches 1 m/s.

(1 mark)

(d) (i) $270 - (12 + 191) = 67$

$\therefore \angle ABC = 67°$

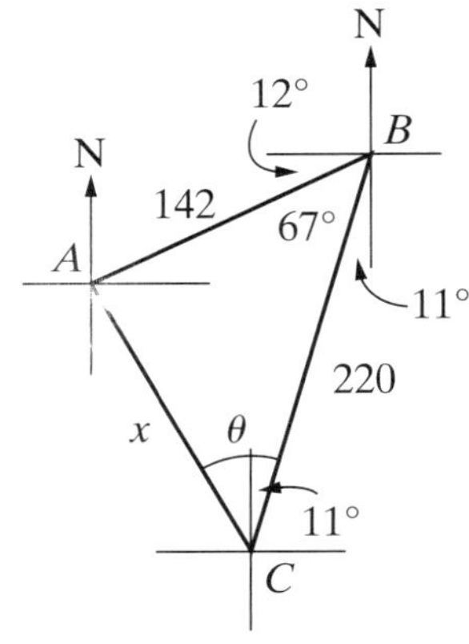

Let $AC = x$:

$x^2 = 142^2 + 220^2 - 2(142)(220)\cos 67°$

$= 44\,151.119\,09\ldots$

$x = 210.121\,6769\ldots$

$= 210$ (nearest whole)

$\therefore$ distance is approx. 210 km.

(2 marks)

(ii) Let $\angle ACB = \theta$:

$\cos\theta = \dfrac{210^2 + 220^2 - 142^2}{2(210)(220)}$

$= 0.782\,857\,142\ldots$

$\theta = 38.477\,077\,75\ldots$

$= 38$ (nearest whole)

Also, $191 - 180 = 11$, $38 - 11 = 27$ and $360 - 27 = 333$.

$\therefore$ bearing is 333°.

(3 marks)

QUESTION 14

(a) $y = e^x - ex$

$\dfrac{dy}{dx} = e^x - e = 0$

$e^x = e$

$x = 1$

$y(1) = e^1 - e(1)$

$= 0 \quad \therefore$ stat. pt at $(1, 0)$

$\dfrac{d^2y}{dx^2} = e^x$

$\dfrac{d^2y}{dx^2}(1) = e^1 > 0 \;\therefore$ minimum

$\therefore$ minimum at $(1, 0)$.

(3 marks)

(b) (i) $\alpha + \beta = \dfrac{-b}{a}$

$= \dfrac{-8}{2}$

$= -4$

(1 mark)

(ii) $\alpha\beta = \dfrac{c}{a}$

$= \dfrac{k}{2}$

$\alpha^2\beta + \alpha\beta^2 = \alpha\beta(\alpha + \beta)$

$= \dfrac{k}{2}(-4)$

$= -2k$

$-2k = 6$

$k = -3$

(2 marks)

(c) $y = 1 + \sqrt{x}$

$$y^2 = (1 + \sqrt{x})^2$$
$$= 1 + 2\sqrt{x} + x$$
$$= 1 + 2x^{\frac{1}{2}} + x$$

$$V = \pi \int_0^4 y^2 \, dx$$
$$= \pi \int_0^4 1 + 2x^{\frac{1}{2}} + x \, dx$$
$$= \pi \left[x + \frac{4x^{\frac{3}{2}}}{3} + \frac{x^2}{2} \right]_0^4$$
$$= \pi \left(4 + \frac{32}{3} + 8 - 0 \right)$$
$$= \frac{68\pi}{3} \quad \therefore \frac{68\pi}{3} \text{ units}^3$$

(3 marks)

(d) (i) Let A_n = amount of drug in body after n doses

$$A_2 = 10 + 10 \times \frac{1}{3}$$
$$= 10 \times 1\frac{1}{3}$$
$$= 13\frac{1}{3}$$

$\therefore 13\frac{1}{3}$ mL

(1 mark)

(ii) $A_3 = 10 + 10 \times \frac{1}{3} + 10 \times \left(\frac{1}{3}\right)^2$

$\therefore A_n = 10 + 10 \times \frac{1}{3} + \ldots + 10 \times \left(\frac{1}{3}\right)^{n-1}$

Total amount is limiting sum:

Using $a = 10, r = \frac{1}{3}$,

limiting sum $= \frac{a}{1 - r}$:

$$A = \frac{10}{1 - \frac{1}{3}}$$
$$= 10 \div \frac{2}{3}$$
$$= 15$$

The maximum amount of the dose is 15 mL.

(2 marks)

(e)

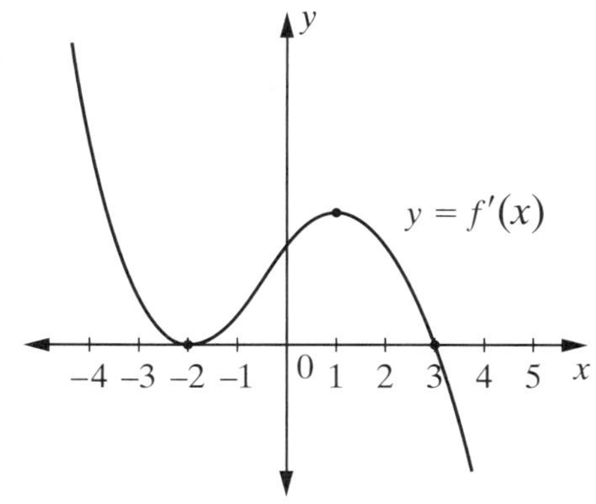

[Note:

- $f(x)$ increasing when $x < -2$.

  $\therefore f'(x)$ is +ve.
- Hor. pt of inflexion at $x = -2$.

  $\therefore f'(-2) = 0$.
- $f(x)$ increasing when $-2 < x < 3$.

  $\therefore f'(x)$ is +ve.
- Maximum at $x = 3$.

  $\therefore f'(3) = 0$.
- $f(x)$ decreasing when $x > 3$.

  $\therefore f'(x)$ is −ve.
- Pt of inflexion at $x = 1$.

  $\therefore f'(1)$ is max value.]

(3 marks)

QUESTION 15

(a)
$$2\sin^2 x + \cos x - 2 = 0$$
$$2(1 - \cos^2 x) + \cos x - 2 = 0$$
$$2 - 2\cos^2 x + \cos x - 2 = 0$$
$$2\cos^2 x - \cos x = 0$$
$$\cos x(2\cos x - 1) = 0$$
$$\cos x = 0 \qquad \cos x = \frac{1}{2}$$
$$\therefore x = \frac{\pi}{2}, \frac{3\pi}{2}, \frac{\pi}{3}, \frac{5\pi}{3}$$

(3 marks)

(b) (i)

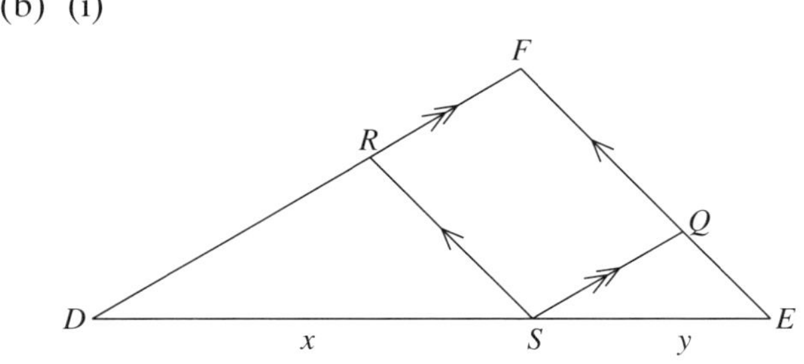

In $\triangle$s DSR, DEF:

$\angle D$ is common

$\angle DRS = \angle DFE$ (corr. $\angle$s, RS $\parallel$ FE)

$\therefore \triangle DSR$ similar to $\triangle DEF$ (equiangular)

(2 marks)

(ii) $\dfrac{DR}{DF} = \dfrac{DS}{DE}$

(matching sides of sim. $\triangle$s are in proportion)

As $DS = x$ and $DE = x + y$,

$$\frac{DR}{DF} = \frac{x}{x+y}$$

(1 mark)

(iii) Area $\triangle DSR = A_1$, Area $\triangle DEF = A$:

Ratio of areas $= A_1 : A$

Ratio of sides $= \sqrt{A_1} : \sqrt{A}$

$$\frac{\sqrt{A_1}}{\sqrt{A}} = \frac{x}{x+y}$$

$$\sqrt{\frac{A_1}{A}} = \frac{x}{x+y}$$

(2 marks)

(iv) Also, it can be proved $\triangle DEF$ similar to $\triangle SEQ$, and hence $\sqrt{\dfrac{A_2}{A}} = \dfrac{y}{x+y}$.

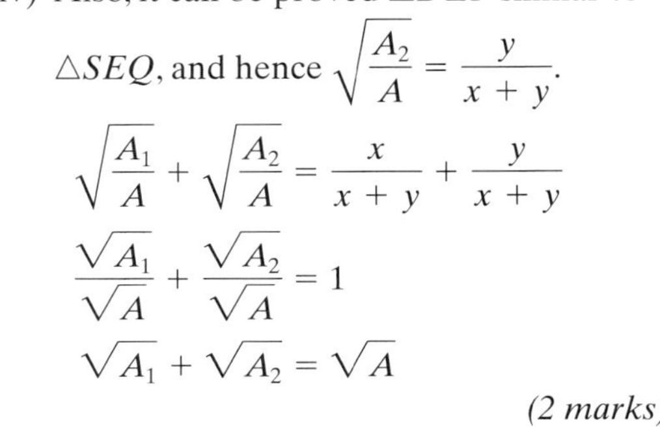

$$\sqrt{\frac{A_1}{A}} + \sqrt{\frac{A_2}{A}} = \frac{x}{x+y} + \frac{y}{x+y}$$

$$\frac{\sqrt{A_1}}{\sqrt{A}} + \frac{\sqrt{A_2}}{\sqrt{A}} = 1$$

$$\sqrt{A_1} + \sqrt{A_2} = \sqrt{A}$$

(2 marks)

(c) (i)

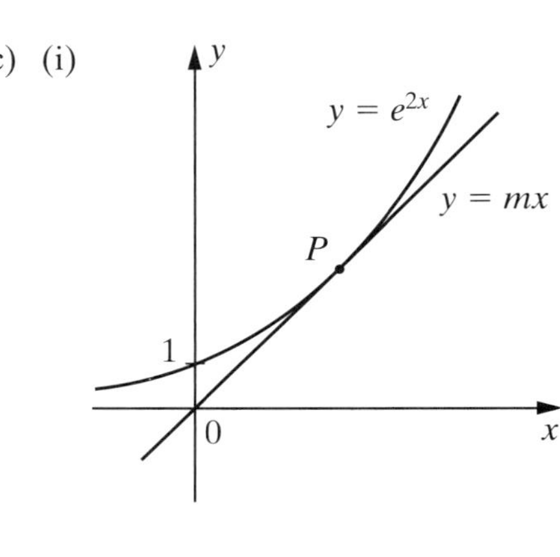

(1 mark)

(ii) For common point, equate y:

$$mx = e^{2x} \text{ .... ①}$$

For common gradient: equate y':

$$m = 2e^{2x} \text{ ...... ②}$$

Subs. ② in ①: $\quad 2e^{2x}x = e^{2x}$

$$2e^{2x}x - e^{2x} = 0$$
$$e^{2x}(2x - 1) = 0$$
$$x = \frac{1}{2} \text{ (as } e^{2x} \neq 0)$$

Subs. in $y = e^{2x}$:

$$y = e^{2\left(\frac{1}{2}\right)}$$
$$= e$$

$$\therefore P\left(\frac{1}{2}, e\right)$$

(3 marks)

(iii) Subs. $P\left(\dfrac{1}{2}, e\right)$ in $y = mx$:

$$e = m\left(\frac{1}{2}\right)$$
$$m = 2e$$

(1 mark)

QUESTION 16

(a)

x	$-\frac{\pi}{3}$	$-\frac{\pi}{6}$	0	$\frac{\pi}{6}$	$\frac{\pi}{3}$
$f(x)$	2	$\frac{2}{\sqrt{3}}$	1	$\frac{2}{\sqrt{3}}$	2

$$\int_{-\frac{\pi}{3}}^{\frac{\pi}{3}} \sec x\,dx$$

$$\approx \frac{\frac{\pi}{6}}{3}\left[2 + 2 + 2(1) + 4\left(\frac{2}{\sqrt{3}} + \frac{2}{\sqrt{3}}\right)\right]$$

$$= \frac{\pi}{18}\left[6 + \frac{16}{\sqrt{3}}\right]$$

$$= \frac{\pi}{9}\left[3 + \frac{8}{\sqrt{3}}\right]$$

(3 marks)

(b) (i) Deposits: 500, 500(1.01), $500(1.01^2)$, …

Let A_n = balance of account after n months

$$A_1 = 500(1.003)$$

$$A_2 = 500(1.003)(1.003) + 500(1.01)(1.003)$$

$$= 500(1.003)^2 + 500(1.01)(1.003)$$

(2 marks)

(ii)

$$A_3 = A_2(1.003) + 500(1.01)^2(1.003)$$

$$= 500(1.003^3) + 500(1.01)(1.003)^2 + 500(1.01)^2(1.003)$$

$$= 500[1.003^3 + (1.01)(1.003)^2 + (1.01)^2(1.003)]$$

$$A_{60} = 500[1.003^{60} + (1.01)(1.003)^{59} + \ldots + (1.01)^{59}(1.003)]$$

Geom. series, $a = 1.003^{60}$, $r = \frac{1.01}{1.003}$, $n = 60$

$$A_{60} = 500\left[\frac{1.003^{60}\left[\left(\frac{1.01}{1.003}\right)^{60} - 1\right]}{\frac{1.01}{1.003} - 1}\right]$$

$= 44\,404.378\,66\ldots$

$= 44\,404$ (nearest whole)

$\therefore$ the balance is \$44 404.

(3 marks)

(c) (i)

$$2x + 2y + \frac{1}{2} \times \pi \times x = 10$$

$$4x + 4y + \pi x = 20$$

$$4y = 20 - x(4 + \pi)$$

$$y = 5 - x(1 + \frac{\pi}{4})$$

(2 marks)

(ii)

$$A = xy + \frac{1}{2} \times \pi \times \left(\frac{x}{2}\right)^2$$

$$L = 3xy + \frac{\pi x^2}{8}$$

$$= 3x\left(5 - x\left(1 + \frac{\pi}{4}\right)\right) + \frac{\pi x^2}{8}$$

$$= 3x\left(5 - x - \frac{\pi x}{4}\right) + \frac{\pi x^2}{8}$$

$$= 15x - 3x^2 - \frac{3\pi x^2}{4} + \frac{\pi x^2}{8}$$

$$= 15x - 3x^2 - \frac{5\pi x^2}{8}$$

$$= 15x - x^2\left(3 + \frac{5\pi}{8}\right)$$

(2 marks)

(iii)

$$L = 15x - x^2\left(3 + \frac{5\pi}{8}\right)$$

$$\frac{dL}{dx} = 15 - 2x\left(3 + \frac{5\pi}{8}\right) = 0$$

$$2x\left(3 + \frac{5\pi}{8}\right) = 15$$

$$x = \frac{15}{2\left(3 + \frac{5\pi}{8}\right)}$$

$$= \frac{15}{6 + \frac{5\pi}{4}}$$

$$= \frac{60}{24 + 5\pi}$$

$$\frac{d^2L}{dx^2} = -2\left(3 + \frac{5\pi}{8}\right) < 0 \therefore \text{maximum}$$

$$\therefore y\left(\frac{60}{24 + 5\pi}\right)$$

$$= 5 - \left(\frac{60}{24 + 5\pi}\right)\left(1 + \frac{\pi}{4}\right)$$

$$= 5 - \left(\frac{60 + 15\pi}{24 + 5\pi}\right)$$

$$= \frac{120 + 25\pi - 60 - 15\pi}{24 + 5\pi}$$

$$= \frac{60 + 10\pi}{24 + 5\pi}$$

$$\therefore x = \frac{60}{24 + 5\pi}, y = \frac{60 + 10\pi}{24 + 5\pi}$$

(3 marks)

2015 **HIGHER SCHOOL CERTIFICATE EXAMINATION**

Mathematics

General Instructions

- Reading time – 5 minutes
- Working time – 3 hours
- Write using black pen
- Board-approved calculators may be used
- A table of standard integrals is provided at the back of this paper
- In Questions 11–16, show relevant mathematical reasoning and/or calculations

Total marks – 100

Section I

10 marks

- Attempt Questions 1–10
- Allow about 15 minutes for this section

Section II

90 marks

- Attempt Questions 11–16
- Allow about 2 hours and 45 minutes for this section

Section I

10 marks
Attempt Questions 1–10
Allow about 15 minutes for this section

Use the multiple-choice answer sheet for Questions 1–10.

---

1 What is 0.005 233 59 written in scientific notation, correct to 4 significant figures?

(A) 5.2336×10^{-2}

(B) 5.234×10^{-2}

(C) 5.2336×10^{-3}

(D) 5.234×10^{-3}

2 What is the slope of the line with equation $2x - 4y + 3 = 0$?

(A) -2

(B) $-\frac{1}{2}$

(C) $\frac{1}{2}$

(D) 2

3 The first three terms of an arithmetic series are 3, 7 and 11.

What is the 15th term of this series?

(A) 59

(B) 63

(C) 465

(D) 495

4 The probability that Mel's soccer team wins this weekend is $\frac{5}{7}$.

The probability that Mel's rugby league team wins this weekend is $\frac{2}{3}$.

What is the probability that neither team wins this weekend?

(A) $\frac{2}{21}$

(B) $\frac{10}{21}$

(C) $\frac{13}{21}$

(D) $\frac{19}{21}$

5 Using the trapezoidal rule with 4 subintervals, which expression gives the approximate area under the curve $y = xe^x$ between $x = 1$ and $x = 3$?

(A) $\frac{1}{4}\left(e^1 + 6e^{1.5} + 4e^2 + 10e^{2.5} + 3e^3\right)$

(B) $\frac{1}{4}\left(e^1 + 3e^{1.5} + 4e^2 + 5e^{2.5} + 3e^3\right)$

(C) $\frac{1}{2}\left(e^1 + 6e^{1.5} + 4e^2 + 10e^{2.5} + 3e^3\right)$

(D) $\frac{1}{2}\left(e^1 + 3e^{1.5} + 4e^2 + 5e^{2.5} + 3e^3\right)$

6 What is the value of the derivative of $y = 2\sin 3x - 3\tan x$ at $x = 0$?

(A) -1

(B) 0

(C) 3

(D) -9

7 The diagram shows the parabola $y = 4x - x^2$ meeting the line $y = 2x$ at $(0, 0)$ and $(2, 4)$.

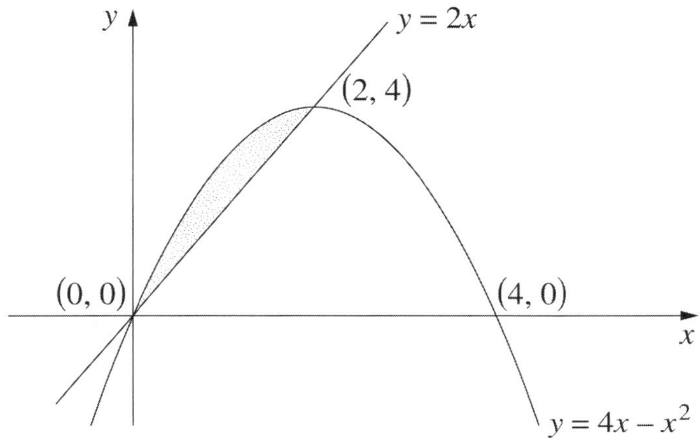

Which expression gives the area of the shaded region bounded by the parabola and the line?

(A) $\int_0^2 x^2 - 2x\,dx$

(B) $\int_0^2 2x - x^2\,dx$

(C) $\int_0^4 x^2 - 2x\,dx$

(D) $\int_0^4 2x - x^2\,dx$

8 The diagram shows the graph of $y = e^x(1 + x)$.

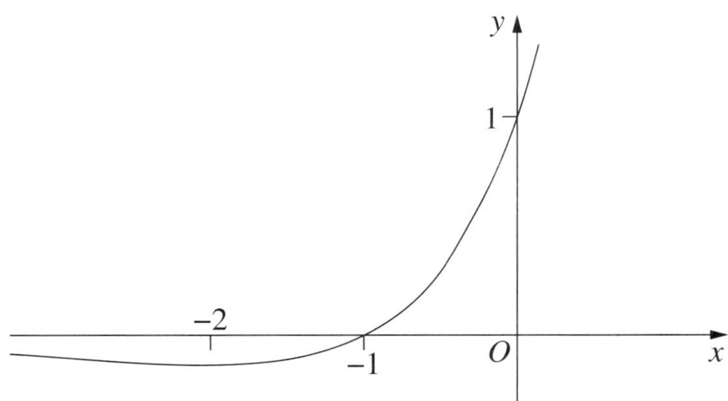

How many solutions are there to the equation $e^x(1 + x) = 1 - x^2$?

(A) 0

(B) 1

(C) 2

(D) 3

9 A particle is moving along the x-axis. The graph shows its velocity v metres per second at time t seconds.

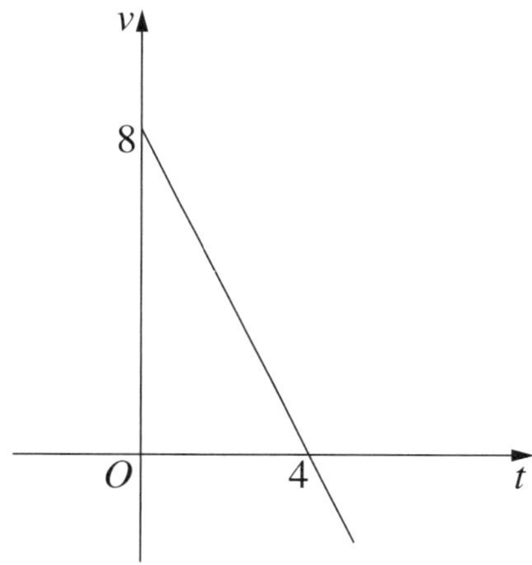

When $t = 0$ the displacement x is equal to 2 metres.

What is the maximum value of the displacement x?

(A) 8 m

(B) 14 m

(C) 16 m

(D) 18 m

10 The diagram shows the area under the curve $y = \frac{2}{x}$ from $x = 1$ to $x = d$.

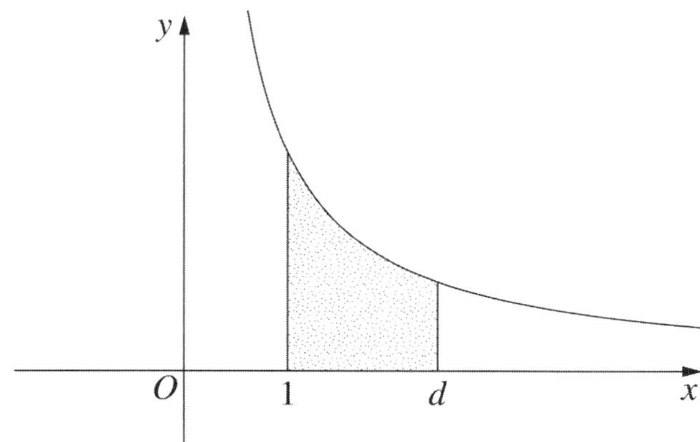

What value of d makes the shaded area equal to 2?

(A) e

(B) $e + 1$

(C) $2e$

(D) e^2

Section II

90 marks
Attempt Questions 11–16
Allow about 2 hours and 45 minutes for this section

Answer each question in the appropriate writing booklet. Extra writing booklets are available.

In Questions 11–16, your responses should include relevant mathematical reasoning and/or calculations.

Question 11 (15 marks) Use the Question 11 Writing Booklet.

(a) Simplify $4x - (8 - 6x)$. **1**

(b) Factorise fully $3x^2 - 27$. **2**

(c) Express $\dfrac{8}{2+\sqrt{7}}$ with a rational denominator. **2**

(d) Find the limiting sum of the geometric series $1 - \dfrac{1}{4} + \dfrac{1}{16} - \dfrac{1}{64} + \cdots$. **2**

(e) Differentiate $\left(e^x + x\right)^5$. **2**

(f) Differentiate $y = (x+4)\ln x$. **2**

(g) Evaluate $\displaystyle\int_0^{\frac{\pi}{4}} \cos 2x \, dx$. **2**

(h) Find $\displaystyle\int \frac{x}{x^2 - 3} \, dx$. **2**

Question 12 (15 marks) Use the Question 12 Writing Booklet.

(a) Find the solutions of $2\sin\theta = 1$ for $0 \le \theta \le 2\pi$. **2**

(b) The diagram shows the rhombus $OABC$.

The diagonal from the point $A(7, 11)$ to the point C lies on the line ℓ_1.

The other diagonal, from the origin O to the point B, lies on the line ℓ_2 which has equation $y = -\frac{x}{3}$.

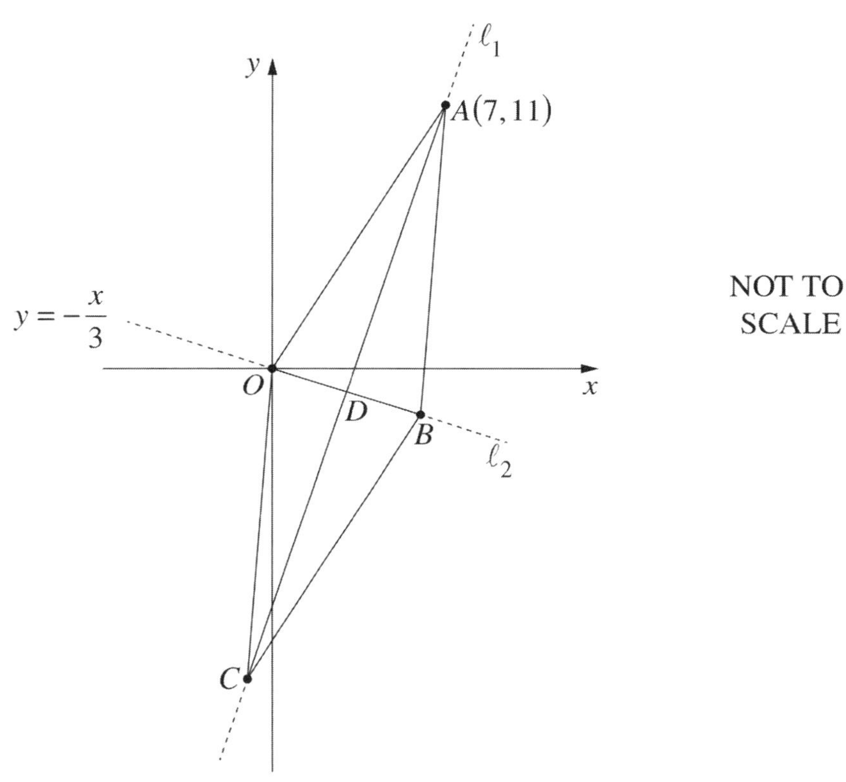

(i) Show that the equation of the line ℓ_1 is $y = 3x - 10$. **2**

(ii) The lines ℓ_1 and ℓ_2 intersect at the point D. **2**

Find the coordinates of D.

Question 12 continues on the following page

Question 12 (continued)

(c) Find $f'(x)$, where $f(x) = \dfrac{x^2 + 3}{x - 1}$. **2**

(d) For what values of k does the quadratic equation $x^2 - 8x + k = 0$ have real roots? **2**

(e) The diagram shows the parabola $y = \dfrac{x^2}{2}$ with focus $S\left(0, \dfrac{1}{2}\right)$. A tangent to the parabola is drawn at $P\left(1, \dfrac{1}{2}\right)$.

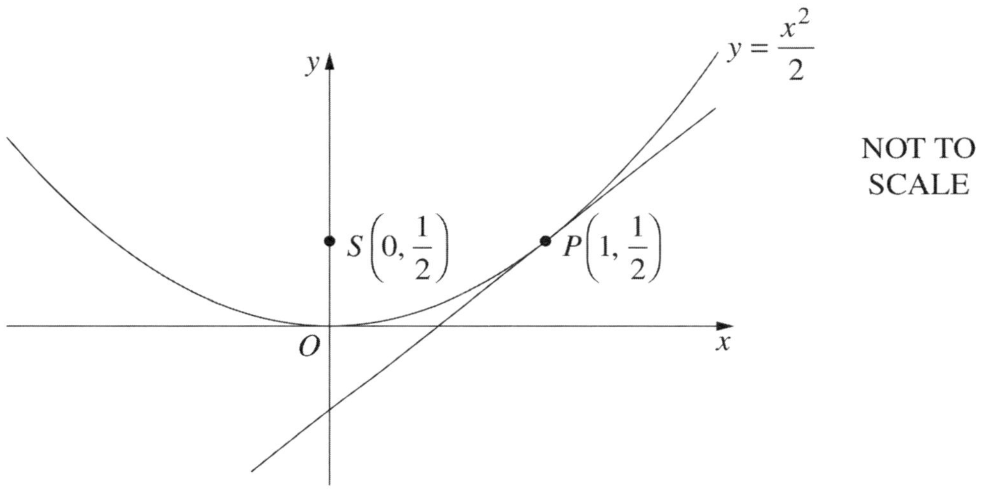

NOT TO SCALE

(i) Find the equation of the tangent at the point P. **2**

(ii) What is the equation of the directrix of the parabola? **1**

(iii) The tangent and directrix intersect at Q.

Show that Q lies on the y-axis. **1**

(iv) Show that $\triangle PQS$ is isosceles. **1**

End of Question 12

Question 13 (15 marks) Use the Question 13 Writing Booklet.

(a) The diagram shows $\triangle ABC$ with sides $AB = 6$ cm, $BC = 4$ cm and $AC = 8$ cm.

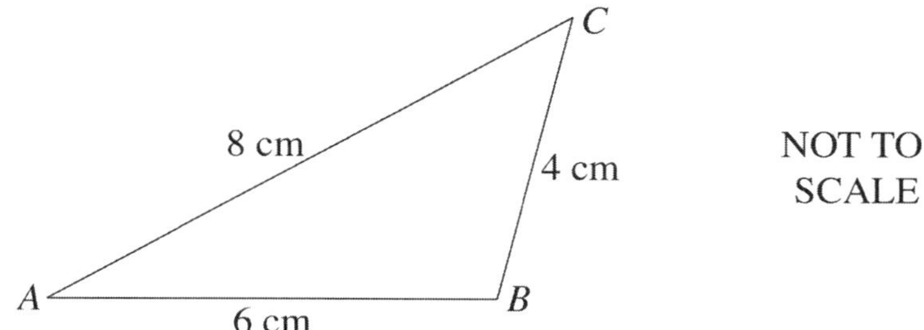

(i) Show that $\cos A = \dfrac{7}{8}$. **1**

(ii) By finding the exact value of $\sin A$, determine the exact value of the area of $\triangle ABC$. **2**

(b) (i) Find the domain and range for the function $f(x) = \sqrt{9 - x^2}$. **2**

(ii) On a number plane, shade the region where the points (x, y) satisfy both of the inequalities $y \le \sqrt{9 - x^2}$ and $y \ge x$. **2**

(c) Consider the curve $y = x^3 - x^2 - x + 3$.

(i) Find the stationary points and determine their nature. **4**

(ii) Given that the point $P\left(\dfrac{1}{3}, \dfrac{70}{27}\right)$ lies on the curve, prove that there is a point of inflexion at P. **2**

(iii) Sketch the curve, labelling the stationary points, point of inflexion and y-intercept. **2**

Question 14 (15 marks) Use the Question 14 Writing Booklet.

(a) In a theme park ride, a chair is released from a height of 110 metres and falls vertically. Magnetic brakes are applied when the velocity of the chair reaches −37 metres per second.

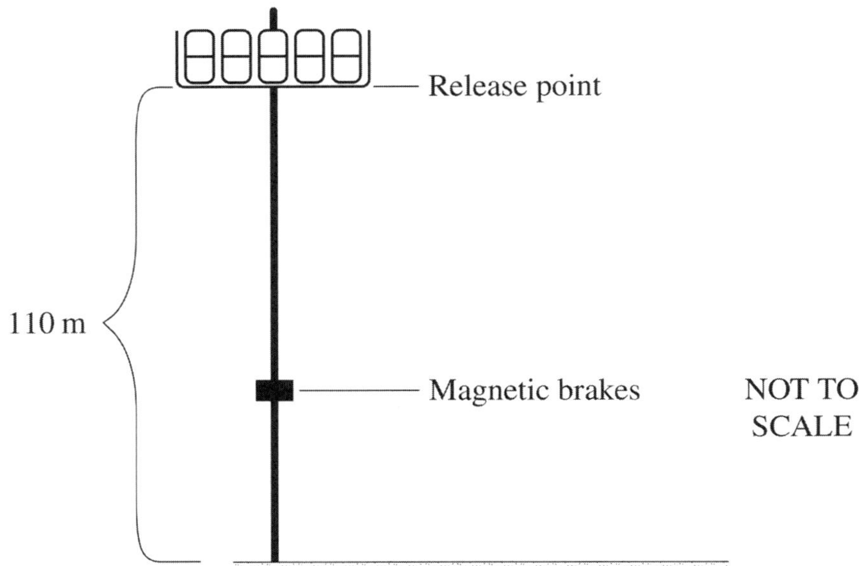

The height of the chair at time t seconds is x metres. The acceleration of the chair is given by $\ddot{x} = -10$. At the release point, $t = 0$, $x = 110$ and $\dot{x} = 0$.

(i) Using calculus, show that $x = -5t^2 + 110$. **2**

(ii) How far has the chair fallen when the magnetic brakes are applied? **2**

Question 14 continues on the following page

Question 14 (continued)

(b) Weather records for a town suggest that:

- if a particular day is wet (W), the probability of the next day being dry is $\frac{5}{6}$
- if a particular day is dry (D), the probability of the next day being dry is $\frac{1}{2}$.

In a specific week Thursday is dry. The tree diagram shows the possible outcomes for the next three days: Friday, Saturday and Sunday.

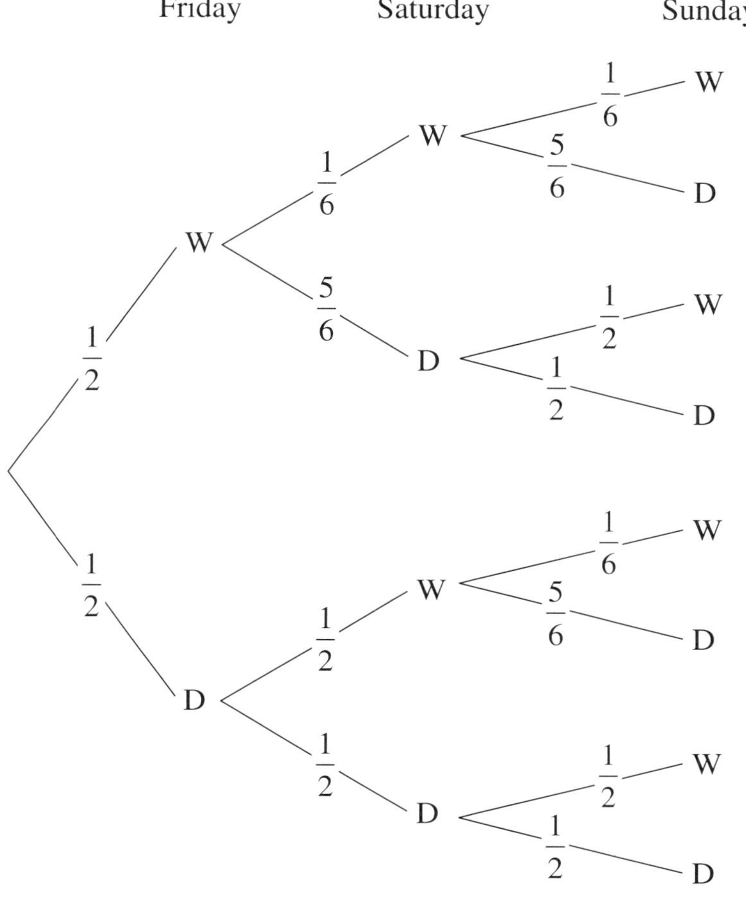

(i) Show that the probability of Saturday being dry is $\frac{2}{3}$. **1**

(ii) What is the probability of both Saturday and Sunday being wet? **2**

(iii) What is the probability of at least one of Saturday and Sunday being dry? **1**

Question 14 continues on the following page

Question 14 (continued)

(c) Sam borrows \$100 000 to be repaid at a reducible interest rate of 0.6% per month. Let $\$A_n$ be the amount owing at the end of n months and $\$M$ be the monthly repayment.

(i) Show that $A_2 = 100\,000(1.006)^2 - M(1 + 1.006)$. **1**

(ii) Show that $A_n = 100\,000(1.006)^n - M\left(\dfrac{(1.006)^n - 1}{0.006}\right)$. **2**

(iii) Sam makes monthly repayments of \$780. **1**

Show that after making 120 monthly repayments the amount owing is \$68 500 to the nearest \$100.

(iv) Immediately after making the 120th repayment, Sam makes a one-off payment, reducing the amount owing to \$48 500. The interest rate and monthly repayment remain unchanged. **3**

After how many more months will the amount owing be completely repaid?

End of Question 14

Question 15 (15 marks) Use the Question 15 Writing Booklet.

(a) The amount of caffeine, C, in the human body decreases according to the equation

$$\frac{dC}{dt} = -0.14C,$$

where C is measured in mg and t is the time in hours.

(i) Show that $C = Ae^{-0.14t}$ is a solution to $\frac{dC}{dt} = -0.14C$, where A is a constant. **1**

When $t = 0$, there are 130 mg of caffeine in Lee's body.

(ii) Find the value of A. **1**

(iii) What is the amount of caffeine in Lee's body after 7 hours? **1**

(iv) What is the time taken for the amount of caffeine in Lee's body to halve? **2**

Question 15 continues on the following page

Question 15 (continued)

(b) The diagram shows $\triangle ABC$ which has a right angle at C. The point D is the midpoint of the side AC. The point E is chosen on AB such that $AE = ED$. The line segment ED is produced to meet the line BC at F.

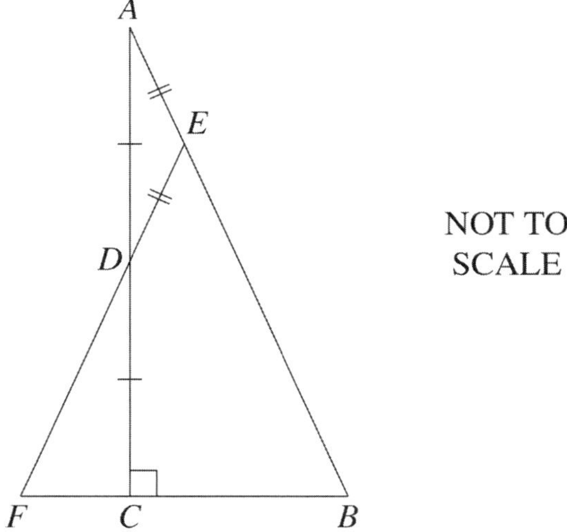

Copy or trace the diagram into your writing booklet.

(i) Prove that $\triangle ACB$ is similar to $\triangle DCF$. **2**

(ii) Explain why $\triangle EFB$ is isosceles. **1**

(iii) Show that $EB = 3AE$. **2**

(c) Water is flowing in and out of a rock pool. The volume of water in the pool at time t hours is V litres. The rate of change of the volume is given by

$$\frac{dV}{dt} = 80\sin(0.5t).$$

At time $t = 0$, the volume of water in the pool is 1200 litres and is increasing.

(i) After what time does the volume of water first start to decrease? **2**

(ii) Find the volume of water in the pool when $t = 3$. **2**

(iii) What is the greatest volume of water in the pool? **1**

End of Question 15

Question 16 (15 marks) Use the Question 16 Writing Booklet.

(a) The diagram shows the curve with equation $y = x^2 - 7x + 10$. The curve intersects the x-axis at points A and B. The point C on the curve has the same y-coordinate as the y-intercept of the curve.

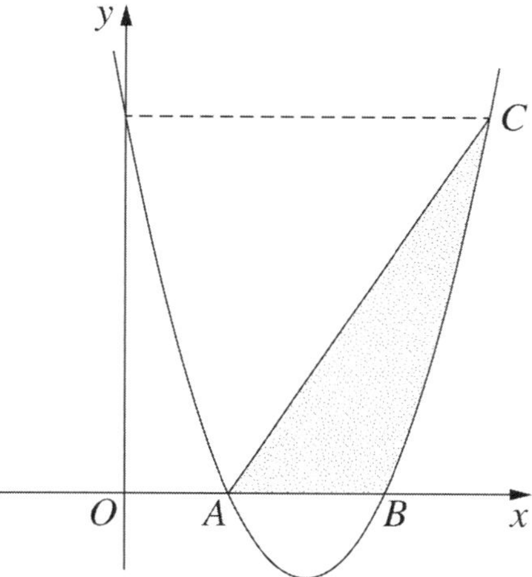

(i) Find the x-coordinates of points A and B. **1**

(ii) Write down the coordinates of C. **1**

(iii) Evaluate $\int_0^2 \left(x^2 - 7x + 10\right) dx$. **1**

(iv) Hence, or otherwise, find the area of the shaded region. **2**

Question 16 continues on the following page

Question 16 (continued)

(b) A bowl is formed by rotating the curve $y = 8\log_e(x-1)$ about the y-axis for $0 \le y \le 6$. **3**

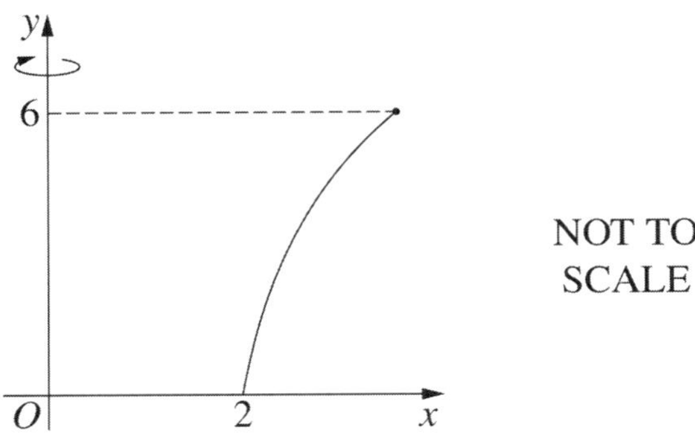

NOT TO SCALE

Find the volume of the bowl. Give your answer correct to 1 decimal place.

(c) The diagram shows a cylinder of radius x and height y inscribed in a cone of radius R and height H, where R and H are constants.

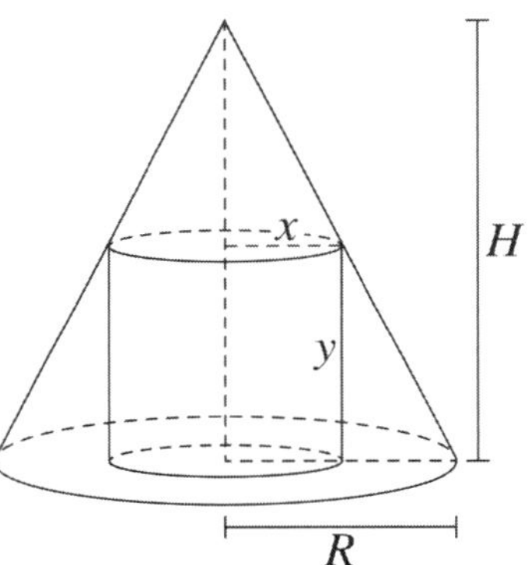

The volume of a cone of radius r and height h is $\frac{1}{3}\pi r^2 h$.

The volume of a cylinder of radius r and height h is $\pi r^2 h$.

(i) Show that the volume, V, of the cylinder can be written as $V = \frac{H}{R}\pi x^2(R-x)$. **3**

(ii) By considering the inscribed cylinder of maximum volume, show that the volume of any inscribed cylinder does not exceed $\frac{4}{9}$ of the volume of the cone. **4**

End of paper

2015 Higher School Certificate

Worked Answers

Section I

QUESTION 1

$0.005\ 233\ 59 = 5.233\ 59 \times 10^{-3}$

$= 5.234 \times 10^{-3}$ (4 sig. figs.)

Answer D

(1 mark)

QUESTION 2

$$2x - 4y + 3 = 0$$
$$4y = 2x + 3$$
$$y = \frac{2x}{4} + \frac{3}{4}$$
$$y = \frac{x}{2} + \frac{3}{4}$$

$\therefore$ slope is $\frac{1}{2}$.

Answer C

(1 mark)

QUESTION 3

3, 7, 11 ...

$a = 3, d = 4, n = 15$:

$$T_n = a + (n - 1)d$$
$$T_{15} = 3 + (15 - 1) \times 4$$
$$= 59$$

$\therefore$ the 15th term is 59.

Answer A

(1 mark)

QUESTION 4

P(Soccer team does not win) $= 1 - \frac{5}{7} = \frac{2}{7}$

P(Rugby League team does not win)

$$= 1 - \frac{2}{3} = \frac{1}{3}$$

P(neither team wins) $= \frac{2}{7} \times \frac{1}{3}$

$$= \frac{2}{21}$$

Answer A

(1 mark)

QUESTION 5

x	1	1.5	2	2.5	3
$f(x)$	e^1	$1.5e^{1.5}$	$2e^2$	$2.5e^{2.5}$	$3e^3$

$$A \approx \frac{0.5}{2}(e^1 + 2\{1.5e^{1.5} + 2e^2 + 2.5e^{2.5}\} + 3e^3)$$
$$= \frac{1}{4}(e^1 + 3e^{1.5} + 4e^2 + 5e^{2.5} + 3e^3)$$

Answer B

(1 mark)

QUESTION 6

$$y = 2\sin 3x - 3\tan x$$
$$y' = 6\cos 3x - 3\sec^2 x$$
$$= 6\cos 3x - \frac{3}{\cos^2 x}$$
$$y'(0) = 6\cos 3(0) - \frac{3}{\cos^2(0)}$$
$$= 6 \times 1 - \frac{3}{1}$$
$$= 3$$

Answer C

(1 mark)

QUESTION 7

$$\text{Area} = \int_0^2 4x - x^2 - 2x \ dx$$
$$= \int_0^2 2x - x^2 \ dx$$

Answer B

(1 mark)

QUESTION 8

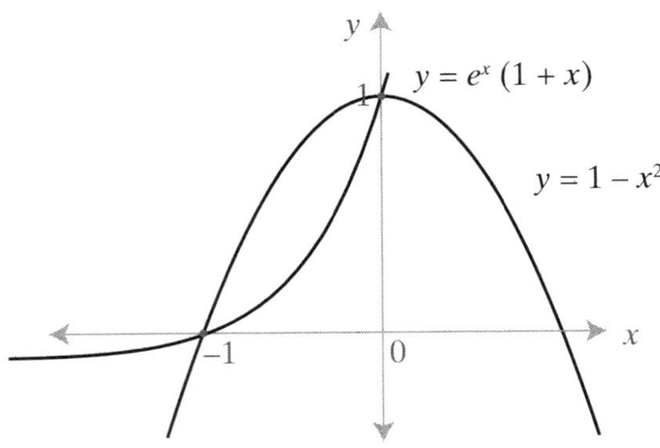

$\therefore$ 2 points of intersection of $y = e^x(1 + x)$ and $y = 1 - x^2$.

$\therefore$ 2 solutions of $e^x(1 + x) = 1 - x^2$.

Answer C

(1 mark)

QUESTION 9

Maximum displacement when $v = 0$.

$$\begin{aligned}\text{Distance travelled} &= \text{area of triangle}\\ &= \frac{1}{2} \times 4 \times 8\\ &= 16\end{aligned}$$

$$\begin{aligned}\text{Displacement} &= 2 + 16\\ &= 18\end{aligned}$$

$\therefore$ the maximum displacement is 18 m.

Alternative method:

The graph has gradient of –2 and v-intercept of 8.

$\therefore v = -2t + 8$.

Now, $x = -t^2 + 8t + c$

Subs, $t = 0, x = 2$:

$$2 = -(0)^2 + 8(0) + c$$

$$\therefore c = 2$$

$$x = -t^2 + 8t + 2$$

Subs. $t = 4$:

$$\begin{aligned}x &= -(4)^2 + 8(4) + 2\\ &= 18\end{aligned}$$

$\therefore$ the maximum displacement is 18 m.

Answer D

(1 mark)

QUESTION 10

$$\int_1^d \frac{2}{x}\,dx = 2$$

$$2\int_1^d \frac{1}{x}\,dx = 2$$

$$\int_1^d \frac{1}{x}\,dx = 1$$

$$\begin{aligned}\log_e d - \log_e 1 &= 1\\ \log_e d &= 1\\ d &= e\end{aligned}$$

Answer A

(1 mark)

Section II

QUESTION 11

(a) $$\begin{aligned}4x - (8 - 6x) &= 4x - 8 + 6x\\ &= 10x - 8\end{aligned}$$

(1 mark)

(b) $$\begin{aligned}3x^2 - 27 &= 3(x^2 - 9)\\ &= 3(x - 3)(x + 3)\end{aligned}$$

(2 marks)

(c) $$\begin{aligned}\frac{8}{2+\sqrt{7}} &= \frac{8}{2+\sqrt{7}} \times \frac{2-\sqrt{7}}{2-\sqrt{7}}\\ &= \frac{8(2-\sqrt{7})}{4-7}\\ &= \frac{8(2-\sqrt{7})}{-3}\\ &= \frac{8(\sqrt{7}-2)}{3}\end{aligned}$$

(2 marks)

(d) $a = 1, r = -\frac{1}{4}$

$$S_\infty = \frac{a}{1-r}$$

$$= \frac{1}{1+\frac{1}{4}}$$

$$= 1 \div \frac{5}{4}$$

$$= \frac{4}{5}$$

(2 marks)

(e) Using the function of a function rule:

$$\frac{d}{dx}[(e^x + x)^5] = 5(e^x + x)^4.(e^x + 1)$$

(2 marks)

(f) $y = (x + 4)\ln x$

Using the product rule,

Let $u = x + 4$ $\quad u' = 1$

Let $v = \ln x$ $\quad v' = \frac{1}{x}$

$$\frac{dy}{dx} = u'v + v'u$$

$$= 1.\ln x + \frac{1}{x}(x+4)$$

$$= \ln x + \frac{x+4}{x}$$

(2 marks)

(g) $\int_0^{\frac{\pi}{4}} \cos 2x \, dx = \left[\frac{1}{2}\sin 2x\right]_0^{\frac{\pi}{4}}$

$$= \frac{1}{2}\left[\sin\frac{\pi}{2} - \sin 0\right]$$

$$= \frac{1}{2}(1-0)$$

$$= \frac{1}{2}$$

(2 marks)

(h) $\int \frac{x}{x^2-3}\,dx = \frac{1}{2}\int \frac{2x}{x^2-3}\,dx$

$$= \frac{1}{2}\log_e(x^2-3) + c$$

(2 marks)

QUESTION 12

(a) $2\sin\theta = 1$

$$\sin\theta = \frac{1}{2}$$

$$\theta = \frac{\pi}{6} \text{ or } \frac{5\pi}{6}.$$

(2 marks)

(b) (i) For ℓ_2: $y = -\frac{x}{3}$, gradient is $-\frac{1}{3}$.

As diagonals of rhombus are perpendicular, then gradient of ℓ_1 is 3.

For line ℓ_1, using (7, 11) and $m = 3$:

$$y - y_1 = m(x - x_1)$$

$$y - 11 = 3(x-7)$$

$$y - 11 = 3x - 21$$

$$y = 3x - 10$$

(2 marks)

(ii) $y = 3x - 10$ ..... ①

$y = -\frac{x}{3}$ .... ②

Let ① = ② : $3x - 10 = -\frac{x}{3}$

$$9x - 30 = -x$$

$$10x = 30$$

$$x = 3$$

Subs. in ② : $y = -\frac{3}{3}$

$$= -1$$

$\therefore D(3, -1)$

(2 marks)

(c) $f(x) = \dfrac{x^2+3}{x-1}$

Using the quotient rule,

Let $u = x^2 + 3$ $\quad u' = 2x$

Let $v = x - 1$ $\quad v' = 1$

$$\frac{dy}{dx} = \frac{vu' - uv'}{v^2}$$
$$= \frac{(x-1).2x - (x^2+3).1}{(x-1)^2}$$
$$= \frac{2x^2 - 2x - x^2 - 3}{(x-1)^2}$$
$$= \frac{x^2 - 2x - 3}{(x-1)^2}$$

(2 marks)

(d) $\Delta \geq 0$ for real roots.
If $x^2 - 8x + k = 0$,

$$\Delta = b^2 - 4ac$$
$$= (-8)^2 - 4 \times 1 \times k$$
$$= 64 - 4k \geq 0$$
$$4k \leq 64$$
$$k \leq 16$$

(2 marks)

(e) (i)
$$y = \frac{x^2}{2}$$
$$y' = \frac{2x}{2}$$
$$= x$$

$y'(1) = 1$

Using $(1, \frac{1}{2})$, $m = 1$:

$$y - y_1 = m(x - x_1)$$
$$y - \frac{1}{2} = 1(x - 1)$$
$$y - \frac{1}{2} = x - 1$$
$$y = x - \frac{1}{2}$$

$\therefore$ the equation of tangent is $y = x - \dfrac{1}{2}$.

(2 marks)

(ii) For $y = \dfrac{x^2}{2}$, vertex $(0,0)$ and focus is $(0, \frac{1}{2})$.

$\therefore$ focal length: $a = \dfrac{1}{2}$.

$\therefore$ equation of the directrix is $y = -\dfrac{1}{2}$.

(1 mark)

(iii) Subs. $y = -\dfrac{1}{2}$ in $y = x - \dfrac{1}{2}$:

$$-\frac{1}{2} = x - \frac{1}{2}$$
$$\therefore \ x = 0$$

$\therefore$ $Q(0, -\frac{1}{2})$, which lies on y-axis.

(1 mark)

(iv) $PS = 1$

$$QS = \frac{1}{2} + \frac{1}{2}$$
$$= 1$$

$\therefore$ $PS = QS$

$\therefore$ ΔPQS is isosceles (2 sides equal).

(1 mark)

QUESTION 13

(a) (i)
$$\cos A = \frac{8^2 + 6^2 - 4^2}{2 \times 8 \times 6}$$
$$= \frac{84}{96}$$
$$= \frac{7}{8}$$

(1 mark)

(ii) Draw a triangle with $\cos A = \dfrac{7}{8}$:

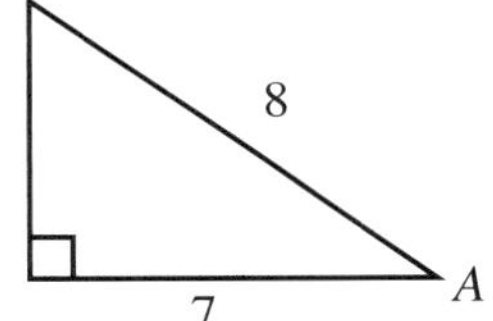

Using Pythagoras' theorem, and letting unknown side be x:

$$x^2 = 8^2 - 7^2$$
$$= 15$$
$$x = \sqrt{15} \quad (x > 0)$$

$\therefore$ $\sin A = \dfrac{\sqrt{15}}{8}$

Now, area of $\Delta ABC = \frac{1}{2}bc \sin A$

$= \frac{1}{2} \times 8 \times 6 \times \frac{\sqrt{15}}{8}$

$= 3\sqrt{15}$

$\therefore$ the area is $3\sqrt{15}$ cm^2.

(2 marks)

(b) (i) Domain: $-3 \le x \le 3$

Range: $0 \le f(x) \le 3$

(2 marks)

(ii)

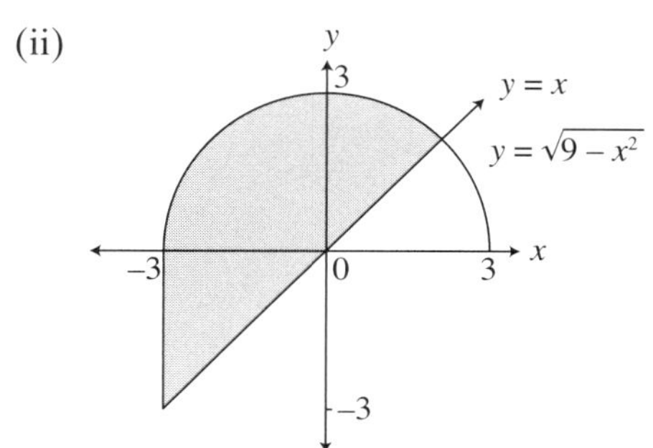

(2 marks)

(c) (i) $y = x^3 - x^2 - x + 3$

$y' = 3x^2 - 2x - 1 = 0$

$(3x + 1)(x - 1) = 0$

$x = -\frac{1}{3}$ or 1

$y(-\frac{1}{3}) = (-\frac{1}{3})^3 - (-\frac{1}{3})^2 - (-\frac{1}{3}) + 3$

$= 3\frac{5}{27}$

$y(1) = 1^3 - 1^2 - 1 + 3$

$= 2$

$\therefore$ stationary points at $(-\frac{1}{3}, 3\frac{5}{27})$ and $(1, 2)$.

$y'' = 6x - 2$

$y''(-\frac{1}{3}) = 6(-\frac{1}{3}) - 2 < 0$

$\therefore$ maximum at $(-\frac{1}{3}, 3\frac{5}{27})$.

$y''(1) = 6(1) - 2 > 0$

$\therefore$ minimum at $(1, 2)$.

(4 marks)

(ii) $y'' = 6x - 2 = 0$

$6x = 2$

$x = \frac{1}{3}$

$\therefore$ possible point of inflexion at $(\frac{1}{3}, \frac{70}{27})$.

Consider neighbourhood:

x	$\frac{1}{4}$	$\frac{1}{3}$	$\frac{1}{2}$
y''	< 0	0	> 0

As change in concavity, then the point of inflexion is at $(\frac{1}{3}, \frac{70}{27})$.

(2 marks)

(iii) Substitute $x = 0$ in $y = x^3 - x^2 - x + 3$:

$\therefore y = 3$.

$\therefore$ y-intercept of 3.

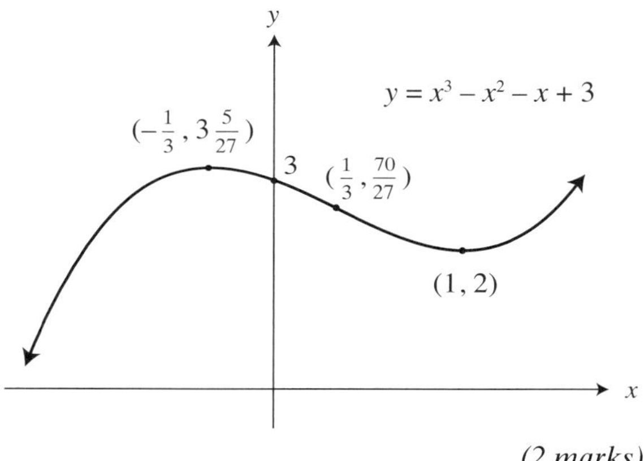

(2 marks)

QUESTION 14

(a) (i) $\ddot{x} = -10$

$\dot{x} = -10t + c$

Substitute $\dot{x} = 0$ and $t = 0$:

$0 = -10(0) + c$

$c = 0$

$\therefore \dot{x} = -10t$

$x = -5t^2 + k$

Substitute $x = 110$ and $t = 0$:

$110 = -5(0)^2 + k$

$k = 110$

$\therefore$ $x = -5t^2 + 110$

(2 marks)

(ii) Substitute $\dot{x} = -37$ in $\dot{x} = -10t$:

$-37 = -10t$

$\therefore\ t = 3.7$

Substitute $t = 3.7$ in $x = -5t^2 + 110$:

$x = -5(3.7)^2 + 110$

$= 41.55$

$\therefore$ the brakes are applied at a height of 41.55 m.

As $110 - 41.55 = 68.45$, the chair has fallen 68.45 m.

(2 marks)

(b) (i) P(Sat. dry)

$= P(WD) + P(DD)$

$= \frac{1}{2} \times \frac{5}{6} + \frac{1}{2} \times \frac{1}{2}$

$= \frac{2}{3}$

(1 mark)

(ii) P(Sat. and Sun. wet)

$= P(WWW) + P(DWW)$

$= \frac{1}{2} \times \frac{1}{6} \times \frac{1}{6} + \frac{1}{2} \times \frac{1}{2} \times \frac{1}{6}$

$= \frac{1}{18}$

$\therefore$ the probability that Saturday and Sunday are both wet is $\frac{1}{18}$.

(2 marks)

(iii) P(at least one of Sat. and Sun. dry)

$= 1 - $ P(Sat. and Sun. wet)

$= 1 - \frac{1}{18}$ (from part (ii))

$= \frac{17}{18}$

$\therefore$ the probability that at least one of Saturday and Sunday being dry is $\frac{17}{18}$.

(1 mark)

(c) (i) $A_1 = 100\,000 \times 1.006 - M$

$A_2 = [100\,000 \times 1.006 - M] \times 1.006 - M$

$= 100\,000 \times 1.006^2 - 1.006M - M$

$= 100\,000 \times 1.006^2 - M(1 + 1.006)$

(1 mark)

(ii) $A_n = 100\,000 \times 1.006^n - M(1 + 1.006 + 1.006^2 + \ldots + 1.006^{n-1})$

Considering the geometric series with $a = 1,\ r = 1.006, n = n$, and using $S_n = \frac{a(r^n - 1)}{r - 1}$:

$$A_n = 100\,000 \times 1.006^n - M\left[\frac{1(1.006^n - 1)}{1.006 - 1}\right]$$

$$= 100\,000 \times 1.006^n - M\left[\frac{1.006^n - 1}{0.006}\right]$$

(2 marks)

(iii) Let $M = 780, n = 120$:

$$A_{120} = 100\,000 \times 1.006^{120} - 780\left[\frac{1.006^{120} - 1}{0.006}\right]$$

$= 68\,499.4583\ldots$

$= 68\,500$ (nearest 500)

$\therefore$ the amount owing is \$68 500.

(1 mark)

(iv) Let amount be 48 500 and $A_n = 0$:

$$0 = 48\,500 \times 1.006^n - 780\left[\frac{1.006^n - 1}{0.006}\right]$$

$48\,500 \times 1.006^n = 130\,000(1.006^n - 1)$

$48\,500 \times 1.006^n = 130\,000 \times 1.006^n - 130\,000$

$81\,500 \times 1.006^n = 130\,000$

$1.006^n = \frac{1300}{815}$

$1.006^n = \frac{260}{163}$

$\log_e 1.006^n = \log_e \frac{260}{163}$

$$n = \frac{\log_e \frac{260}{163}}{\log_e 1.006}$$

$= 78.055\,137\,98\ldots$

$\therefore$ Sam will need 79 repayments.

(3 marks)

QUESTION 15

(a) (i) $C = Ae^{-0.14t}$

$$\frac{dC}{dt} = -0.14 \times Ae^{-0.14t}$$
$$= -0.14C$$

$\therefore C = Ae^{-0.14t}$ is a solution.

(1 mark)

(ii) Substitute $C = 130, t = 0$ in $C = Ae^{-0.14t}$

$$130 = Ae^{-0.14(0)}$$
$$130 = A$$
$$\therefore \quad A = 130$$

(1 mark)

(iii) Substitute $t = 7$ in $C = 130e^{-0.14t}$

$$C = 130e^{-0.14(7)}$$
$$= 48.790\,442\,85\ldots$$
$$= 48.79 \text{ (2 dec. pl.)}$$

$\therefore$ there is 48.79 mg of caffeine in Lee's body.

(1 mark)

(iv) Substitute $C = 65$ in $C = 130e^{-0.14t}$

$$65 = 130e^{-0.14t}$$
$$e^{-0.14t} = 0.5$$
$$\log_e e^{-0.14t} = \log_e 0.5$$
$$-0.14t = \log_e 0.5$$
$$t = \frac{\log_e 0.5}{-0.14}$$
$$= 4.951\,051\,29\ldots$$
$$= 4.95 \text{ (2 dec. pl.)}$$

$\therefore$ it will take 4.95 hours.

(2 marks)

(b)

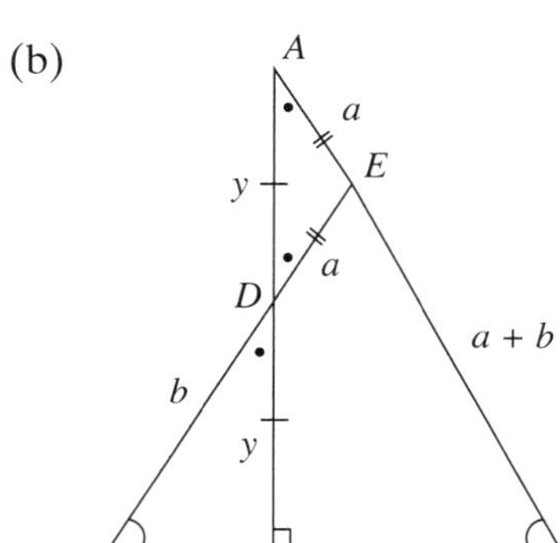

(i) $\angle BAC = \angle ADE$ (base $\angle$s of isosceles Δ)

$\angle ADE = \angle FDC$ (vertically opposite $\angle$s)

$\therefore \angle BAC = \angle FDC$

Also, $\angle BCA = \angle FCD$ (given)

$\therefore$ ΔACB is similar to ΔDCF (2 $\angle$s equal)

(2 marks)

(ii) $\angle ABC = \angle DFC$ (matching $\angle$s of similar Δs)

$\therefore$ ΔEFB is isosceles (two $\angle$s equal)

(1 mark)

(iii) Let $AE = ED = a$, $AD = DC = y$.

Also, let $DF = b$, then $EF = a + b$ and hence $EB = a + b$.

Now, $\frac{2a+b}{b} = \frac{2y}{y}$ (matching sides of similar Δs in proportion)

$$\frac{2a+b}{b} = 2$$

$$2a + b = 2b$$

$$b = 2a$$

But $EB = a + b$

$= a + 2a$

$= 3a$

$\therefore$ $EB = 3AE$

(2 marks)

(c) (i) 'First starts to decrease' when $\frac{dV}{dt} = 0$:

$$80 \sin (0.5t) = 0$$

$$\sin (0.5t) = 0$$

$$0.5t = 0, \pi, 2\pi, \ldots$$

$$t = 0, 2\pi, 4\pi, \ldots$$

As $t > 0$, then $t = 2\pi, 4\pi, \ldots$

$$\frac{d^2V}{dt^2} = 40 \cos (0.5t)$$

$$\frac{d^2V}{dt^2}(2\pi) = 40 \cos (0.5(2\pi)) < 0$$

$\therefore$ maximum volume when $t = 2\pi$.

$\therefore$ the volume starts to decrease after 2π hours.

(2 marks)

(ii) $\dfrac{dV}{dt} = 80\sin(0.5t)$

$V = -160\cos(0.5t) + c$

Substitute $V = 1200$ and $t = 0$:

$1200 = -160\cos(0.5(0)) + c$

$1200 = -160 + c$

$c = 1360$

$\therefore V = 1360 - 160\cos(0.5t)$

Substitute $t = 3$:

$V = 1360 - 160\cos(0.5(3))$

$= 1348.682\,048...$

$= 1349$ (nearest whole)

$\therefore$ the volume is 1349 L.

(2 marks)

(iii) Greatest volume when $t = 2\pi$ (from part (i)).

$V = 1360 - 160\cos(0.5(2\pi))$

$= 1360 - 160\cos\pi$

$= 1360 + 160$

$= 1520$

$\therefore$ the greatest volume is 1520 L.

(1 mark)

QUESTION 16

(a) (i) $x^2 - 7x + 10 = 0$

$(x-2)(x-5) = 0$

$x = 2, 5$

$\therefore$ the x-coordinates are 2 and 5.

(1 mark)

(ii) y-intercept of $x^2 - 7x + 10$ is 10.

Substitute $y = 10$ in $y = x^2 - 7x + 10$:

$10 = x^2 - 7x + 10$

$x^2 - 7x = 0$

$x(x-7) = 0$

$x = 0, 7$

$\therefore C(7, 10)$

(1 mark)

(iii) $\displaystyle\int_0^2 (x^2 - 7x + 10)\ dx$

$= \left[\dfrac{x^3}{3} - \dfrac{7x^2}{2} + 10x\right]_0^2$

$= \dfrac{2^3}{3} - \dfrac{7(2)^2}{2} + 10(2) - 0$

$= 8\dfrac{2}{3}$

(1 mark)

(iv)

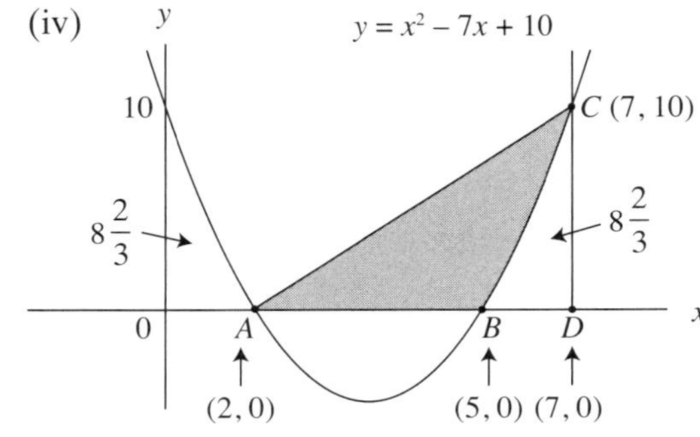

Parabola is symmetrical:

$$\text{Area} = \text{Area of } \Delta ACD - \int_0^2 (x^2 - 7x + 10)\,dx$$

$$= \frac{1}{2} \times 5 \times 10 - 8\frac{2}{3}$$

$$= 16\frac{1}{3}$$

$\therefore$ shaded area is $16\frac{1}{3}$ units^2.

(2 marks)

(b)

$$y = 8\log_e(x-1)$$

$$\frac{y}{8} = \log_e(x-1)$$

$$x - 1 = e^{\frac{y}{8}}$$

$$x = e^{\frac{y}{8}} + 1$$

$$x^2 = (e^{\frac{y}{8}} + 1)^2$$

$$= e^{\frac{y}{4}} + 2e^{\frac{y}{8}} + 1$$

$$V = \pi\int_0^6 x^2\,dy$$

$$= \pi\int_0^6 \left(e^{\frac{y}{4}} + 2e^{\frac{y}{8}} + 1\right) dy$$

$$= \pi\left[4e^{\frac{y}{4}} + 16e^{\frac{y}{8}} + y\right]_0^6$$

$$= \pi(4e^{1.5} + 16e^{0.75} + 6 - (4e^0 + 16e^0 + 0))$$

$$= 118.748\,2959\ldots$$

$$= 118.7 \text{ (1 dec. pl.)}$$

$\therefore$ the volume is 118.7 units^3.

(3 marks)

(c) (i)

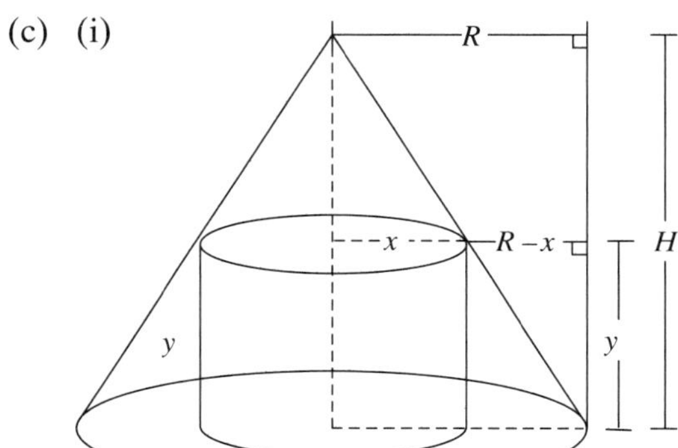

Consider sides of similar triangles:

$$\frac{y}{H} = \frac{R-x}{R}$$

$$y = \frac{H(R-x)}{R}$$

Now, volume of cylinder:

$$V = \pi x^2 y$$

$$= \pi x^2\left[\frac{H(R-x)}{R}\right]$$

$$= \frac{H\pi}{R}x^2(R-x)$$

(3 marks)

(ii) $V = \frac{H\pi}{R}[x^2(R-x)]$

$$\frac{dV}{dx} = \frac{H\pi}{R}[2x(R-x) + x^2(-1)]$$

$$= \frac{H\pi}{R}[2xR - 2x^2 - x^2]$$

$$= \frac{H\pi}{R}[2xR - 3x^2]$$

Maximum volume occurs when $\frac{dV}{dx} = 0$

$$\therefore 2xR - 3x^2 = 0$$

$$x(2R - 3x) = 0$$

$$x = 0, \frac{2R}{3}$$

$$x = \frac{2R}{3} \quad (\text{as } x > 0)$$

$$\frac{d^2V}{dx^2} = \frac{H\pi}{R}[2R - 6x]$$

$$\frac{d^2V}{dx^2}\left[\frac{2R}{3}\right] = \frac{H\pi}{R}[2R - 6(\frac{2R}{3})] < 0$$

$$= \frac{H\pi}{R}[-2R] < 0$$

(as $H > 0, R > 0$)

$\therefore$ maximum volume when $x = \frac{2R}{3}$.

Substitute $x = \frac{2R}{3}$ in

$$V = \frac{H\pi}{R}x^2(R-x):$$

$$V = \frac{H\pi}{R}(\frac{2R}{3})^2(R - \frac{2R}{3})$$

$$= \frac{H\pi}{R} \times \frac{4R^2}{9} \times \frac{R}{3}$$

$$\therefore V_{\text{cylinder}} = \frac{4H\pi R^2}{27}$$

Now, volume of cone with radius R and height H:

$$V_{\text{cone}} = \frac{1}{3}\pi R^2 H$$

But, $\frac{4}{9} \times \frac{1}{3}\pi R^2 H = \frac{4H\pi R^2}{27}$.

$\therefore$ volume of cylinder does not exceed $\frac{4}{9}$ of volume of cone.

(4 marks)

2016 **HIGHER SCHOOL CERTIFICATE EXAMINATION**

Mathematics

General Instructions

- Reading time – 5 minutes
- Working time – 3 hours
- Write using black pen
- Board-approved calculators may be used
- A reference sheet is provided at the back of this paper
- In Questions 11–16, show relevant mathematical reasoning and/or calculations

Total marks – 100

Section I

10 marks

- Attempt Questions 1–10
- Allow about 15 minutes for this section

Section II

90 marks

- Attempt Questions 11–16
- Allow about 2 hours and 45 minutes for this section

Section I

10 marks
Attempt Questions 1–10
Allow about 15 minutes for this section

Use the multiple-choice answer sheet for Questions 1–10.

---

1 For the angle θ, $\sin\theta = \dfrac{7}{25}$ and $\cos\theta = -\dfrac{24}{25}$.

Which diagram best shows the angle θ?

(A)

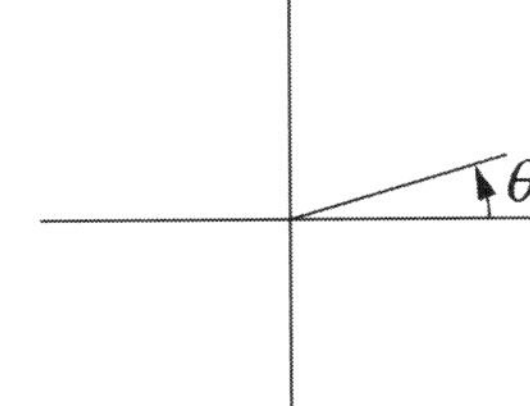

(B)

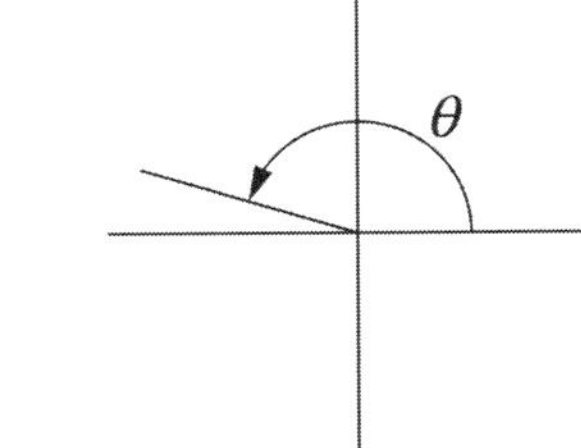

(C)

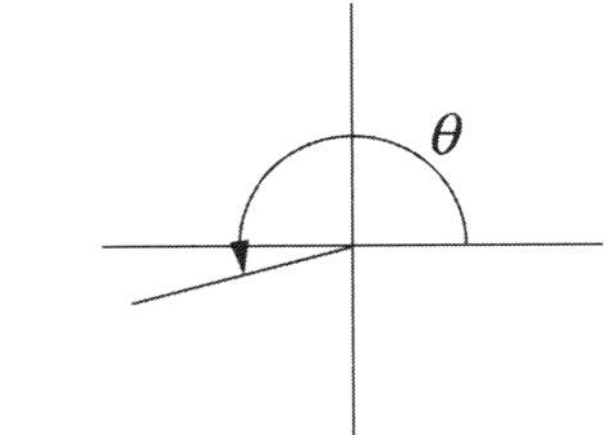

(D)

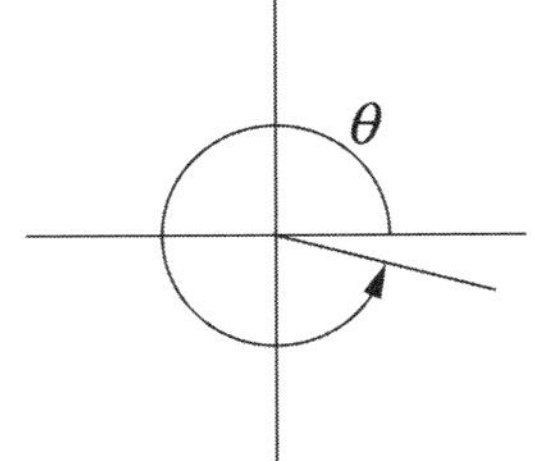

2 In a raffle, 30 tickets are sold and there is one prize to be won.

What is the probability that someone buying 6 tickets wins the prize?

(A) $\dfrac{1}{30}$

(B) $\dfrac{1}{6}$

(C) $\dfrac{1}{5}$

(D) $\dfrac{1}{4}$

3 Which diagram best shows the graph of the parabola $y = 3 - (x - 2)^2$?

(A)

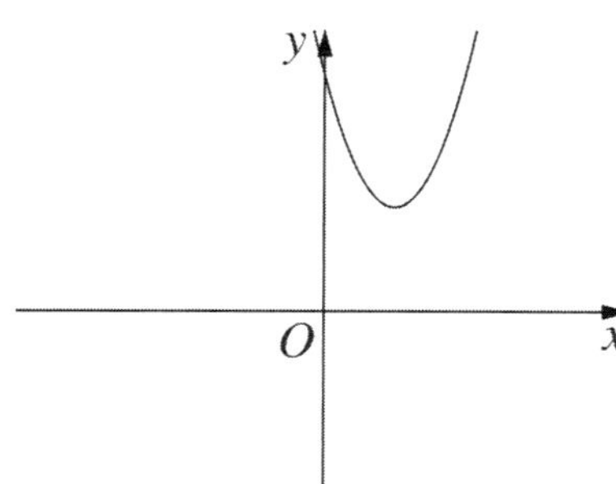

(B)

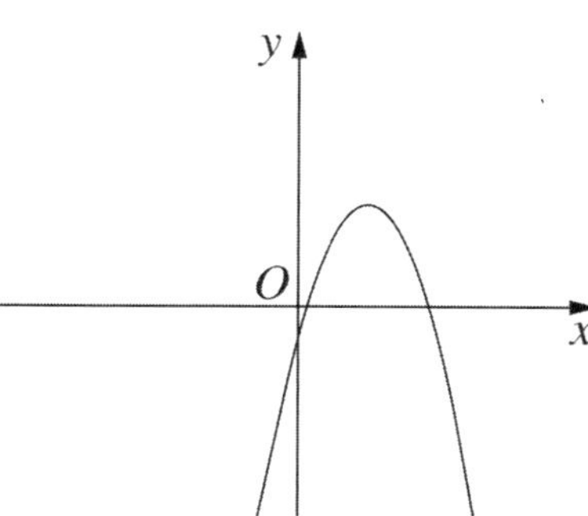

(C)

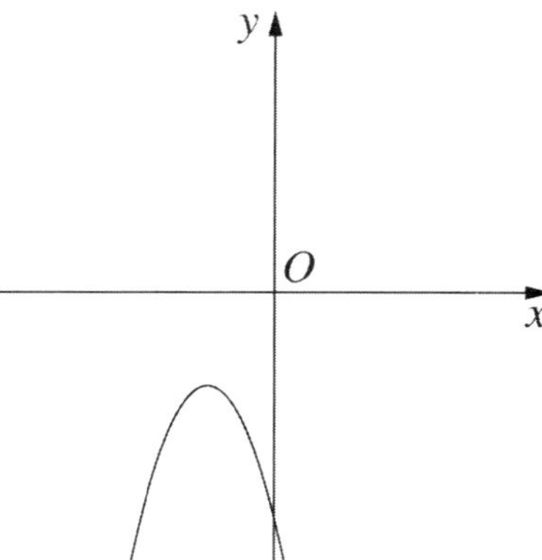

(D)

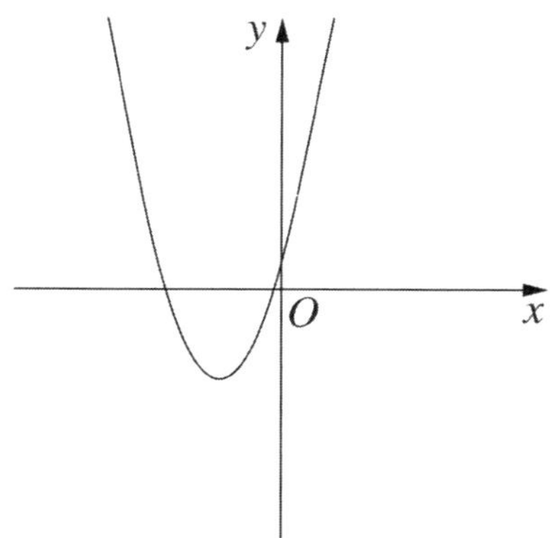

4 Which diagram shows the graph of an odd function?

(A)

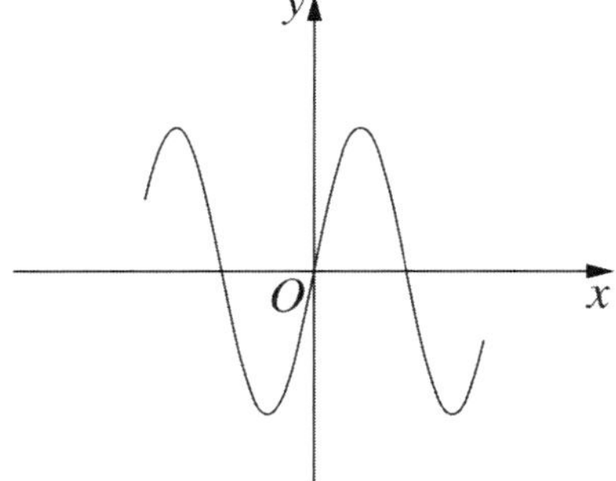

(B)

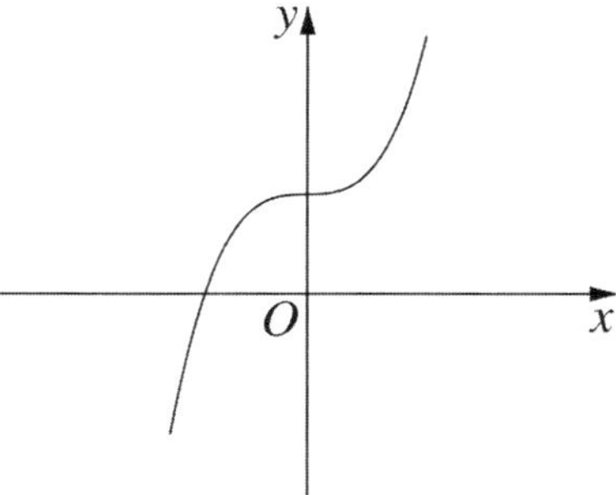

(C)

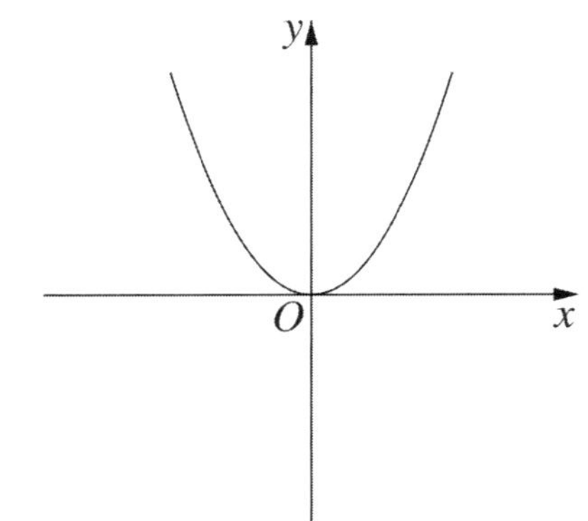

(D)

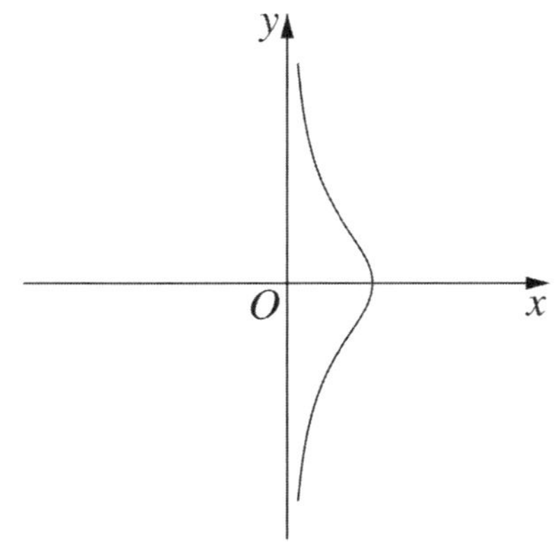

5 What is the derivative of $\ln(\cos x)$?

(A) $-\sec x$

(B) $-\tan x$

(C) $\sec x$

(D) $\tan x$

6 What is the period of the function $f(x) = \tan(3x)$?

(A) $\dfrac{\pi}{3}$

(B) $\dfrac{2\pi}{3}$

(C) 3π

(D) 6π

7 The circle centred at O has radius 5. Arc AB has length 7 as shown in the diagram.

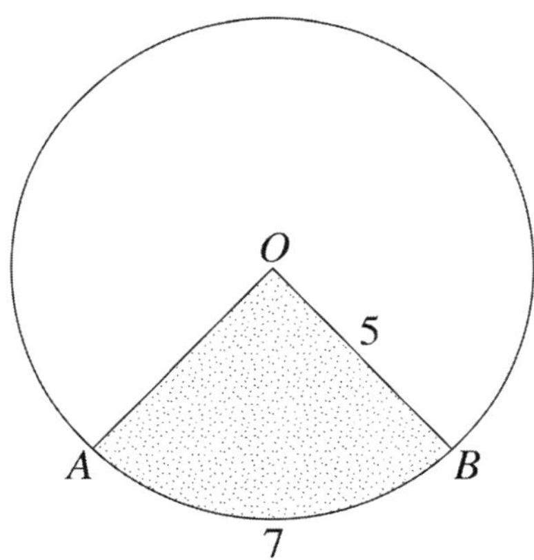

What is the area of the shaded sector OAB?

(A) $\dfrac{35}{2}$

(B) $\dfrac{35}{2}\pi$

(C) $\dfrac{125}{14}$

(D) $\dfrac{125}{14}\pi$

8 How many solutions does the equation $|\cos(2x)| = 1$ have for $0 \le x \le 2\pi$?

(A) 1

(B) 3

(C) 4

(D) 5

9 What is the value of $\displaystyle\int_{-3}^{2} |x+1|\,dx$?

(A) $\frac{5}{2}$

(B) $\frac{11}{2}$

(C) $\frac{13}{2}$

(D) $\frac{17}{2}$

10 Which expression is equivalent to $4 + \log_2 x$?

(A) $\log_2(2x)$

(B) $\log_2(16 + x)$

(C) $4\log_2(2x)$

(D) $\log_2(16x)$

Section II

90 marks
Attempt Questions 11–16
Allow about 2 hours and 45 minutes for this section

Answer each question in the appropriate writing booklet. Extra writing booklets are available.

In Questions 11–16, your responses should include relevant mathematical reasoning and/or calculations.

Question 11 (15 marks) Use the Question 11 Writing Booklet.

(a) Sketch the graph of $(x-3)^2 + (y+2)^2 = 4$. **2**

(b) Differentiate $\dfrac{x+2}{3x-4}$. **2**

(c) Solve $|x-2| \le 3$. **2**

(d) Evaluate $\displaystyle\int_0^1 (2x+1)^3\,dx$. **2**

(e) Find the points of intersection of $y = -5 - 4x$ and $y = 3 - 2x - x^2$. **3**

(f) Find the gradient of the tangent to the curve $y = \tan x$ at the point where $x = \dfrac{\pi}{8}$. **2**

Give your answer correct to 3 significant figures.

(g) Solve $\sin\left(\dfrac{x}{2}\right) = \dfrac{1}{2}$ for $0 \le x \le 2\pi$. **2**

Question 12 (15 marks) Use the Question 12 Writing Booklet.

(a) The diagram shows points $A(1, 0)$, $B(2, 4)$ and $C(6, 1)$. The point D lies on BC such that $AD \perp BC$.

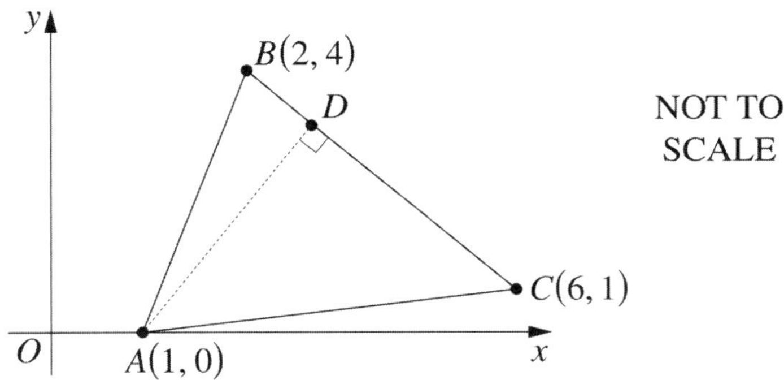

(i) Show that the equation of BC is $3x + 4y - 22 = 0$. **2**

(ii) Find the length of AD. **2**

(iii) Hence, or otherwise, find the area of $\triangle ABC$. **2**

(b) The diagram shows a semicircle with centre O. It is given that $AB = OB$, $\angle COD = 87°$ and $\angle BAO = x°$.

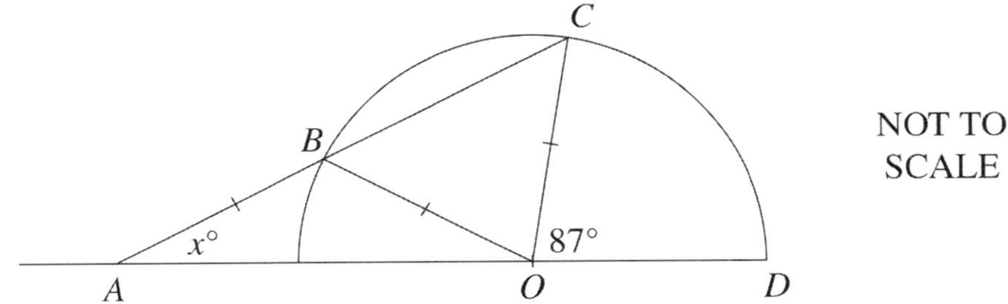

(i) Show that $\angle CBO = 2x°$, giving reasons. **1**

(ii) Find the value of x, giving reasons. **2**

Question 12 continues on the following page

Question 12 (continued)

(c) Square tiles of side length 20 cm are being used to tile a bathroom. **3**

The tiler needs to drill a hole in one of the tiles at a point P which is 8 cm from one corner and 15 cm from an adjacent corner.

To locate the point P the tiler needs to know the size of the angle θ shown in the diagram.

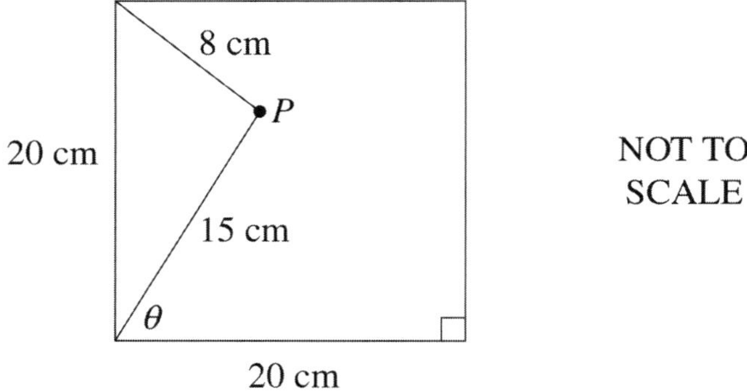

Find the size of the angle θ to the nearest degree.

(d) (i) Differentiate $y = xe^{3x}$. **1**

(ii) Hence find the exact value of $\int_0^2 e^{3x}(3+9x)\,dx$. **2**

End of Question 12

Question 13 (15 marks) Use the Question 13 Writing Booklet.

(a) Consider the function $y = 4x^3 - x^4$.

(i) Find the two stationary points and determine their nature. **4**

(ii) Sketch the graph of the function, clearly showing the stationary points and the x and y intercepts. **2**

(b) Consider the parabola $x^2 - 4x = 12y + 8$.

(i) By completing the square, or otherwise, find the focal length of the parabola. **2**

(ii) Find the coordinates of the focus. **1**

(c) A radioactive isotope of Curium has a half-life of 163 days. Initially there are 10 mg of Curium in a container.

The mass $M(t)$ in milligrams of Curium, after t days, is given by

$$M(t) = Ae^{-kt},$$

where A and k are constants.

(i) State the value of A. **1**

(ii) Given that after 163 days only 5 mg of Curium remain, find the value of k. **2**

(d) The curve $y = \sqrt{2}\cos\left(\frac{\pi}{4}x\right)$ meets the line $y = x$ at $P(1, 1)$, as shown in the diagram. **3**

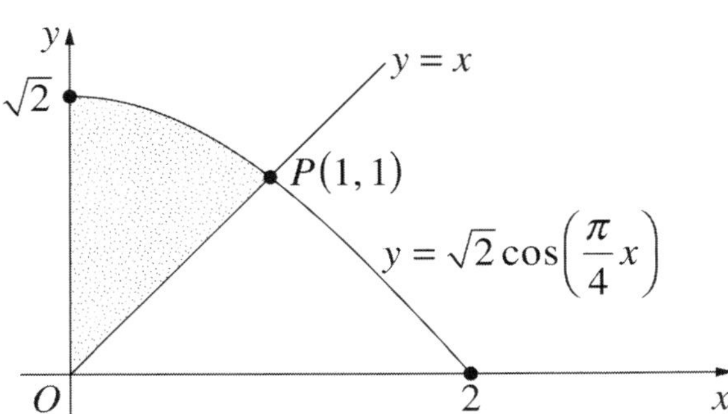

Find the exact value of the shaded area.

Question 14 (15 marks) Use the Question 14 Writing Booklet.

(a) The diagram shows the cross-section of a tunnel and a proposed enlargement. **3**

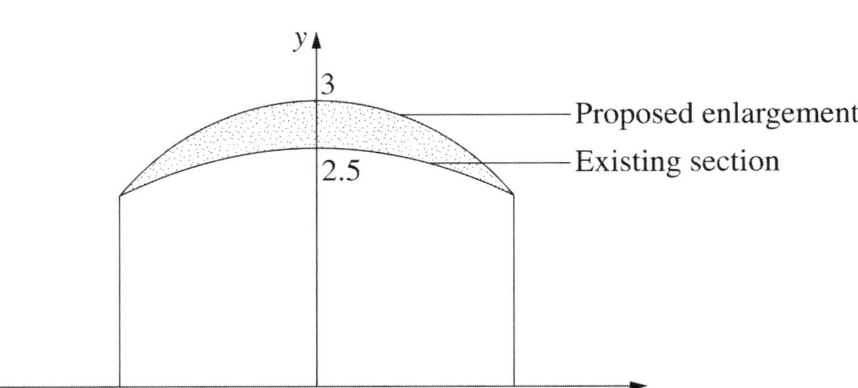

The heights, in metres, of the existing section at 1 metre intervals are shown in Table A.

Table A: Existing heights

x	−2	−1	0	1	2
y	2	2.38	2.5	2.38	2

The heights, in metres, of the proposed enlargement are shown in Table B.

Table B: Proposed heights

x	−2	−1	0	1	2
y	2	2.78	3	2.78	2

Use Simpson's rule with the measurements given to calculate the approximate increase in area.

Question 14 continues on the following page

Question 14 (continued)

(b) A gardener develops an eco-friendly spray that will kill harmful insects on fruit trees without contaminating the fruit. A trial is to be conducted with 100 000 insects. The gardener expects the spray to kill 35% of the insects each day and that exactly 5000 new insects will be produced each day.

The number of insects expected at the end of the nth day of the trial is A_n.

(i) Show that $A_2 = 0.65(0.65 \times 100\,000 + 5000) + 5000$. **2**

(ii) Show that $A_n = 0.65^n \times 100\,000 + 5000\dfrac{\left(1 - 0.65^n\right)}{0.35}$. **1**

(iii) Find the expected insect population at the end of the fourteenth day, correct to the nearest 100. **1**

(c) A farmer wishes to make a rectangular enclosure of area 720 m^2. She uses an existing straight boundary as one side of the enclosure. She uses wire fencing for the remaining three sides and also to divide the enclosure into four equal rectangular areas of width x m as shown.

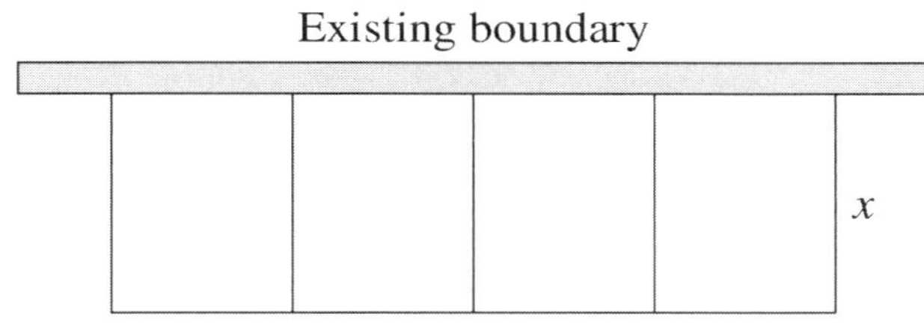

(i) Show that the total length, ℓ m, of the wire fencing is given by **1**

$$\ell = 5x + \frac{720}{x}.$$

(ii) Find the minimum length of wire fencing required, showing why this is the minimum length. **3**

(d) By summing the geometric series $1 + x + x^2 + x^3 + x^4$, or otherwise, **2**

find $\lim\limits_{x \to 1} \dfrac{x^5 - 1}{x - 1}$.

(e) Write $\log 2 + \log 4 + \log 8 + \cdots + \log 512$ in the form $a \log b$ where a and b are integers greater than 1. **2**

End of Question 14

Question 15 (15 marks) Use the Question 15 Writing Booklet.

(a) The diagram shows two curves C_1 and C_2. The curve C_1 is the semicircle $x^2 + y^2 = 4$, $-2 \le x \le 0$. The curve C_2 has equation $\dfrac{x^2}{9} + \dfrac{y^2}{4} = 1$, $0 \le x \le 3$. **4**

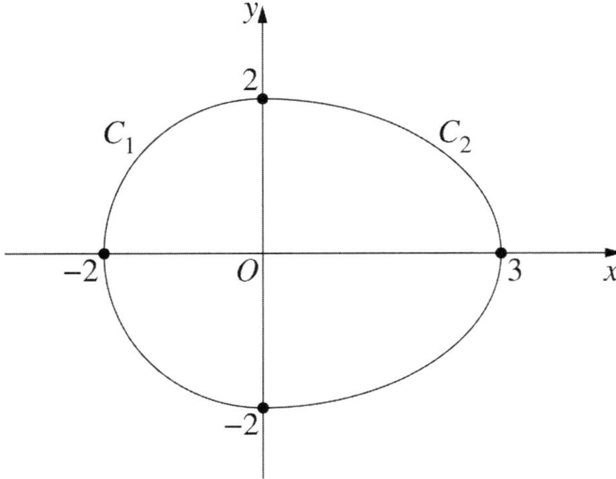

An egg is modelled by rotating the curves about the x-axis to form a solid of revolution.

Find the exact value of the volume of the solid of revolution.

(b) An eight-sided die is marked with numbers 1, 2, ..., 8. A game is played by rolling the die until an 8 appears on the uppermost face. At this point the game ends.

(i) Using a tree diagram, or otherwise, explain why the probability of the game ending before the fourth roll is **2**

$$\frac{1}{8} + \frac{7}{8} \times \frac{1}{8} + \left(\frac{7}{8}\right)^2 \times \frac{1}{8}.$$

(ii) What is the smallest value of n for which the probability of the game ending before the nth roll is more than $\dfrac{3}{4}$? **3**

Question 15 continues on the following page

Question 15 (continued)

(c) Maryam wishes to estimate the height, h metres, of a tower, ST, using a square, $ABCD$, with side length 1 metre.

She places the point A on the horizontal ground and ensures that the point D lies on the line joining A to the top of the tower T. The point F is the intersection of the line joining B and T and the side CD. The point E is the foot of the perpendicular from B to the ground. Let CF have length x metres and AE have length y metres.

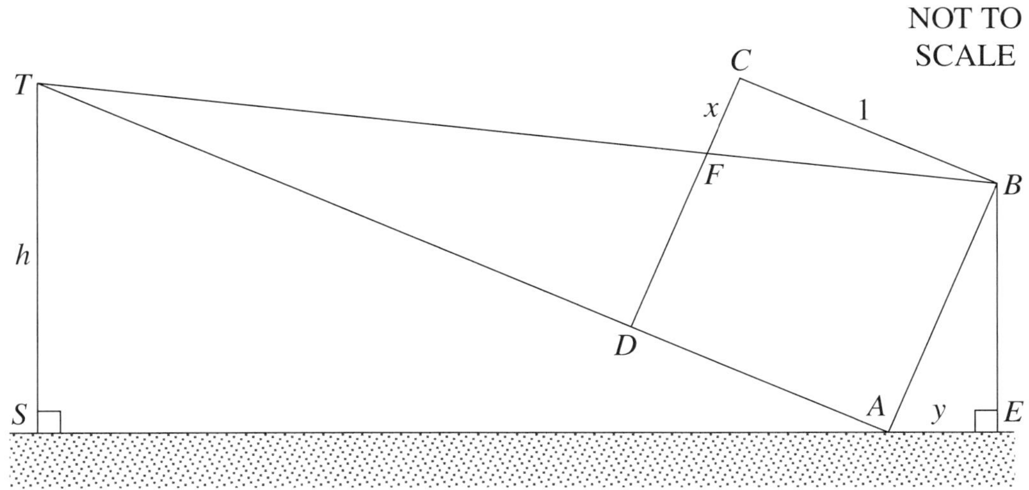

Copy or trace the diagram into your writing booklet.

(i) Show that $\triangle FCB$ and $\triangle BAT$ are similar. **2**

(ii) Show that $\triangle TSA$ and $\triangle AEB$ are similar. **2**

(iii) Find h in terms of x and y. **2**

End of Question 15

Question 16 (15 marks) Use the Question 16 Writing Booklet.

(a) A particle moves in a straight line. Its velocity v m s^{-1} at time t seconds is given by

$$v = 2 - \frac{4}{t+1}.$$

(i) Find the initial velocity. **1**

(ii) Find the acceleration of the particle when the particle is stationary. **2**

(iii) By considering the behaviour of v for large t, sketch a graph of v against t for $t \geq 0$, showing any intercepts. **2**

(iv) Find the exact distance travelled by the particle in the first 7 seconds. **3**

(b) Some yabbies are introduced into a small dam. The size of the population, y, of yabbies can be modelled by the function

$$y = \frac{200}{1+19e^{-0.5t}},$$

where t is the time in months after the yabbies are introduced into the dam.

(i) Show that the rate of growth of the size of the population is **2**

$$\frac{1900e^{-0.5t}}{\left(1+19e^{-0.5t}\right)^2}.$$

(ii) Find the range of the function y, justifying your answer. **2**

(iii) Show that the rate of growth of the size of the population can be rewritten as **1**

$$\frac{y}{400}(200-y).$$

(iv) Hence, find the size of the population when it is growing at its fastest rate. **2**

End of paper

2016 Higher School Certificate

Worked Answers

Section I

QUESTION 1

$\sin\theta > 0$ and $\cos\theta < 0$.

$\therefore$ 2nd quadrant

Answer B

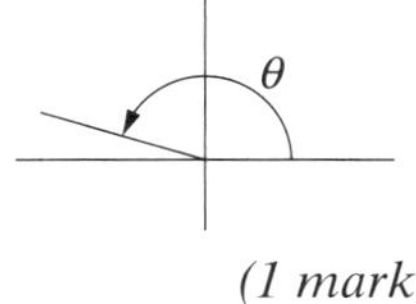

(1 mark)

QUESTION 2

$$\text{P(wins)} = \frac{6}{30} = \frac{1}{5}$$

Answer C

(1 mark)

QUESTION 3

$y = (x-2)^2$ has an axis of symmetry $x = 2$, and $y = 3 - (x-2)^2$ is concave down.

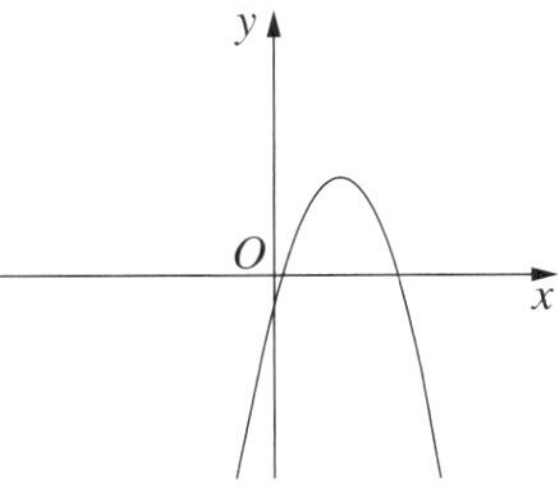

Answer B

(1 mark)

QUESTION 4

The graph of an odd function has point symmetry about the origin.

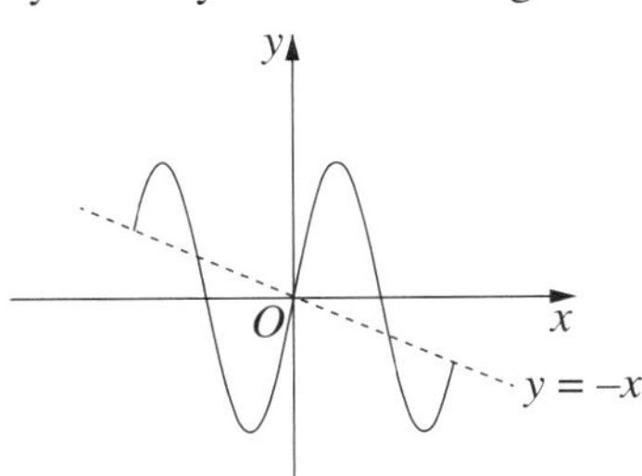

Answer A

(1 mark)

QUESTION 5

$$\frac{d}{dx}[\ln(\cos x)] = \frac{1}{\cos x}.-\sin x$$
$$= -\tan x$$

Answer B

(1 mark)

QUESTION 6

The period is $\frac{\pi}{3}$.

Answer A

(1 mark)

QUESTION 7

Use $\ell = r\theta$:

$$7 = 5\theta$$
$$\theta = \frac{7}{5}$$

Use $A = \frac{1}{2}r^2\theta$:

$$= \frac{1}{2} \times 5^2 \times \frac{7}{5}$$
$$= \frac{35}{2}$$

Answer A

(1 mark)

QUESTION 8

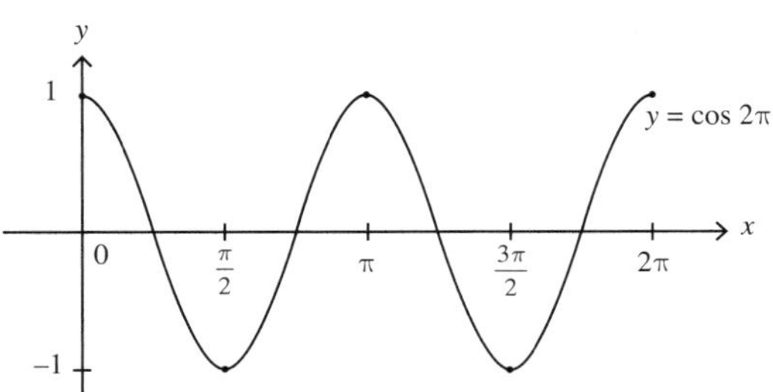

cos $2x = 1$ three times and cos $2x = -1$ twice.

$\therefore$ a total of 5 times.

Answer D

(1 mark)

QUESTION 9

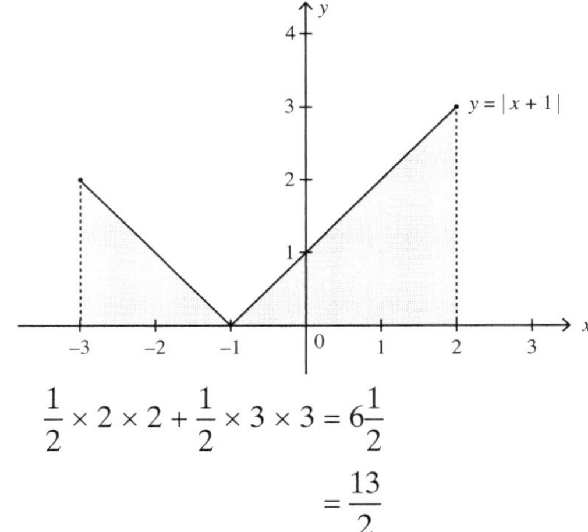

$$\frac{1}{2} \times 2 \times 2 + \frac{1}{2} \times 3 \times 3 = 6\frac{1}{2}$$
$$= \frac{13}{2}$$

Answer C

(1 mark)

QUESTION 10

Use $4 = \log_2 2^4$.

$$\begin{aligned} 4 + \log_2 x &= \log_2 2^4 + \log_2 x \\ &= \log_2 16 + \log_2 x \\ &= \log_2 16x \end{aligned}$$

Answer D

(1 mark)

Section II

QUESTION 11

(a) Circle, centre $(3, -2)$ and radius 2.

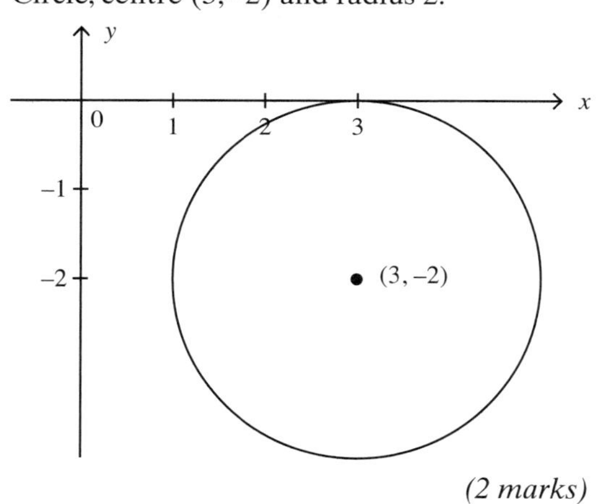

(2 marks)

(b) Using the quotient rule,

Let $u = x + 2$ $\qquad \frac{du}{dx} = 1$

Let $v = 3x - 4$ $\qquad \frac{dv}{dx} = 3$

$$\begin{aligned} \frac{d}{dx}\left[\frac{x+2}{3x-4}\right] &= \frac{v\frac{du}{dx} - u\frac{dv}{dx}}{v^2} \\ &= \frac{(3x-4).1-(x+2).3}{(3x-4)^2} \\ &= \frac{3x-4-3x-6}{(3x-4)^2} \\ &= \frac{-10}{(3x-4)^2} \end{aligned}$$

(2 marks)

(c) $|x-2| \leq 3$

$x-2 \leq 3 \qquad x-2 \geq -3$

$x \leq 5 \qquad x \geq -1$

$\therefore \quad -1 \leq x \leq 5$

[or: $-3 \leq x-2 \leq 3$

$-1 \leq x \leq 5$]

(2 marks)

(d) $\int_0^1 (2x+1)^3\,dx = \left[\frac{(2x+1)^4}{4\times 2}\right]_0^1$

$= \left[\frac{(2x+1)^4}{8}\right]_0^1$

$= \frac{1}{8}[(2(1)+1)^4 - (2(0)+1)^4]$

$= \frac{1}{8}[81-1]$

$= 10$

(2 marks)

(e) $y = -5-4x \ \ldots$ ①

$y = 3-2x-x^2 \ \ldots$ ②

Let ① = ②:

$-5-4x = 3-2x-x^2$

$x^2-2x-8=0$

$(x-4)(x+2)=0$

$x = 4, -2$

Subs. in ①:

$y(4) = -5-4(4)$

$= -21$

$y(-2) = -5-4(-2)$

$= 3$

$\therefore$ (4, –21) and (–2, 3)

(3 marks)

(f) $y = \tan x$

$\frac{dy}{dx} = \sec^2 x$

$\frac{dy}{dx}\left(\frac{\pi}{8}\right) = \sec^2\frac{\pi}{8}$

$= \frac{1}{\cos^2\frac{\pi}{8}}$

$= 1.171\,572\,857\ldots$

$= 1.17$ (3 sig. figs.)

(2 marks)

(g) $\sin\frac{x}{2} = \frac{1}{2}$

$\therefore \quad \frac{x}{2} = \frac{\pi}{6}, \frac{5\pi}{6}$

$\therefore \quad x = \frac{\pi}{3}, \frac{5\pi}{3}$

(2 marks)

QUESTION 12

(a) (i) grad. $BC = \frac{4-1}{2-6}$

$= -\frac{3}{4}$

Using (2, 4) and $m = -\frac{3}{4}$:

$y-4 = -\frac{3}{4}(x-2)$

$4y-16 = -3x+6$

$3x+4y-22=0$

(2 marks)

(ii) Using (1, 0) and $3x+4y-22=0$:

$d = \left|\frac{3(1)+4(0)-22}{\sqrt{3^2+4^2}}\right|$

$= \left|\frac{-19}{5}\right|$

$= \frac{19}{5}$

$\therefore$ the length of AD is $\frac{19}{5}$ units.

(2 marks)

(iii) Length $BC = \sqrt{(6-2)^2+(1-4)^2}$

$= \sqrt{25}$

$= 5$

$\therefore$ Area $= \frac{1}{2}\times 5\times\frac{19}{5}$

$= 9\frac{1}{2}$

$\therefore$ the area is $9\frac{1}{2}$ units^2.

(2 marks)

(b) (i) $\angle AOB = x^\circ$ (base $\angle$s of isos. Δ)

$\therefore \angle CBO = 2x^\circ$ (ext. $\angle$ of Δ result)

(1 mark)

(ii)

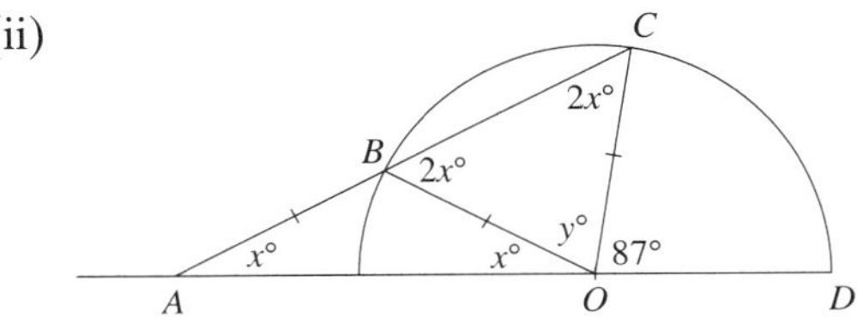

$\angle BCO = 2x^\circ$ (base $\angle$s of isos. Δ)

Let $\angle BOC = y^\circ$

$2x+2x+y = 180$ ($\angle$ sum of Δ)

$x+87+y = 180$ (straight $\angle$)

$\therefore 2x+2x = x+87$

$3x = 87$

$x = 29$

(2 marks)

(c) Consider α. Using the cosine rule:

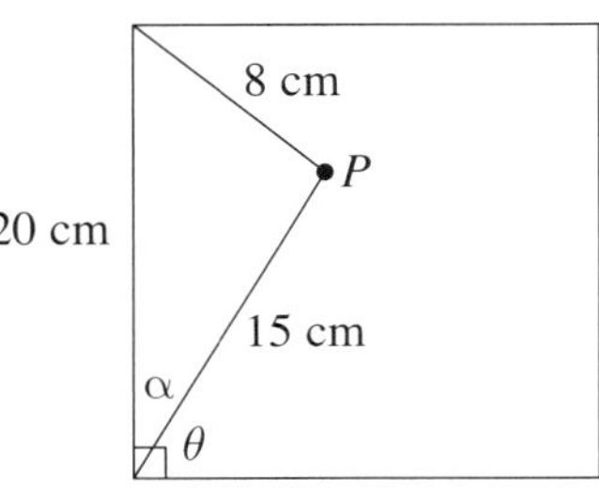

$$\cos\alpha = \frac{20^2 + 15^2 - 8^2}{2(20)(15)}$$
$$= 0.935$$
$$\alpha = 20.771\,855\,05\ldots$$
$$= 21 \text{ (nearest whole)}$$
$$\therefore \quad \theta = 90 - 21 \ (\angle \text{ in square})$$
$$= 69$$

(3 marks)

(d) (i) $y = xe^{3x}$

Using the product rule,

Let $u = x$ $\quad \dfrac{du}{dx} = 1$

Let $v = e^{3x}$ $\quad \dfrac{dv}{dx} = 3e^{3x}$

$$\frac{dy}{dx} = u\frac{dv}{dx} + v\frac{du}{dx}$$
$$= x.3e^{3x} + e^{3x}.1$$
$$= e^{3x} + 3xe^{3x}$$
$$= e^{x}(1 + 3x)$$

(1 mark)

(ii) $\displaystyle\int_0^2 e^{3x}(3+9x)dx = 3\int_0^2 e^{3x}(1+3x)dx$

$$= 3\left[xe^{3x}\right]_0^2 dx \qquad \text{(from part (i))}$$
$$= 3[2e^6 - 0]$$
$$= 6e^6$$

(2 marks)

QUESTION 13

(a) (i)
$$y = 4x^3 - x^4$$
$$y' = 12x^2 - 4x^3 = 0$$
$$4x^2(3 - x) = 0$$
$$x = 0 \text{ or } 3$$
$$y(0) = 4(0)^3 - (0)^4$$
$$= 0$$
$$y(3) = 4(3)^3 - (3)^4$$
$$= 27$$

$\therefore$ stationary points at $(0, 0)$ and $(3, 27)$.

$$y'' = 24x - 12x^2$$

$y''(0) = 24(0) - 12(0)^2 = 0$ $\quad\therefore$ possible horizontal point of inflexion

Now, consider the slope of curve near $x = 0$:

x	-1	0	1
y'	>0	0	>0

$\therefore$ horizontal point of inflexion at $(0, 0)$

Also, $y''(3) = 24(3) - 12(3)^2 < 0$ $\quad\therefore$ maximum at $(3, 27)$

$\therefore$ horizontal point of inflexion at $(0, 0)$ and maximum at $(3, 27)$

(4 marks)

(ii)

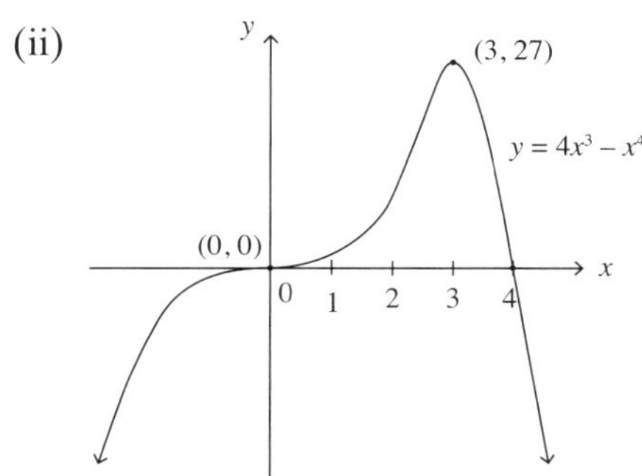

(2 marks)

(b) (i)

$$x^2 - 4x = 12y + 8$$
$$x^2 - 4x + 4 = 12y + 8 + 4$$
$$(x - 2)^2 = 12y + 12$$
$(x - 2)^2 = 12(y + 1)$, which is of the form $(x - h)^2 = 4a(y - k)$.

$\therefore a = 3$

$\therefore$ the focal length is 3 units.

(2 marks)

(ii) The parabola is concave up with vertex $(2, -1)$ and focal length 3.

$\therefore$ focus $(2, 2)$

(1 mark)

(c) (i) Consider $M = Ae^{-kt}$

Let $t = 0$, $M = 10$:

$\therefore 10 = Ae^{-k(0)}$

$\therefore A = 10$

(1 mark)

(ii)

$$M = 10e^{-kt}$$
Let $t = 163$, $M = 5$:
$$\therefore 5 = 10e^{-k(163)}$$
$$e^{-163k} = 0.5$$
$$-163k = \ln 0.5$$
$$k = -\frac{\ln 0.5}{163} \quad \left(\text{or } k = \frac{\ln 2}{163}\right)$$

(2 marks)

(d) Shaded area $= \int_0^1 \sqrt{2}\cos\left(\frac{\pi}{4}x\right)dx - \frac{1}{2} \times 1 \times 1$

$$= \left[\frac{4\sqrt{2}}{\pi}\sin\frac{\pi}{4}x\right]_0^1 - \frac{1}{2}$$
$$= \frac{4\sqrt{2}}{\pi}\left[\sin\frac{\pi}{4} - \sin 0\right] - \frac{1}{2}$$
$$= \frac{4\sqrt{2}}{\pi}\left[\frac{1}{\sqrt{2}} - 0\right] - \frac{1}{2}$$
$$= \frac{4}{\pi} - \frac{1}{2}$$
$$= \frac{8 - \pi}{2\pi}$$

$\therefore$ the area is $\frac{8 - \pi}{2\pi}$ units^2 *(3 marks)*

QUESTION 14

(a) Form a table using the differences in the y-values:

x	-2	-1	0	1	2
Difference in y	0	0.4	0.5	0.4	0

Using two applications of Simpson's Rule:

$$\text{Increase} = \frac{0-(-2)}{6}[0+4(0.4)+0.5] + \frac{2-0}{6}[0.5+4(0.4)+0]$$
$$= 1.4$$

$\therefore$ the increase is 1.4 m^2.

[OR: $\text{Increase} = \frac{1}{3}[0+0+2(0.5)+4(0.4+0.4)]$
$= 1.4$]

[OR: $\text{Increase} = \frac{1}{3}[2+2+2(3)+4(2.78+2.78)] - \frac{1}{3}[2+2+2(2.5)+4(2.38+2.38)]$
$= 1.4$]

(3 marks)

(b) (i) As 35% of insects die, then 65% of insects survive.

$\therefore A_1 = 0.65 \times 100\,000 + 5000$

$A_2 = 0.65 \times A_1 + 5000$

$= 0.65(0.65 \times 100\,000 + 5000) + 5000$

(2 marks)

(ii) $A_2 = 0.65^2 \times 100\,000 + 0.65 \times 5000 + 5000$

$= 0.65^2 \times 100\,000 + 5000(1 + 0.65)$

$\therefore A_n = 0.65^n \times 100\,000 + 5000(1 + 0.65 + 0.65^2 + \ldots + 0.65^{n-1})$

Using a geometric sum with $a = 1, r = 0.65, n = n, S_n = \frac{a(1-r^n)}{1-r}$:

$$A_n = 0.65^n \times 100\,000 + 5000\left[\frac{1(1-0.65^n)}{1-0.65}\right]$$
$$= 0.65^n \times 100\,000 + 5000\frac{(1-0.65^n)}{0.35}$$

(1 mark)

(iii) Let $n = 14$:

$$A_{14} = 0.65^{14} \times 100\,000 + 5000\frac{(1-0.65^{14})}{0.35}$$
$$= 14\,491.701\,47\ldots$$
$$= 14\,500 \text{ (nearest hundred)}$$

$\therefore$ the expected population is 14 500.

(1 mark)

(c) (i) Let the length of the enclosure be y m.

$\therefore \text{Area} = xy = 720$

$\therefore y = \frac{720}{x}$

Now, $\ell = 5x + y$

$= 5x + \frac{720}{x}$

(1 mark)

(ii) $\ell = 5x + 720x^{-1}$

$$\frac{d\ell}{dx} = 5 - 720x^{-2}$$

$$= 5 - \frac{720}{x^2} = 0$$

$5x^2 = 720$

$x^2 = 144$

$x = 12$ (as $x > 0$)

$$\frac{d^2\ell}{dx^2} = 1440x^{-3}$$

As $\frac{d^2\ell}{dx^2} > 0$, for all positive x, then minimum.

Subs. $x = 12$:

$$\ell = 5(12) + \frac{720}{12}$$

$$= 120$$

$\therefore$ the minimum length is 120 m.

(3 marks)

(d) For the geometric series, $1 + x + x^2 + x^3 + x^4$,

$a = 1, r = x, n = 5, S_n = \dfrac{a(r^n - 1)}{r - 1}$:

$$S_5 = \frac{1(x^5 - 1)}{x - 1}$$

$$\therefore \quad \lim_{x\to 1} \frac{x^5 - 1}{x - 1} = \lim_{x\to 1}(1 + x + x^2 + x^3 + x^4)$$

$$= 1 + 1 + 1 + 1 + 1$$

$$= 5$$

[OR: $\displaystyle\lim_{x\to 1} \frac{x^5 - 1}{x - 1} = \lim_{x\to 1} \frac{(x-1)(x^4 + x^3 + x^2 + x + 1)}{x - 1}$

$\displaystyle = \lim_{x\to 1}(x^4 + x^3 + x^2 + x + 1)$

$= 5$]

(2 marks)

(e) Consider $\log 2 + \log 4 + \log 8 + \ldots + \log 512$

$= \log 2^1 + \log 2^2 + \log 2^3 + \ldots + \log 2^9$

$= \log 2 + 2 \log 2 + 3 \log 2 + \ldots + 9 \log 2$

$= \log 2(1 + 2 + 3 + \ldots + 9)$

As $1 + 2 + 3 + \ldots + 9$ is arithmetic series with $a = 1, d = 1, n = 9, S_n = \dfrac{n}{2}[2a + (n - 1)d]$:

Sum $= \log 2 \times \dfrac{9}{2}[2(1) + (9 - 1)1]$

$= \log 2 \times 45$

$= 45 \log 2$

$\therefore \log 2 + \log 4 + \log 8 + \ldots + \log 512 = 45 \log 2$

(2 marks)

QUESTION 15

(a) Consider two volumes:

For C_1, rotated around x-axis, use $y^2 = 4 - x^2$. The solid is a hemisphere with radius 2.

For C_2, rotated around x-axis, use $\frac{y^2}{4} = 1 - \frac{x^2}{9}$, i.e. $y^2 = 4 - \frac{4x^2}{9}$.

$$\begin{aligned}
\text{Total volume} &= \frac{1}{2} \times \frac{4}{3}\pi r^3 + \pi\int_a^b y^2\,dx \\
&= \frac{1}{2} \times \frac{4}{3} \times \pi \times 2^3 + \pi\int_0^3 \left(4 - \frac{4x^2}{9}\right)dx \\
&= \frac{16\pi}{3} + \pi\left[4x - \frac{4x^3}{27}\right]_0^3 \\
&= \frac{16\pi}{3} + \pi\left[4(3) - \frac{4(3)^3}{27} - (4(0) - \frac{4(0)^3}{27})\right] \\
&= \frac{16\pi}{3} + \pi[12 - 4] \\
&= \frac{16\pi}{3} + 8\pi \\
&= \frac{40\pi}{3}
\end{aligned}$$

$\therefore$ the volume is $\frac{40\pi}{3}$ units^3.

(4 marks)

(b) (i) Let A = appear and N = not appear:

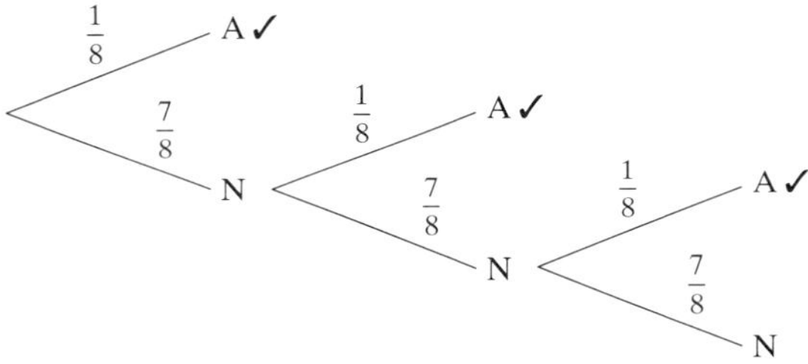

$\therefore$ P(ending before 4th roll)

= P(ending after 1st roll or after 2nd roll or after 3rd roll)

= P(ending after 1st roll) + P(ending after 2nd roll) + P(ending after 3rd roll)

$= \frac{1}{8} + \frac{7}{8} \times \frac{1}{8} + \frac{7}{8} \times \frac{7}{8} \times \frac{1}{8}$

$= \frac{1}{8} + \frac{7}{8} \times \frac{1}{8} + \left(\frac{7}{8}\right)^2 \times \frac{1}{8}$

(2 marks)

(ii) P(ending before 4th roll) $= \frac{1}{8} + \frac{7}{8} \times \frac{1}{8} + \left(\frac{7}{8}\right)^2 \times \frac{1}{8}$

$\therefore$ P(ending before nth roll) $= \frac{1}{8} + \frac{7}{8} \times \frac{1}{8} + \left(\frac{7}{8}\right)^2 \times \frac{1}{8} + \ldots + \left(\frac{7}{8}\right)^{n-2} \times \frac{1}{8}$

which is geometric sum with $a = \frac{1}{8}$, $r = \frac{7}{8}$, $n = n - 1$, $S_n = \frac{a(1 - r^n)}{1 - r}$:

$$\frac{\frac{1}{8}\left[1-\left(\frac{7}{8}\right)^{n-1}\right]}{1-\frac{7}{8}} > \frac{3}{4}$$

$$1-\left(\frac{7}{8}\right)^{n-1} > \frac{3}{4}$$

$$\left(\frac{7}{8}\right)^{n-1} < \frac{1}{4}$$

$$(n-1)\ln\left(\frac{7}{8}\right) < \ln\left(\frac{1}{4}\right)$$

$$n-1 > \ln\left(\frac{1}{4}\right) \div \ln\left(\frac{7}{8}\right) \quad \left(\text{as } \ln\left(\frac{7}{8}\right) < 0\right)$$

$$n-1 > 10.381\,786\,14\ldots$$

$$n > 11.381\,786\,14\ldots$$

$\therefore$ before the 12th roll.

(3 marks)

(c)

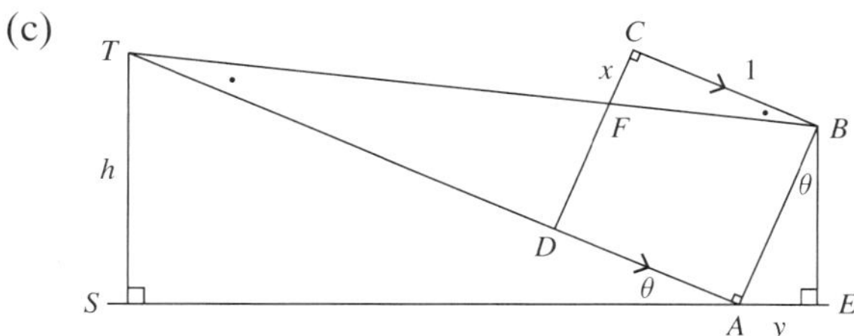

(i) $\angle FCB = \angle BAT$ ($\angle$s in square)

Now, $BC \parallel AT$ (opp. sides of square),

$\therefore \angle CBF = \angle ATB$ (alt. $\angle$s, $BC \parallel AT$)

$\therefore \Delta FCB \equiv \Delta BAT$ (equiangular)

(2 marks)

(ii) $\angle TSA = \angle AEB$ (given)

Let $\angle SAT = \theta$

As $\angle BAD = 90^\circ$ ($\angle$s in square)

$\therefore \angle BAE = (90-\theta)^\circ$ (straight $\angle$)

$\therefore \angle EBA = \theta$ ($\angle$ sum of Δ)

$\therefore \angle SAT = \angle EBA$

$\therefore \Delta TSA \equiv \Delta AEB$ (equiangular)

(2 marks)

(iii) Using $\Delta FCB \,|||\, \Delta BAT$:

$\frac{1}{x} = \frac{AT}{1}$ (matching sides of sim. Δs in proportion)

$AT = \frac{1}{x}$ … ①

Using $\Delta TSA \,|||\, \Delta AEB$:

$\frac{h}{y} = \frac{AT}{1}$ (matching sides of sim. Δs in proportion)

$h = y \times AT$

$= y \times \frac{1}{x}$ (from ①)

$\therefore h = \frac{y}{x}$

(2 marks)

QUESTION 16

(a) (i) Let $t = 0$:

$\therefore v(0) = 2 - \frac{4}{1}$

$= -2$

$\therefore$ the initial velocity is -2 ms^{-1}.

(1 mark)

(ii) Let $v = 0$:

$\therefore 2 - \frac{4}{t+1} = 0$

$\frac{4}{t+1} = 2$

$2t + 2 = 4$

$2t = 2$

$t = 1$

As $v = 2 - 4(t+1)^{-1}$

$\therefore a = 4(t+1)^{-2}.1$

$= \frac{4}{(t+1)^2}$

Subs. $t = 1$:

$a(1) = \frac{4}{(1+1)^2}$

$= 1$

$\therefore$ the acceleration is 1 ms^{-2}.

(2 marks)

(iii)

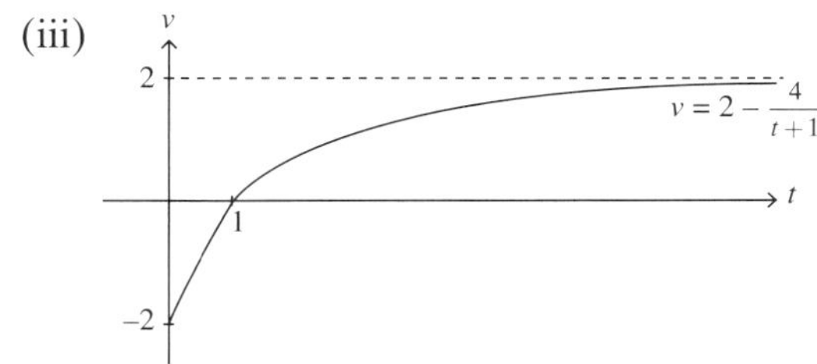

(2 marks)

(iv) $d = \left|\int_0^1 \left(2 - \frac{4}{t+1}\right) dt\right| + \int_1^7 \left(2 - \frac{4}{t+1}\right) dt$

$= \left|\left[2t - 4\ln(t+1)\right]_0^1\right| + \left[2t - 4\ln(t+1)\right]_1^7$

$= |2 - 4\ln 2 - 0| + 2(7) - 4\ln(7+1) - [2(1) - 4\ln(1+1)]$

$= 4\ln 2 - 2 + 14 - 4\ln 8 - 2 + 4\ln 2$ (as $|2 - 4\ln 2| = 4\ln 2 - 2$)

$= 8\ln 2 + 10 - 4\ln 8$

$= 8\ln 2 + 10 - 4\ln 2^3$

$= 8\ln 2 + 10 - 12\ln 2$

$= 10 - 4\ln 2$

$\therefore$ the distance is $(10 - 4\ln 2)$ m.

(3 marks)

(b) (i) $y = 200(1 + 19e^{-0.5t})^{-1}$

$\frac{dy}{dt} = -200(1 + 19e^{-0.5t})^{-2} \times -9.5e^{-0.5t}$

$= \frac{1900e^{-0.5t}}{(1+19e^{-0.5t})^2}$

(2 marks)

(ii) As $t \geq 0$, consider $t = 0$:

$y(0) = \frac{200}{1+19e^{-0.5(0)}}$

$= \frac{200}{1+19}$

$= 10$

As $t \to \infty$, $y(\infty) = \frac{200}{1+19e^{-0.5(\infty)}}$

$= \frac{200}{1+0}$

$= 200$

As y is monotonic increasing, then range is $10 \leq y < 200$.

(2 marks)

(iii) From $y = \frac{200}{1+19e^{-0.5t}}$

$\therefore 1 + 19e^{-0.5t} = \frac{200}{y}$

$19e^{-0.5t} = \frac{200}{y} - 1$

$19e^{-0.5t} = \frac{200-y}{y}$

$\therefore 1900e^{-0.5t} = 100\left[\frac{200-y}{y}\right]$

As $\frac{dy}{dt} = \frac{1900e^{-0.5t}}{(1+19e^{-0.5t})^2}$

$= 100\left[\frac{200-y}{y}\right] \div \left[\frac{200}{y}\right]^2$

$= 100\left[\frac{200-y}{y}\right] \times \frac{y^2}{40\,000}$

$= \frac{y(200-y)}{400}$

$= \frac{y}{400}(200-y)$

(1 mark)

(iv) Let $R = \frac{y}{400}(200-y)$

$\therefore R = \frac{y}{2} - \frac{y^2}{400}$

$\frac{dR}{dy} = \frac{1}{2} - \frac{y}{200} = 0$

$\therefore \frac{y}{200} = \frac{1}{2}$

$y = 100$

$\frac{d^2R}{dy^2} = \frac{1}{200} < 0 \therefore$ maximum.

$\therefore$ fastest rate when 100 yabbies.

(2 marks)

2017 | HIGHER SCHOOL CERTIFICATE EXAMINATION

Mathematics

General Instructions

- Reading time – 5 minutes
- Working time – 3 hours
- Write using black pen
- NESA approved calculators may be used
- A reference sheet is provided at the back of this paper
- In Questions 11–16, show relevant mathematical reasoning and/or calculations

Total marks: 100

Section I – 10 marks

- Attempt Questions 1–10
- Allow about 15 minutes for this section

Section II – 90 marks

- Attempt Questions 11–16
- Allow about 2 hours and 45 minutes for this section

Section I

10 marks
Attempt Questions 1–10
Allow about 15 minutes for this section

Use the multiple-choice answer sheet for Questions 1–10.

---

1 What is the gradient of the line $2x + 3y + 4 = 0$?

A. $-\frac{2}{3}$

B. $\frac{2}{3}$

C. $-\frac{3}{2}$

D. $\frac{3}{2}$

2 Which expression is equal to $3x^2 - x - 2$?

A. $(3x - 1)(x + 2)$

B. $(3x + 1)(x - 2)$

C. $(3x - 2)(x + 1)$

D. $(3x + 2)(x - 1)$

3 What is the derivative of e^{x^2}?

A. $x^2 e^{x^2 - 1}$

B. $2xe^{2x}$

C. $2xe^{x^2}$

D. $2e^{x^2}$

4 The function $f(x)$ is defined for $a \le x \le b$.

On this interval, $f'(x) > 0$ and $f''(x) < 0$.

Which graph best represents $y = f(x)$?

A. 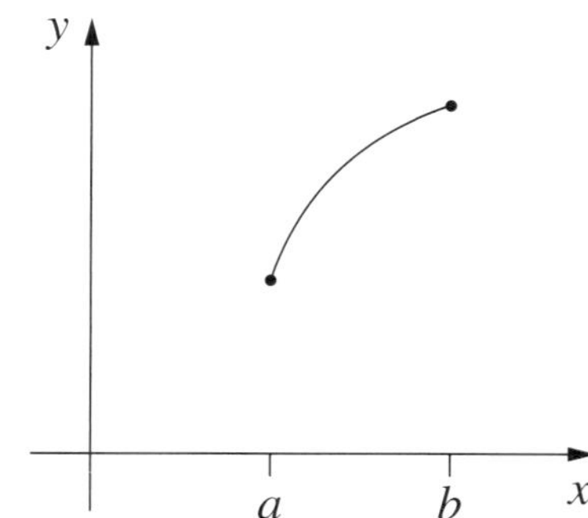

B. 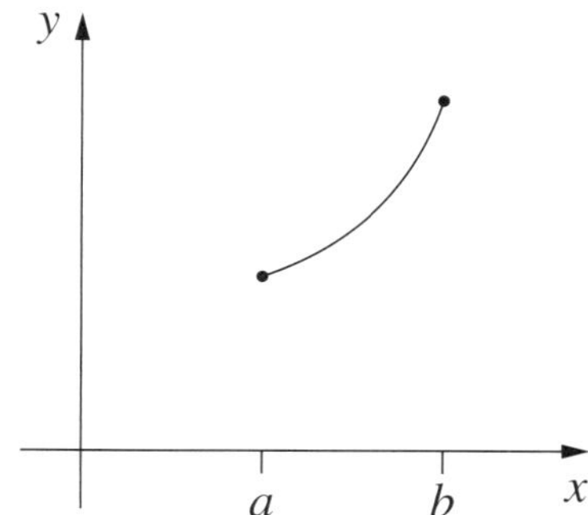

C. 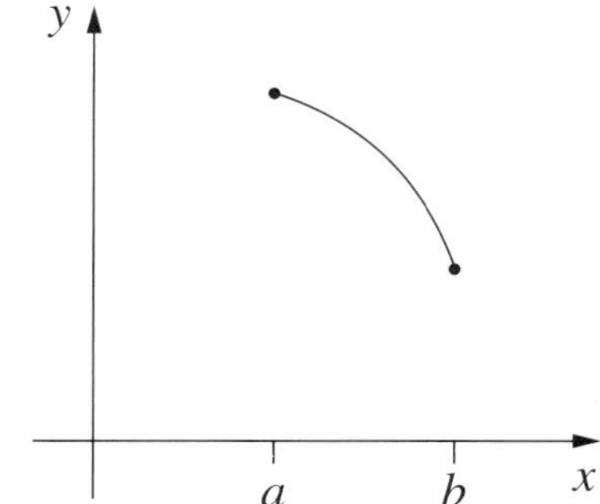

D. 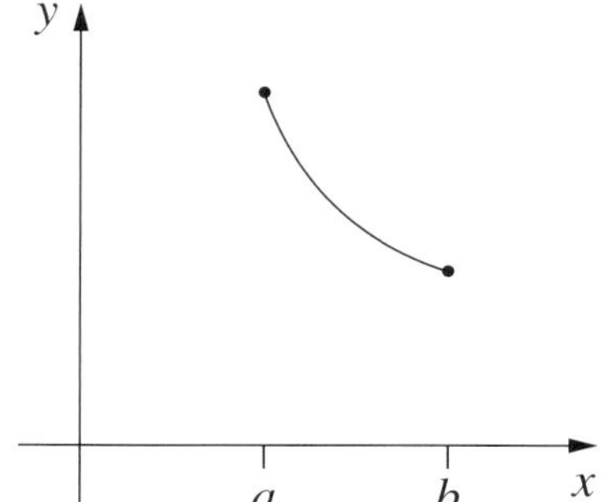

5 It is given that $\ln a = \ln b - \ln c$, where $a, b, c > 0$.

Which statement is true?

A. $a = b - c$

B. $a = \dfrac{b}{c}$

C. $\ln a = \dfrac{b}{c}$

D. $\ln a = \dfrac{\ln b}{\ln c}$

6 The point P moves so that it is always equidistant from two fixed points, A and B.

What best describes the locus of P?

A. A point

B. A circle

C. A parabola

D. A straight line

7 Which expression is equivalent to $\tan\theta + \cot\theta$?

A. $\operatorname{cosec}\theta + \sec\theta$

B. $\sec\theta \operatorname{cosec}\theta$

C. 2

D. 1

8 The region enclosed by $y = 4 - x$, $y = x$ and $y = 2x + 1$ is shaded in the diagram below.

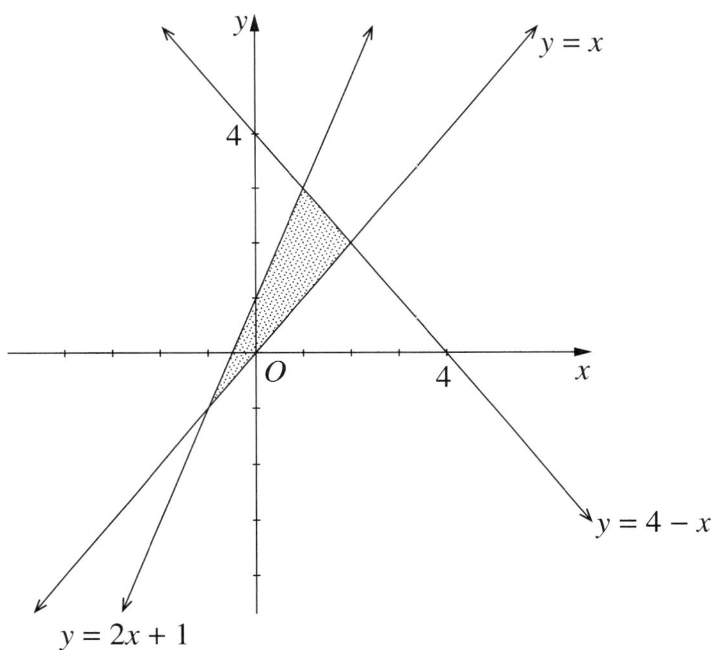

Which of the following defines the shaded region?

A. $y \le 2x + 1, \quad y \le 4 - x, \quad y \ge x$

B. $y \ge 2x + 1, \quad y \le 4 - x, \quad y \ge x$

C. $y \le 2x + 1, \quad y \ge 4 - x, \quad y \ge x$

D. $y \ge 2x + 1, \quad y \ge 4 - x, \quad y \ge x$

9 The graph of $y = f'(x)$ is shown.

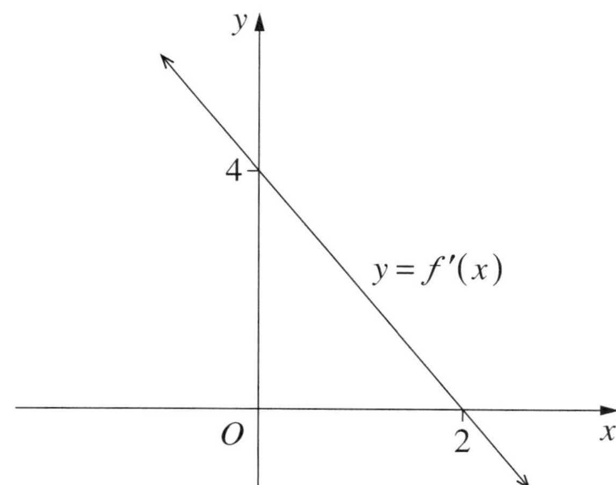

The curve $y = f(x)$ has a maximum value of 12.

What is the equation of the curve $y = f(x)$?

A. $y = x^2 - 4x + 12$

B. $y = 4 + 4x - x^2$

C. $y = 8 + 4x - x^2$

D. $y = x^2 - 4x + 16$

10 A particle is moving along a straight line.

The graph shows the velocity, v, of the particle for time $t \geq 0$.

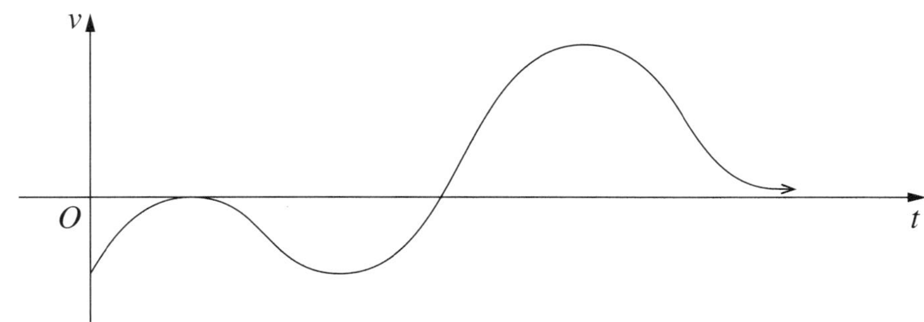

How many times does the particle change direction?

A. 1

B. 2

C. 3

D. 4

Section II

90 marks
Attempt Questions 11–16
Allow about 2 hours and 45 minutes for this section

Answer each question in the appropriate writing booklet. Extra writing booklets are available.

In Questions 11–16, your responses should include relevant mathematical reasoning and/or calculations.

Question 11 (15 marks) Use the Question 11 Writing Booklet.

(a) Rationalise the denominator of $\dfrac{2}{\sqrt{5}-1}$. **2**

(b) Find $\displaystyle\int (2x+1)^4\,dx$. **1**

(c) Differentiate $\dfrac{\sin x}{x}$. **2**

(d) Differentiate $x^3 \ln x$. **2**

(e) In the diagram, OAB is a sector of the circle with centre O and radius 6 cm, where $\angle AOB = 30°$.

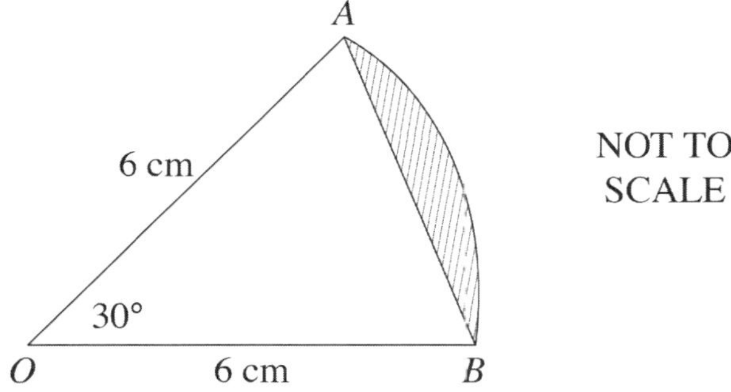

(i) Find the exact value of the area of the triangle OAB. **1**

(ii) Find the exact value of the area of the shaded segment. **1**

Question 11 continues on the following page

Question 11 (continued)

(f) Determine the equation of the parabola shown. Write your answer in the form $(x-h)^2 = 4a(y-k)$. **2**

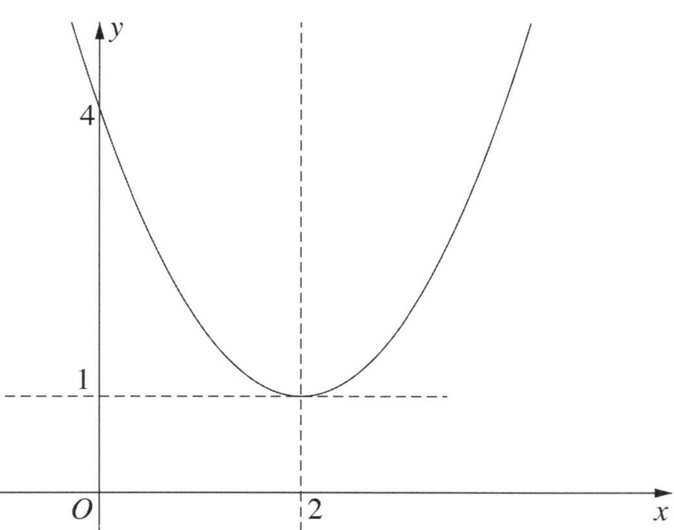

(g) Solve $|3x-1| = 2$. **2**

(h) Find the domain of the function $f(x) = \sqrt{3-x}$. **2**

End of Question 11

Question 12 (15 marks) Use the Question 12 Writing Booklet.

(a) Find the equation of the tangent to the curve $y = x^2 + 4x - 7$ at the point $(1, -2)$. **2**

(b) The diagram shows the region bounded by $y = \sqrt{16 - 4x^2}$ and the x-axis. **3**

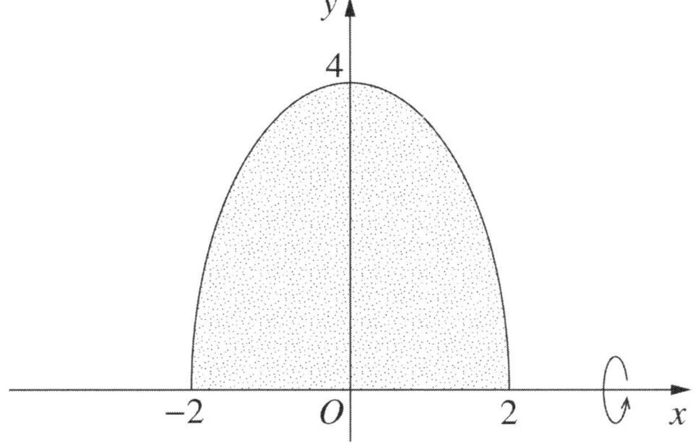

The region is rotated about the x-axis to form a solid.

Find the exact volume of the solid formed.

(c) In an arithmetic series, the fifth term is 200 and the sum of the first four terms is 1200. **3**

Find the value of the tenth term.

Question 12 continues on the following page

Question 12 (continued)

(d) The points $A(-4, 0)$ and $B(1, 5)$ lie on the line $y = x + 4$.

The length of AB is $5\sqrt{2}$.

The points $C(0, -2)$ and $D(3, 1)$ lie on the line $x - y - 2 = 0$.

The points A, B, D, C form a trapezium as shown.

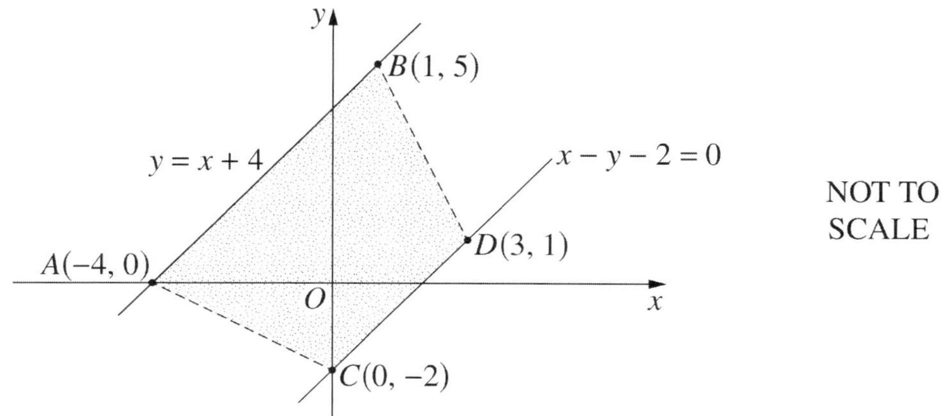

(i) Find the perpendicular distance from point $A(-4, 0)$ to the line $x - y - 2 = 0$. **1**

(ii) Calculate the area of the trapezium. **2**

(e) A spinner is marked with the numbers 1, 2, 3, 4 and 5. When it is spun, each of the five numbers is equally likely to occur.

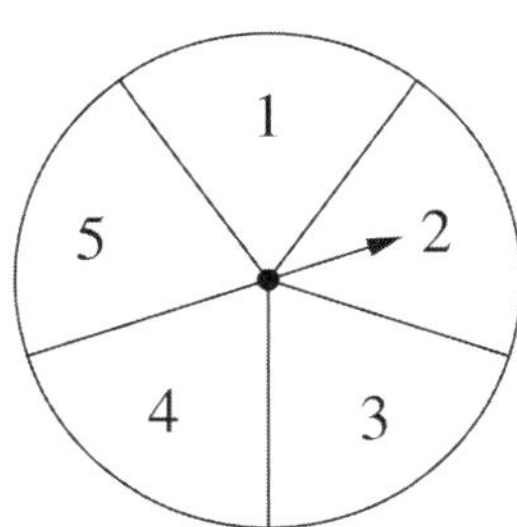

The spinner is spun three times.

(i) What is the probability that an even number occurs on the first spin? **1**

(ii) What is the probability that an even number occurs on at least one of the three spins? **1**

(iii) What is the probability that an even number occurs on the first spin and odd numbers occur on the second and third spins? **1**

(iv) What is the probability that an even number occurs on exactly one of the three spins? **1**

End of Question 12

Question 13 (15 marks) Use the Question 13 Writing Booklet.

(a) Using the cosine rule, find the value of x in the following diagram. **3**

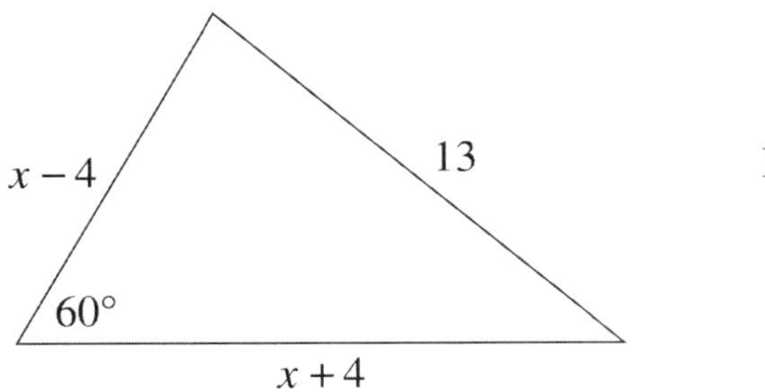

(b) Consider the curve $y = 2x^3 + 3x^2 - 12x + 7$.

(i) Find the stationary points of the curve and determine their nature. **4**

(ii) Sketch the curve, labelling the stationary points. **2**

(iii) Hence, or otherwise, find the values of x for which $\frac{dy}{dx}$ is positive. **1**

(c) By letting $m = t^{\frac{1}{3}}$, or otherwise, solve $t^{\frac{2}{3}} + t^{\frac{1}{3}} - 6 = 0$. **2**

(d) The rate at which water flows into a tank is given by **3**

$$\frac{dV}{dt} = \frac{2t}{1+t^2},$$

where V is the volume of water in the tank in litres and t is the time in seconds.

Initially the tank is empty.

Find the exact amount of water in the tank after 10 seconds.

Question 14 (15 marks) Use the Question 14 Writing Booklet.

(a) Sketch the curve $y = 4 + 3\sin 2x$ for $0 \le x \le 2\pi$. **3**

(b) (i) Find the exact value of $\int_0^{\frac{\pi}{3}} \cos x \, dx$. **1**

(ii) Using Simpson's rule with one application, find an approximation to the integral **2**

$$\int_0^{\frac{\pi}{3}} \cos x \, dx,$$

leaving your answer in terms of π and $\sqrt{3}$.

(iii) Using parts (i) and (ii), show that **1**

$$\pi \approx \frac{18\sqrt{3}}{3 + 4\sqrt{3}}.$$

Question 14 continues on the following page

Question 14 (continued)

(c) Carbon-14 is a radioactive substance that decays over time. The amount of carbon-14 present in a kangaroo bone is given by

$$C(t) = Ae^{kt},$$

where A and k are constants, and t is the number of years since the kangaroo died.

(i) Show that $C(t)$ satisfies $\dfrac{dC}{dt} = kC$. **1**

(ii) After 5730 years, half of the original amount of carbon-14 is present. **2**

Show that the value of k, correct to 2 significant figures, is -0.00012.

(iii) The amount of carbon-14 now present in a kangaroo bone is 90% of the original amount. **2**

Find the number of years since the kangaroo died. Give your answer correct to 2 significant figures.

(d) The shaded region shown is enclosed by two parabolas, each with x-intercepts at $x = -1$ and $x = 1$. **3**

The parabolas have equations $y = 2k\left(x^2 - 1\right)$ and $y = k\left(1 - x^2\right)$, where $k > 0$.

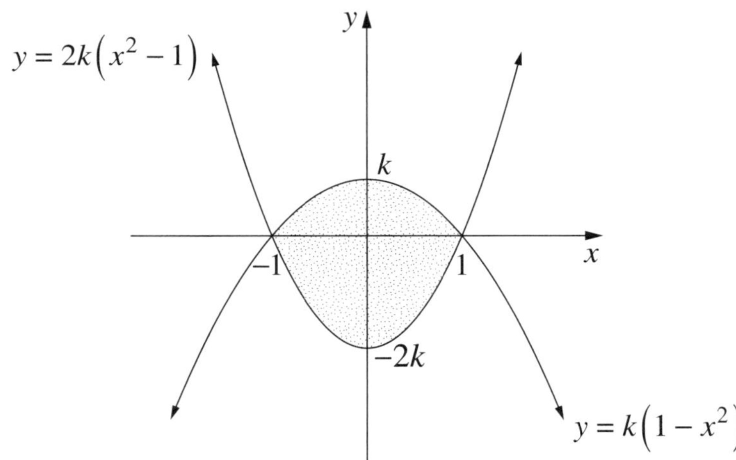

Given that the area of the shaded region is 8, find the value of k.

End of Question 14

Question 15 (15 marks) Use the Question 15 Writing Booklet.

(a) The triangle ABC is isosceles with $AB = AC$ and the size of $\angle BAC$ is $x°$. Points D and E are chosen so that $\triangle ABC$, $\triangle ACD$ and $\triangle ADE$ are congruent, as shown in the diagram. **3**

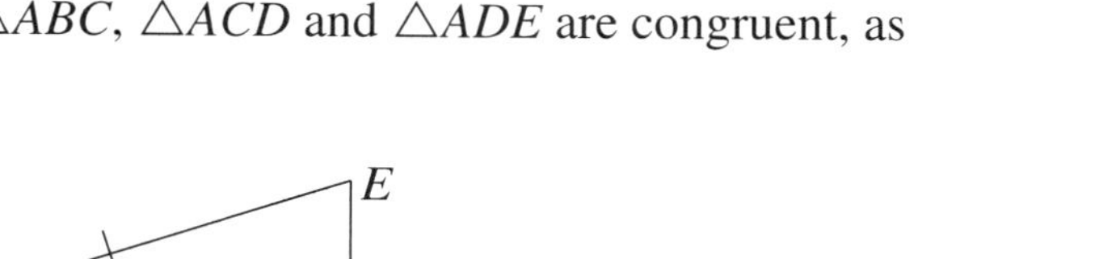

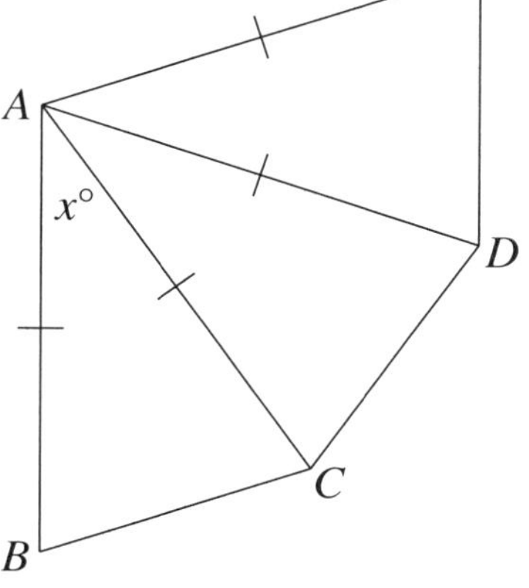

Find the value of x for which AB is parallel to ED, giving reasons.

(b) Anita opens a savings account. At the start of each month she deposits $\$X$ into the savings account. At the end of each month, after interest is added into the savings account, the bank withdraws \$2500 from the savings account as a loan repayment. Let M_n be the amount in the savings account after the nth withdrawal.

The savings account pays interest of 4.2% per annum compounded monthly.

(i) Show that after the second withdrawal the amount in the savings account is given by **2**

$$M_2 = X\left(1.0035^2 + 1.0035\right) - 2500\left(1.0035 + 1\right).$$

(ii) Find the value of X so that the amount in the savings account is \$80 000 after the last withdrawal of the fourth year. **3**

Question 15 continues on the following page

Question 15 (continued)

(c) Two particles move along the x-axis.

When $t = 0$, particle P_1 is at the origin and moving with velocity 3.

For $t \geq 0$, particle P_1 has acceleration given by $a_1 = 6t + e^{-t}$.

(i) Show that the velocity of particle P_1 is given by $v_1 = 3t^2 + 4 - e^{-t}$. **2**

When $t = 0$, particle P_2 is also at the origin.

For $t \geq 0$, particle P_2 has velocity given by $v_2 = 6t + 1 - e^{-t}$.

(ii) When do the two particles have the same velocity? **2**

(iii) Show that the two particles do not meet for $t > 0$. **3**

End of Question 15

Question 16 (15 marks) Use the Question 16 Writing Booklet.

(a) John's home is at point A and his school is at point B. A straight river runs nearby.

The point on the river closest to A is point C, which is 5 km from A.

The point on the river closest to B is point D, which is 7 km from B.

The distance from C to D is 9 km.

To get some exercise, John cycles from home directly to point E on the river, x km from C, before cycling directly to school at B, as shown in the diagram.

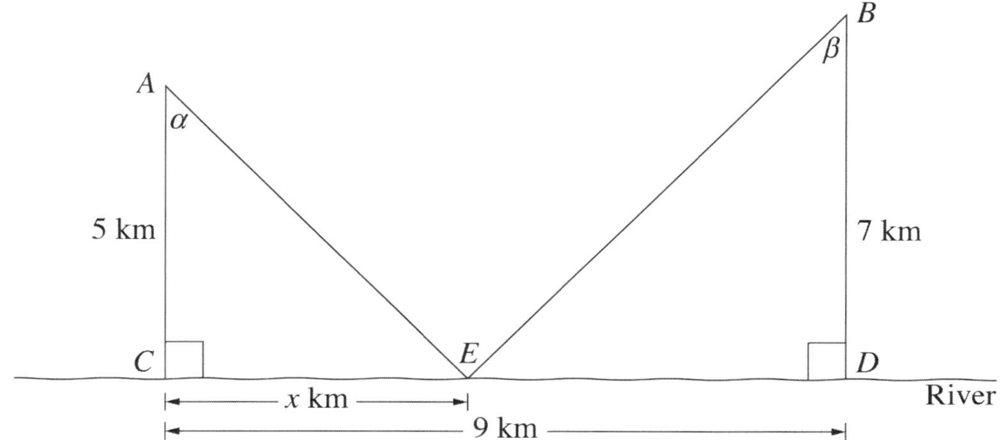

The total distance John cycles from home to school is L km.

(i) Show that $L = \sqrt{x^2 + 25} + \sqrt{49 + (9 - x)^2}$. **1**

(ii) Show that if $\dfrac{dL}{dx} = 0$, then $\sin\alpha = \sin\beta$. **3**

(iii) Find the value of x that makes $\sin\alpha = \sin\beta$. **2**

(iv) Explain why this value of x gives a minimum for L. **1**

Question 16 continues on the following page

Question 16 (continued)

(b) A geometric series has first term a and limiting sum 2. **3**

Find all possible values for a.

(c) In the triangle ABC, the point M is the mid-point of BC. The point D lies on AB and

$$BD = DA + AC.$$

The line that passes through the point C and is parallel to MD meets BA produced at E.

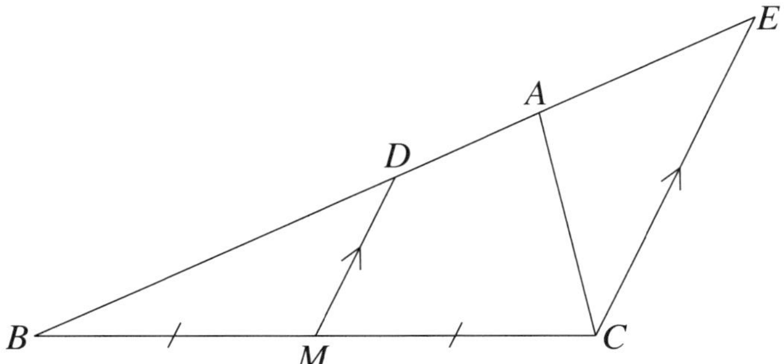

Copy or trace this diagram into your writing booklet.

(i) Prove that $\triangle ACE$ is isosceles. **3**

(ii) The point F is chosen on BC so that AF is parallel to DM. **2**

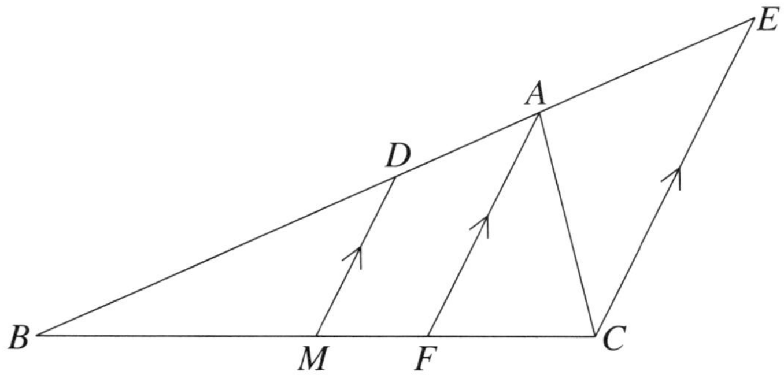

Show that AF bisects $\angle BAC$.

End of paper

2017 Higher School Certificate

Worked Answers

Section I

QUESTION 1

$$2x + 3y + 4 = 0$$
$$3y = -2x - 4$$
$$y = \frac{-2x-4}{3}$$
$$\therefore m = -\frac{2}{3}$$

Answer A

(1 mark)

QUESTION 2

$3x^2 - x - 2 = (3x + 2)(x - 1)$

Answer D

(1 mark)

QUESTION 3

$$\frac{d}{dx}\left[e^{x^2}\right] = 2xe^{x^2}$$

Answer C

(1 mark)

QUESTION 4

As $f'(x) > 0$, function is increasing, and $f''(x) < 0$, function is concave down.

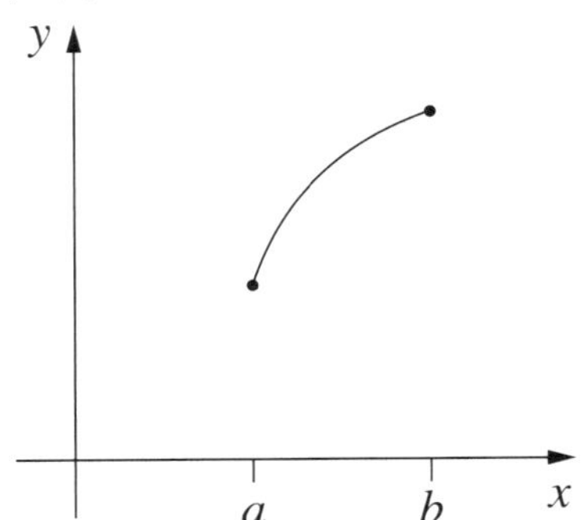

Answer A

(1 mark)

QUESTION 5

$$\ln a = \ln b - \ln c$$
$$\ln a = \ln \frac{b}{c}$$
$$a = \frac{b}{c}$$

Answer B

(1 mark)

QUESTION 6

The locus of P is a straight line.

Answer D

(1 mark)

QUESTION 7

$$\tan\theta + \cot\theta = \frac{\sin\theta}{\cos\theta} + \frac{\cos\theta}{\sin\theta}$$
$$= \frac{\sin^2\theta + \cos^2\theta}{\cos\theta\sin\theta}$$
$$= \frac{1}{\cos\theta\sin\theta}$$
$$= \sec\theta\ \text{cosec}\ \theta$$

Answer B

(1 mark)

QUESTION 8

The point $(1, 2)$ is in the shaded region and satisfies each inequality. Substitute $(1, 2)$ in:

$y = 2x + 1$: as $2 \leq 2(1) + 1$, then we require $y \leq 2x + 1$.

$y = 4 - x$: as $2 \leq 4 - 1$, then we require $y \leq 4 - x$.

$y = x$: as $2 \geq 1$, then we require $y \geq x$.

The region is defined by $y \leq 2x + 1, y \leq 4 - x, y \geq x$.

Answer A

(1 mark)

QUESTION 9

Using gradient-intercept form, the line is $y' = 4 - 2x$.

$\therefore y = 4x - x^2 + c$

Now, maximum of $y = 12$ when $y' = 0$, i.e. when $x = 2$:

$\therefore 12 = 4(2) - (2)^2 + c$

$12 = 8 - 4 + c$

$c = 8$

$\therefore y = 4x - x^2 + 8$

i.e. $y = 8 + 4x - x^2$

Answer C

(1 mark)

QUESTION 10

Let the times when the particle stops ($v = 0$) be t_1 and t_2.

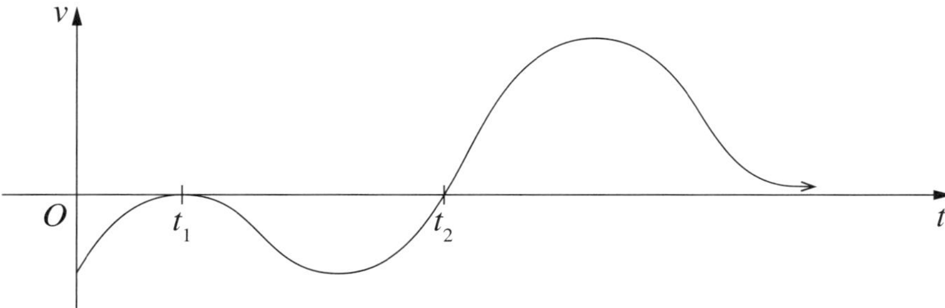

The particle changes direction when the particle's velocity changes sign. This occurs only once at t_2 seconds. (At time t_1 the particle stops but continues in the same direction: moving to the left.)

Answer A

(1 mark)

Section II

QUESTION 11

(a) $$\frac{2}{\sqrt{5}-1} = \frac{2}{\sqrt{5}-1} \times \frac{\sqrt{5}+1}{\sqrt{5}+1}$$
$$= \frac{2\sqrt{5}+2}{5-1}$$
$$= \frac{2(\sqrt{5}+1)}{4}$$
$$= \frac{\sqrt{5}+1}{2}$$

(2 marks)

(b) $$\int (2x+1)^4\, dx = \frac{(2x+1)^5}{5\times 2} + c$$
$$= \frac{(2x+1)^5}{10} + c$$

(1 mark)

(c) Using the quotient rule,

Let $u = \sin x$ $\quad \frac{du}{dx} = \cos x$

Let $v = x$ $\quad \frac{dv}{dx} = 1$

$$\frac{d}{dx}\left[\frac{\sin x}{x}\right] = \frac{v\frac{du}{dx} - u\frac{dv}{dx}}{v^2}$$
$$= \frac{x.\cos x - \sin x.1}{x^2}$$
$$= \frac{x\cos x - \sin x}{x^2}$$

(2 marks)

(d) Using the product rule,

Let $u = x^3$ $\quad \frac{du}{dx} = 3x^2$

Let $v = \ln x$ $\quad \frac{dv}{dx} = \frac{1}{x}$

$$\frac{d}{dx}\left[x^3 \ln x\right] = u\frac{dv}{dx} + v\frac{du}{dx}$$
$$= x^3.\frac{1}{x} + \ln x.3x^2$$
$$= x^2 + 3x^2 \ln x$$

(2 marks)

(e) (i) $$\text{Area of triangle} = \frac{1}{2}ab\sin C$$
$$= \frac{1}{2} \times 6 \times 6 \times \sin 30^{\circ}$$
$$= 9$$

$\therefore$ the area is 9 cm^2.

(1 mark)

(ii) $$\text{Area} = \text{Area of sector} - \text{Area of triangle}$$
$$= \frac{1}{2}r^2\theta - 9$$
$$= \frac{1}{2} \times 6^2 \times \frac{\pi}{6} - 9$$
$$= 3\pi - 9$$

$\therefore$ the area is $(3\pi - 9)$ cm^2.

(1 mark)

(f) Vertex: (2, 1)

$\therefore$ Equation: $(x-2)^2 = 4a(y-1)$

Also, subs. (0, 4):

$$\therefore (0-2)^2 = 4a(4-1)$$
$$4 = 4a(3)$$
$$4a = \frac{4}{3}$$

$\therefore$ the equation is $(x-2)^2 = \frac{4}{3}(y-1)$

(2 marks)

(g) $3x - 1 = 2$ $\qquad$ $3x - 1 = -2$

$3x = 2 + 1$ $\qquad$ $3x = -2 + 1$

$3x = 3$ $\qquad$ $3x = -1$

$x = 1$ $\qquad$ $x = -\frac{1}{3}$

$\therefore x = 1, -\frac{1}{3}$

(2 marks)

(h) As $3 - x \geq 0$, then
$x \leq 3$

(2 marks)

QUESTION 12

(a) $$\frac{dy}{dx} = 2x + 4$$
$$\frac{dy}{dx}(1) = 2(1) + 4$$
$$= 6$$

Using $(1, -2)$ and $m = 6$,
$$y - y_1 = m(x - x_1)$$
$$y + 2 = 6(x - 1)$$
$$y + 2 = 6x - 6$$
$$6x - y - 8 = 0$$

(2 marks)

(b) $y = \sqrt{16-4x^2}$

$y^2 = 16 - 4x^2$

$V = 2\pi \int_0^2 y^2 \ dx$

$= 2\pi \int_0^2 16 - 4x^2 \ dx$

$= 2\pi \left[16x - \frac{4x^3}{3}\right]_0^2$

$= 2\pi \left(16(2) - \frac{4(2)^3}{3} - 0\right)$

$= 2\pi \left(32 - \frac{32}{3} - 0\right)$

$= \frac{128\pi}{3}$

$\therefore \frac{128\pi}{3}$ units^3

(3 marks)

(c) $n = 5$, $T_5 = 200$:

$T_n = a + (n-1)d$

$200 = a + (5-1)d$

$a + 4d = 200 \ldots ①$

$n = 4$, $S_4 = 1200$:

$S_n = \frac{n}{2}(2a + (n-1)d)$

$1200 = \frac{4}{2}(2a + (4-1)d)$

$2(2a + 3d) = 1200$

$2a + 3d = 600 \ldots ②$

$2 \times ① : 2a + 8d = 400 \ldots ③$

$③ - ② : 5d = -200$

$d = -40$

Subs. in ②:

$2a + 3(-40) = 600$

$2a - 120 = 600$

$2a = 720$

$a = 360$

$\therefore a = 360, d = -40, n = 10$:

$T_{10} = 360 + (10-1)(-40)$

$= 0$

(3 marks)

(d) (i) $(-4, 0)$ and $x - y - 2 = 0$:

$d = \frac{|ax_1 + by_1 + c|}{\sqrt{a^2+b^2}}$

$= \frac{|1(-4) - 1(0) - 2|}{\sqrt{{}^2 + (-1)^2}}$

$= \frac{|-6|}{\sqrt{2}}$

$= \frac{6}{\sqrt{2}}$

$\therefore$ distance is $\frac{6}{\sqrt{2}}$ units.

(1 mark)

(ii) As $C(0, -2)$ and $D(3, 1)$:

Length of CD:

$d = \sqrt{(3-0)^2 + (1+2)^2}$

$= \sqrt{9+9}$

$= \sqrt{18}$

$= 3\sqrt{2}$

$\therefore$ length is $3\sqrt{2}$ units

Area $= \frac{1}{2}h(a+b)$

$= \frac{1}{2} \times \frac{6}{\sqrt{2}}(3\sqrt{2} + 5\sqrt{2})$

$= \frac{1}{2} \times \frac{6}{\sqrt{2}}(8\sqrt{2})$

$\therefore$ the area is 24 units^2.

(2 marks)

(e) (i) P(even number on first spin) $= \frac{2}{5}$

(1 mark)

(ii) P(at least one even number)

$= 1 -$ P(no even numbers)

$= 1 - \frac{3}{5} \times \frac{3}{5} \times \frac{3}{5}$

$= \frac{98}{125}$

(1 mark)

(iii) P(even, then odd, then odd)

$= \frac{2}{5} \times \frac{3}{5} \times \frac{3}{5}$

$= \frac{18}{125}$

(1 mark)

(iv) P(even, then odd, then odd) + P(odd, then even, then odd) + P(odd, then odd, then even)

$= 3 \times \frac{2}{5} \times \frac{3}{5} \times \frac{3}{5}$

$= \frac{54}{125}$

(1 mark)

QUESTION 13

(a) $c^2 = a^2 + b^2 - 2ab\cos C$

$13^2 = (x+4)^2 + (x-4)^2 - 2(x+4)(x-4)\cos 60^{\circ}$

$169 = x^2 + 8x + 16 + x^2 - 8x + 16 - 2(x^2 - 16) \times \frac{1}{2}$

$169 = 2x^2 + 32 - x^2 + 16$

$x^2 = 169 - 48$

$x^2 = 121$

$x = 11 \qquad (x > 4)$

(3 marks)

(b) $y = 2x^3 + 3x^2 - 12x + 7$

(i) $y' = 6x^2 + 6x - 12 = 0$

$6(x^2 + x - 2) = 0$

$(x+2)(x-1) = 0$

$x = -2, 1$

$y(-2) = 2(-2)^3 + 3(-2)^2 - 12(-2) + 7$

$= 27$

$y(1) = 2(1)^3 + 3(1)^2 - 12(1) + 7$

$= 0$

$\therefore$ stationary points at $(-2, 27)$ and $(1, 0)$.

$y'' = 12x + 6$

$y''(-2) = 12(-2) + 6 < 0 \therefore$ maximum

$y''(1) = 12(1) + 6 > 0 \therefore$ minimum

$\therefore$ maximum at $(-2, 27)$ and minimum at $(1, 0)$.

(4 marks)

(ii)

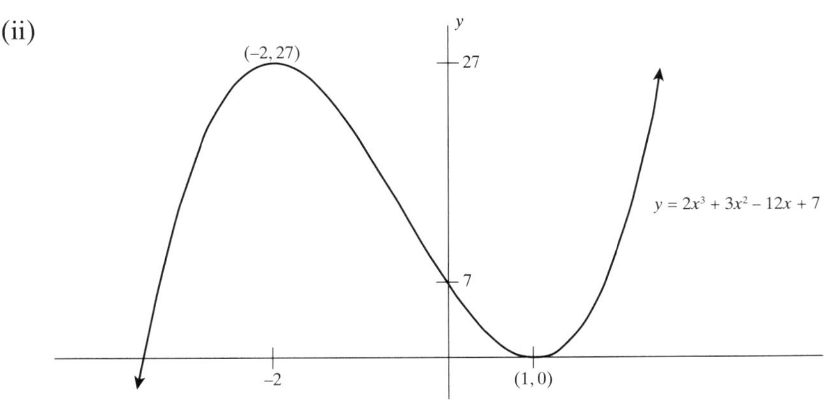

(2 marks)

(iii) From the graph, $y' > 0$ when the curve is increasing: $x < -2$ and $x > 1$.

(1 mark)

(c) Let $m = t^{\frac{1}{3}}$:

$$m^2 + m - 6 = 0$$

$$(m + 3)(m - 2) = 0$$

$$m = -3, 2$$

$\therefore\ t^{\frac{1}{3}} = -3 \qquad t^{\frac{1}{3}} = 2$

$t = -27 \qquad t = 8$

$\therefore t = -27$ or 8

(2 marks)

(d) $\dfrac{dV}{dt} = \dfrac{2t}{1+t^2}$

$$V = \ln(1 + t^2) + c$$

Subs. $t = 0$, $V = 0$:

$$0 = \ln(1 + 0^2) + c$$

$$c = 0$$

$$\therefore V = \ln(1 + t^2)$$

$$V(10) = \ln(1 + 10^2)$$

$$= \ln(101)$$

$\therefore$ the volume is $\ln(101)$ litres.

(3 marks)

QUESTION 14

(a)

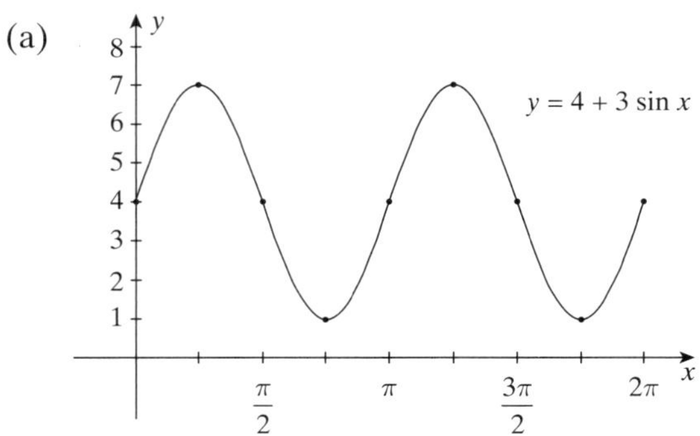

(3 marks)

(b) (i) $\displaystyle\int_0^{\frac{\pi}{3}} \cos x \; dx = \left[\sin x\right]_0^{\frac{\pi}{3}}$

$$= \sin\frac{\pi}{3} - \sin 0$$

$$= \frac{\sqrt{3}}{2} - 0$$

$$= \frac{\sqrt{3}}{2}$$

(1 mark)

(ii) $\displaystyle\int_a^b f(x)\; dx = \frac{b-a}{6}\left[f(a) + 4f\left(\frac{a+b}{2}\right) + f(b)\right]$

$$\int_0^{\frac{\pi}{3}} \cos x \; dx = \frac{\frac{\pi}{3} - 0}{6}\left[\cos 0 + 4\cos\frac{\pi}{6} + \cos\frac{\pi}{3}\right]$$

$$= \frac{\pi}{18}\left[1 + \frac{4\sqrt{3}}{2} + \frac{1}{2}\right]$$

$$= \frac{\pi}{18}\left[\frac{3 + 4\sqrt{3}}{2}\right]$$

$$= \frac{\pi(3 + 4\sqrt{3})}{36}$$

(2 marks)

(iii) $\dfrac{\pi(3+4\sqrt{3})}{36} \approx \dfrac{\sqrt{3}}{2}$

$$\pi \approx \frac{36\sqrt{3}}{2(3+4\sqrt{3})}$$

$$\pi \approx \frac{18\sqrt{3}}{3+4\sqrt{3}}$$

(1 mark)

(c) (i) $C = Ae^{kt}$

$$\frac{dC}{dt} = k.Ae^{kt} = kC$$

$\therefore C = Ae^{kt}$ satisfies $\dfrac{dC}{dt} = kC$

(1 mark)

(ii) Let $t = 5730$, $A = 1$, $C = 0.5$:

$$0.5 = 1e^{k(5730)}$$

$$e^{5730k} = 0.5$$

$$\ln e^{5730k} = \ln 0.5$$

$$5730k \ln e = \ln 0.5$$

$$k = \frac{\ln 0.5}{5730} = -0.000\,120\,968\ldots$$

$$= -0.000\,12 \text{ (2 sig. figs)}$$

(2 marks)

(iii) Let $A = 1$, $C = 0.9$:

$$0.9 = 1e^{kt}$$

$$e^{kt} = 0.9$$

$$\ln e^{kt} = \ln 0.9$$

$$kt \ln e = \ln 0.9$$

$$t = \frac{\ln 0.9}{k} = 870.977\,7254\ldots$$

$$= 870 \text{ (2 sig. figs)}$$

$\therefore$ the kangaroo died 870 years ago.

(or 880 years ago if using $k = -0.000\,12$)

(2 marks)

(d) $A = 2\int_0^1 k(1-x^2) - 2k(x^2-1)\,dx = 8$

$$2\int_0^1 k - kx^2 - 2kx^2 + 2k\,dx = 8$$

$$\int_0^1 3k - 3kx^2\,dx = 4$$

$$\left[3kx - kx^3\right]_0^1 = 4$$

$$3k(1) - k(1)^3 - 0 = 4$$

$$3k - k = 4$$

$$2k = 4$$

$$k = 2$$

(3 marks)

QUESTION 15

(a) In ΔAED, $\angle EAD = x^\circ$ (matching $\angle$s of congruent Δs AED & BAC equal)

Similarly, $\angle CAD = x^\circ$.

$\therefore \angle BAE = 3x^\circ$

$\angle AED + \angle ADE = (180 - x)^\circ$ ($\angle$ sum of Δ)

$\angle AED = \left(\dfrac{180-x}{2}\right)^\circ$ (base $\angle$s of isosceles Δ)

$\therefore\ 3x + \dfrac{180-x}{2} = 180$ (co-interior $\angle$s, $AB \| ED$)

$$6x + 180 - x = 360$$

$$5x = 180$$

$$x = 36$$

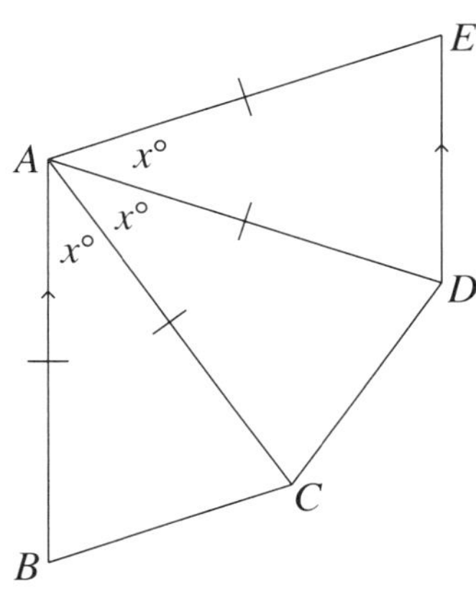

(3 marks)

(b) (i) 4.2% p.a. = 0.35% per month = 0.0035 per month, 4 years = 48 months

$M_1 = X \times 1.0035 - 2500$

$M_2 = M_1 \times 1.0035 + X \times 1.0035 - 2500$

$= (X \times 1.0035 - 2500) \times 1.0035 + X \times 1.0035 - 2500$

$= X \times 1.0035^2 - 2500 \times 1.0035 + X \times 1.0035 - 2500$

$= X(1.0035^2 + 1.0035) - 2500(1.0035 + 1)$

(2 marks)

(ii) $M_{48} = X(1.0035^{48} + 1.0035^{47} + \ldots + 1.0035) - 2500(1.0035^{47} + 1.0035^{46} + \ldots + 1)$

$= X(1.0035 + 1.0035^2 + \ldots + 1.0035^{48}) - 2500(1 + 1.0035^2 + \ldots + 1.0035^{47})$

Using geometric series and $S_n = \dfrac{a(r^n - 1)}{r - 1}$:

$$M_{48} = X\left[\frac{1.0035(1.0035^{48} - 1)}{1.0035 - 1}\right] - 2500\left[\frac{1(1.0035^{48} - 1)}{1.0035 - 1}\right] = 80\,000$$

$$52.351X - 130\,421 = 80\,000$$

$X = 4019$ (nearest whole)

(3 marks)

(c) (i) $a_1 = 6t + e^{-t}$

$v_1 = 3t^2 - e^{-t} + c$

Subs. $t = 0, v_1 = 3$:

$3 = 3(0)^2 - e^0 + c$

$3 = 0 - 1 + c$

$c = 4$

$v_1 = 3t^2 + 4 - e^{-t}$

(2 marks)

(ii) Let $v_1 = v_2$

$3t^2 + 4 - e^{-t} = 6t + 1 - e^{-t}$

$3t^2 - 6t + 3 = 0$

$t^2 - 2t + 1 = 0$

$(t - 1)^2 = 0$

$t = 1$

$\therefore$ after 1 second.

(2 marks)

(iii) $v_1 = 3t^2 + 4 - e^{-t}$

$\therefore x_1 = t^3 + 4t + e^{-t} + k_1$

Subs. $t = 0, x_1 = 0$:

$0 = (0)^3 + 4(0) + e^0 + k_1$

$0 = 1 + k_1$

$k_1 = -1$

$x_1 = t^3 + 4t + e^{-t} - 1$

Also $v_2 = 6t + 1 - e^{-t}$

$\therefore x_2 = 3t^2 + t + e^{-t} + k_2$

Subs. $t = 0, x_2 = 0$:

$0 = 3(0)^2 + 0 + e^0 + k_2$

$0 = 1 + k_2$

$k_2 = -1$

$x_2 = 3t^2 + t + e^{-t} - 1$

Let $x_1 = x_2$:

$t^3 + 4t + e^{-t} - 1 = 3t^2 + t + e^{-t} - 1$

$t^3 - 3t^2 + 3t = 0$

$t(t^2 - 3t + 3) = 0$

For $t^2 - 3t + 3$, $\Delta = 9 - 4(1)(3) < 0$, and $a = 1 > 0$.

Hence $t^2 - 3t + 3 > 0$ for all t, and so particles do not meet for $t > 0$.

(3 marks)

QUESTION 16

(a)

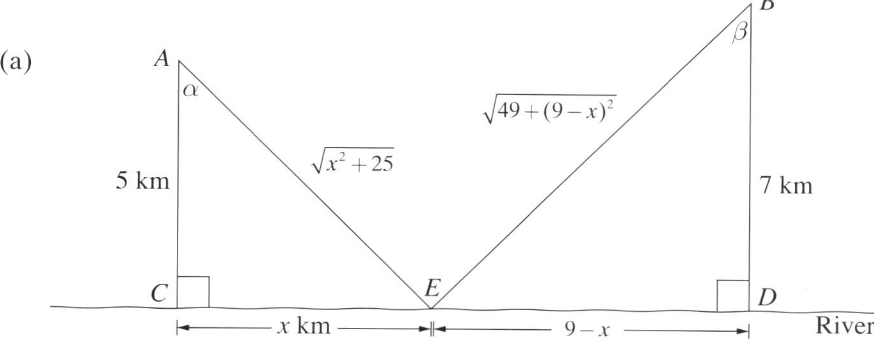

(i) $AE = \sqrt{x^2 + 5^2}$ (by Pythagoras)

$= \sqrt{x^2 + 25}$

As $DE = 9 - x$,

$EB = \sqrt{7^2 + (9-x)^2}$ (by Pythagoras)

$= \sqrt{49 + (9-x)^2}$

As $L = AE + EB$,

$\therefore L = \sqrt{x^2 + 25} + \sqrt{49 + (9-x)^2}$

(1 mark)

(ii) $L = (x^2+25)^{\frac{1}{2}} + (49 + (9-x)^2)^{\frac{1}{2}}$

$$\frac{dL}{dx} = \frac{1}{2}(x^2+25)^{-\frac{1}{2}}.2x + \frac{1}{2}(49+(9-x)^2)^{-\frac{1}{2}}.(-18+2x) = 0$$

$$x(x^2+25)^{-\frac{1}{2}} - \left[49+(9-x)^2\right]^{-\frac{1}{2}}.(9-x) = 0$$

$$\frac{x}{\sqrt{x^2+25}} - \frac{9-x}{\sqrt{49+(9-x)^2}} = 0$$

But, from ΔACE, $\dfrac{x}{\sqrt{x^2+25}} = \sin\alpha$, and from ΔBDE, $\dfrac{9-x}{\sqrt{49+(9-x)^2}} = \sin\beta$,

$\therefore \sin\alpha - \sin\beta = 0$

$\therefore \sin\alpha = \sin\beta$

(3 marks)

(iii) If $\sin\alpha = \sin\beta$, then $\alpha = \beta$ (for acute α, β) and hence ΔACE and ΔBDE are similar (2 $\angle$s equal).

$\therefore \dfrac{x}{5} = \dfrac{9-x}{7}$ (matching sides of similar Δs)

$7x = 45 - 5x$

$12x = 45$

$x = 3.75$

(2 marks)

(iv) As $\dfrac{dL}{dx}(3.75) = 0$, then check nature of stationary point:

x	3	3.75	4
L'	<0	0	>0

As sign changes, then $x = 3.75$ gives a minimum value for L.

(1 mark)

(b) Using $S = \dfrac{a}{1-r}$:

$$2 = \frac{a}{1-r}$$
$$2 - 2r = a$$
$$2r = 2 - a$$
$$r = \frac{2-a}{2}$$

Now, limiting sum when $-1 < r < 1$:

$$-1 < \frac{2-a}{2} < 1$$
$$-2 < 2 - a < 2$$
$$-4 < -a < 0$$
$$0 < a < 4$$

(3 marks)

(c) (i) $\dfrac{BM}{MC} = \dfrac{BD}{DE}$ (ratio of intercepts of parallel lines)

As $\dfrac{BM}{MC} = 1$ (given), then $\dfrac{BD}{DE} = 1$.

$\therefore BD = DE$

$DA + AC = DE$ (as $BD = DA + AC$)

$AC = DE - DA$

$\therefore AC = AE$

$\therefore \Delta ACE$ is isosceles.

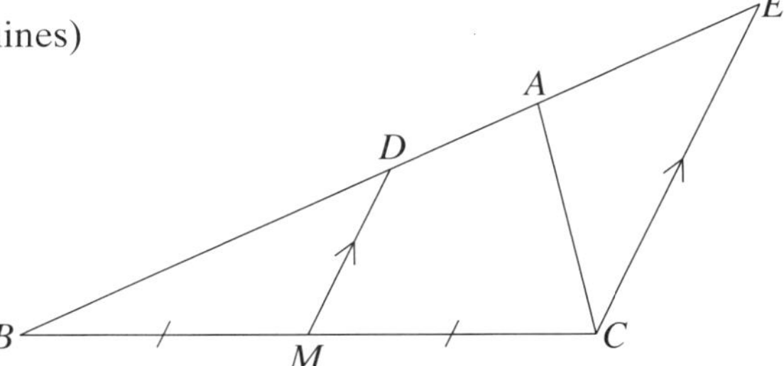

(3 marks)

(ii) Let $\angle FAC = \alpha$

$\angle ACE = \alpha$ (alternate $\angle$s, $AF \| EC$)

$\angle AEC = \alpha$ (base $\angle$s of isosceles Δ)

$\angle BAF = \alpha$ (corresponding $\angle$s, $AF \| EC$)

$\therefore \angle FAC = \angle BAF$

$\therefore AF$ bisects $\angle BAC$.

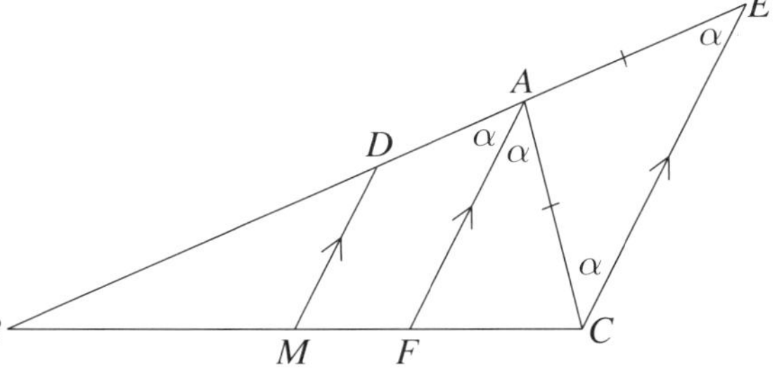

(2 marks)

2017 | HIGHER SCHOOL CERTIFICATE **EXAMINATION**

REFERENCE SHEET

– Mathematics –

– Mathematics Extension 1 –

– Mathematics Extension 2 –

Mathematics

Factorisation

$$a^2 - b^2 = (a+b)(a-b)$$

$$a^3 + b^3 = (a+b)(a^2 - ab + b^2)$$

$$a^3 - b^3 = (a-b)(a^2 + ab + b^2)$$

Angle sum of a polygon

$$S = (n-2) \times 180°$$

Equation of a circle

$$(x-h)^2 + (y-k)^2 = r^2$$

Trigonometric ratios and identities

$$\sin\theta = \frac{\text{opposite side}}{\text{hypotenuse}}$$

$$\cos\theta = \frac{\text{adjacent side}}{\text{hypotenuse}}$$

$$\tan\theta = \frac{\text{opposite side}}{\text{adjacent side}}$$

$$\operatorname{cosec}\theta = \frac{1}{\sin\theta}$$

$$\sec\theta = \frac{1}{\cos\theta}$$

$$\tan\theta = \frac{\sin\theta}{\cos\theta}$$

$$\cot\theta = \frac{\cos\theta}{\sin\theta}$$

$$\sin^2\theta + \cos^2\theta = 1$$

Exact ratios

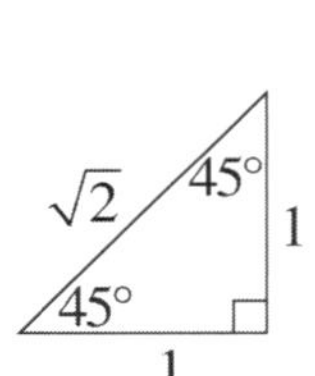

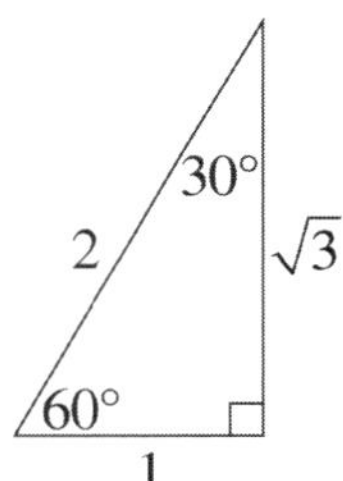

Sine rule

$$\frac{a}{\sin A} = \frac{b}{\sin B} = \frac{c}{\sin C}$$

Cosine rule

$$c^2 = a^2 + b^2 - 2ab\cos C$$

Area of a triangle

$$\text{Area} = \frac{1}{2}ab\sin C$$

Distance between two points

$$d = \sqrt{(x_2 - x_1)^2 + (y_2 - y_1)^2}$$

Perpendicular distance of a point from a line

$$d = \frac{|ax_1 + by_1 + c|}{\sqrt{a^2 + b^2}}$$

Slope (gradient) of a line

$$m = \frac{y_2 - y_1}{x_2 - x_1}$$

Point–gradient form of the equation of a line

$$y - y_1 = m(x - x_1)$$

***n*th term of an arithmetic series**

$$T_n = a + (n-1)d$$

Sum to *n* terms of an arithmetic series

$$S_n = \frac{n}{2}[2a + (n-1)d] \quad \text{or} \quad S_n = \frac{n}{2}(a + l)$$

***n*th term of a geometric series**

$$T_n = ar^{n-1}$$

Sum to *n* terms of a geometric series

$$S_n = \frac{a(r^n - 1)}{r - 1} \quad \text{or} \quad S_n = \frac{a(1 - r^n)}{1 - r}$$

Limiting sum of a geometric series

$$S = \frac{a}{1-r}$$

Compound interest

$$A_n = P\left(1 + \frac{r}{100}\right)^n$$

Differentiation from first principles

$$f'(x) = \lim_{h \to 0} \frac{f(x+h) - f(x)}{h}$$

Derivatives

If $y = x^n$, then $\frac{dy}{dx} = nx^{n-1}$

If $y = uv$, then $\frac{dy}{dx} = u\frac{dv}{dx} + v\frac{du}{dx}$

If $y = \frac{u}{v}$, then $\frac{dy}{dx} = \frac{v\frac{du}{dx} - u\frac{dv}{dx}}{v^2}$

If $y = F(u)$, then $\frac{dy}{dx} = F'(u)\frac{du}{dx}$

If $y = e^{f(x)}$, then $\frac{dy}{dx} = f'(x)e^{f(x)}$

If $y = \log_e f(x) = \ln f(x)$, then $\frac{dy}{dx} = \frac{f'(x)}{f(x)}$

If $y = \sin f(x)$, then $\frac{dy}{dx} = f'(x)\cos f(x)$

If $y = \cos f(x)$, then $\frac{dy}{dx} = -f'(x)\sin f(x)$

If $y = \tan f(x)$, then $\frac{dy}{dx} = f'(x)\sec^2 f(x)$

Solution of a quadratic equation

$$x = \frac{-b \pm \sqrt{b^2 - 4ac}}{2a}$$

Sum and product of roots of a quadratic equation

$\alpha + \beta = -\frac{b}{a}$ $\qquad$ $\alpha\beta = \frac{c}{a}$

Equation of a parabola

$$(x-h)^2 = \pm 4a(y-k)$$

Integrals

$$\int (ax+b)^n \, dx = \frac{(ax+b)^{n+1}}{a(n+1)} + C$$

$$\int e^{ax+b} \, dx = \frac{1}{a}e^{ax+b} + C$$

$$\int \frac{f'(x)}{f(x)} \, dx = \ln|f(x)| + C$$

$$\int \sin(ax+b)\,dx = -\frac{1}{a}\cos(ax+b) + C$$

$$\int \cos(ax+b)\,dx = \frac{1}{a}\sin(ax+b) + C$$

$$\int \sec^2(ax+b)\,dx = \frac{1}{a}\tan(ax+b) + C$$

Trapezoidal rule (one application)

$$\int_a^b f(x)\,dx \approx \frac{b-a}{2}\left[f(a) + f(b)\right]$$

Simpson's rule (one application)

$$\int_a^b f(x)\,dx \approx \frac{b-a}{6}\left[f(a) + 4f\left(\frac{a+b}{2}\right) + f(b)\right]$$

Logarithms – change of base

$$\log_a x = \frac{\log_b x}{\log_b a}$$

Angle measure

$180° = \pi$ radians

Length of an arc

$l = r\theta$

Area of a sector

$$\text{Area} = \frac{1}{2}r^2\theta$$

Mathematics Extension 1

Angle sum identities

$\sin(\theta + \phi) = \sin\theta\cos\phi + \cos\theta\sin\phi$

$\cos(\theta + \phi) = \cos\theta\cos\phi - \sin\theta\sin\phi$

$\tan(\theta + \phi) = \dfrac{\tan\theta + \tan\phi}{1 - \tan\theta\tan\phi}$

***t* formulae**

If $t = \tan\dfrac{\theta}{2}$, then

$$\sin\theta = \frac{2t}{1+t^2}$$

$$\cos\theta = \frac{1-t^2}{1+t^2}$$

$$\tan\theta = \frac{2t}{1-t^2}$$

General solution of trigonometric equations

$\sin\theta = a, \quad \theta = n\pi + (-1)^n \sin^{-1} a$

$\cos\theta = a, \quad \theta = 2n\pi \pm \cos^{-1} a$

$\tan\theta = a, \quad \theta = n\pi + \tan^{-1} a$

Division of an interval in a given ratio

$\left(\dfrac{mx_2 + nx_1}{m+n}, \dfrac{my_2 + ny_1}{m+n}\right)$

Parametric representation of a parabola

For $x^2 = 4ay$,

$$x = 2at, \quad y = at^2$$

At $(2at, at^2)$,

tangent: $y = tx - at^2$

normal: $x + ty = at^3 + 2at$

At (x_1, y_1),

tangent: $xx_1 = 2a(y + y_1)$

normal: $y - y_1 = -\dfrac{2a}{x_1}(x - x_1)$

Chord of contact from (x_0, y_0): $xx_0 = 2a(y + y_0)$

Acceleration

$$\frac{d^2x}{dt^2} = \frac{dv}{dt} = v\frac{dv}{dx} = \frac{d}{dx}\left(\frac{1}{2}v^2\right)$$

Simple harmonic motion

$x = b + a\cos(nt + \alpha)$

$\ddot{x} = -n^2(x - b)$

Further integrals

$$\int \frac{1}{\sqrt{a^2 - x^2}}\,dx = \sin^{-1}\frac{x}{a} + C$$

$$\int \frac{1}{a^2 + x^2}\,dx = \frac{1}{a}\tan^{-1}\frac{x}{a} + C$$

Sum and product of roots of a cubic equation

$\alpha + \beta + \gamma = -\dfrac{b}{a}$

$\alpha\beta + \alpha\gamma + \beta\gamma = \dfrac{c}{a}$

$\alpha\beta\gamma = -\dfrac{d}{a}$

Estimation of roots of a polynomial equation

Newton's method

$$x_2 = x_1 - \frac{f(x_1)}{f'(x_1)}$$

Binomial theorem

$$(a+b)^n = \sum_{k=0}^{n}\binom{n}{k}a^k b^{n-k} = \sum_{k=0}^{n}\binom{n}{k}a^{n-k}b^k$$

CHEMISTRY

CHAPTER 1 PROPERTIES OF MATTER

YEAR 11 EXAM-TYPE QUESTIONS

BIOLOGY

YEAR 11 EXAM-TYPE QUESTIONS

PHYSICS

CHAPTER 1 MOTION IN A STRAIGHT LINE

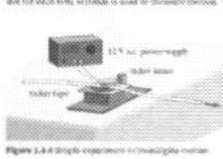

YEAR 11 EXAM-TYPE QUESTIONS

STANDARD MATHEMATICS

CHAPTER 1 ALGEBRA

YEAR 11 EXAM-TYPE QUESTIONS